COULTER LIBRARY ONONDAGA COMM. COLL.
3 0418 00244674 6

Sidney B.
Coulter
Library
ONONDAGA COMMUNITY COLLEGE
SYRACUSE, NEW YORK 13215

D1307196

DIGITAL DESIGNING WITH PROGRAMMABLE LOGIC DEVICES

JOHN W. CARTER
University of North Carolina, Charlotte

Prentice Hall

Upper Saddle River, New Jersey   Columbus, Ohio

Library of Congress Cataloging-in-Publication Data

Carter, John W.
Digital designing with programmable logic devices / by John W. Carter.
p. cm.
Includes index.
ISBN 0-13-373721-7
1. Programmable logic devices—Design. 2. Logic design. I. Title.
TK7872.L64C37 1997 96-1547
621.39'5—dc20 CIP

Editor: Charles E. Stewart, Jr.
Production Editor: Mary Harlan
Cover Designer: Tammy Johnson
Cover photo: Diana Ong/Superstock
Production Supervision: Custom Editorial Productions, Inc.
Production Manager: Patricia A. Tonneman
Marketing Manager: Debbie Yarnell

This book was set in Times Roman by Custom Editorial Productions, Inc., and was printed and bound by Book Press, Inc., a Quebecor America Book Group Company. The cover was printed by Phoenix Color Corp.

Printed in the United States of America

10 9 8 7 6 5 4 3 2

ISBN: 0-13-373721-7

Prentice-Hall International (UK) Limited, *London*
Prentice-Hall of Australia Pty. Limited, *Sydney*
Prentice-Hall of Canada, Inc., *Toronto*
Prentice-Hall Hispanoamericana, S. A., *Mexico*
Prentice-Hall of India Private Limited, *New Delhi*
Prentice-Hall of Japan, Inc., *Tokyo*
Simon & Schuster Asia Pte. Ltd., *Singapore*
Editora Prentice-Hall do Brasil, Ltda., *Rio de Janeiro*

To Ann Marie, Jennifer Ann and John Andrew

PREFACE

WHY PROGRAMMABLE LOGIC?

Digital. The word has become a part of our common vocabulary. Few people who are part of any modern industrial community are unaware of it. This has been true only in the past few years, when the miniaturization of electronic circuits has opened the floodgates of applications for digital electronic technology. Now, one can hardly go through a day without coming in contact with digital technology. To some, the word is synonymous with *reliable,* or *programmable,* or, to others, *complex.* Many applications that were heretofore analog in nature, such as communications and television, are being replaced with functionally equivalent digital systems. Various products in the consumer marketplace today contain electronic controllers that are digital in nature. We often recognize their application when data entry keyboards are present, such as with microwave ovens and digital clocks. In other cases, we may be unaware of the digital circuits controlling our products, as in the electronic ignition system in an automobile.

The "brain" of these devices is the state machine: the electronic circuit that provides control of the device by gathering input information, processing it with some decision-making capability, and providing output in the form of information to the user and stimulus to the appliance under control. The complexity of the state machine varies with the complexity of the application. Simple control applications may require only a few logic circuits. Very complex applications may require the use of a microcontroller or microprocessor. In the center, between the ends of this spectrum of control complexity, lies an entire range of applications that are too complex to be controlled by simple logic devices, yet simple enough that applying a microprocessor is too expensive and inappropriate. The solution lies in the application of the digital state machine described in this text. This circuit, which uses a controlled binary counter and strategically designed output logic, will often fulfill the task of the much more expensive microcontroller. This type of state machine can frequently be implemented with a single programmable logic device (PLD).

The knowledge of and ability to apply state machines to control situations are vital to the overall education of the digital designer. Often the PLD chip, which costs only a few dollars, can be used to replace a microprocessor-based or even a very expensive programmable logic controller (PLC) based system. Furthermore, the state machine can be the basis of designs for new applications.

INDUCTIVE, UNDERSTANDING-BASED EDUCATION

More than ever before it is expedient that the academic industry not fall behind in preparing individuals to be productive in this changing field. It is no longer sufficient to be satisfied with a boilerplate approach to education, which treats students as machines who will operate within a finite set of parameters defined by current (or outdated) technology. Even the current technology will soon be obsolete. It is necessary that the already effective knowledge-based curriculum be supplemented with yet greater understanding-based emphasis.

It is to this end that this text has been prepared. It appears that the proliferation of texts used to teach electronic and computer technology, though excellent, are theory- or knowledge-based and do not require the "hands-on" application and "on the job training" approach that best facilitates understanding. It is the purpose of this text to use that "hands-on" methodology to present state machine design from a viewpoint that will prepare the student for application within the digital design industry. Its primary focus is inductive in nature. The material is presented from a standpoint of application-based design rather than theory-based design and requires student interaction. Following the presentation of a principle, its application in a design is presented as a "practical example." This example is part of the pedagogy of the text and often contains new information and stimulates thought into different directions. This is followed by an opportunity for the reader to exercise an understanding of the information by applying it in a simple design that is similar to the one just presented. Using this methodology, the level of complexity of each subject increases until the subject is covered. The intent is to teach a *skill,* rather than to teach theory. When the assigned tasks have been completed, the student should be able to *do* something rather than just tell how something is done.

TARGET AUDIENCE

This text is intended as the main resource for a single-semester course of study in an upper-division baccalaureate program in Electronics or Computer Engineering Technology. It could also be used as an elective course in an Engineering curriculum. Such a course might be entitled "Digital Design II," "Advanced Digital Design," "Digital State Machine Design," or even "Digital Designing with Programmable Logic."

PREREQUISITES

This text presupposes that the reader has an introductory background in digital logic, including logic gates, combinational and sequential logic, flip-flops, and registers, as well as an introduction to microprocessors and assembly language. It also assumes the background in binary arithmetic and Boolean algebra common to such a digital course sequence. The ability to reduce Boolean expressions using a Karnaugh map is a must.

TEXT OVERVIEW

Each chapter includes an introductory overview, a set of objectives, an in-depth discussion that includes worked problems, and a set of questions and problems at the end of the chapter.

Chapter 1, "An Introduction to Programmable Logic Devices," describes the architecture of programmable logic devices from an applications viewpoint. The form of Boolean expressions to be implemented is first established, and common methods of applying digital logic are investigated. This is followed by a description of programmable logic circuits and the methods of using them to apply Boolean logic expressions. The purpose of this chapter is to lay the foundation for the applications of programmable logic devices encountered later in the text.

Chapter 2, "Synchronous Binary Counter Design," introduces the application of the synchronous binary counter as the engine that drives the state machine, which will be presented in the next few chapters. In order to apply the state machine that resides in the center of the aforementioned spectrum, the ability to design a synchronous binary counter is vital. The design methodology is systematically presented, and followed by opportunities for the reader to apply that method.

Chapter 3, "Decision Making with Binary Counters," adds to the design methodology of Chapter 2 the capability for the synchronous binary counter to make decisions based on input from an outside source.

Chapter 4, "Output-Forming Logic Design," adds output control capability to the state machine. The very purpose of the state machine is to control some peripheral object. This chapter presents a systematic method of designing circuitry that can take the output of the synchronous counter of Chapter 3 and create the control signals needed by external devices.

As discussed in Chapter 5, "Programming PLD Circuits with CUPL," when programmable logic devices are used to implement the state machine function, the design process critically changes. Rather than dealing with the encumbrances of state tables, Karnaugh maps, and excessively long logic equations, PAL programming languages can be used. The PAL program design method replaces much of the logic design previously discussed. Though other common PAL programming languages are readily available and commonly used in the field, CUPL has been selected because of its simplicity, its power, and its similarity to most of the other languages. An individual who can program a PAL in the CUPL language can easily translate that ability to other languages such as PALASM or Texas Instrument's "ProLogic." This chapter describes a subset of the CUPL programming language as it is applied to digital state machine design.

Chapter 6, "Design Examples Using PLDs," contains several completely documented circuit designs that utilize the principles and methods from the first five chapters. It illlustrates some of the variety of applications of state machines and programmable logic devices within the context of practical use.

Chapter 7, "ROM-Based Controller Architecture," moves us further up the spectrum of complexity to those applications too complex for a PAL or FPLA, yet simple enough that a microcontroller is still an extravagant solution. This gap in the spectrum is caused by the limit of complexity of PALs and FPLAs that are currently available on the commercial market. For those applications, a ROM chip can be used to replace the PAL, and a binary register is included to replace the function of the synchronous binary counter. With this simple architecture, very complex controllers can be designed.

Chapter 8, "Microprocessor Architecture," completes the design spectrum between discrete logic and microprocessor applications. In this chapter the architecture of Chapter 7 is enhanced with capabilities of doing arithmetic processes and managing the movement of data between internal resources, external memory, and external peripheral devices. The MC68HC11 microcontroller is then used to apply the principles of the previous two chapters using a common commercial microcontroller. Its programmer's model and instruction set are introduced. Examples are provided that describe the use of the microcontroller for a variety of applications.

Chapter 9, "The MC68HC11 Microcontroller," is a relatively intense study of the technical properties of this common and quite useful microcontroller. Its basic architecture is presented, followed by a detailed description of the architecture and programming of each of its I/O ports. Finally, a series of examples are provided that illustrate the application of the microcontroller in a variety of interfacing scenarios. Some of the more advanced features of the microcontroller are not presented.

Chapter 10, "Address Bus Decoding and Logic Design," describes a very detailed methodology for designing circuitry to interface memory-mapped devices to a microcontroller or microprocessor. The method uses the design principles of Chapters 1 through 5 to apply programmable logic devices to the solution. The target microprocessor is the MC68HC11 microcontroller, configured in the expanded mode of operation, though the principles apply to any processor architecture.

Appendix A, "Selected PLD Architectures," includes the technical documentation necessary to apply a variety of commercial PAL and FPLA chips to designs consistent with the material of this text. Included are the DIP pin layouts, supplier information, and technical information when necessary.

Appendix B, "CUPL PAL/PROM Device Library," is a list of the PAL and PROM devices, sorted by supplier, supported by the CUPL PAL Expert® software.

ACKNOWLEDGMENTS

Research and development for this text began several years before the first manuscript page was typed. It is a product of the contribution of effort from several sources. I would first like to thank the Computer and Electronic Engineering Technology faculty at the University of North Carolina, Charlotte (Go Forty-Niners!), and the SUNY Institute of

Technology, Utica, New York (Go Wildcats!), for their assistance and contributions. Particular thanks are extended to those students at these campuses who have been subjected to the myriad of ideas that are presented here, and for their input, which helped to shape this presentation. I wish to thank those who reviewed the manuscript for their helpful comments and suggestions: Jaspal Attrey, Central Piedmont Community College; Michael A. Miller, DeVry Technical Institute, Phoenix; Gregory S. Romine, Indiana University/Purdue University–Indianapolis; and Asad Yousuf, Savannah State College.

I would like to thank Logical Devices, Inc., Philips Electronics North America Corporation, and Motorola, Inc. for their permission to use their devices and data sheets in this text. Much of the material in Chapter 9 is taken directly from Motorola, Inc. technical literature.

I would also like to thank those faculty, now mostly retired, in the Oklahoma State University Department of Engineering Technology for guiding the creative interests of a young student into the field of digital electronics. On that staff I pay most gratitude to Mr. Charles B. Harrison. Thanks to his friendship and encouragement in those early years, I am still in academia eighteen years later. Thank you, Mr. Harrison, wherever you are.

Finally, and most important, I would like to thank my computer widow, Ann Marie, and my children, who patiently endured the hours that I dedicated to producing this book.

BRIEF CONTENTS

CONTENTS

CHAPTER 1

An Introduction to Programmable Logic Devices

OBJECTIVES

After completing this chapter you should be able to:

- Describe sum of products (SOP) Boolean logic function expressions in a variety of ways, including:
    - Conventional logical expressions
    - Typographic logical expressions
    - Arithmetic summations
- Use discrete combinational logic circuits to implement logical function expressions.
- Describe how a read-only memory (ROM) can be used to implement logical function expressions.
- Describe how a programmable array logic (PAL) circuit can be used to implement logical function expressions.
- Describe how a field programmable logic array (FPLA) circuit can be used to implement logical function expressions.
- Describe how a field programmable logic sequencer (FPLS) circuit can be used to implement logical function expressions.
- Describe the following programmable logic output macrocell configurations:
    - Logic-type macrocell
    - Register-type macrocell
    - Variable-type macrocell
- Describe the characteristics and use of generic array logic (GAL) devices.
- Describe the architecture and use of several PROM technologies, including mask-programmed ROMs, field-programmable ROMs, UV-erasable PROMs, and electrically erasable PROMs.

1.1 OVERVIEW

In this chapter we will introduce the concept of programmable logic devices and their use. Programmable logic devices include relatively complex medium-scale integration (MSI) and more complex large-scale integration (LSI) circuits that can be configured to emulate most discrete logic circuits, as well as complex interconnections of those discrete circuits. Because of this characteristic, many would-be prophets once predicted that the age of discrete logic was over because all designers would aggressively replace it with programmable logic. Such a decision would reduce inventory and provide a more flexible design environment. Though such reasoning was adequate, the multimillion dollar market still held by discrete logic circuits proves this prophesy false. However, the failure of this prediction may be due to the ignorance of the digital design market as well as several other factors. Indeed, programmable logic circuits can be easily configured to emulate most of the circuits included in digital systems, producing very cost-effective design solutions.

The most significant aspect of this concept, which we will take advantage of, is that the programmable logic circuit can be configured to replace more than one discrete logic circuit. For example, assume that you, as a digital designer, were assigned the task of creating a combinational logic circuit that generates 10 logic functions, each comprised of up to eight minterms of variables coming from a set of 12 different inputs. Such a task would be formidable if implemented with discrete logic gates. A minimum of 10 eight-input OR gates would be summing together eight minterms, each created by a 12-input AND gate. This requires a total of 96 12-input AND gates. This problem can be solved using a single programmable logic device that is commercially available. Furthermore, programmable logic devices are available that contain the flip-flops that are part of common sequential circuits.

As we develop the designs for digital state machines we will discover that programmable logic devices are particularly well suited for this application. A moderately complex state machine would require so much discrete logic that its construction would be impractical. However, the same design may require only one or two programmable logic devices.

If the application of programmable logic devices is so advantageous, it is worth our time to understand them and develop an ability to apply them.

1.2 SUM OF PRODUCTS EXPRESSIONS

Though logical **expressions** can be presented in a variety of forms, we are going to represent all expressions in the **sum of products** (**SOP**) form. Expressions in this form are particularly well suited for minimization using the most common methods. Furthermore, they are well suited for direct application in the programming languages that apply them. Finally, dealing with only one form of an expression will simplify our discussion.

1.2.1 Conventional Form

Consider the following logical expression (Expression Set 1.1):

$$Y = A\overline{B}C + \overline{A}\,\overline{B}\,\overline{C} + \overline{A}B\overline{C} + \overline{A}\,\overline{B}C \quad \textbf{1.1}$$

We will refer to this as the **conventional form** of an SOP expression. Note that the structure of the expression shows four groups of products, each referred to as a **minterm.** These minterms are ORed together in order to define the expression. It is implied (and true) that the AND logic has first operation precedence, followed by the OR operator. Therefore, this expression contains four minterms with three variables in each.

The list of the variables in the entire expression is referred to as its **domain.** That is, the domain of Expression Set 1.1 is the set of variables {A,B,C}. It would be correct to state that, according to Expression Set 1.1, "*Y* is a function of A, B, and C," with the definition of the function following the equals sign. Since there are three variables in the domain, it is also correct to say that the expression has a domain of three.

$$Y = f(A, B, C) \quad \textbf{1.2}$$

It is certainly reasonable that this domain definition can become part of the entire expression. This is particularly valuable when one or more variables in the domain are not otherwise part of the expression. Expression Set 1.3 illustrates this complete form of an SOP expression:

$$Y = f(A, B, C) = A\overline{B}C + \overline{A}\,\overline{B}\,\overline{C} + \overline{A}B\overline{C} + \overline{A}\,\overline{B}C \quad \textbf{1.3}$$

1.2.2 Typographic Form

If we are relegated to using a conventional text editor to describe these expressions, we run into a bit of a problem. It is difficult in most typing environments to use the overscore to indicate the complement of a variable. For this reason, another convention will be used in addition to that shown above. This form, referred to as the **typographic SOP form,** replaces all of the logical operators with characters commonly found on a typewriter. Table 1.1 illustrates the characters that will be used to form logical expressions.

Using the definitions of Table 1.1, Expression Set 1.3 could be written:

$$Y = f(A, B, C) = (A\&!B\&C)\#(!A\&!B\&!C)\#(!A\&B\&!C)\#(!A\&!B\&C) \quad \textbf{1.4}$$

Note that the precedence of operators has not changed. However, parentheses were added to the expression in order to make it easier to read. Though this form may be more cumbersome

TABLE 1.1 Typographical logical operators.

Operator	*Character*
NOT	!
AND	&
OR	#
XOR	$

to read when compared with the conventional form, it is this form that we will be using to write programs for programmable devices. Therefore, we must become familiar with it.

1.2.3 Summation Form

An improved form of a logical expression can be obtained by referring to each minterm in the expression by its numerical equivalent in the domain, rather than by the list of variables. Again, consider Expression Set 1.3 and observe the method used to convert this expression to its equivalent **summation SOP form:**

$$
\begin{aligned}
&1.\quad Y = f(A,B,C) = A\overline{B}C + \overline{A}\,\overline{B}\,\overline{C} + \overline{A}B\overline{C} + \overline{A}\,\overline{B}C \\
&2.\quad Y = f(A,B,C) = 101 + 000 + 010 + 001 \\
&3.\quad Y = f(A,B,C) = \sum(5,0,2,1) \\
&4.\quad Y = f(A,B,C) = \sum(0,1,2,5)
\end{aligned}
\qquad \textbf{1.5}
$$

The original conventional expression is shown as line number 1, and the conversion process takes three steps:

- Each minterm is converted to its binary equivalent in order to produce the expression of line number 2. This is done by treating each complemented variable as a 0 digit and each non-complemented variable as a 1 digit.
- Again, this form is a bit cumbersome, so each minterm is converted from binary to decimal, realizing the expression of line number 3.
- Finally, for convention purposes, the minterms are ordered in ascending sequence in line number 4.

1.2.4 Advantages of the Different Forms

We have reviewed three forms of logical expressions: conventional, typographic, and summation. Each form has predominant advantages that are realized in specific situations, all of which will be utilized in state machine design. The conventional form is most useful for implementation with discrete logic gates. The typographical form is most useful when implementing with programmable logic when the summation form is not used. The summation form is most useful when implementing the expression in programmable logic.

One advantage of the summation form of a logical expression is that very long and complex expressions can be described easily. Another advantage is the ease of conversion between this and other forms. Finally, the most important advantage of this form concerns its implementation with hardware devices.

Practical Example 1.1: Converting Forms of SOP Expressions

Convert the following SOP expressions to typographic and summation form:

$$
\begin{aligned}
W &= f(A,B,C) = \overline{A}\,\overline{B}C + A\overline{B}\,\overline{C} + ABC \\
X &= f(A,B,C,D) = \overline{A}CD + AB\overline{C} + \overline{C}\,\overline{D} \\
Y &= f(A,B,C) = \overline{A} + B\overline{C} + \overline{A}C
\end{aligned}
\qquad \textbf{1.6}
$$

Conversion from Conventional to Typographic Form. The conversions to typographic form are relatively straightforward. Simply replace each logical operator with the equivalent typographic operator. Note that parentheses are placed around minterms only to make the expression more readable.

$$W = (!A\&!B\&C)\#(A\&!B\&!C)\#(A\&B\&C) \qquad \textbf{1.7}$$
$$X = (!A\&C\&D)\#(A\&B\&!C)\#(!C\&!D)$$
$$Y = !A\ \#(B\&!C)\#(!A\&C)$$

Conversion from Conventional to Summation Form. The conversion from conventional to summation form is not quite as straightforward as the conversion to typographic form. The summation form lists all true minterms of the expression without regard to any minimization of the expression. Therefore, if the expression has been minimized in any way, it must be expanded to include all of the individual minterms. For example, anytime a variable that is part of the domain is missing from the minterm, that variable is **redundant.** An example of a redundant variable is as shown:

1. $Y = AB\overline{C}D + AB\overline{C}\,\overline{D}$ **1.8**
2. $Y = AB\overline{C}(D + \overline{D})$
3. $Y = AB\overline{C}(1)$
4. $Y = AB\overline{C}$

Observing the first equation, we can see that the set of variables $AB\overline{C}$ is repeated in both minterms. Therefore, we can factor out these variables, resulting in the second equation. The remaining term, D, is revealed to be redundant since the term within the parentheses is an OR operation between the variable and its complement. Therefore, the following expression is true:

$$Y = f(A, B, C, D) = AB\overline{C} = AB\overline{C}D + AB\overline{C}\,\overline{D} \qquad \textbf{1.9}$$

When a minterm does not contain all of the variables defined within the domain, it can be expanded to include those variables using the method shown in Expression Set 1.9. The original minterm is logically multiplied by all of the possible permutations of the missing variable, in this case D, which has two permutations, D and $\overline{D}$.

Consider another example, one with two missing terms: $Y = f$ (A,B,C,D) = AB. In this case, the missing terms are C and D. There are four possible permutations of the two variables: $\overline{C}\,\overline{D}$, $\overline{C}\,D$, $C\,\overline{D}$, and C D. Therefore, the following is the correct expansion of the expression:

$$Y = f(A, B, C, D) = AB \qquad \textbf{1.10}$$
$$Y = f(A, B, C, D) = AB\overline{C}\,\overline{D} + AB\overline{C}D + ABC\overline{D} + ABCD$$

This expansion step is necessary to convert a conventional expression to summation form. Let us return to our original problem. Convert the following conventional SOP expressions to summation form:

$$W = f(A, B, C) = \overline{A}\,\overline{B}C + A\overline{B}\,\overline{C} + ABC \qquad \textbf{1.11}$$
$$X = f(A, B, C, D) = \overline{A}CD + AB\overline{C} + \overline{C}\,\overline{D}$$
$$Y = f(A, B, C) = \overline{A} + B\overline{C} + \overline{A}C$$

The first expression, *W*, contains no minterms with missing variables, so its conversion is straightforward:

$$W = f(A,B,C) = \bar{A}\bar{B}C + A\bar{B}\bar{C} + ABC \quad \textbf{1.12}$$
$$W = f(A,B,C) = \sum(001, 100, 111)$$
$$W = f(A,B,C) = \sum(1,4,7)$$

The second expression, *X*, contains minterms with missing variables, so, prior to conversion, each of these minterms must be expanded:

1. $X = f(A,B,C,D) = \bar{A}CD + AB\bar{C} + \bar{C}\bar{D}$ **1.13**
2. $X = f(A,B,C,D) = \bar{A}CD(B+\bar{B}) + AB\bar{C}(D+\bar{D}) + \bar{C}\bar{D}(\bar{A}\bar{B} + \bar{A}B + A\bar{B} + AB)$
3. $X = f(A,B,C,D) = \bar{A}BCD + \bar{A}\bar{B}CD + AB\bar{C}D + AB\bar{C}\bar{D} + \bar{A}\bar{B}\bar{C}\bar{D} + \bar{A}B\bar{C}\bar{D} + A\bar{B}\bar{C}\bar{D} + AB\bar{C}\bar{D}$
4. $X = f(A,B,C,D) = \sum(0111, 0011, 1101, 1100, 0000, 0100, 1000, 1100)$
5. $X = f(A,B,C,D) = \sum(7, 3, 13, 12, 0, 4, 8, 12)$
6. $X = f(A,B,C,D) = \sum(0, 3, 4, 7, 8, 12, 13)$

Note that in line 2 each redundant term was added to each minterm to create the expression of line 3. Then each minterm was converted to its binary equivalent in line 4 and its decimal equivalent in line 5. Finally, the minterms were ordered and redundant terms eliminated, resulting in the solution on line 6.

The solution of the third expression, *Y*, in this example is similar:

1. $Y = f(A,B,C) = \bar{A} + B\bar{C} + \bar{A}C$ **1.14**
2. $Y = f(A,B,C) = \bar{A}(\bar{B}\bar{C} + \bar{B}C + B\bar{C} + BC) + B\bar{C}(A+\bar{A}) + \bar{A}C(B+\bar{B})$
3. $Y = f(A,B,C) = \bar{A}\bar{B}\bar{C} + \bar{A}\bar{B}C + \bar{A}B\bar{C} + \bar{A}BC + AB\bar{C} + \bar{A}B\bar{C} + \bar{A}BC + \bar{A}\bar{B}C$
4. $Y = f(A,B,C) = \sum(000, 001, 010, 011, 110, 010, 011, 001)$
5. $Y = f(A,B,C) = \sum(0, 1, 2, 3, 6, 2, 3, 1)$
6. $Y = f(A,B,C) = \sum(0, 1, 2, 3, 6)$

A simpler method for converting to summation form involves using a Karnaugh map. Note that the final solution simply lists the true minterms in the domain of the expression. This is also a basic function of the Karnaugh map. Again, consider the third expression *Y* in the preceding problem. If we take each minterm of the original expression of line 1 and place it on a Karnaugh map, we get the results illustrated in Figure 1.1.

Note that the placement of the minterms onto the map effectively expands the expression for us. It also removes redundant terms from consideration. All that is needed is to extract the map cell numbers of the true terms, 0,1,2,3,6. This matches the results of the system of preceding equations.

If expressions are maintained in their summation form, their placement on and extraction from a Karnaugh map is a trivial matter. We will be using Karnaugh maps frequently in the solution of state machine design problems, and our data will often be in numeric rather than logical form. Consequently, we will be making frequent use of the summation form of an SOP expression.

FIGURE 1.1 Karnaugh map for expression *Y*.

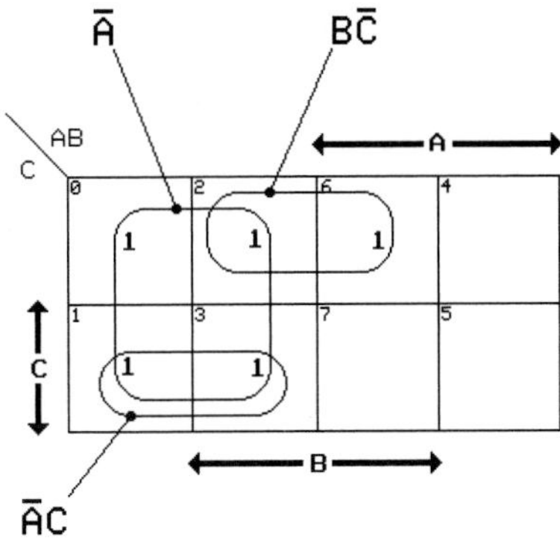

Practical Exercise 1.1: SOP Expression Conversions

Consider the following SOP expressions:

$$W = f(A, B, C) = \overline{A}\,\overline{B}\,\overline{C} + \overline{A}\,\overline{B}C + A\overline{B}\,\overline{C} + A\overline{B}C \qquad \textbf{1.15}$$

$$X = f(A, B, C, D) = \overline{A}B + A\overline{B}C\overline{D} + \overline{B}CD$$

$$Y = f(A, B, C) = A\overline{B} + \overline{B}C$$

1. Convert each expression to its equivalent typographic form.
2. Use Boolean algebra to convert each expression to its equivalent summation form.
3. Use Karnaugh maps to convert each expression to its equivalent summation form.

Consider the following SOP expressions:

$$T = f(A, B, C) = \sum(0, 3, 4, 5) \qquad \textbf{1.16}$$

$$U = f(A, B, C, D) = \sum(0, 1, 2, 8, 9, 11, 14, 15)$$

$$V = f(A, B, C) = \sum(1, 3, 4, 6, 7)$$

1. Convert each expression to its equivalent conventional form.
2. Reduce each expression to its simplest SOP form.

1.3 IMPLEMENTATION WITH COMBINATIONAL LOGIC

The purpose of the preceding text was primarily to define the expression vocabulary we would need in order to produce real solutions to real problems. If our problem originates with a logical argument, that argument can be placed into SOP form using Boolean algebra. Once in this form, implementation using the methods to follow is a relatively straightforward task. First, note that we are going to look at only a few methods of implementation. The methods we will consider are those that directly relate to the application of programmable logic in the solution. We will be focusing primarily on the use of programmable logic to solve state machine design problems.

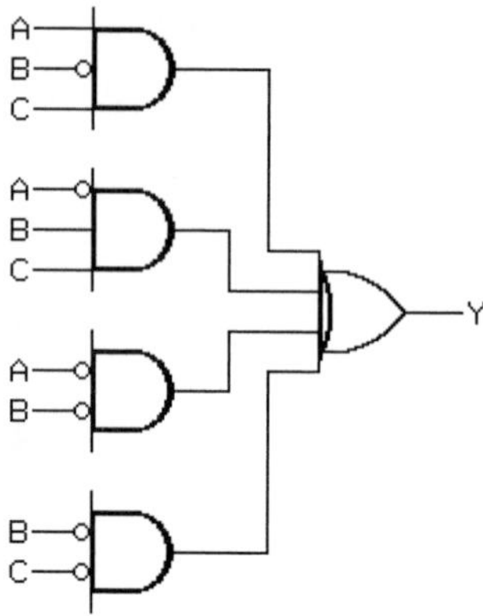

FIGURE 1.2 Discrete logic implementation of SOP expression.

1.3.1 Discrete Logic

Consider a given SOP expression:

$$Y = f(A, B, C) = A\overline{B}C + \overline{A}BC + \overline{A}\,\overline{B} + \overline{B}\,\overline{C} \qquad \textbf{1.17}$$

Without minimization, this expression could be implemented using discrete combinational logic gates as shown in Figure 1.2.

It might be noted that to actually construct this circuit would require three discrete logic chips, one that contains a multiple of two-input AND gates, one that contains a multiple of three-input AND gates, and one that contains a four-input OR gate. It is usually more practical to implement circuits using negative logic, so the following conversion to negative logic is offered.

First, note that the output gate in this circuit, the four-input OR, can be converted to a NAND gate by utilizing **DeMorgan's Theorem:**

$$A + B = \overline{\overline{A}\,\overline{B}} \qquad \textbf{1.18}$$

If this theorem is applied to the output buffer of the circuit shown in Figure 1.2, the circuit shown in Figure 1.3 is realized.

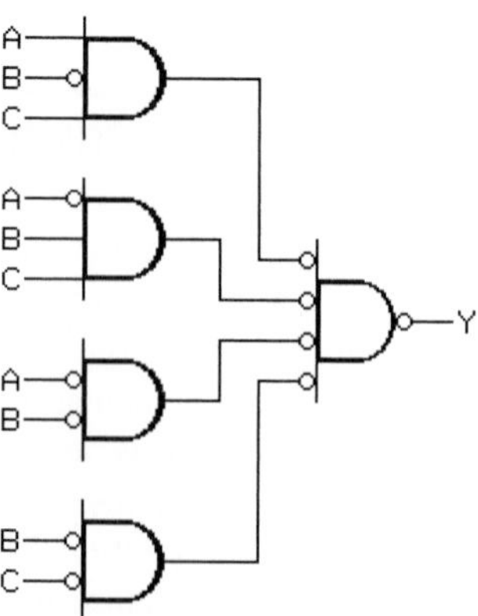

FIGURE 1.3 DeMorgan's equivalent NAND output buffer.

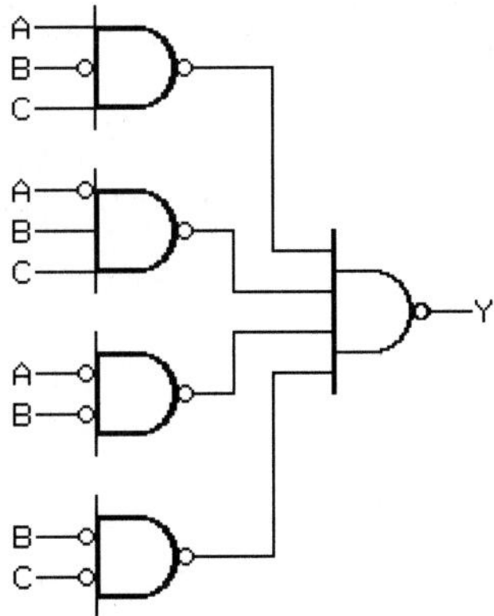

FIGURE 1.4 NAND implementation of conventional summation expression.

If we "slide" the inverters from the inputs to the four-input NAND gate back to the outputs of the AND gates, we will end up with an implementation that uses entirely NAND gates. This is illustrated in Figure 1.4.

As we can see from these examples, the implementation of SOP expressions using discrete logic circuits is relatively straightforward. Each minterm is generated with an AND gate, and the minterms are summed with an OR gate. It is also possible to implement the same expression with only NAND gates if the preceding sequence of design is adhered to.

The application of discrete logic circuits to a real problem may become impractical if the problem is anything but very simple. When the problem becomes more complex, the number of devices required for the solution becomes impractical. There are other methods to be used that implement much more efficiently. This includes the use of multiplexers, ROMs, PALs, and FPLAs.

1.3.2 Digital Multiplexers

A more effective solution to the problem may be to apply a multiplexer instead of discrete gates. The **digital multiplexer** can be used to provide both the minterm generation and the summation needed to implement SOP expressions. Consider this SOP expression again, which was used in the previous example:

$$Y = f(A,B,C) = A\overline{B}C + \overline{A}BC + \overline{A}\,\overline{B} + \overline{B}\,\overline{C} \qquad \textbf{1.19}$$

To implement this expression with a multiplexer, it must first be expanded into each of its unique minterms as described previously in this chapter. It is advantageous to express the argument in summation form:

$$\begin{aligned}
Y &= f(A,B,C) = A\overline{B}C + \overline{A}BC + \overline{A}\,\overline{B} + \overline{B}\,\overline{C} \qquad \textbf{1.20}\\
Y &= f(A,B,C) = A\overline{B}C + \overline{A}BC + \overline{A}\,\overline{B}(C + \overline{C}) + \overline{B}\,\overline{C}(A + \overline{A})\\
Y &= f(A,B,C) = A\overline{B}C + \overline{A}BC + \overline{A}\,\overline{B}C + \overline{A}\,\overline{B}\,\overline{C} + A\overline{B}\,\overline{C} + \overline{A}\,\overline{B}\,\overline{C}\\
Y &= f(A,B,C) = \sum(101, 011, 001, 000, 100, 000)\\
Y &= f(A,B,C) = \sum(5, 3, 1, 0, 4, 0)\\
Y &= f(A,B,C) = \sum(0, 1, 3, 4, 5)
\end{aligned}$$

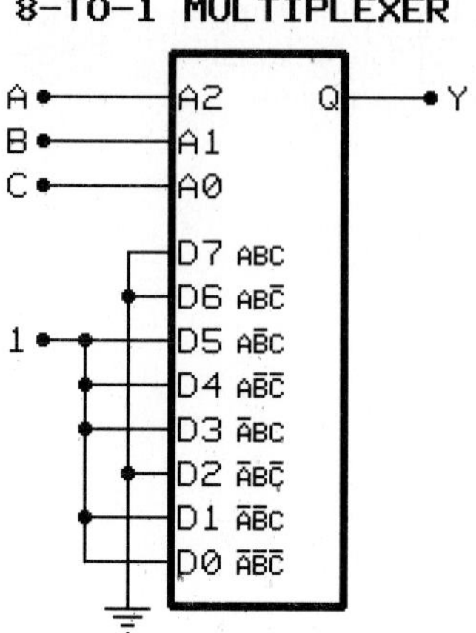

FIGURE 1.5 Multiplexer implementation of SOP expression.

Armed with the summation expression, it is a straightforward matter to configure an eight-input digital multiplexer to generate the minterms. This configuration is shown in Figure 1.5.

Note that the inputs to the multiplexer that are identified as TRUE in the summation expression are tied to a logic 1. The remaining inputs are tied to a logic 0. When used in this configuration, the action of the multiplexer is to act as a "rotary switch," providing on the Q output the value on the addressed D input. Therefore, if ABC = 000, the input on D0 will be present on the output of the multiplexer. According to our expression, minterm 0 is to be TRUE, and it has been tied to a logic 1, consistent with our definition.

Practical Example 1.2: Multiplexer Implementation of SOP Expressions

Use a digital multiplexer to implement the following expression:

$$X = f(A,B,C,D) = \overline{A}CD + AB\overline{C} + \overline{C}\,\overline{D} \qquad \textbf{1.21}$$

Note first that the expression has four variables in the domain. Therefore, there are 16 possible minterms in the domain, and a 16-to-1 multiplexer will be used. First, the expression must be converted to summation form. This was done in Expression Set 1.13 and yielded the following results:

1. $X = f(A,B,C,D) = \overline{A}CD + AB\overline{C} + \overline{C}\,\overline{D}$ **1.22**
2. $X = f(A,B,C,D) = \overline{A}CD(B+\overline{B}) + AB\overline{C}(D+\overline{D}) + \overline{C}\,\overline{D}(\overline{A}\,\overline{B} + \overline{A}B + A\overline{B} + AB)$
3. $X = f(A,B,C,D) = \overline{A}BCD + \overline{A}\,\overline{B}CD + AB\overline{C}D + AB\overline{C}\,\overline{D} + \overline{A}\,\overline{B}\,\overline{C}\,\overline{D} + \overline{A}B\overline{C}\,\overline{D} + A\overline{B}\,\overline{C}\,\overline{D} + AB\overline{C}\,\overline{D}$
4. $X = f(A,B,C,D) = \sum(0111, 0011, 1101, 1100, 0000, 0100, 1000, 1100)$
5. $X = f(A,B,C,D) = \sum(7, 3, 13, 12, 0, 4, 8, 12)$
6. $X = f(A,B,C,D) = \sum(0, 3, 4, 7, 8, 12, 13)$

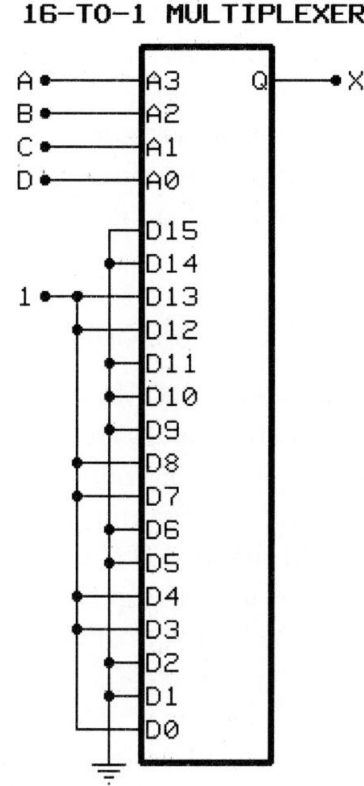

FIGURE 1.6 Practical Example 1.2, multiplexer implementation of SOP expression.

According to the preceding summation expression, minterms 0, 3, 4, 7, 8, 12, and 13 are TRUE, so these inputs to the multiplexer will be tied to a logic 1. Figure 1.6 illustrates the correct solution to this problem.

Practical Exercise 1.2: Multiplexer Implementation of SOP Expressions

Draw the schematic diagram of a digital circuit that uses multiplexers to realize each of the following expressions:

$$W = f(A,B,C) = \overline{A}\,\overline{B}\,\overline{C} + \overline{A}\,\overline{B}\,C + A\overline{B}\,\overline{C} + A\overline{B}\,C \qquad \textbf{1.23}$$

$$X = f(A,B,C,D) = \overline{A}B + A\overline{B}C\overline{D} + \overline{B}CD$$

$$Y = f(A,B,C) = A\overline{B} + \overline{B}C$$

1.4 IMPLEMENTATION WITH PROGRAMMABLE LOGIC

Programmable logic devices are digital logic circuits that can be programmed by the user to implement a wide variety of combinational and sequential logic functions.

The use of a multiplexer to realize a logical argument is sufficient as long as that argument is simple. However, if the domain of the expression exceeds four, the multiplexer solution becomes too complex, and a more powerful solution is needed. Also, if the digital circuit to be implemented requires more than one function, more than one multiplexer is needed. Again, a more powerful solution is required. This solution can come in the form of the application of a ROM, a PAL, or an FPLA.

1.4.1 Read-Only Memory (ROM)

A **read-only memory (ROM)** is a logic circuit that can generate all of the possible minterms of its inputs. Diodes are used to summate minterms into an SOP expression. The summation diodes in a ROM are placed in the device during fabrication.

When we were configuring the multiplexer in the previous section, we were actually emulating a ROM that had *N* one-bit words, where *N* was the number of multiplexer inputs. That is, the 8-to-1 multiplexer was emulating the function of an eight-word ROM, where each word had a length of one bit. Figure 1.7 illustrates such a ROM circuit.

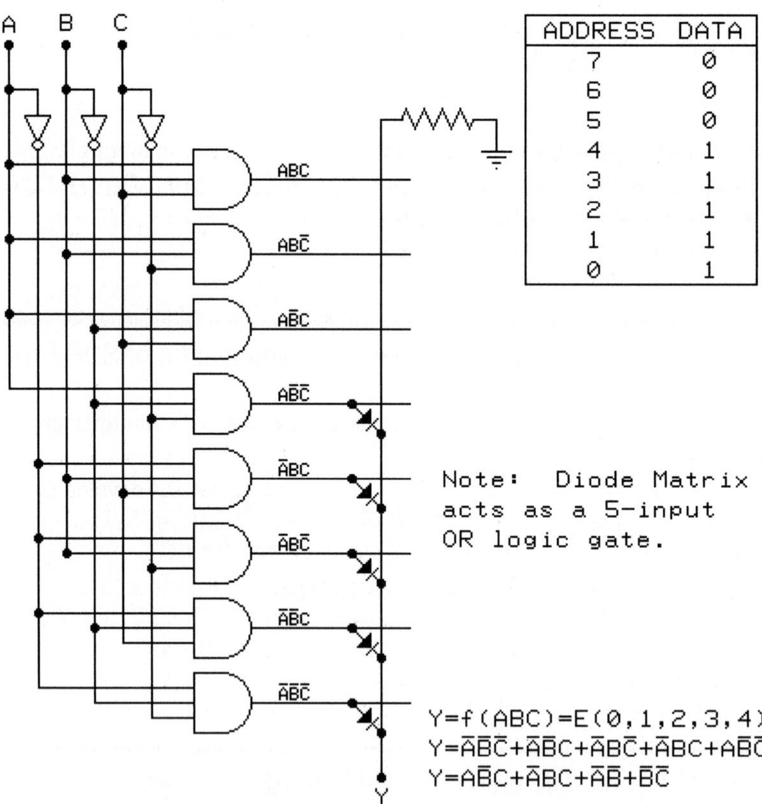

ADDRESS	DATA
7	0
6	0
5	0
4	1
3	1
2	1
1	1
0	1

FIGURE 1.7 Eight-word by one-bit read-only memory.

Each of the AND gates in Figure 1.7 is used to decode one of the eight possible minterms generated by three variables. When configured in this manner, the action of the output diodes is to logically OR together all minterms where a diode is present. In this example, diodes are present for minterms 0, 1, 2, 3, and 4. Consequently, the output function of the circuit is equal to $\Sigma(0,1,2,3,4)$.

Observe the following characteristic of the ROM circuit: *Input* products are *hard-wired* and include all possible minterms. *Output* summation circuitry is *programmable.*

Practical Example 1.3: ROM Implementation of SOP Functions

Draw the schematic diagram of a ROM implementation of the following set of logic equations:

$$\begin{aligned} W &= AB\overline{C} + A\overline{B}C + \overline{A}\,\overline{B}\,\overline{C} \\ X &= A\overline{B} + \overline{A}C \\ Y &= AB + \overline{A}\,\overline{B} + AC \\ Z &= A\overline{B}\,\overline{C} + BC \end{aligned} \qquad \textbf{1.24}$$

First, the expressions must be converted to summation form. Let's use Karnaugh maps to accomplish the conversion this time. By extracting the minterm cell numbers from the Karnaugh maps that are represented in Figure 1.8, we obtain the following summation equations:

$$\begin{aligned} W &= f(A,B,C) = \sum(0,5,6) \\ X &= f(A,B,C) = \sum(1,3,4,5) \\ Y &= f(A,B,C) = \sum(0,1,5,6,7) \\ Z &= f(A,B,C) = \sum(3,4,7) \end{aligned} \qquad \textbf{1.25}$$

Now, armed with these summation equations, we can develop the ROM schematic diagram by generating an output bit for each expression and inserting diodes for the true minterms. (See Figure 1.9.)

The operation of the ROM is a little bit more obvious with this example. The input lines A, B, and C provide the addresses for the ROM. Since there are three address lines, the ROM decodes them to address eight separate locations. At each location is a four-bit word, defined by the presence of the diodes.

If this system of equations were implemented with multiplexers, how many multiplexers would be required? The answer is four, one for each expression.

Since each logic function is generated by a ROM data bit, additional functions can be generated by adding additional bits. As the domain of the function increases, the number of inputs increases. For example, if the domain of an expression is four (A, B, C, D), then the ROM will have four address lines decoded into 16 unique minterms.

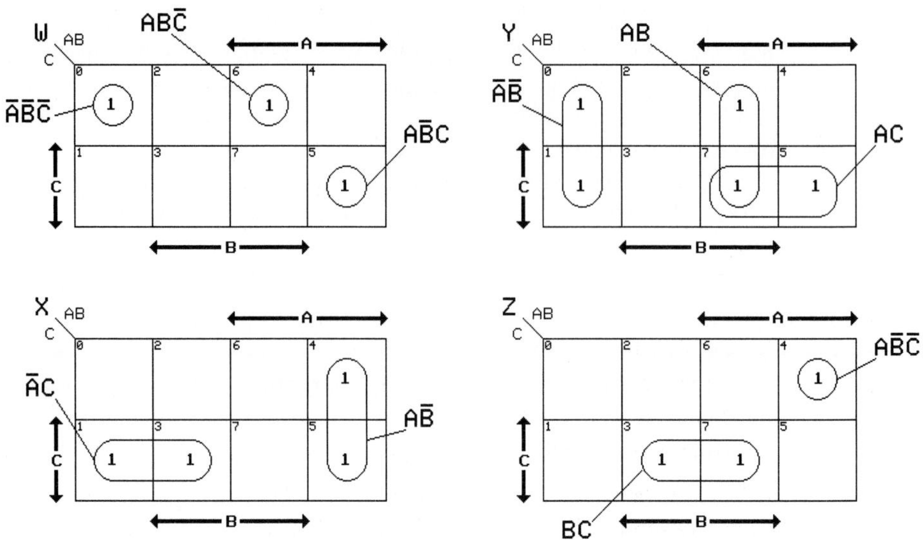

FIGURE 1.8 Practical Example 1.3, Karnaugh maps of logical functions W, X, Y, Z.

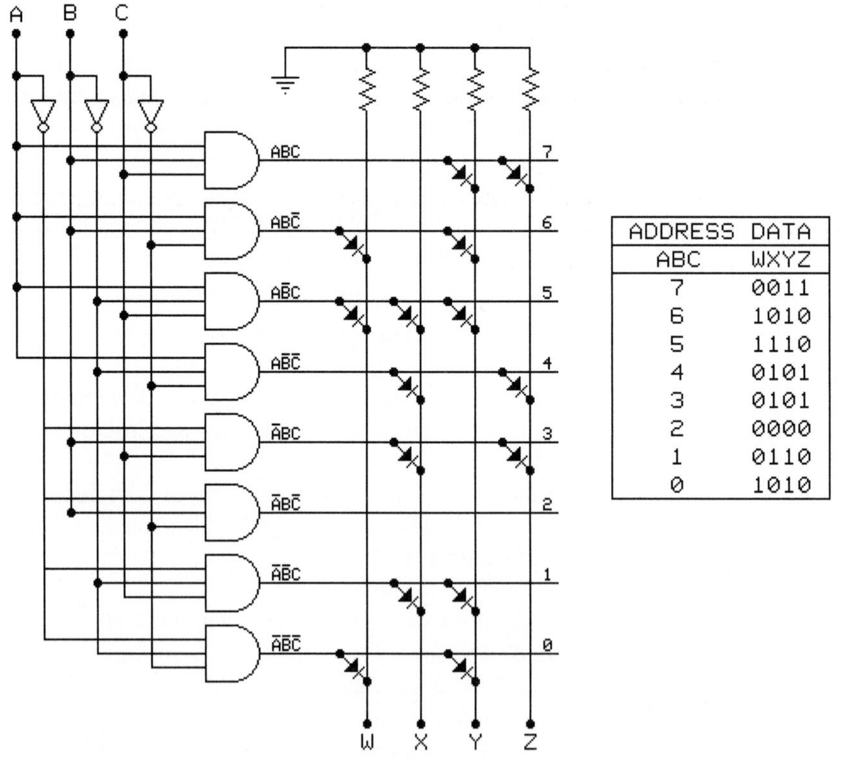

ADDRESS	DATA
ABC	WXYZ
7	0011
6	1010
5	1110
4	0101
3	0101
2	0000
1	0110
0	1010

FIGURE 1.9 Practical Example 1.3, ROM circuit.

Practical Exercise 1.3: ROM Implementation of SOP Functions

Draw the schematic diagram of a ROM implementation of the following functions:

$$\begin{aligned} W &= \overline{A}\,\overline{C} + A\overline{C}\,\overline{D} + A\overline{B}D \\ X &= A\overline{C} + BD + AB + \overline{A}\,\overline{B}C\overline{D} \\ Y &= \overline{C}\,\overline{D} + \overline{A}B + B\overline{C} + A\overline{B}D \\ Z &= \overline{C}D + AB\overline{C} + \overline{A}B\overline{D} + A\overline{B}C\overline{D} \end{aligned} \tag{1.26}$$

The ROM circuit is useful for generation of logical functions and is particularly suited to situations where those functions are numerous and/or quite complex. For example, if a combinational logic environment has a domain of eight, and there are 16 different logic functions generated, a ROM with an eight-bit address and a 16-bit data word (256 words by 16 bits/word) would fulfill the need.

ROM Specification. One common way of specifying a ROM is to refer to it by its architectural limits. For example, the ROM we are observing has three input lines generating *eight* (2^3) unique minterms, and it has *four* output lines. This represents eight words of four bits each. This device can be specified as an 8 x 4 ROM. A typical commercial device may be specified as a 4096 x 8 device. This device could be used to generate eight output expressions with a domain of twelve ($2^{12} = 4096$).

In most applications requiring such complex logical functions, not all of the minterms that can be generated in the domain are needed. When this is the case, there are two other circuits, each based on the ROM architecture, that will fill the need: the PAL and the FPLA.

1.4.2 Field Programmable Logic Array (FPLA)

This section will differentiate between **field programmable logic array (FPLA)** and **programmable array logic (PAL)** devices. Though similar in design, each device has distinctive characteristics and implementation advantages. We will start by looking at the FPLA. First, consider the following set of logical functions:

$$\begin{aligned} W &= B\overline{C} + ACD + \overline{A}\,\overline{B}C \\ X &= \overline{A}\,\overline{B}C + BD + AB\overline{C} \\ Y &= AC + \overline{A}\,\overline{C} \\ Z &= B\overline{C} + BD \end{aligned} \tag{1.27}$$

The circuit that will produce this output will have four inputs, A, B, C, and D, and four outputs, W, X, Y, and Z. Figure 1.10 illustrates a detailed schematic diagram of an FPLA implementation of these functions. Note the similarity to the ROM circuit. The input circuit uses three-input AND logic gates driven by a *programmable* diode matrix to generate each of the unique minterms in the expression. In the ROM implementation, all possible minterms are generated in the input circuit. In the FPLA circuit, only those minterms that are needed are generated. Also, each is generated only once, even though it may appear multiple times in the output expressions. The output circuit uses a programmable diode matrix to combine the generated minterms into the complete output expression. The output diodes are inserted at each output bit in a logical OR arrangement, generating the required SOP expression.

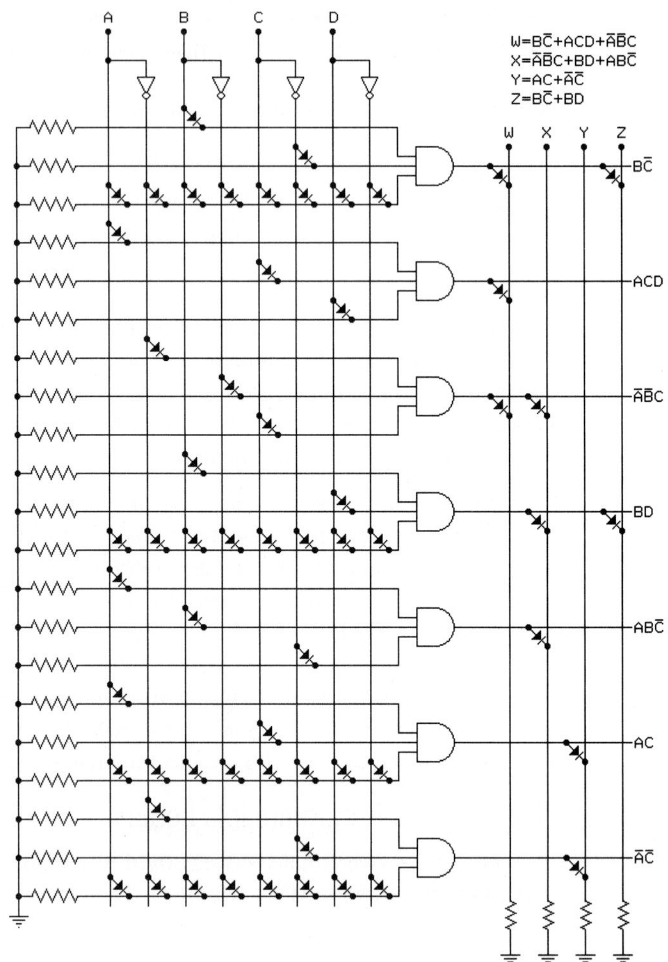

FIGURE 1.10 FPLA implementation of logical functions.

The FPLA can be viewed as a ROM with many of its locations deleted. Only those locations that are defined by the input logic are required. Consequently, an FPLA is usually smaller than the ROM that would be required to solve the same problem. For example, the ROM solution to the current problem would require decoding of 16 unique minterms. However, when an FPLA is used, only seven are needed.

Observe the following characteristic of the FPLA circuit: *Input* product circuitry is *programmable. Output* summation circuitry is *programmable.*

Note that the unused inputs to the AND logic gates of Figure 1.10 are tied to all of the possible input lines. This ensures a logical 1 on the input to the AND gate. This was done to allow the use of three-input AND gates for each product. Also, it may be of interest to note that an unprogrammed device would typically have diodes placed at each intersection. The programming procedure would burn out diodes that are not desired.

FPLA Specification. One common way of specifying an FPLA device is to refer to it by its architectural limits. For example, the FPLA we are observing has *four* input lines, it can generate *seven* minterms, and it has *four* output lines. This device can be specified as a 4 x 7 x 4 FPLA. A typical commercial device may be specified as a 12 x 96 x 16 device. This device could be used to generate 16 output expressions with a domain of 12, as long as no more than 96 unique minterms are generated. The smallest ROM that could be used would have 4096 16-bit words, or an architecture of 12 x 4096 x 16. Therefore, the FPLA implementation would be 96/4096, or 2.3% of the size of the ROM, which is quite an improvement.

1.4.3 Programmable Array Logic (PAL)

A more common programmable solution uses **programmable array logic (PAL)** devices. The input circuitry to the PAL is similar to that of the FPLA. However, the output circuitry includes hardwired OR logic and is not programmable. Figure 1.11 illustrates the solution of the previous logical expressions using a PAL circuit. Note that the circuit shown can create up to four outputs. Each output, by virtue of the hardwired OR logic, can create an expression with three minterms, and each minterm is limited to three arguments, which are selected from the available inputs.

Note that the limitation of this device is its inflexibility of the output logic. Since the output is hardwired, it is easier than the FPLA to program. However, the device must be selected with care to make sure that enough minterms are generated by each output and that enough variables are available for each minterm so that the device can be programmed to generate the desired functions.

PAL Specification. One common way of specifying a PAL device is to refer to it by its architectural limits. The enumeration method is not quite as straightforward as with ROM and FPLA devices. A typical PAL specification of a common commercial device is 16L8. This means that there are 16 inputs to the programmable diode matrix, and there are eight outputs. In this device six of the outputs are fed back to the diode matrix. If that is the case in this example, there are 10 inputs to the diode matrix left. These are the external inputs to the device. The "L" refers to the type of output buffer used; three types of output buffer will be investigated in this text. The "L" type refers to a combinational logic output buffer. This is the SOP logic circuit we have been observing. Commercial devices usually invert the output. We will look at these output buffers in more detail later.

Observe the following characteristic of the PAL circuit: *Input* product circuitry is *programmable. Output* summation circuitry is *hardwired.*

1.5 PROGRAMMABLE LOGIC SCHEMATIC FORMS

The logic circuits of Figures 1.9, 1.10, and 1.11 are quite detailed, illustrating each programmable junction by the insertion of a diode. A simpler convention will be used in the remainder of this text. Instead of using diodes, a cross will be placed at the junction where a diode is to be placed. A dot will be used to identify a hardwired junction. Figure 1.12 illustrates a ROM schematic where the diodes have been replaced with dots.

Figure 1.13 illustrates the simplified form of the FPLA schematic. Note its similarity to the ROM circuit. A comparison at this point may help in understanding the differences between them. Again, the PAL and FPLA input matrices are programmable, reducing the number of minterms generated, resulting in a smaller device.

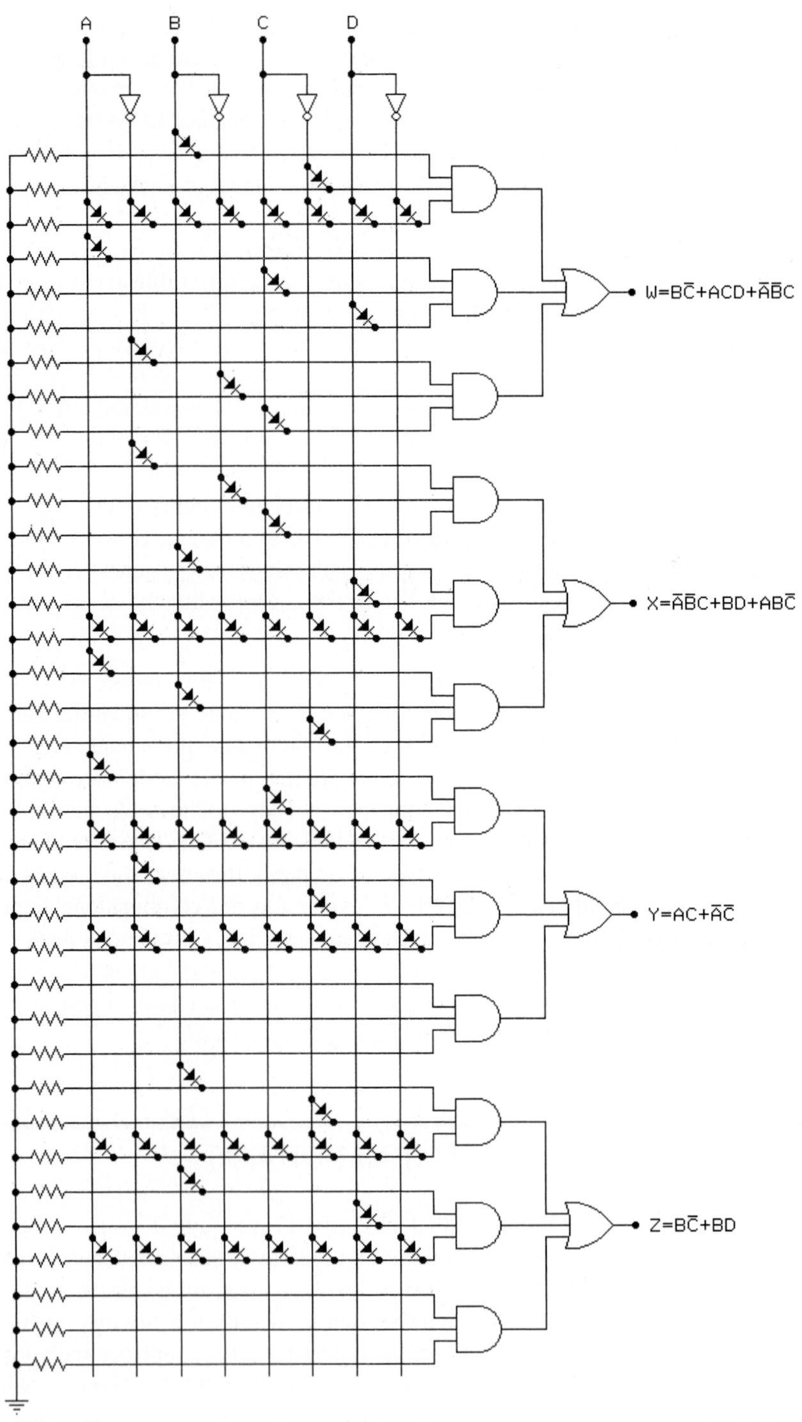

FIGURE 1.11 PAL implementation of logical functions.

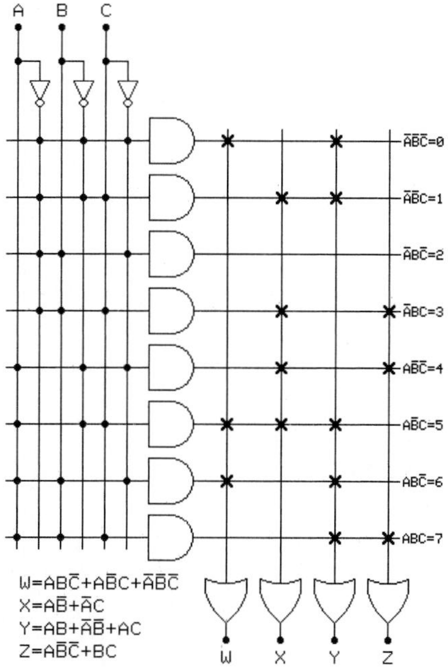

FIGURE 1.12 ROM schematic, simplified form.

Figure 1.14 illustrates a simplified form of the PAL schematic. The function of this device can be compared with that of the ROM and the FPLA. Figure 1.15 illustrates a second form of the PAL schematic diagram. Note that these two schematic diagrams are functionally identical, though the second may be simpler to draw. Also, note that the diode matrix in the output circuit of Figure 1.15 is hardwired, as it should be.

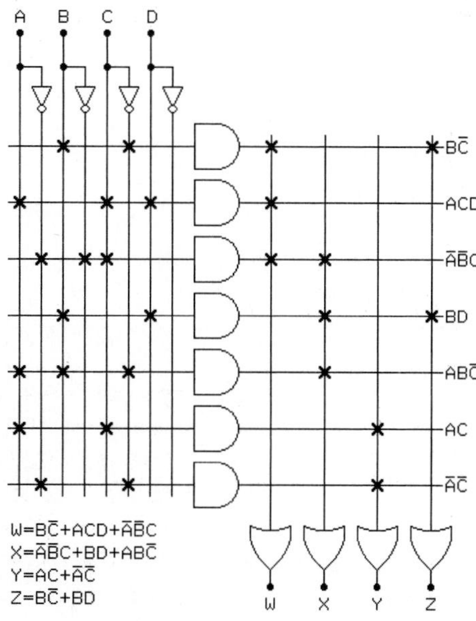

FIGURE 1.13 FPLA schematic, simplified form.

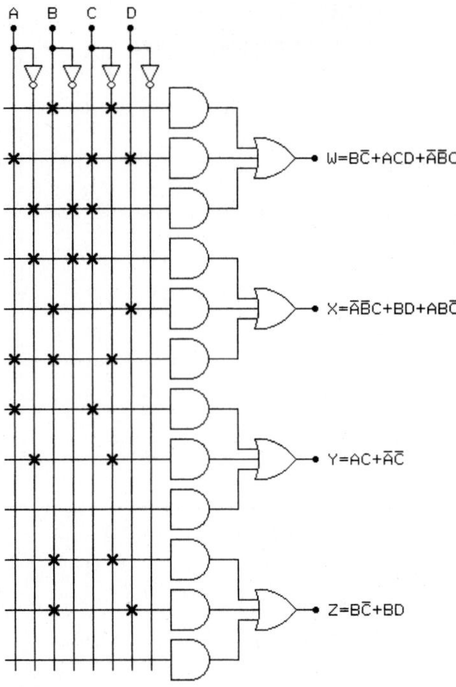

FIGURE 1.14 PAL schematic, first form.

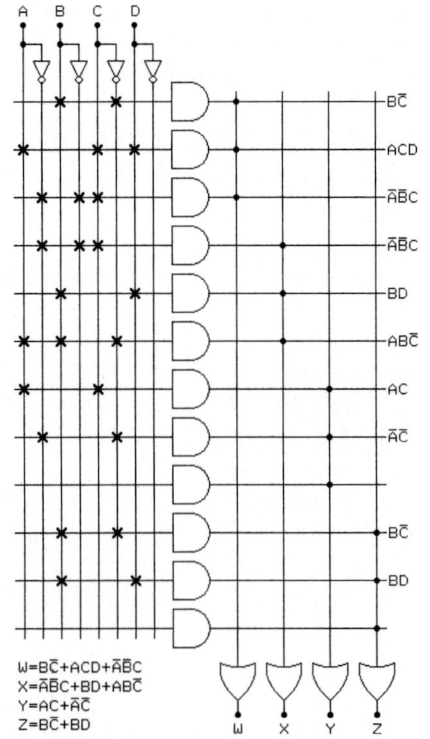

FIGURE 1.15 PAL schematic, second form.

TABLE 1.2 ROM/FPLA/PAL device summary.

Device	*Input Matrix*	*Output Matrix*	*Output Logic*
ROM	Hardwired	Programmable	Combinational, SOP only
FPLA	Programmable	Programmable	Several output buffer types available
PAL	Programmable	Hardwired	Several output buffer types available

1.5.1 Device Summary

Table 1.2 provides a summary of the distinctions associated with each of the three programmable devices described here, the ROM, the PAL, and the FPLA.

1.6 PAL OUTPUT MACROCELLS

The output circuit of the PAL device is hardwired and is commercially available in a variety of different forms. The most common forms are described here. This description is by no means complete, but it does represent a significant portion of the devices available. There are a variety of additional output circuits available.

The output circuit of a PAL device is referred to as a **macrocell.** This is the portion of the circuit that is driven by the OR logic gates that generate the SOP expression.

1.6.1 Logic-Type Macrocell

Figure 1.16 is a detailed schematic diagram of a commercial-grade PAL that utilizes the **L-type** or **logic-type macrocell.** This macrocell is identified by the output buffer following the OR logic gate. The output logic is combinational.

Observe the following characteristics of this circuit:

1. The output buffer is enabled from logic derived from the input diode matrix. This is not always the case. Often, all of the output buffers in the device are enabled by a single externally applied logical signal. When the output buffer is enabled from the diode matrix, any single minterm that can be generated as a function of any of the inputs to the matrix can be used to enable the output. If the output is always to be enabled, all of the diodes driving the AND logic gate for the buffer would be left intact. **If the output is disabled, it can be used as an additional input.**
2. The outputs from the PAL are fed back to the input diode matrix. Again, this is not always the case. When the outputs are fed back, they can be used as arguments in the logical expressions generated by other outputs. This will allow the PAL to be programmed to execute logical functions that have more than one level of SOP terms. It also allows the PAL to be programmed to execute logical functions that have feedback, such as flip-flops and other sequential circuits.
3. The input buffers have been drawn slightly differently from those of the previous figures. Instead of appearing at the top of the drawing, they appear at the left side. Though

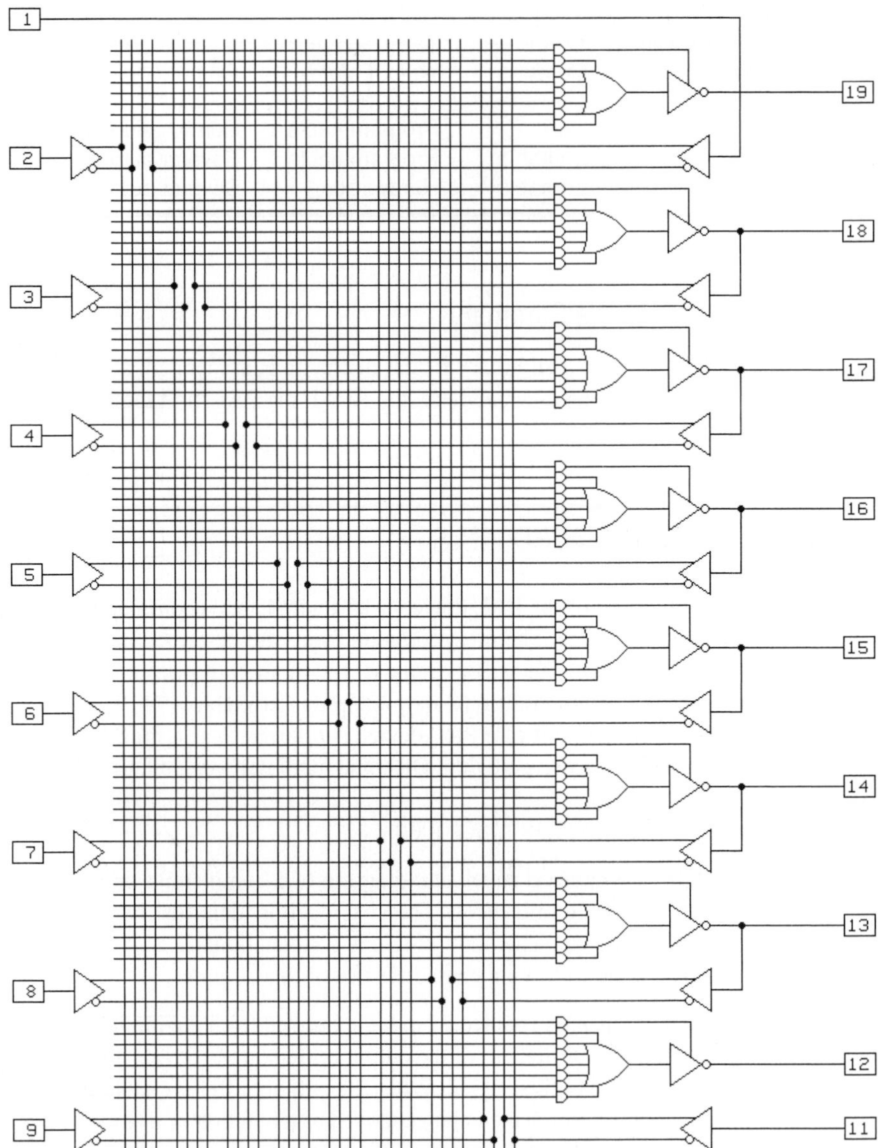

FIGURE 1.16 Unprogrammed 16L8 PAL circuit diagram.

their function is identical, it is simpler to draw them this way when the outputs are fed back to the diode matrix. This is also a convention used by several device manufacturers.

4. Each output is limited to seven minterms, though each can have a domain of 16.

Figure 1.17 shows the pin layout of a typical 16L8 PAL device. This is a convenient way of viewing the device, as it shows both the pin layout and the output macrocell circuits.

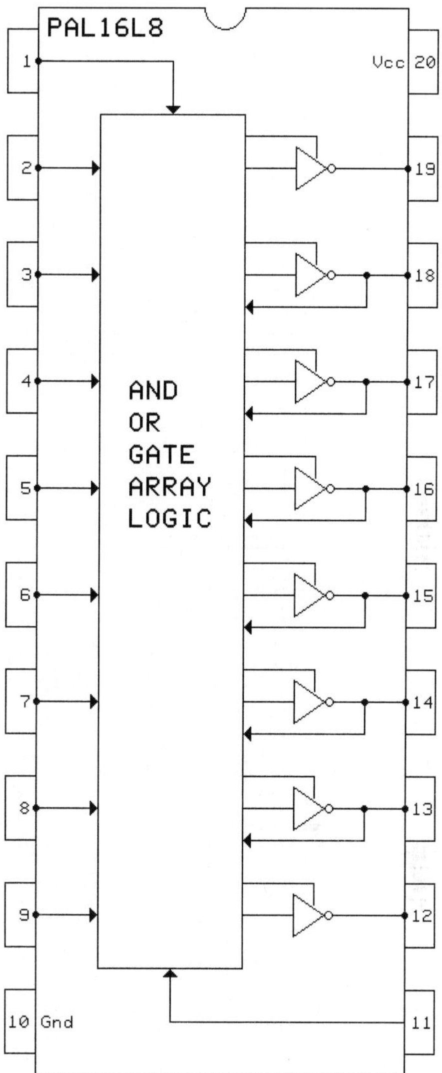

FIGURE 1.17 PAL16L8 device diagram.

1.6.2 Registered-Type Macrocell

Figure 1.18 illustrates the detailed schematic diagram of a commercial grade PAL which utilizes the **R-type** or **registered-type macrocell.** This macrocell is identified by the insertion of the D-type flip-flop between the OR logic gate and the output buffer. The clock signal for each flip-flop is driven by an externally applied signal. This feature makes this device especially well suited for use in synchronous counter design. A typical synchronous counter is a register with common clocks. The count sequence is determined by logical arguments fed back to the flip-flop inputs. Note that these facilities are available in this circuit. Some PAL chips are available that have individual clock inputs driven by a minterm generated from the diode matrix.

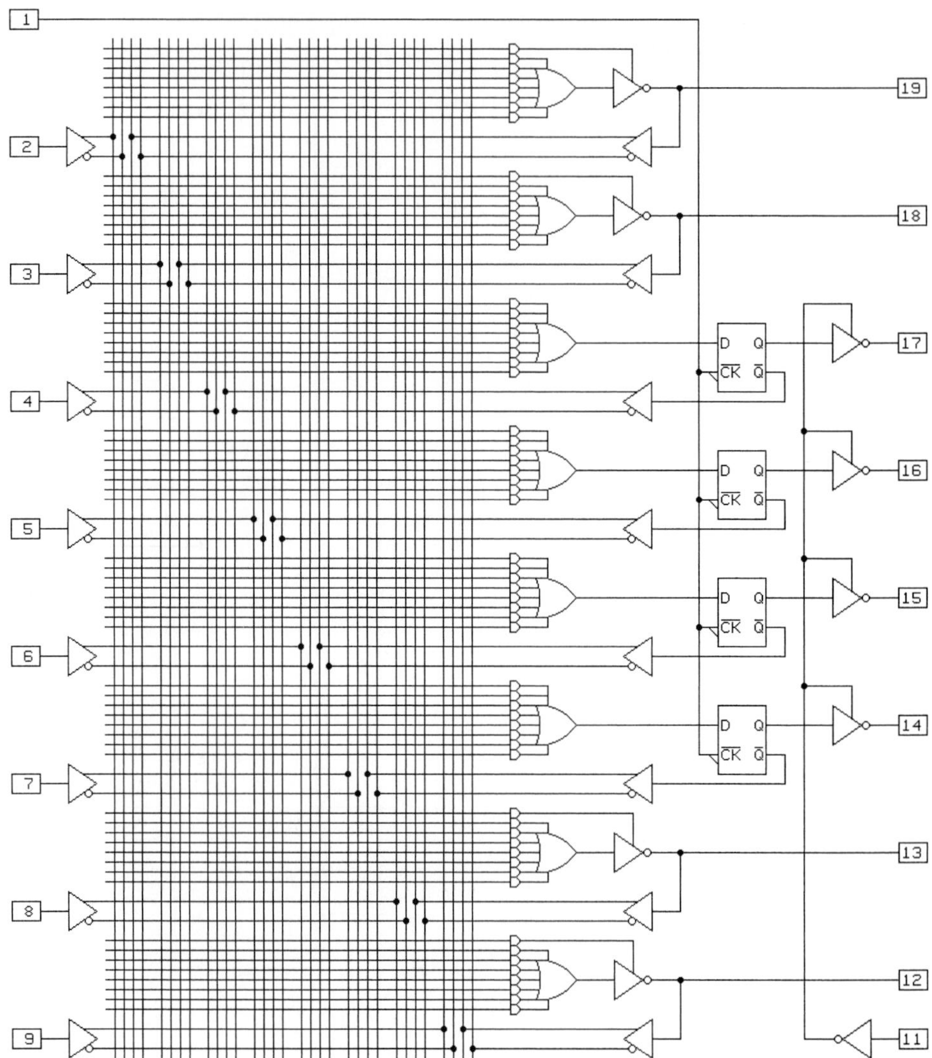

FIGURE 1.18 Unprogrammed PAL 16R4 circuit diagram.

The device illustrated in Figure 1.18 is referred to as a 16R4 device. This implies that there are 16 input arguments to the diode matrix (eight from external inputs and eight fed back from the outputs), and there are four *R*egistered outputs. This device is particularly well suited to applications that combine logical and registered arguments, such as a four-bit counter-type state machine. The registered outputs can be used to implement sequential portions of the problem, such as a synchronous feedback counter, and the logic outputs can be used to generate other combinational arguments. Each logical output expression generated by this device is limited to seven minterms. Each registered output can be driven by an expression with eight minterms. While the logical output buffers can be programmed from the diode matrix, the registered output buffers are enabled by an externally applied signal.

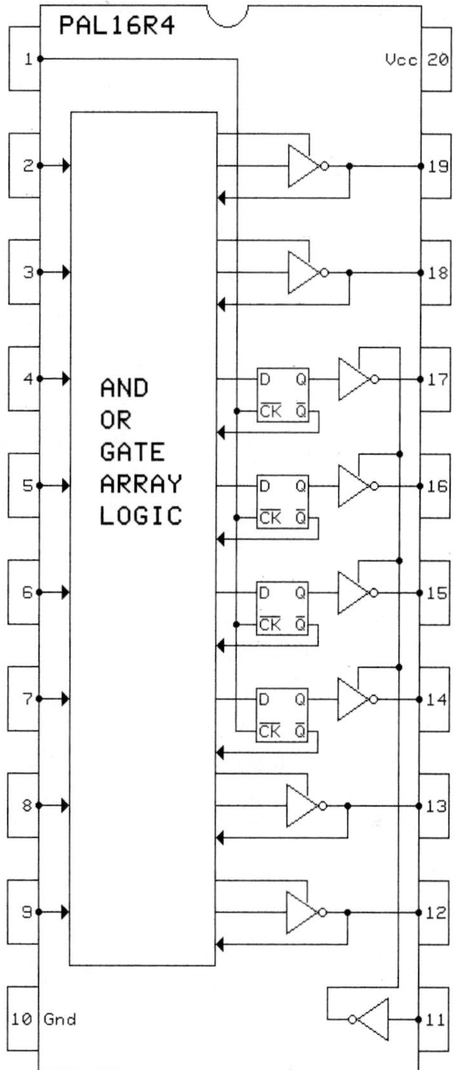

FIGURE 1.19 PAL16R4 device diagram.

Figure 1.19 shows the device diagram for the 16R4.

1.6.3 Variable-Type Macrocell

The **V-type** or **variable-type macrocell** is a circuit that can be programmed to execute the function of either an L or an R type. Furthermore, the assertion level of the output can be programmed to be either active-high or active- low. Figure 1.20 illustrates a detailed logic diagram of the V-type macrocell. As with the L-type, it is placed between the OR gate that generates the SOP expression and the output of the device.

The flexibility of this device is provided by the use of a 4-to-1 multiplexer, which is shown in the center of the schematic. One of four modes is selected by programming the

V-TYPE MACROCELL

A1	A0	OUTPUT CONFIGURATION
0	0	REGISTERED/ACTIVE-LOW
0	1	REGISTERED/ACTIVE-HIGH
1	0	COMBINATORIAL/ACTIVE-LOW
1	1	COMBINATORIAL/ACTIVE-HIGH

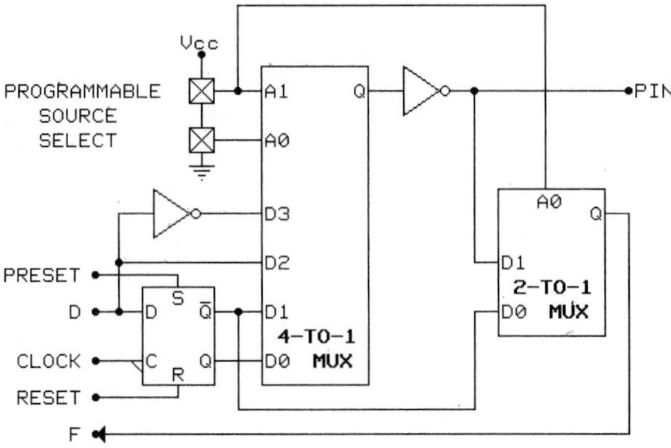

FIGURE 1.20 V-type macrocell.

fusible links associated with the A_1 and A_0 inputs. For example, if $A_1A_0 = 11$, the D input (which comes from the OR gate generating the SOP expression) is passed through to the output pin without change. If $A_1A_0 = 10$, the D input is inverted prior to being passed through to the output pin. These two modes, where $A_1 = 1$, select the L-type output buffer modes of the device. If $A_1A_0 = 00$, the Q output of the D-type flip-flop is selected. If $A_1A_0 = 01$, the $\overline{Q}$ output of the D-type flip-flop is selected. These two modes, where $A_1 = 0$, select the R-type output buffer modes of the device.

The V-type macrocell is a popular choice for digital system designers because of its ability to function as either an L or an R type.

Figure 1.21 illustrates the device diagram for the PAL16V8.

1.7 EXAMPLE PROGRAMMABLE LOGIC DEVICE ARCHITECTURES

Recall from our previous discussion the fundamental differences between PAL and FPLA devices: the summation logic in the gate array of PAL devices is hardwired, while the FPLA is programmable. This creates a limitation of the number of minterms summed in expressions generated by PAL devices. The most common PAL devices limit the number of minterms to seven or eight per expression. A few larger PALs increase that limit to as high as 16. The FPLA typically places little or no restriction on the number of minterms generated in each expression. This requires considerably more circuitry. Consequently, FPLA devices are often more expensive.

FIGURE 1.21 PAL16V8 device diagram.

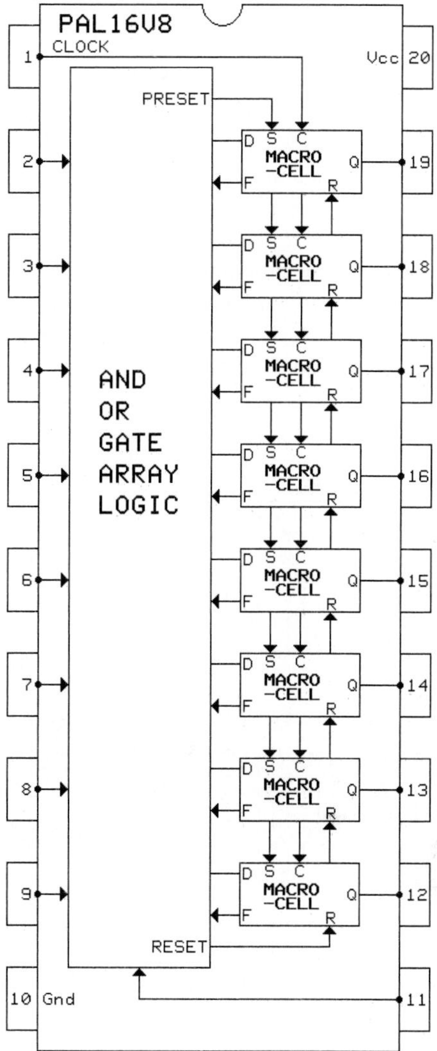

1.7.1 Field Programmable Logic Array (FPLA)

Consider the FPLA schematic illustrated in Figure 1.22. Careful examination of it will reveal the following features:

- The device has eight logic inputs, I0 through I7.
- The device has 10 outputs, B0 through B9. Each output assertion level is programmable using an XOR gate.
- All outputs are fed back to the gate array, providing a total of 18 inputs to the array. If the outputs are not enabled, they may be used as inputs.
- Each of 10 outputs is combinational and can generate expressions of up to 32 minterms using any or all of the eight inputs or 10 outputs as operands.

This device and others like it are useful for application as code converters, encoders, arithmetic circuits, or small ROMs. When the PLS153 device is applied as a ROM, it looks like a 32-word device with ten bits per word. Only five of the input lines would be used.

Note that the outputs of the PLS153 are individually enabled from expressions generated by the gate array. Since the devices are shipped with all diodes in place, the outputs are enabled until programmed. Also, note the method used to program the assertion level of each output. At the lower center of the schematic diagram we see that the output of the summation circuitry (OR gate) drives an input to an XOR gate. The other input to the XOR gate can be programmed to a logic 1 or a logic 0. When programmed to a logic 0, the data passes through it unchanged. However, if the programmable input of the XOR gate is tied to a logic 1, the data passing through it is complemented by virtue of the effect of the XOR function on the data.

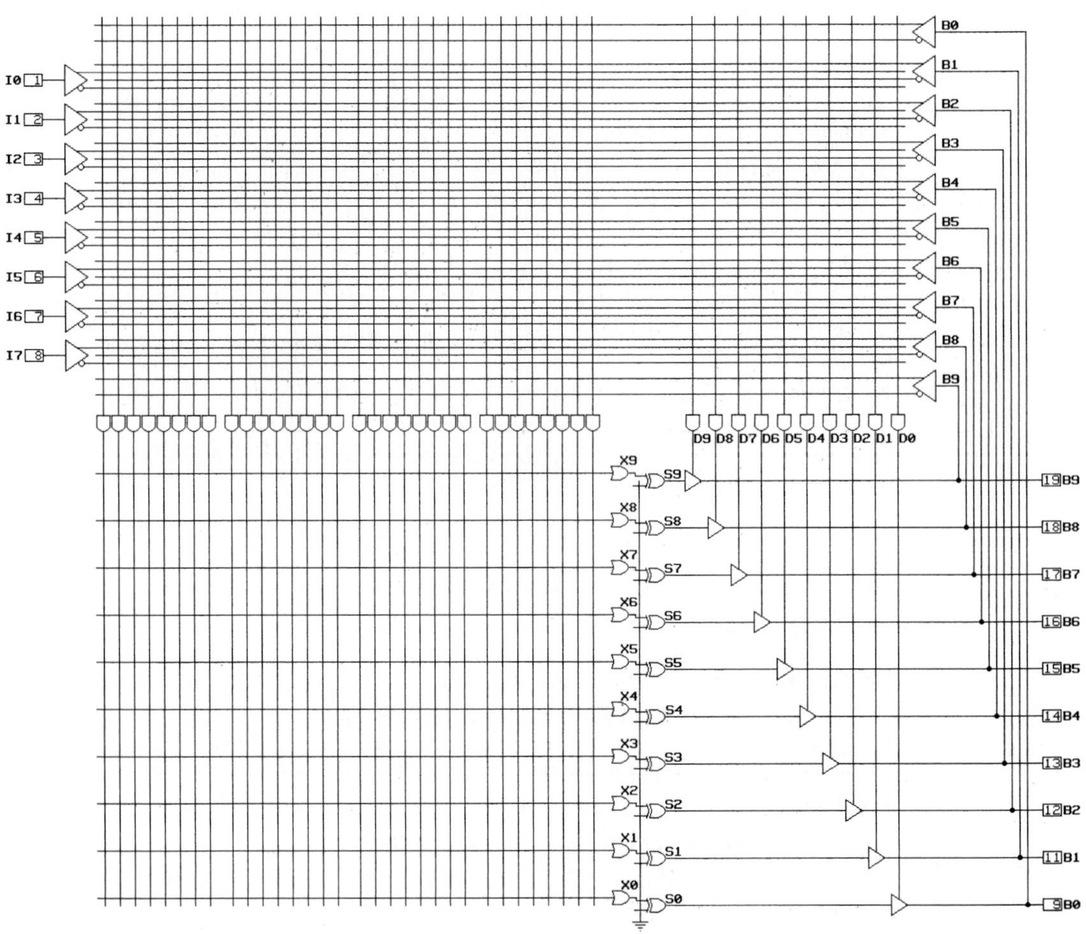

FIGURE 1.22 Philips PLS153 FPLA (18 x 32 x 10).
Source: Philips, "Programmable Logic Devices (PLD) Data Handbook," Philips Semiconductor, Sunnyvale, CA, 1994.

1.7.2 Field Programmable Logic Sequencer (FPLS)

In this text we will find that a common use of programmable logic is in the design of control circuits. We will often be designing state machines with relatively complex logic that is used to control a defined count sequence. Similar to the use of the register-type macrocells of the PAL devices, the field programmable logic sequencer (FPLS) circuit will provide such counter control capabilities. When combined with an FPLA circuit, the use of an FPLS can result in an elaborate control circuit.

Consider the FPLS device illustrated in Figure 1.23. This device has 16 external inputs (I0 through I15) to the gate array, and the gate array is capable of generating functions of up to 48 minterms. Each minterm can have up to 23 terms, 16 coming from the input pins, and seven coming from internally generated functions. The eight outputs are enabled by a single externally applied signal. Each output macrocell is comprised of a single SR-type flip-flop. The S and R inputs of each flip-flop are driven by the gate array. None of the outputs are fed back to the gate array.

An additional register of six flip-flops (the outputs are referred to as P0–P5 in the schematic diagram) have their outputs fed back to the gate array. This provides an important function. Often the state machine to be implemented contains a binary counter that may count in any variety of sequences. It is not necessary for the output of the counter to be available to the external circuitry, but it is needed to control the state machine sequence. These six flip-flops provide such a capability. A synchronous six-bit feedback counter can be implemented using these flip-flops without tying up output pins.

The register of flip-flops can be preset by a logic 1 on pin 19, the pin used to enable the outputs. That is, when the pin is a logic 0, the output of the register is enabled. When the pin is a logic 1, the register is preset such that each flip-flop is a logic 1. This is an asynchronous input. Presetting to a logic 1 provides a starting point for a state machine design. When used, it is often possible to design simpler feedback circuitry where unused binary counter states can be treated as "don't cares." Otherwise, if there are missing counts in the sequence, it is possible for such a count to be loaded at power-up, placing the state machine in an undefined state.

Located in the schematic diagram of Figure 1.23, immediately above the register of SR-type flip-flops, is a single summation circuit labeled C. This allows for the generation of a single logical expression that is fed back to the gate array for use by all of the other circuits.

Finally, note that it is possible to configure any of the SR-type flip-flops as a D-type output. If the complement of the SET function is assigned to the RESET input of an SR-type flip-flop, the output of the flip-flop simply follows its input on the asserted edge of the clock cycle, similar to the definition of a D-type flip-flop. It is not possible to configure any of the outputs of the PLS105 as a logical circuit. If needed, it will be necessary to add an FPLA or a PAL to the overall design when such a capability is needed.

Consider the PLS167 device illustrated in Figure 1.24. Though similar to the PLS105 device, note that this device has two fewer inputs and two fewer outputs. It contains the internal register six-bit register of SR flip-flops labeled P2–P7, and two outputs labeled P0 and P1 are connected back to the gate array. This device can also generate a single logical expression, C, which is also connected back to the gate array.

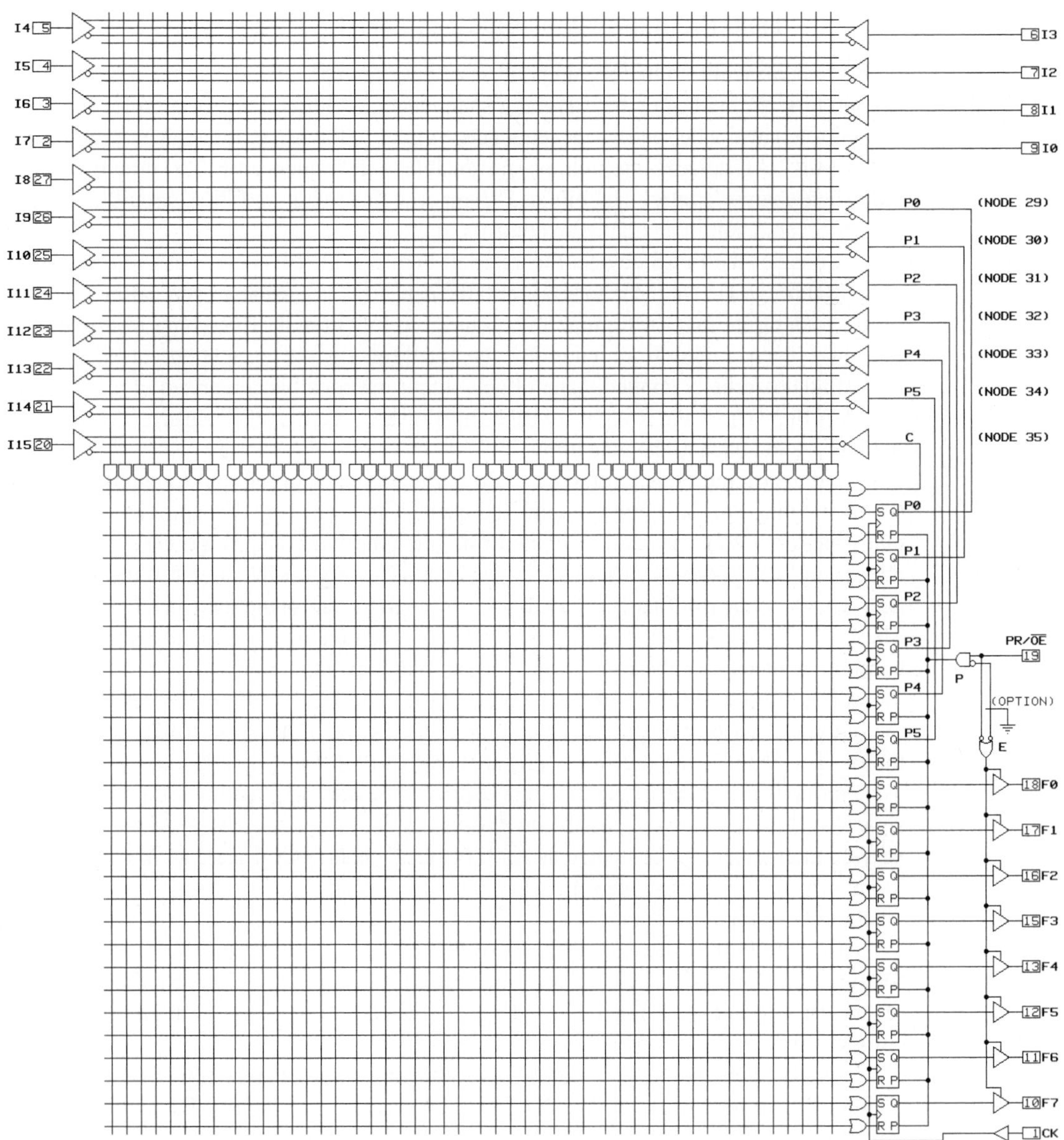

FIGURE 1.23 Philips PLS105 FPLS (16 x 48 x 8).
Source: Philips, "Programmable Logic Devices (PLD) Data Handbook," Philips Semiconductor, Sunnyvale, CA, 1994.

Consider the PLS168 device illustrated in Figure 1.25. This device has 12 inputs to the gate array and eight registered outputs. Four of those outputs are connected back to the gate array. It also contains the internal six-bit register and an internally generated logic expression, C.

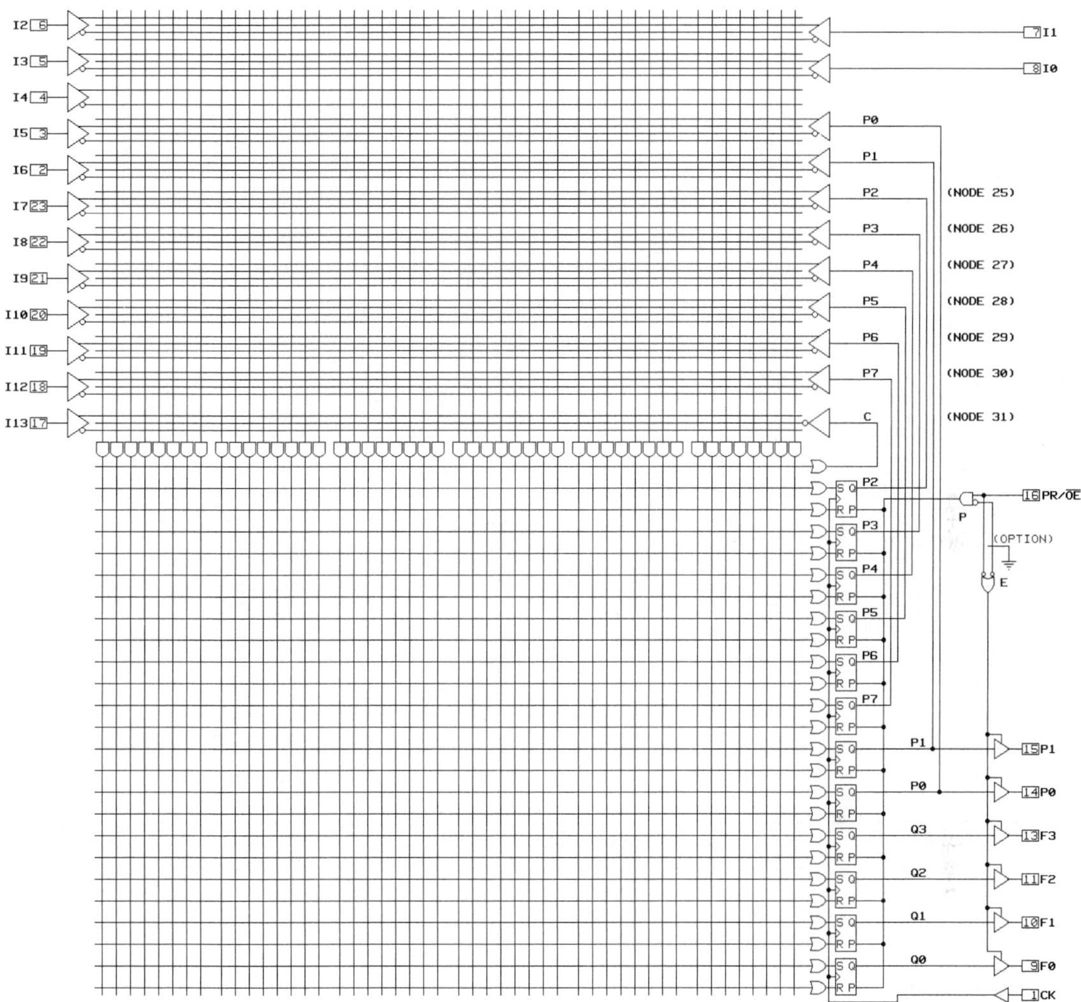

FIGURE 1.24 Philips PLS167 FPLS (14 x 48 x 6).
Source: Philips, "Programmable Logic Devices (PLD) Data Handbook," Philips Semiconductor, Sunnyvale, CA, 1994.

We will be able to see in some of the examples and exercises later in this text that it is often necessary to look for devices that generate more minterms than we might expect. A common error in design is to select a device that cannot generate enough minterms. Consider the programmable logic sequencer illustrated in Figure 1.26. At this point in our study we should be able to identify the capabilities of the PLS405 device. Note that it is quite an enhancement of the previous devices. It has 16 external inputs to the gate array and eight registered outputs. It also contains an internal eight-bit register and can generate two internal logical expressions, C0 and C1. It has the ability to generate logical expressions of 64 minterms, where each minterm can have any of 29 variables.

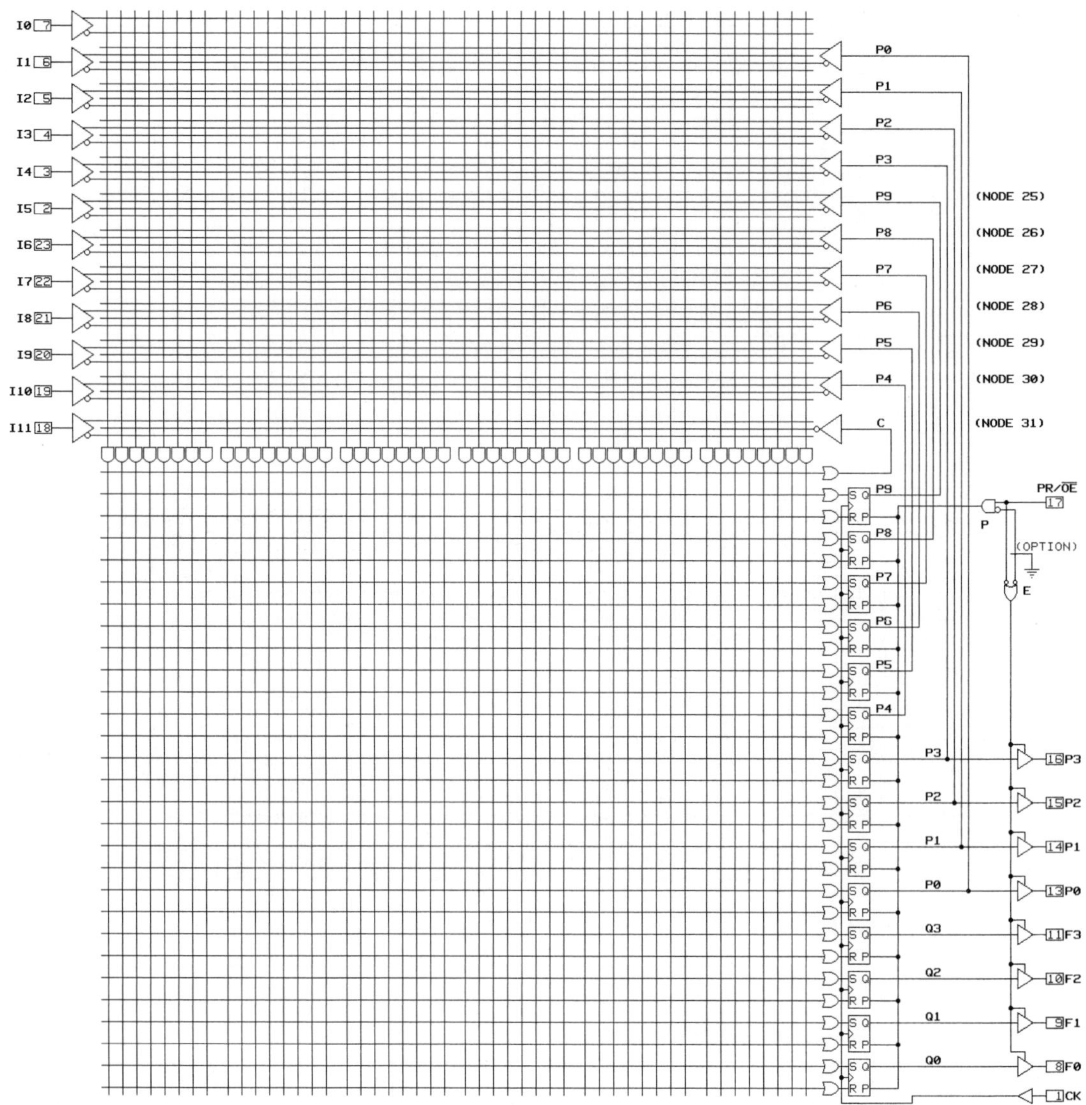

FIGURE 1.25 Philips PLS168 FPLS (12 x 48 x 8).
Source: Philips, "Programmable Logic Devices (PLD) Data Handbook," Philips Semiconductor, Sunnyvale, CA, 1994.

1.7.3 Generic Array Logic (GAL)

Lattice Semiconductor Corporation has established the GAL® family of products as a result of pioneering efforts at applying E^2CMOS (electrically erasable floating gate technology) to programmable logic. These devices, once programmed, can be bulk-erased for the purpose of reprogramming. This makes these devices ideal for the prototyping of programmable logic devices. The devices use a flexible output microcell design so that a single GAL device can be configured to emulate several bipolar PAL devices.

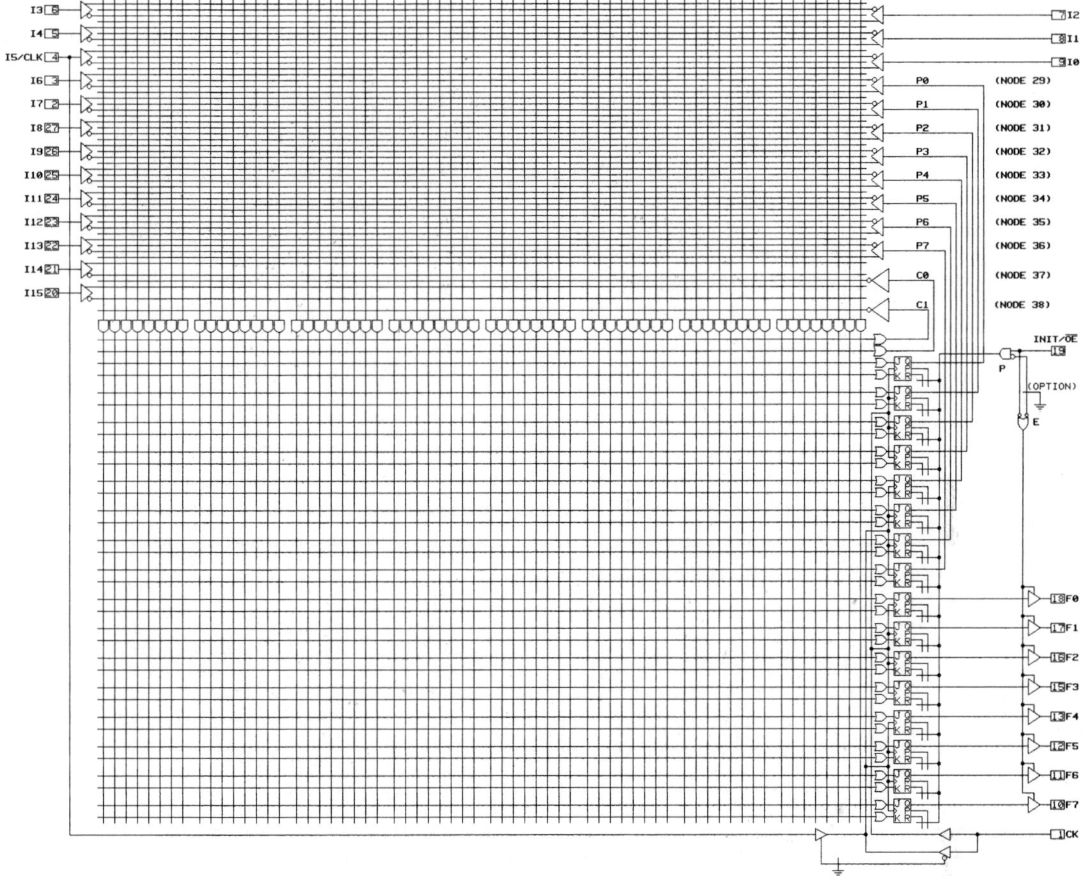

FIGURE 1.26 Philips PLS405 FPLS (16 x 64 x 8).
Source: Philips, "Programmable Logic Devices (PLD) Data Handbook," Philips Semiconductor, Sunnyvale, CA, 1994.

Table 1.3 lists bipolar PALs that can be emulated by one or more of the family of five GAL devices. The output macrocell of the GAL device can be programmed to emulate both logical and registered outputs in a variety of configurations.

Product versions are available that have from 25 down to five nanoseconds total propagation delay, allowing operating speeds up to 200 megahertz.

1.7.4 Programmable Read-Only Memory (PROM)

The number of minterms per logical expression generated by PAL devices is typically less than a dozen. FPLA devices, though more expensive, extend that number to 64 or more. Some applications require expressions of more minterms than are available from PAL and FPLA devices. Recall the previous argument that a PLA device is essentially the same in architectural form as a ROM, except that fewer minterms are available. If we design a gate array which fan generate 2^N minterms, where N is the number of inputs, we have a ROM.

TABLE 1.3 GAL/PAL replacement table.

Lattice GAL Devices

16V8
20V8
20RA10
20XV10
22V10

replace the following *Bipolar Programmable Logic Devices*

10H8	14H4	16H2	16P2	16RP4	18P8	20L10*	20RP6
10L8	14H8	16H6	16P6	16RP6	18U8	20P8	20RP8
10P8	14L4	16H8	16P8	16RP8	20H2	20R4	20RA10
12H6	14L8	16L2	16R4	18H4	20H8	20R6	20X4*
12L6	14P4	16L6	16R6	18L4	20L2	20R8	20X8*
12P6	14P8	16L8	16R8	18P4	20L8	20RP4	20X10*
							22V10

* GAL20XV10 replaces 20L10, 20X10, 20X8 and 20X4.

A PROM, or programmable read-only memory, has the characteristics of a ROM, and it is field-programmable, as is a PAL or FPLA. Consider Figure 1.27.

Unlike the PAL or FPLA, the input to the AND logic is not programmable. However, each available product of input terms is decoded to an individual minterm. Therefore,

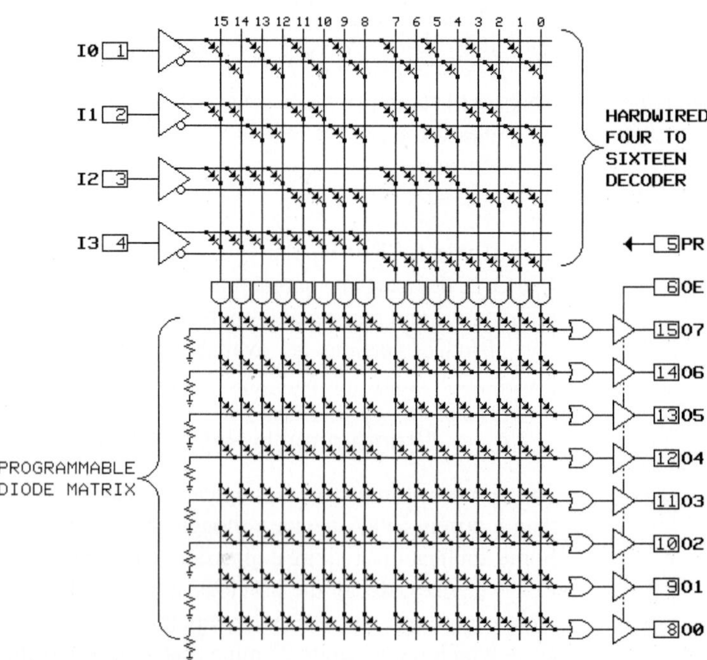

FIGURE 1.27
Unprogrammed 16 x 8 programmable ROM (PROM).

there are always 2^N AND logic gates, generating all of the possible minterms of the N input arguments. When a new PROM device is shipped, the diodes between the AND and OR logic gates are all intact, as shown in Figure 1.27. Programming is accomplished by burning out fuses that are in series with each diode, leaving only the desired diodes intact. This programming is usually accomplished by placing the desired data on the data output lines, placing the storage address on the input lines, and placing a series of pulses on the program line. The voltage level of these pulses typically ranges from –12 to –25 volts.

It is common to refer to a memory device using the two numbers that define its architecture. The number of words is stated first, and the word size is stated second. Since this example PROM has 16 words, and each word is eight bits in length, this can be referred to as a 16 x 8 PROM.

Practical Example 1.4: PROM-Based Binary to ASCII Code Converter

Consider the following problem: it is desired to design a "black box" that converts a four-bit binary nibble to its equivalent eight-bit ASCII representation. Table 1.4 illustrates the relationship between these two arguments.

This example has defined an eight-bit output for each possible permutation of input patterns. Since there are four input bits, there are 2^4, or 16, output patterns. Each output pattern is a function of a unique input pattern. Also, each input pattern represents a single minterm. Therefore, the function of each output bit is simply the logical sum of the true minterms. In similar examples shown throughout this text, the design methodology is to calculate these functions, minimize using Boolean algebra or Karnaugh maps, and implement the logic. When implementing with a PROM, no such reduction is necessary, since

TABLE 1.4 Practical Example 1.4, binary to ASCII code conversion.

Binary	*ASCII*
0000	00110000
0001	00110001
0010	00110010
0011	00110011
0100	00110100
0101	00110101
0110	00110110
0111	00110111
1000	00111000
1001	00111001
1010	01000001
1011	01000010
1100	01000011
1101	01000100
1110	01000101
1111	01000110

all possible minterms are generated by the input logic of the device. It is necessary only to treat the device as a memory unit with the input logic used to define addresses with the desired output data stored at each respective address. Consider Table 1.5, which modifies the information in Table 1.4 to include this address and data information.

Table 1.5 describes the contents of the PROM at address 0000_2 to be the binary pattern 00110000_2. Therefore, when the binary pattern 0000_2 is presented at the inputs of the PROM, the binary pattern 00110000_2 is presented at its outputs, and the code conversion has taken place. Note that this method of design does not require any Boolean calculations or minimizations. The output is simply expressed as a function of each input minterm and stored in the PROM accordingly. The actual programming of the PROM requires that a diode be left intact for each output bit that is to remain at a logical 1. Look carefully at Figure 1.28 to see the solution to this example. The AND logic gate in the least significant position is located on the far right. The output of this AND gate is TRUE when the input data pattern is 0000_2. Also, since the AND logic circuits are arranged as a decoder, *only* the least significant AND gate output is TRUE at this time. The diodes have been left intact on bits O5 and O4 on the output of this AND gate, resulting in an output pattern of 00110000_2. Note that the pull-down resistors on the inputs to the OR logic cause those inputs to drop low when the diode has been removed.

Carefully examine the placement of each of the diodes on the output matrix (between the AND and OR logic gates) to determine the relationship between the diode pattern and the data of Table 1.5.

In the last several years PROMs have been manufactured in a wide variety of sizes. Let's look at a slightly larger PROM illustrated in Figure 1.29. This PROM has an architecture of 32 eight-bit words. That is, there are 32 addresses, and each address contains eight data bits.

TABLE 1.5 Practical Example 1.4, PROM-based binary to ASCII code conversion.

Binary	*ASCII*
Address I 3210	*Data O 76543210*
0000	00110000
0001	00110001
0010	00110010
0011	00110011
0100	00110100
0101	00110101
0110	00110110
0111	00110111
1000	00111000
1001	00111001
1010	01000001
1011	01000010
1100	01000011
1101	01000100
1110	01000101
1111	01000110

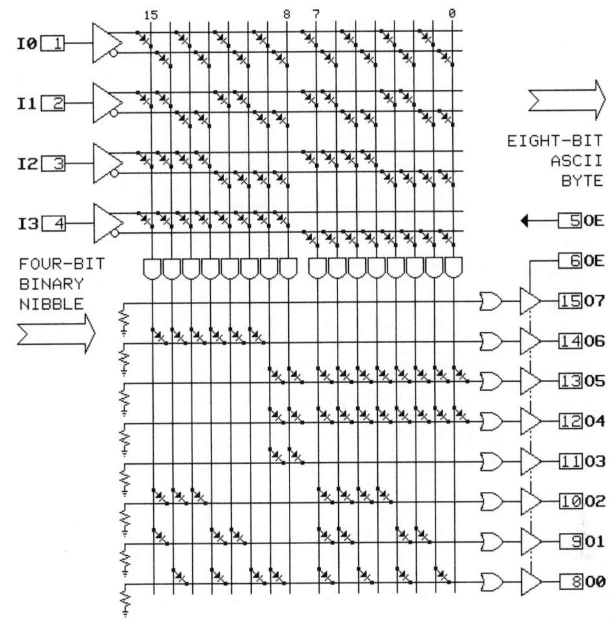

FIGURE 1.28 Practical Example 1.4, ROM-based binary to ASCII code converter.

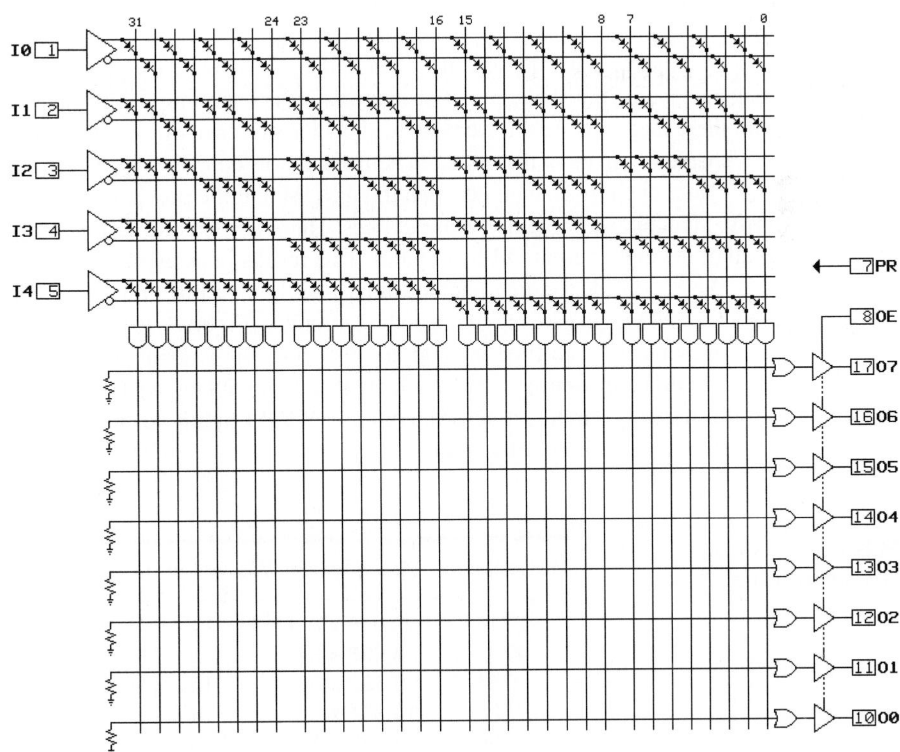

FIGURE 1.29 32 x 8 PROM schematic diagram.

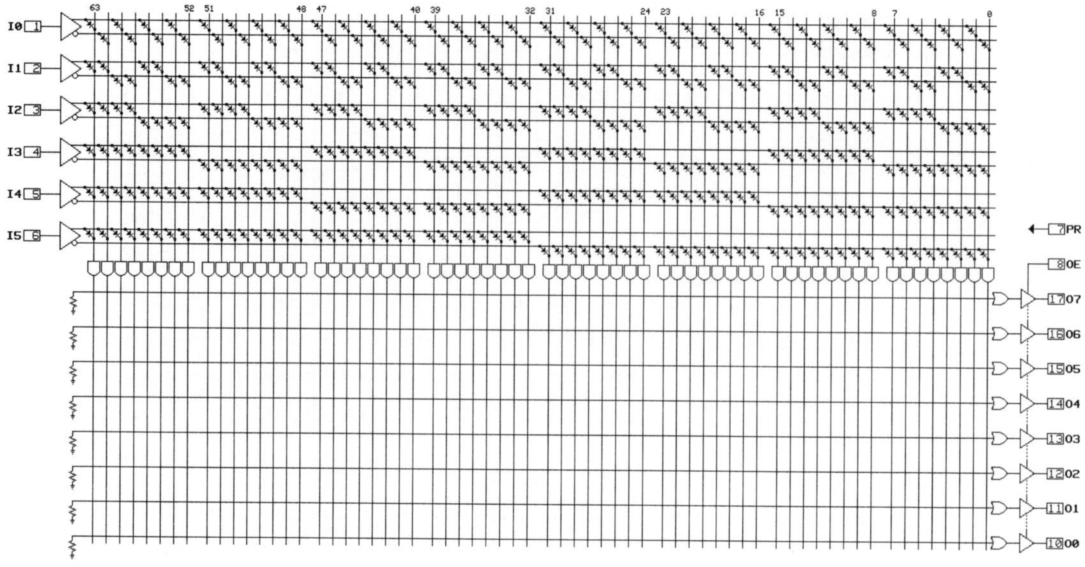

FIGURE 1.30 64 x 8 PROM schematic diagram.

In order to make the schematic diagram easier to read, the programmable diodes have been removed from the summation matrix. This is the most convenient way of working with these diagrams. When we wish to show the presence of a diode at any intersection, an "X" character can be drawn on the schematic as was done in previous examples in this text.

Note the only change between this PROM and the previous one was the addition of a fifth input bit, resulting in five input lines. These decode into 32 minterms, each available to all of the eight output OR logic gates. This results in an eight-bit by 32-word PROM.

Let's consider adding another input line. Such a PROM, with six input lines, must have 64 AND logic gates in the decoding circuitry. Figure 1.30 illustrates such a device.

Obviously, each time we add an input line, the amount of decoding circuitry doubles. PROMs are commercially available that are much larger than these examples. Table 1.6 illustrates the architecture of a common PROM family of devices. These devices are available from a variety of vendors.

TABLE 1.6 Example PROM architectures.

Designation	*Architecture*	*No. of Inputs*	*Total Bits*
2716	2,048 x 8	11	16,384
2732	4,096 x 8	12	32,768
2764	8,192 x 8	13	65,536
27128	16,384 x 8	14	131,072
27256	32,768 x 8	15	262,144
27512	65,536 x 8	16	524,288

There are several programming methods employed in PROM design. Consider the following types.

Mask-Programmed ROM. When many similar PROMs are to be manufactured, it may be advantageous to order them preprogrammed from the manufacturer. A data table is sent to the manufacturer, typically on a floppy diskette, and this table is used to create a mask that is used during fabrication to remove the unwanted diodes. These devices cannot be reprogrammed and are typically the least expensive to purchase when bought in large quantities.

Field-Programmable ROM (PROM or FPROM). These PROMs are programmed by the user by placing them into a PROM programming device. The programming method burns out fuses that are placed in series with each programmable diode. Once the fuses have been burned out, the junction cannot be restored, so each bit can be programmed to a logical 0 only once. This type of device is typically more expensive than the masked-programmed ROM but less expensive than the reprogrammable types.

UV-Erasable PROM (EPROM). This type of device uses a floating gate CMOS transistor in the place of the diode. The junction is programmed in a fashion very similar to the other PROMs, but in this case a charge (or lack of a charge) is placed on the floating gate. Because of the very high resistance of the insulator beneath the floating gate, the gate can hold its charge for a very long time. The hold time of an EPROM is measured in years. The charge on the floating gate can be dissipated by exposure to ultraviolet light. The effect of the light is to reduce the resistance of the insulator, allowing the charge to dissipate. The entire EPROM is exposed to the light, resulting in a bulk-erase of the entire device. Erasure under a bright ultraviolet lamp may take 10 to 20 minutes. Precautions must be taken to avoid looking at a UV lamp, as the eyes can be easily sunburned, resulting in blindness.

Electrically Erasable PROM (EEPROM, E^2PROM, EAROM). This device is very similar in operation to the EPROM, except that the floating gate CMOS technology is designed so that each addressed word can be erased (set to a logical 1) by applying an external stimulus. EEPROMs cannot typically be erased in a bit-by-bit fashion because of the complex decoding circuitry that would be required to do so. Usually EEPROMs are erased in bulk, but under software control they can be erased one addressable word at a time. The GAL devices described earlier use this type of technology.

1.8 PROGRAMMABLE LOGIC DEVICE (PLD) PROGRAMMING

It is probably evident by this time that the programming of these devices is typically not done by hand. Two predominant methods are available.

The first programming method, and the more cumbersome, is to manually create a "fuse plot." This device is simply a table which defines the binary pattern to be programmed into each address of the PLD. Recall that diodes are "removed" when programmed by the blowing of a fuse that is placed in series with it. The simplest example of this is the application of PROM programming. Consider a table that identifies the binary number to be programmed into each PROM address. Such a table is literally a fuse plot, since binary 1s and 0s in the table represent present and absent fuses.

TABLE 1.7 PLD programming languages.

Language	*Source*	*Application*
ABLE	Data I/O Corporation	Generic
AMAZE	Signetics	Vendor-specific
CUPL	Logical Devices, Inc.[1]	Generic
MAX+	Altera	Vendor-specific
PALASM	Advanced Micro Devices	Generic
PLDesigner	Minc, Inc.	Generic
Pro-Logic	Texas Instruments	Generic

[1]Programming these devices with CUPL language will be discussed later in this text.

The second method uses programming language compilers to generate the fuse plot. CAD software systems, both generic and vendor-specific, have been developed that allow the user to write a program using a typical text editor. This program includes information that describes the device to be programmed, the pin assignments, any logical functions to be generated, and a variety of other statements that make the design of common circuits a reasonably straightforward task. Some representative PLD programming languages are listed in Table 1.7.

The programming language generates a fuse plot list as a text data file in some standard format such as ASCII hex, Intel Hex, Motorola S-format, or JEDEC format. A PLD programmer is then used to load the PLD with the fuse plot data. These devices are often referred to as EPROM programmers.

Practical Example 1.5: PAL Implementation of SOP Functions

Consider the following set of logic equations:

$$W = AB\overline{C} + A\overline{B}C + \overline{A}\,\overline{B}\,\overline{C} \qquad \textbf{1.28}$$

$$X = A\overline{B} + \overline{A}C$$

$$Y = AB + \overline{A}\,\overline{B} + AC$$

$$Z = A\overline{B}\,\overline{C} + BC$$

1. What size PAL is needed to implement these functions?
   The PAL to be chosen must produce at least four output functions. Each output function must be able to have up to three minterms since the largest expression has three minterms. Finally, each minterm must be able to select arguments from a domain of three input variables.

   Number of logical outputs: 4
   Number of input arguments: 3
   Number of minterms generated for each output: 3
   Number of arguments for each minterm: 3

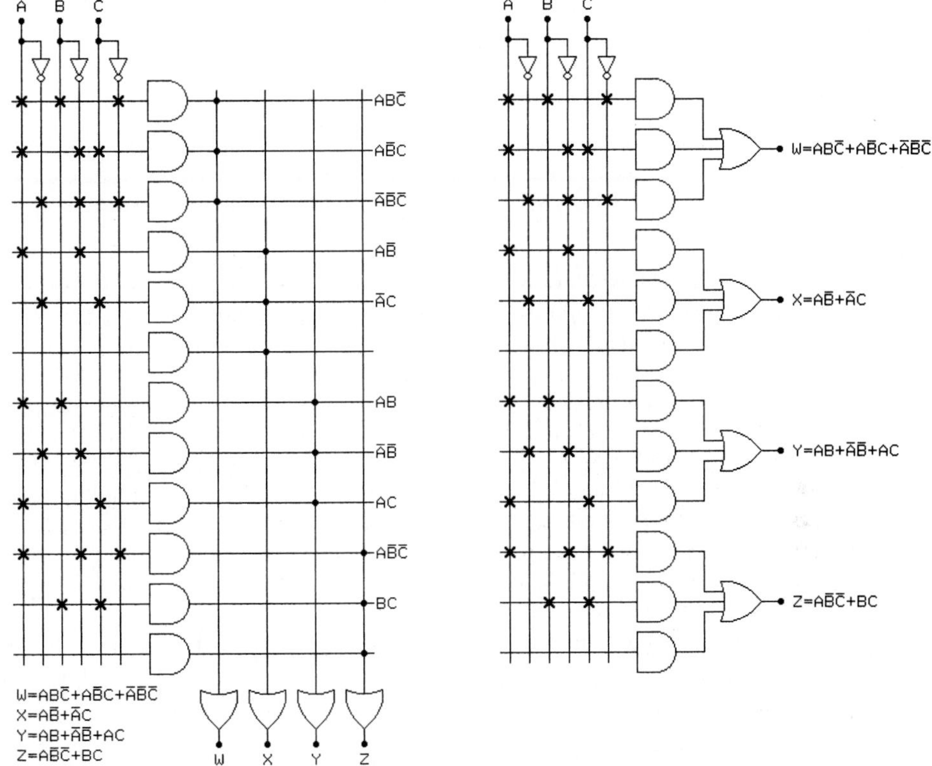

FIGURE 1.31 Practical Example 1.4, solutions.

2. Draw a schematic diagram of the programmed PAL.
   Figure 1.31 illustrates the minimum solution. Note that both simpler forms of the schematic diagram are shown, though only one form is needed to obtain a solution. The design is relatively straightforward, as a dot (or diode) is inserted at each point where a minterm is to be generated.

Practical Exercise 1.4: PAL Implementation of SOP Functions

Consider the following functions:

$$W=\overline{A}\,\overline{C}+A\overline{C}\,\overline{D}+A\overline{B}D \tag{1.29}$$

$$X=A\overline{C}+BD+AB+\overline{A}\,\overline{B}C\overline{D}$$

$$Y=\overline{C}\,\overline{D}+\overline{A}B+B\overline{C}+A\overline{B}D$$

$$Z=\overline{C}D+AB\overline{C}+\overline{A}B\overline{D}+A\overline{B}C\overline{D}$$

1. Determine the minimum size PAL that can be used to generate these functions.
2. Draw a schematic diagram (similar to one of those in Figure 1.31) of a PAL circuit that generates these functions.

Practical Exercise 1.5: BCD to Seven-Segment Code Converter

Seven-segment displays are used quite frequently in order to display decimal numbers on devices such as calculators, instrumentation, clocks, etc. In order to be readable, it is necessary to illuminate specific segments as a function of the number to be displayed. This PROM-based code converter is to accept four input bits which may be any of the 10 binary patterns 0000_2 through 1001_2, as illustrated in the top portion of Figure 1.32.

The output of the code converter will be seven lines, labeled a, b, c, d, e, f, g. Each represents the signal for the respective seven-segment display bit illustrated in Figure 1.32. Bold lines in the character displays across the top of the figure represent a logical 1 output.

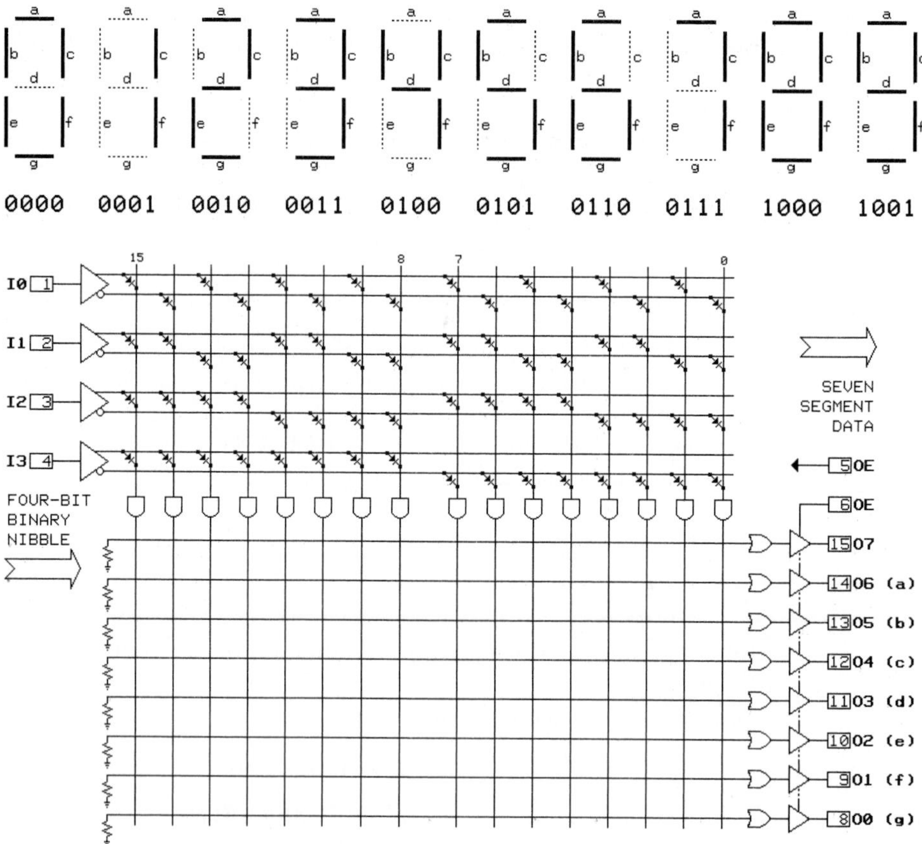

FIGURE 1.32 Practical Exercise 1.5, BCD to seven-segment code definitions and PROM.

Dotted lines represent a logical 0 output. That is, when a 0000_2 is presented to the inputs, the output pattern a, b, c, d, e, f, g should be the binary values 1,1,1,0,1,1,1 respectively. When a 0001_2 is presented to the inputs, the output pattern a, b, c, d, e, f, g should be the binary values 0,0,1,0,0,1,0 respectively. The same concept is repeated for each of the input patterns.

Your task in this exercise is to describe the diode pattern to be programmed into a 16 x 8 EPROM similar to that in Figure 1.32 that will generate the defined code conversion. Use Practical Example 1.4 as a guide. Place either Xs or diodes at the junctions of the summation matrix as required.

1.9 SUMMARY: PROGRAMMABLE LOGIC, FILLING THE GAP

Consider Figure 1.33. Without a knowledge of programmable logic devices, a digital designer may be inclined to use a microcontroller or microprocessor-based solution to a problem when a simpler and more cost-effective alternative may be available. PAL devices can take the place of many discrete logic circuits, but are limited in the number of minterms that can be generated in an expression. FPLA and FPLS devices can generate considerably more complex expressions than PAL devices. ROMs can be used to provide the optimum expression complexity as well as to use table look-up methods for code conversion and signal generation. When the problem is too complex for a ROM solution, or extensive I/O operations are to be performed, it is time to move up to the microcontroller. Only when all of these methods have been determined to be insufficient for the problem at hand is it reasonable to go to the expense of a microprocessor-based system.

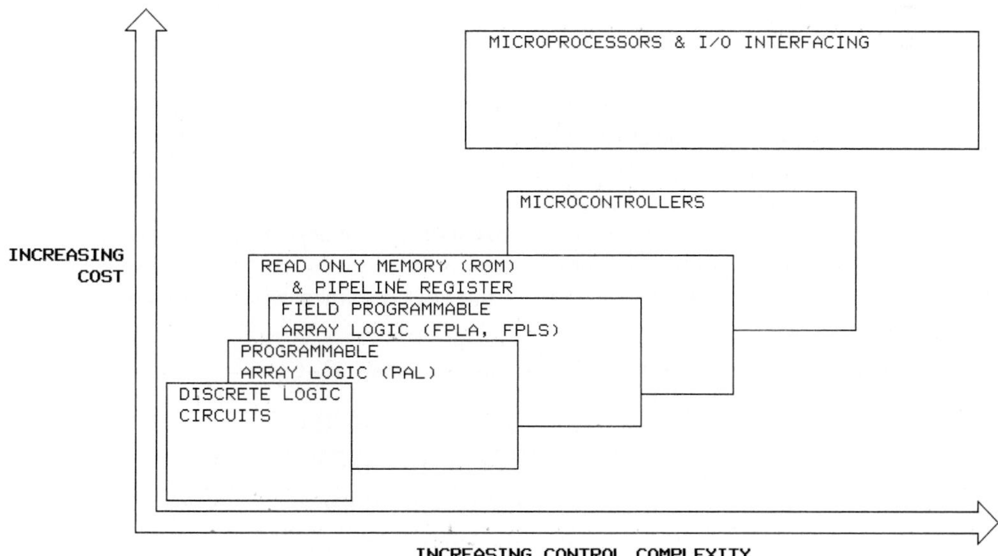

FIGURE 1.33 Spectrum of digital design complexity.

QUESTIONS AND PROBLEMS

Define the terms in 1–23.

1. Expression
2. Sum of products
3. Conventional SOP form
4. Minterm
5. Domain of an SOP expression
6. Typographic SOP form
7. Summation SOP form
8. Redundant variable
9. DeMorgan's Theorem
10. Digital multiplexer
11. Programmable logic devices
12. Read-only memory (ROM)
13. Field programmable logic array (FPLA)
14. Programmable array logic (PAL)
15. PAL output macrocell
16. L-type macrocell
17. R-type macrocell
18. V-type macrocell
19. Programmable ROM (PROM)
20. EPROM
21. Mask-programmed ROM
22. Field-programmable ROM
23. EEPROM or E^2ROM

24. Describe the three forms used to define an SOP expression.
25. What is the advantage of maintaining an SOP expression in conventional form?
26. What is the advantage of maintaining an SOP expression in typographic form?
27. What is the advantage of maintaining an SOP expression in summation form?
28. Which SOP expression form is most useful when designing using multiplexer circuits?
29. Which SOP expression form is most useful when designing using ROMs?
30. Which SOP expression form is most useful when designing using discrete logic?
31. Which SOP expression form is most useful when writing programs using a text editor?

Use Expression Sets 1.30–1.32 to answer Problems 32–63.

$$W = A\overline{B} + A\overline{B}\,\overline{D} + A\overline{B}D + ACD$$ **1.30**

$$X = A\overline{B} + CD + \overline{A}C + \overline{A}B\overline{C}\,\overline{D}$$

$$Y = \overline{B}\,\overline{D} + \overline{A}C + \overline{B}C + ACD$$

$$Z = \overline{B}\,\overline{D} + A\overline{B}C + \overline{A}C\overline{D} + \overline{A}B\overline{C}\,\overline{D}$$

32. Convert each of the expressions in Expression Set 1.30 to their typographic form.

33. Convert each of the expressions in Expression Set 1.30 to summation form using Boolean algebra.
34. Convert each of the expressions in Expression Set 1.30 to summation form using Karnaugh maps.

$$R = (A \& B \& !D) \# (B \& !C) \# (!A \& B) \# (!C \& !D)$$
$$S = (!A \& !B \& !D) \# (A \& B) \# (B \& D) \# (A \& !C)$$
$$T = (A \& !B \& C \& !D) \# (!A \& B \& !D) \# (A \& B \& !C) \# (!C \& D)$$
$$U = (A \& !B \& C) \# (!A \& B \& C) \# (!A \& !B) \# (!B \& !C)$$
1.31

35. Convert each expression of Expression Set 1.31 to conventional form.
36. Convert each expression of Expression Set 1.31 to summation form using Boolean algebra.
37. Convert each expression of Expression Set 1.31 to summation form using Karnaugh maps.

$$L = f(ABC) = \sum(0,2,4,5,7)$$
$$M = f(ABCD) = \sum(1,2,5,6,7,8,10,11,13,15)$$
$$N = f(ABC) = \sum(2,3,4,5,6)$$
$$P = f(ABCDE) = \sum(1,2,3,4,9,10,11,12,21,22,23,24)$$
1.32

38. Convert each expression of Expression Set 1.32 to conventional form.
39. Convert each expression of Expression Set 1.32 to typographic form.
40. Draw a schematic diagram which uses discrete logic devices to implement the expressions of Expression Set 1.30.
41. What would be the smallest ROM that could be used to implement the functions of Expression Set 1.30?
42. What would be the smallest ROM that could be used to implement the functions of Expression Set 1.31?
43. What would be the smallest ROM that could be used to implement the functions of Expression Set 1.32?
44. What would be the smallest FPLA that could be used to implement the functions of Expression Set 1.30?
45. What would be the smallest FPLA that could be used to implement the functions of Expression Set 1.31?
46. What would be the smallest FPLA that could be used to implement the functions of Expression Set 1.32?
47. What would be the smallest PAL that could be used to implement the functions of Expression Set 1.30?
48. What would be the smallest PAL that could be used to implement the functions of Expression Set 1.31?
49. What would be the smallest PAL that could be used to implement the functions of Expression Set 1.32?
50. Draw a schematic diagram that uses multiplexers to implement the expressions of Expression Set 1.30.

51. Draw a schematic diagram that uses a ROM to implement the expressions of Expression Set 1.30.
52. Draw a schematic diagram that uses a PAL to implement the expressions of Expression Set 1.30.
53. Draw a schematic diagram that uses an FPLA to implement the expressions of Expression Set 1.30.
54. Draw a schematic diagram that uses discrete logic devices to implement the expresions of Expression Set 1.31.
55. Draw a schematic diagram that uses multiplexers to implement the expressions of Expression Set 1.31.
56. Draw a schematic diagram that uses a ROM to implement the expressions of Expression Set 1.31.
57. Draw a schematic diagram that uses a PAL to implement the expressions of Expression Set 1.31.
58. Draw a schematic diagram that uses an FPLA to implement the expressions of Expression Set 1.31.
59. Draw a schematic diagram that uses discrete logic devices to implement the expressions of Expression Set 1.32.
60. Draw a schematic diagram that uses multiplexers to implement the expressions of Expression Set 1.32.
61. Draw a schematic diagram that uses a ROM to implement the expressions of Expression Set 1.32.
62. Draw a schematic diagram that uses a PAL to implement the expressions of Expression Set 1.32.
63. Draw a schematic diagram that uses an FPLA to implement the expressions of Expression Set 1.32.

64. Describe in detail the L-type macrocell as implemented in a PAL.
65. Describe in detail the R-type macrocell as implemented in a PAL.
66. Describe in detail the V-type macrocell as implemented in a PAL.
67. In what situation or situations would the use of discrete logic gates be more applicable for generating functions than the use of programmable logic?
68. In what situation or situations would the use of a ROM be more applicable for generating functions than discrete logic, an FPLA or a PAL?
69. In what situation or situations would the use of an FPLA be more applicable for generating logic functions than discrete logic, a ROM or a PAL?
70. In what situation or situations would the use of a PAL be more applicable for generating logic functions than discrete logic, a ROM or an FPLA?
71. By what parameters is a ROM commonly specified?
72. By what parameters is an FPLA commonly specified?
73. By what parameters is a PAL commonly specified?
74. Describe how each of the following devices is programmed: masked-programmed ROM, field-programmable ROM, EPROM, EEPROM.
75. Describe how each of the following memory devices is erased: EPROM, EEPROM.

CHAPTER 2

Synchronous Binary Counter Design

OBJECTIVES

After completing this chapter you should be able to:

- Describe a step-by-step method of designing a synchronous binary counter.
- Describe the excitation tables for a variety of clocked flip-flops.
- Describe the count sequence of a synchronous binary counter as a state diagram.
- Design a state table that defines the count sequence and excitation requirements of a synchronous binary counter using JK-, D-, or SR-type flip-flops.
- Extract from a state table those sum-of-product expressions that form the input forming logic of a synchronous binary counter.
- Draw a detailed logic diagram of a synchronous binary counter using discrete logic devices including flip-flops and logic gates.
- Define and describe the need for hazard protection.
- Assign state numbers to a state diagram to assure glitch-free operation.
- Describe the purpose and use of delayed-state timing.
- Describe the purpose and use of alternate-state timing.

2.1 OVERVIEW

The overall objective of this book is to attain the skills necessary to design digital state machines. These devices may be implemented using discrete logic gates, programmable logic, or, in complex applications, a microcontroller or microprocessor. The state machines we will study all have a timing element that requires the storage of state information. The timing element in those state machines that use discrete or programmable logic is essentially a synchronous binary counter that steps from one count to another in a sequence defined by the current state and any input stimulus. Consequently, the synchronous binary counter is a fundamental part of state machine design, and a comprehensive understanding of its design is vital.

This chapter will present the design of a synchronous binary counter as a multistep process where each step can be easily mastered until the design is complete. Subsequent chapters will be concerned with controlling the counter from external stimulus and using the counter to control external devices.

2.2 THE SYNCHRONOUS BINARY COUNTER

A **synchronous binary counter** is made up of two parts: a synchronous binary register and input-forming logic. A synchronous **binary register** is formed by combining a set of one or more flip-flops in a linear array. This connection is attained by connecting their edge-triggered clock inputs together. Figure 2.1 illustrates a synchronous binary register.

A synchronous binary register can be formed using any of a variety of flip-flops, provided that they use an **edge-triggered** clock. This feature allows the current state information to be used in determining the next state. That is, the outputs of the synchronous binary register can be fed back to its own inputs to define the state to which the register will advance when the next clock edge comes.

The **input-forming logic** in a synchronous binary counter is the combinational circuitry that drives the control inputs of its flip-flops. For example, if JK-type flip-flops are used, the combinational circuitry that drives the J and K inputs would be the input-forming logic. Those inputs are used in a synchronous binary counter to define the next state following the clock edge. The source for the input forming logic will come from two places: the counter outputs (which define, and are unique for, the current state), and any external stimulus that may be used to modify the count sequence. Figure 2.2 illustrates a block diagram of a generalized three-bit synchronous binary counter. Note that the clock inputs of the flip-flops are interconnected and are defined to be asserted active-low. The J and K inputs are driven by the combinational input-forming logic. This logic is used to determine the state to which the counter will change following the next clock edge. Note also that external stimulus may be provided to the input-forming logic to modify the count sequence. This particular feature will be explored in Chapter 3.

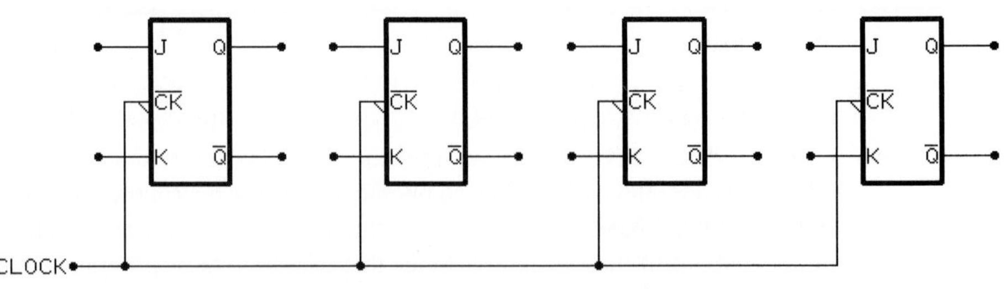

FIGURE 2.1 Synchronous binary register using JK flip-flops.

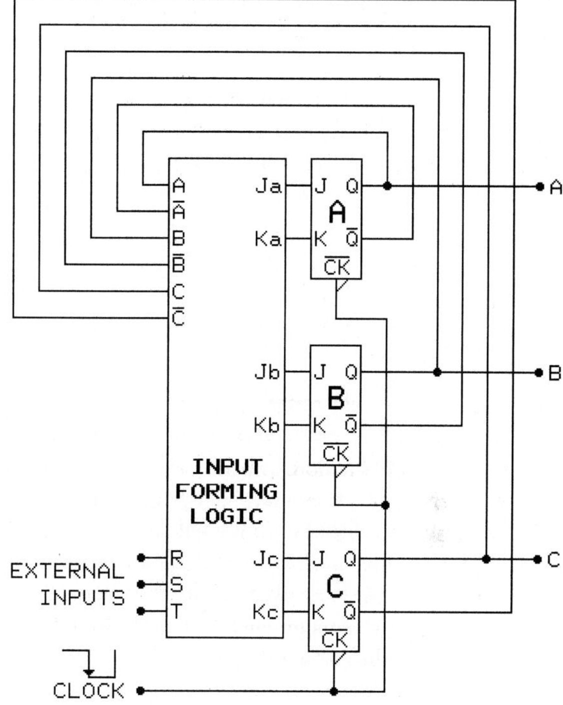

FIGURE 2.2 Synchronous binary counter.

2.3 FLIP-FLOP EXCITATION LOGIC

The first step in the design of the synchronous binary counter is the selection of the flip-flop to be used. This will determine the input-forming logic required. Again, recall that the flip-flops must use edge-triggered clocks. For this discussion we will consider the design of counters that use D-type, JK-type, and Set-Reset (SR-type) flip-flops. Consider Figure 2.3. The **flip-flop excitation table** describes the logic level that must be placed on the inputs of the flip-flop to cause it to attain a known logic level following the clock edge. For example, if the present state (Q_t) of a D-type flip-flop is a logical 0 and it is desired to have the flip-flop at a zero state following the clock edge (next state, Q_{t+1}), it is necessary to place a logical 0 on the D input.

For example, if the present state (Q_t) of a JK-type flip-flop is a logical 0 and it is desired to have the flip-flop at a zero state following the clock edge (next state, Q_{t+1}), it is necessary to place a logical 0 on the J input. It doesn't matter what logic is on the K input since a logical 0 would cause the flip-flop to remain stable at a logical 0, and a logical 1 would reset the flip-flop, also resulting in a logical 0. Spend a few moments to review the excitation table of Figure 2.3.

What states would be placed on the inputs of an SR-type flip-flop to cause it to change from a logical 1 to a logical 0 state? This represents a $Q_t = 0$ and $Q_{t+1} = 1$. Referring to the excitation table, we find out that S should be set to a logical 0 and R to a logical 1.

FIGURE 2.3 Flip-flop excitation tables.

PRESENT STATE Q_t	NEXT STATE Q_{t+1}	POSITIVE D-TYPE D_t	POSITIVE JK-TYPE J_t	K_t	POSITIVE SR TYPE S_t	R_t
0	0	0	0	*	0	*
0	1	1	1	*	1	0
1	0	0	*	1	0	1
1	1	1	*	0	*	0

* = Don't Care State

Caution: The excitation tables of Figure 2.3 define the responses of specific devices. When specifying a logic device, its particular excitation logic must be considered. For example, the excitation for the JK flip-flop just discussed is true for a 7476 dual JK flip-flop chip. However, a 74107 dual JK flip-flop chip has a different response. For the purposes of this discussion, we will use the responses shown in Figure 2.3. When applying actual devices, the data sheets for the devices should be used.

We will be using the excitation tables when we design the input-forming logic for a synchronous binary counter, so familiarity with them is vital. We will see in examples to follow how they are used.

2.4 STATE DIAGRAMS

It will be necessary for us to be able to describe the sequence of the binary counter in some reasonable fashion. We can do it using literal statements, logical expressions, or state diagrams. For example, we would like to design a binary counter that resets to zero, and then steps through the repeating sequence 4, 2, 1, 0. We would expect the counter to count through the following sequence:

0, 4, 2, 1, 0, 4, 2, 1, 0, 4, 2, 1, 0, 4, 2, 1, 0, 4, 2, 1, 0, 4, 2, 1, ...

We could use a logical expression to define the count sequence. This could be done by listing the sequence and showing the repetition of the pattern. For example:

Count = (0, 4, 2, 1, 0, 4, 2, 1, ...)

These two methods may be sufficient for very simple applications, but they fall far short of the need for defining the count sequence in a state machine. A state machine uses a binary counter that progresses in a sequence that may not repeat the same pattern but rather in a sequence that changes based on external logical inputs. For this purpose, the third method of sequence description is needed, the state diagram.

A **state diagram** is simply a logical flowchart that describes all of the counter states, their interrelationships, the effect of external stimuli, and the output characteristics of each

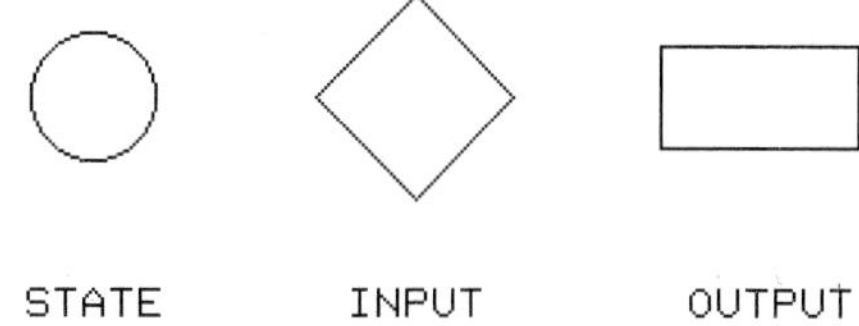

FIGURE 2.4 State diagram components.

state in a simple graphical manner. Figure 2.4 illustrates the three primary components of the state diagram: the state box, the input box, and the output box.

2.4.1 The State Box

The **state box** is used to describe a stable state of the counter. Usually this is done by placing either the count number in the box, a variable representing the count number, or both. Also, it is common to place one of these objects inside the box and the other outside. This is the convention we will use in this text. For example, a presettable counter that repeats a four-count sequence could be described as shown in Figure 2.5.

In the example shown in Figure 2.5, no count sequence numbers have been assigned and only variables are shown. This is a typical starting point for a state machine design. Often the sequence is defined first and the count numbers for each state in the sequence are assigned later. Such an assignment could be documented as shown in Figure 2.6.

There are no hard and fast rules as to the methods used to document the state assignments. The example shown in Figure 2.7 is as applicable and reasonable as that shown in Figure 2.6. This text will maintain the convention illustrated in Figure 2.6 for the remainder of the design examples.

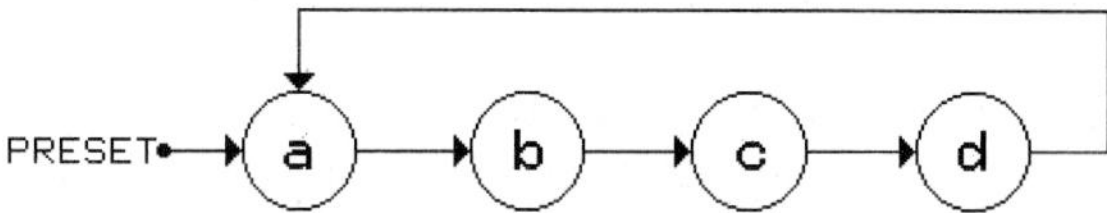

FIGURE 2.5 State diagram of a variable four-count sequence.

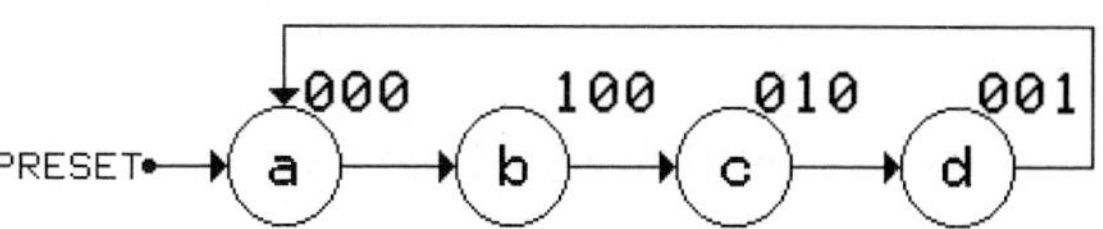

FIGURE 2.6 State diagram of a four-count sequence, first form.

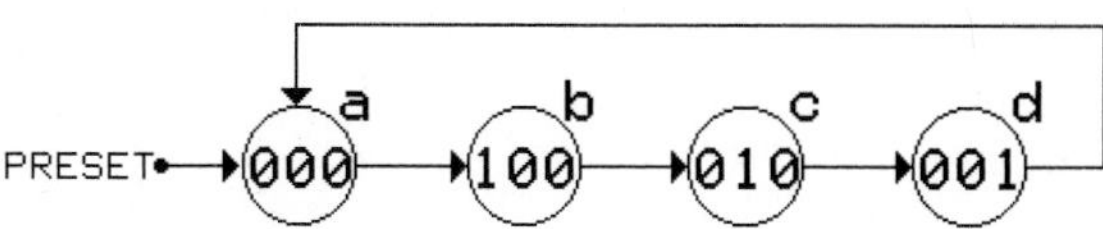

FIGURE 2.7 State diagram of a four-count sequence, second form.

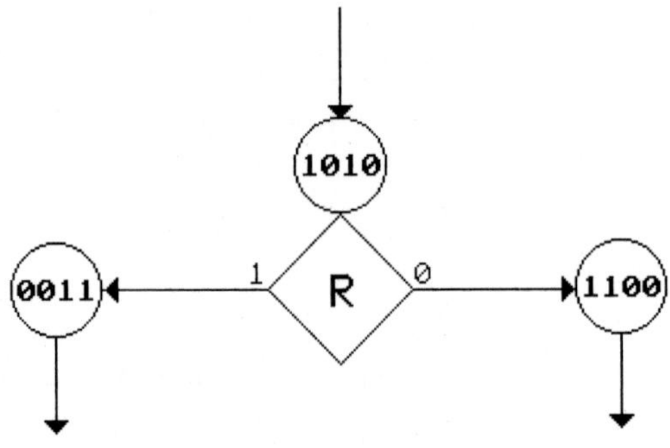

FIGURE 2.8 State diagram input box example.

2.4.2 The Input Box

The **input box** shown in Figure 2.4 is used to document the effect on the count sequence caused by the logical state of inputs presented to the counter. This is the mechanism by which the counter can make decisions. Consider the following example. A synchronous four-bit binary counter is in a stable state with the count equal to 1010_2. If the logic level on input line R is equal to a logical 0, the count sequence is to change to 1100_2 following the clock edge. If the logic level on line R is equal to a logical 1, the count sequence is to change to 0011_2 following the clock edge. Figure 2.8 illustrates the state diagram for this example. Note how the input box is used to document the decision. A convention used in this text is to place the argument inside the input box and the permutations of that argument around the perimeter. This method will provide clear documentation of the count sequence.

This chapter will not be concerned with decision making in the state machine. There are various forms of decision making in a state machine, and these forms will also be described in Chapter 3.

2.4.3 The Output Box

The **output box** is used to describe any outputs from the state machine that are defined during the present state. For example, a four-bit synchronous counter is supposed to turn on a lamp whenever it is in state 0101_2. The lamp is driven by a logic signal on line X. Figure 2.9 illustrates an output box that defines this state. In the example shown, the contents of the box states: "X goes high at the state beginning and low at the state end."

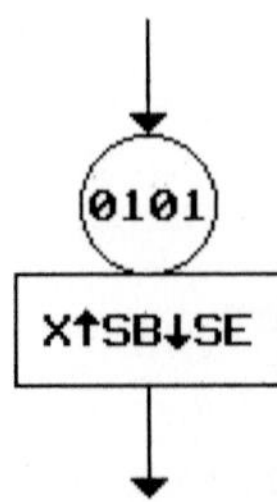

FIGURE 2.9 State diagram output box example.

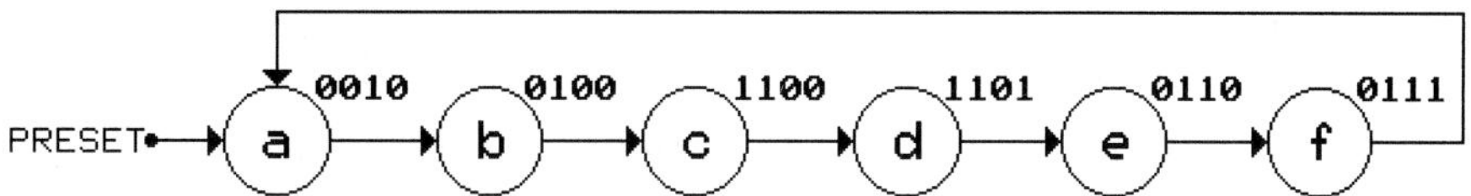

FIGURE 2.10 Practical Example 2.1, solution.

Chapter 4 will deal with output arguments. The remainder of this chapter will be concerned only with the state box, as we will not be concerned with a synchronous binary counter with inputs and outputs until later chapters.

Practical Example 2.1: Synchronous Counter State Diagram

Consider that a four-bit synchronous binary counter is to be designed that presets to the value 0010_2 and progresses through the repeating sequence 0100_2, 1100_2, 1101_2, 0110_2, 0111_2, 0010_2 when presented with clock pulses. Draw a state diagram that describes this count sequence. (See Figure 2.10)

2.5 STATE TABLES

The design of a synchronous binary counter is a multistep process. The first step is to determine the count sequence. With this information in hand, we have seen that we can document that sequence using a state diagram. The next step in the design is to develop the state table. The **state table** defines the digital levels to be generated by the input-forming logic at each state in the count sequence. In order to illustrate the development of the state table, consider the counter illustrated in Figure 2.11.

Let's assume that we are going to implement the counter using D-type flip-flops. According to the state diagram, when the counter is in the binary state 000_2, it is to go to state 100_2 after the clock edge. That is, if the present state is 000_2, the next state is 100_2. The fundamental question to be answered is: "What logic levels must be placed on the flip-flop inputs to cause the counter to change from state 000_2 to state 100_2?" The answer to that question depends on the type of flip-flop used.

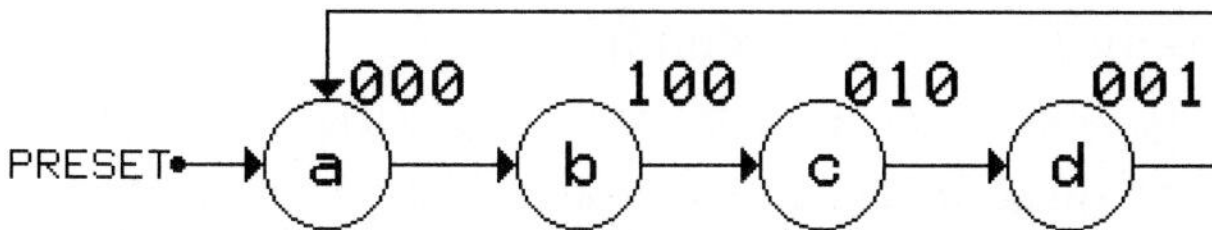

FIGURE 2.11 State diagram of example counter.

PRESENT STATE ABC	NEXT STATE ABC	Input Forming Logic D_A	D_B	D_C
000	100	1	0	0
001	000	0	0	0
010	001	0	0	1
011	***	*	*	*
100	010	0	1	0
101	***	*	*	*
110	***	*	*	*
111	***	*	*	*

D FF Excitation Q_t	Q_{t+1}	D_t
0	0	0
0	1	1
1	0	0
1	1	1

* = Don't Care

FIGURE 2.12 Example state table for D-type counter.

Consider the state table illustrated in Figure 2.12. The state table for this example is the left-hand table in Figure 2.12. (The right-hand table is the excitation table for a D-type flip-flop. It is included here so that the information it contains is readily available.) Note that the state table contains three major columns, the present state column, the next state column, and the input-forming logic column.

2.5.1 Present State Column

Though there are no hard and fast rules about the format of the present state column, it is probably best at this point to create it by writing all of the possible permutations of the counter in an ascending sequence. Once one is familiar with the use of the state table, or in the case of very large state tables, it may be better to list only those states that are actually defined in the count sequence. For now, it is advantageous to maintain the column in ascending sequence since we will be entering this information later onto a Karnaugh map in order to determine the minimum input-forming logic necessary to build the counter.

2.5.2 Next State Column

The next state column lists the corresponding next state for each present state. Recall our previous argument, which stated that if the present state is a logical 000_2, the next state is a logical 100_2. Note that this information is included on the first line of the table where there is recorded a present state of zero. The next state change defined by the state diagram says that if the present state is a logical 100_2, the next state is a logical 010_2. Note that this is recorded on the state table also. Each count sequence definition is annotated on the table using this method. If there are values in the present state column that are not a part of the count sequence, there are no next state definitions for them. Therefore, since these values will never be in the count sequence, it is appropriate to record a "don't care" state for them in the next state column. Note how this is done in Figure 2.12.

2.5.3 Input-Forming Logic Column

The input-forming logic column of the state table will be used to create the logical expressions that define the input-forming logic of the synchronous binary counter. This set of values needs to be formed carefully. The excitation table is used to define the required logical state of the inputs to the flip-flops for each bit in each state. For example, observe the contents of the first

FIGURE 2.13 Example state table entry.

PRESENT STATE ABC	NEXT STATE ABC	Input Forming Logic D_A	D_B	D_C
000	100	1	0	0

row of data in a state table as shown in Figure 2.13. The first row of data describes the logic levels in the counter during the 000_2 state. Take a look at bit A. Its present state is a logical 0, and following the clock edge it is to go to a logical 1. What value has to be placed on the D input of counter bit-A, (referred to as D_A) to cause this to take place? From the excitation table we find that a logical 1 must be placed on the D input to cause the flip-flop to change state to a logical 1. This is the value recorded under D_A for this state. It doesn't take a lot of inspection to note that when D-type flip-flops are used, the logical values generated by the input-forming logic are the same as that in the next state column. This pattern of design is repeated for each line in the state table, completing it as was shown in Figure 2.12.

Note that states 011_2, 101_2, 110_2, and 111_2 are included in the present state column but are not part of the count sequence. Since these values will never be present on the counter outputs, there is no next state definition; therefore, "don't care" values are entered in the next state column. For the same reason, the input-forming logic column bits for those rows are also "don't care" states. Again, the reasoning behind this is that these states will never be encountered in the count sequence, so it doesn't matter what the logical values generated by the input-forming logic are.

Now that the state table is completed, we will need to observe the relationship between each flip-flop input and the present states. If we look at the D_A column, we find that: $D_A = 1$ when ($A = 0$ and $B = 0$ and $C = 0$). That is, $D_A = \overline{A}\,\overline{B}\,\overline{C}$.

We will later be using Karnaugh map reduction techniques to take advantage of the "don't care" states to further minimize this expression. We will look at the extraction of logical arguments from state tables shortly. However, let's first look at examples of state tables that define the same count sequence using other flip-flop types.

Figure 2.14 illustrates a similar state table that uses JK-type flip-flops to implement the previously described counter design. Again, the first row, first bit shows that bit A

PRESENT STATE ABC	NEXT STATE ABC	Input Forming Logic J_A	K_A	J_B	K_B	J_C	K_C
000	100	1	*	0	*	0	*
001	000	0	*	0	*	*	1
010	001	0	*	*	1	1	*
011	***	*	*	*	*	*	*
100	010	*	1	1	*	0	*
101	***	*	*	*	*	*	*
110	***	*	*	*	*	*	*
111	***	*	*	*	*	*	*

JK FF Excitation Q_t	Q_{t+1}	J_t	K_t
0	0	0	*
0	1	1	*
1	0	*	1
1	1	*	0

* = Don't Care

FIGURE 2.14 Example state table for JK-type counter.

PRESENT STATE ABC	NEXT STATE ABC	Input Forming Logic S_A	R_A	S_B	R_B	S_C	R_C
000	100	1	0	0	*	0	*
001	000	0	*	0	*	0	1
010	001	0	*	0	1	1	0
011	***	*	*	*	*	*	*
100	010	0	1	1	0	0	*
101	***	*	*	*	*	*	*
110	***	*	*	*	*	*	*
111	***	*	*	*	*	*	*

SR FF Excitation Q_t	Q_{t+1}	S_t	R_t
0	0	0	*
0	1	1	0
1	0	0	1
1	1	*	0

* = Don't Care

FIGURE 2.15 Example state table for SR-type counter.

must change from a logical 0 to a logical 1. According to the excitation table for the positive JK flip-flop, the J input must be set to a logical 1. The K_A input is recorded as a "don't care" since if $K_A = 0$, the flip-flop will set to a logical 1, and if $K_A = 1$, it will toggle to a logical 1. Take care to observe how each of the bit definitions in the input-forming logic column of Figure 2.14 was determined.

Figure 2.15 illustrates a similar state table that uses Set-Reset (SR-type) flip-flops to implement the above counter design. Again, the first row, first bit shows that Bit A must change from a logical 0 to a logical 1. According to the excitation table for the positive SR flip-flop, the S_A input must be set to a logical 1 and the R_A input must be set to a logical 0. Take care to observe how each of the bit definitions in the input-forming logic column of Figure 2.15 was determined.

We will have plenty of opportunities to develop state tables through the remainder of this text. However, readers should become very familiar with the methodology used to create these tables so that it will not be a stumbling block when the design becomes more complex.

2.6 EXTRACTING SOP EXPRESSIONS

Again, the purpose of the state table is to produce the input-forming logic for the counter. The input-forming logic columns define the relationships between the input-forming logic outputs and the counter states. All that is needed is to treat each subcolumn in the input-forming logic section of the state table as a function of the counter bits. We often use Karnaugh maps to reduce such expressions, and since the functions are already defined in terms of numeric values, the Karnaugh maps work well.

Consider Figure 2.16. It is important that the relationship between the state table and the Karnaugh maps of Figure 2.16 is understood. The first subcolumn of the input-forming logic column defines the relationship between the D input of bit A in the counter (D_A) and the counter states A, B, and C. Each bit in the column is recorded in the D_A Karnaugh map in the box corresponding to its minterm value defined by arguments A, B, and C. That is,

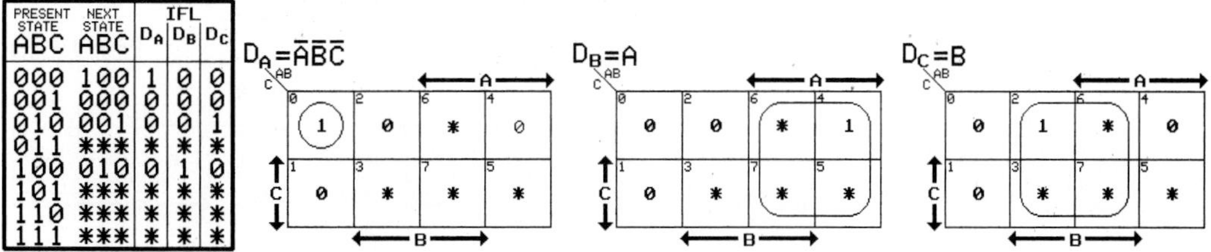

PRESENT STATE ABC	NEXT STATE ABC	IFL D_A	D_B	D_C
000	100	1	0	0
001	000	0	0	0
010	001	0	0	1
011	***	*	*	*
100	010	0	1	0
101	***	*	*	*
110	***	*	*	*
111	***	*	*	*

FIGURE 2.16 Extraction of SOP expressions for the D-type counter.

$D_A = 1$ when $\overline{A}\,\overline{B}\,\overline{C}$, $D_A = 0$ when $\overline{A}\,\overline{B}\,C$, etc. These values can be placed on the Karnaugh map either by using these logical arguments or by simply using the decimal equivalent of those arguments. The use of the numeric arguments is simpler and will be used in this text. Once all of the terms from the subcolumns have been entered into the Karnaugh maps (including the "don't care" states), the minimized expression for each bit can be extracted. From Figure 2.16 we find that: $D_A = \overline{A}\,\overline{B}\,\overline{C}$, $D_B = A$, and $D_C = B$.

Figure 2.17 illustrates the Karnaugh maps used to extract the SOP expressions for the JK implementation described above. Again, each input-forming logic column is placed on a Karnaugh map, and the minimum SOP expression is extracted from the map in order to define the relationship between the individual flip-flop input and the arguments A, B, and C.

Our last example of implementation of this counter used Set-Reset flip-flops. Figure 2.18 illustrates the Karnaugh maps for this design. Armed with the SOP expressions for each flip-flop input, we are ready to build the synchronous counter.

2.7 COUNTER CONSTRUCTION

We are now at the final stage of the synchronous binary counter design. We have defined the counter sequence and documented it in a state diagram. We have used the state diagram

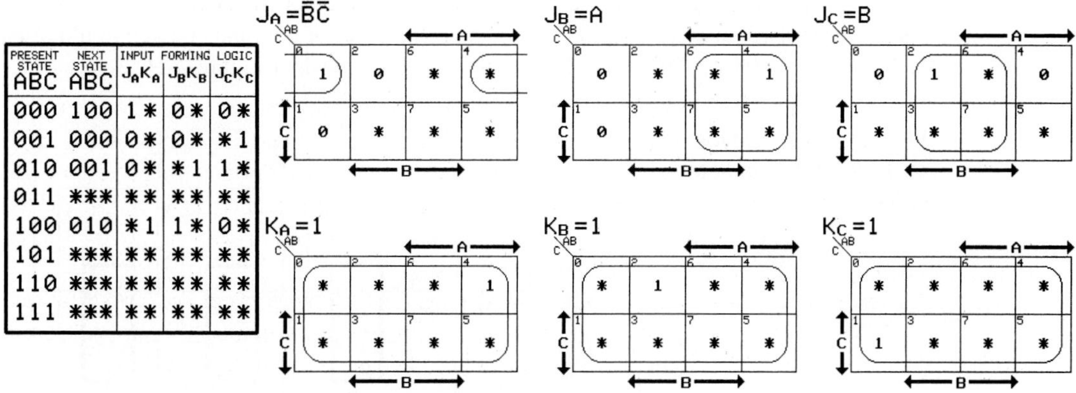

PRESENT STATE ABC	NEXT STATE ABC	INPUT FORMING LOGIC J_AK_A	J_BK_B	J_CK_C
000	100	1*	0*	0*
001	000	0*	0*	*1
010	001	0*	*1	1*
011	***	**	**	**
100	010	*1	1*	0*
101	***	**	**	**
110	***	**	**	**
111	***	**	**	**

FIGURE 2.17 Extraction of SOP expressions for the JK-type counter.

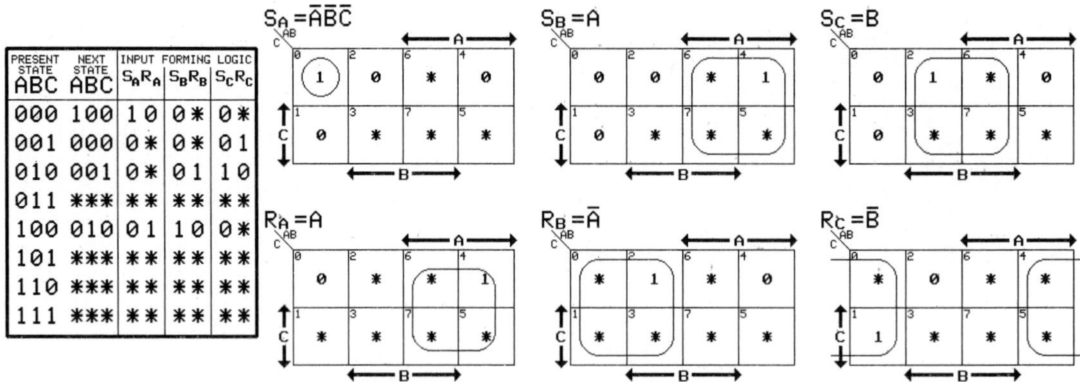

PRESENT STATE ABC	NEXT STATE ABC	S_AR_A	S_BR_B	S_CR_C
		INPUT FORMING LOGIC		
000	100	10	0*	0*
001	000	0*	0*	01
010	001	0*	01	10
011	***	**	**	**
100	010	01	10	0*
101	***	**	**	**
110	***	**	**	**
111	***	**	**	**

FIGURE 2.18 Extraction of SOP expressions for the SR-type counter.

information to create a state table. We have then used the information in the state tables to create the Karnaugh maps and the SOP expressions for each flip-flop input.

Our first example used D-type flip-flops to implement the counter design. We found that the input-forming logic generated the following signals for the flip-flop inputs: $D_A = \overline{A}\,\overline{B}\,\overline{C}$, $D_B = A$, and $D_C = B$. Knowing this, we can construct the synchronous binary counter by placing three D-type flip-flops in a synchronous register format (interconnecting their clock inputs) and interconnecting their D inputs according to the calculated expressions. Figure 2.19 illustrates this completed counter design.

Note that the counter can be preset to a logic 000_2 by a logic 1 pulse on the RESET input to the counter. Then, as clock edges are provided, it can progress through the sequence defined by our design.

Figure 2.20 illustates the completed design of the JK-type counter. Note that this counter requires less input-forming logic than the D-type. This will usually be the case when a JK-type design is compared with a D-type. Observation of the state table will show why this is true. Half of the input logic levels for the JK-type design are "don't care" states. These states will provide more redundant terms in the minimization process.

Since the SR-type design falls between these two other methods in the number of "don't care" states in the excitation tables, the complexity of the final counter would often be between these two. This is true for this design also. Consider Figure 2.21.

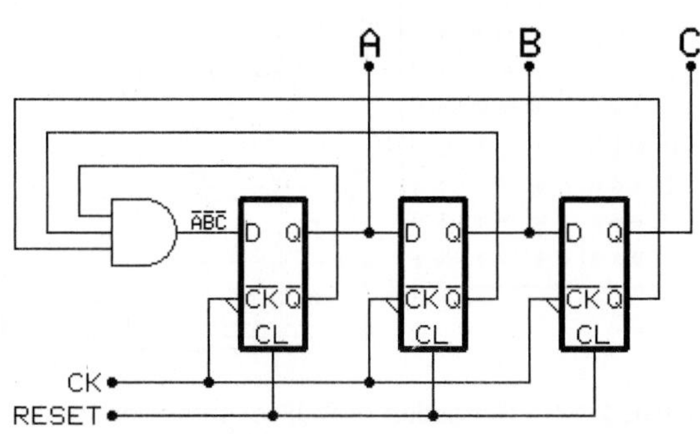

FIGURE 2.19 D-type synchronous binary counter example.

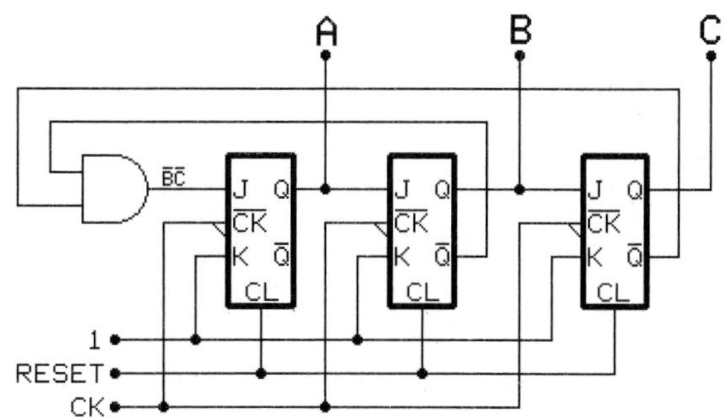

FIGURE 2.20 JK-type synchronous binary counter example.

2.7.1 Design Method Summary

The method used here for designing synchronous counters has five steps:

1. Define the count sequence.
2. Document the sequence as a state diagram.
3. Generate a state table from the state diagram information.
4. Use Karnaugh maps to extract the product terms for each flip-flop control input.
5. Draw the final schematic diagram.

Practical Example 2.2: Synchronous Binary Counter Design

Design a synchronous binary counter that repeats the sequence 1,2,3,7,4,1,2,3,7,4, ... If the counter is in state 0, 5, or 6, cause it to jump into the sequence at state 1.

First, document this sequence as a state diagram, as shown in Figure 2.22. Note that the three states that are outside the count sequence are shown in the state diagram, and in each case, if the state is entered, the next state is 001_2.

Second, generate a state table from the state diagram information, as shown in Figure 2.23. Note that, since the three states outside the count sequence are programmed to jump into the sequence, there are no "don't care" states in this table.

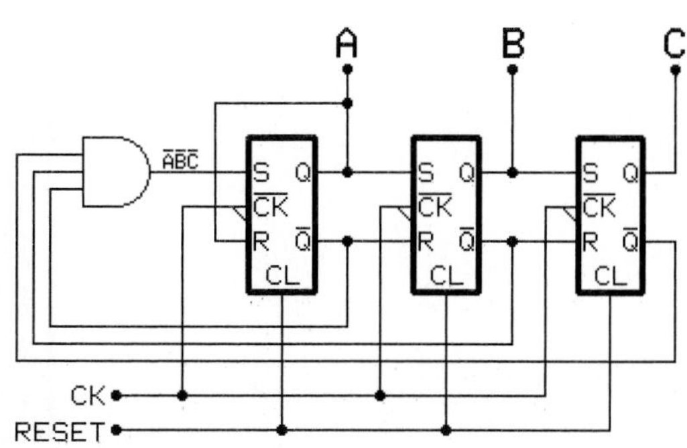

FIGURE 2.21 SR-type sychronous binary counter example.

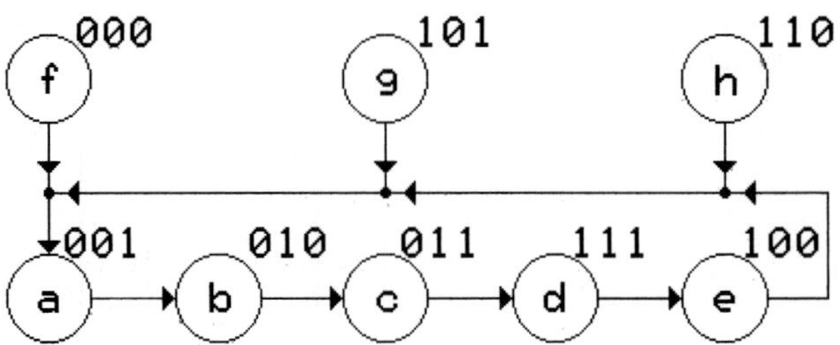

FIGURE 2.22 Practical Example 2.2, state diagram.

Third, use Karnaugh maps to extract the product terms for each flip-flop control input, as shown in Figure 2.24. Observe that the expressions for each flip-flop input are maintained in SOP form. The last function, $D_C = \overline{C} + \overline{A}B + A\overline{B}$, could be expressed as $D_C = C + (A \oplus B)$. Though the second expression is logically simpler, we must consider that ultimately we will be implementing our designs using programmable logic. It is often simpler to manipulate SOP expressions in a programmable logic environment, so it may be best to leave the expression in that form.

Your last task is to draw the final schematic diagram (see Figure 2.25). The logic gates in Figure 2.25 represent the input-forming logic of the counter.

Since all of the possible permutations of the count sequence were defined and there was no preset definition, this circuit does not contain a line to reset the counter. Though this would probably not be a normal circumstance, it is consistent with the definition of the problem. The circuit will power up to some permutation of the three-bit count. If the value is not in the count sequence, it will be after a single clock pulse.

Practical Exercise 2.1: JK-Type Synchronous Binary Counter

Using the method described in this chapter, redesign the problem in Practical Example 2.2 using JK-type flip flops. Show all steps of the design in detail.

FIGURE 2.23 Practical Example 2.2, state table.

PRESENT STATE ABC	NEXT STATE ABC	Input Forming Logic D_A	D_B	D_C
000	001	0	0	1
001	010	0	1	0
010	011	0	1	1
011	111	1	1	1
100	001	0	0	1
101	001	0	0	1
110	001	0	0	1
111	100	1	0	0

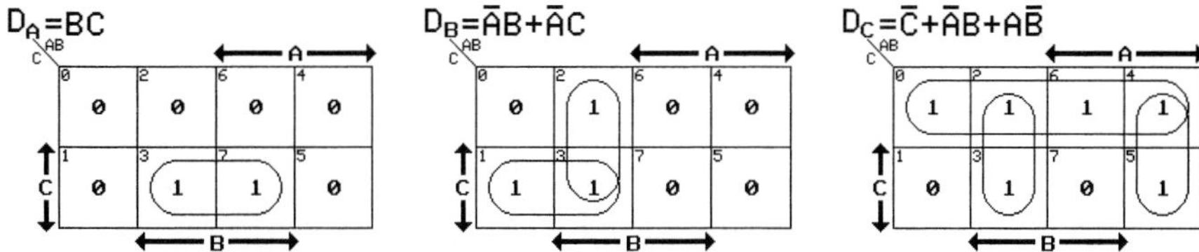

FIGURE 2.24 Practical Example 2.2, Karnaugh map simplification.

Practical Exercise 2.2: SR-Type Synchronous Binary Counter

Using the method described in this chapter, redesign the problem in Practical Example 2.2 using SR-type flip flops. Show all steps of the design in detail.

2.8 HAZARD PROTECTION

The design process we have just discussed is effective in producing reliable binary counters. However, there is one significant problem associated with this, and any other register-type device: glitches.

A **glitch** is an undesired logic level that occurs during a short period of time. It is often caused by unequal propagation delays through devices that are generating multibit data. For example, consider a three-bit synchronous counter that is changing state from a 000_2 to a 111_2. If all of the flip-flops in the counter change state at exactly the same time, there is no problem. However, if one flip-flop lags behind another, a glitch will be generated. Consider Figure 2.26.

Let's assume that we have built our three-bit counter of flip-flops from a variety of sources.

- The flip-flop generating bit A has a propagation delay of 4.0 nanoseconds.
- The flip-flop generating bit B has a propagation delay of 6.0 nanoseconds.
- The flip-flop generating bit C has a propagation delay of 8.0 nanoseconds.

At time 0 the negative edge of the clock initiates the change of the counter from state 000_2 to state 111_2. Bit A changes state first after a delay of 4.0 nanoseconds.

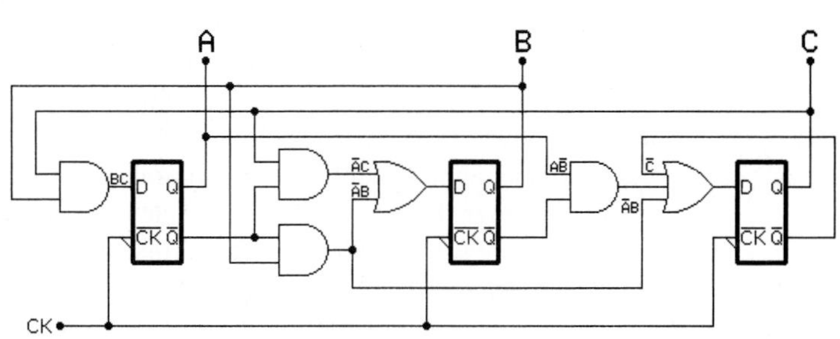

FIGURE 2.25 Practical Example 2.2, schematic diagram.

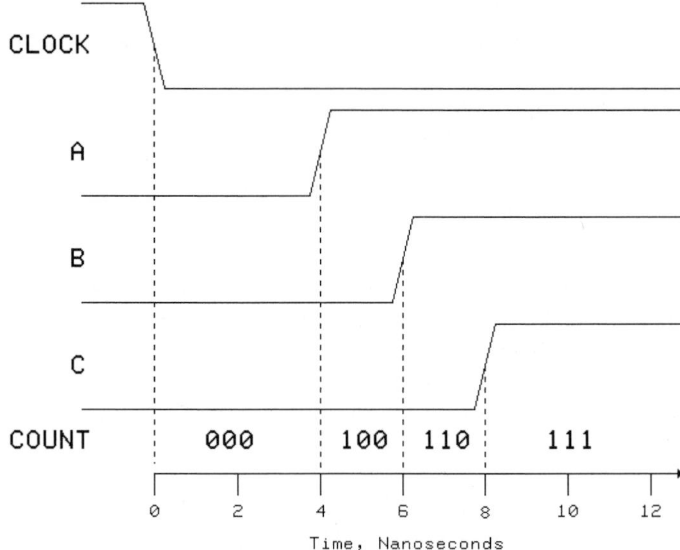

FIGURE 2.26 Glitches caused by unequal propagation delays.

The output of the counter is now 100_2. Two nanoseconds later, bit B changes state, placing a 110_2 on the output of the counter. Finally, eight nanoseconds after the clock edge, bit C changes state, placing a 111_2 on the output of the counter as originally desired.

What was taking place on the output of the counter during the time period from four to eight nanoseconds after the clock edge? In this case two unwanted bit patterns were present for two nanoseconds each. If this counter were driving circuitry that expected a clean transition from one state to the next, a disaster could result. For example, the circuit could be designed to activate a specific output device whenever the counter is in the 100_2 state. Imagine the consternation of the user when the device is activated at the wrong time!

Hazard protection refers to the system of protection designed into a digital circuit to prevent glitches from creating errors. This is no trivial subject and should be considered in any situation where glitches cannot be tolerated.

Glitch-tolerant circuits are common in a control environment. Many devices require pulse widths that are several nanoseconds, or even milliseconds, in length. For example, consider the application of a stepper motor. A common stepper must have a pulse width of at least 200 milliseconds in order to operate. Glitches will not cause problems with this type of device. The same is true for relays and many other electromagnetic devices.

Glitch-sensitive circuits are common in the digital environment. Any circuit that is edge-triggered is going to respond to the edge of a glitch. Any circuit that can respond to a very narrow pulse is also going to respond to the glitch. State machines are used in many situations where such errors cannot be tolerated. One clear example is in the field of medical instrumentation, where life support or medication doses have little room for error. Another could be in the area of weapons systems control.

There are a few methods available that can be used to eliminate the destructive effect of glitches. These fall into three categories. First, **glitch-free design** methods are available that never allow the generation of a glitch in the first place. Second, glitch-suppression methods are available that allow glitch generation, but use added propagation delay methods to remove them from the signal prior to use. Third, the circuit under control can be designed to be glitch-tolerant. Since we are concerned in this study with the state machine design—that is, the device doing the controlling—only the first two of these methods are relevant.

2.8.1 Glitch-Free Counter Design

The most effective method for hazard protection is to design systems without glitches. In the subject of binary counter design this is not a very difficult issue. What caused the glitches in Figure 2.26 in the first place? Three different devices were instructed to change state at the same time, and their delay periods were different. Since we cannot control the propagation delay of the devices, we are left with only one alternative: **Do not allow more than one bit in the counter to change state on the same clock edge**. At first this may seem to be a daunting task. Count sequences must be generated that flip only a single bit from one state to the next. Consider the following binary count sequence:

0001 - 0101 - 0100 - 1100 - 1000 - 1001 - 1011 - 1111 - 1110 - 0110 - 0111 - 0011 - 0001 . . .

Note that as the count progresses either to the left or to the right, only one bit changes state at a time. It is not possible for a glitch to be generated with a count sequence like this. Therefore, it would be useful to be able to come up with count sequences that fulfill this requirement. One way to do this would be to locate a digital fundamentals text that discusses binary Gray codes in detail. However, the approach this text will use takes advantage of an interesting phenomenon that is evident in the construction of a Karnaugh map. Every cell in a Karnaugh map is bordered by four other cells, either directly in the graphic or by "wrapping around" from side to side. As we consider a transition from any cell to any other neighboring cell we find that the binary value of the original minterm is only changed by one bit. Consider Figure 2.27.

Note that the binary equivalent of the minterm value has been recorded in each box of the Karnaugh map. If you observe any single value in the map and look to its right, its left, above it, or below it, you will find that only one bit has changed in the number.

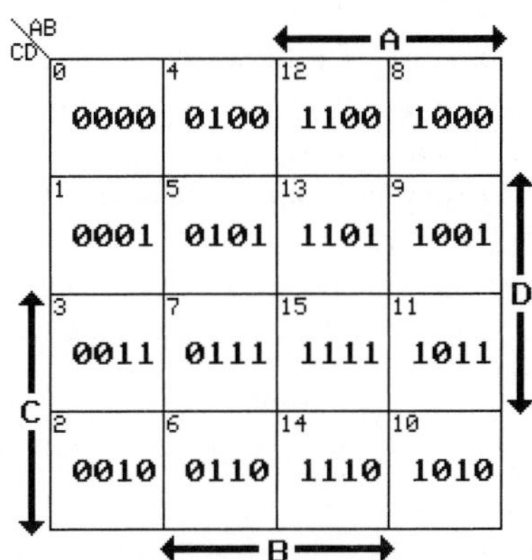

FIGURE 2.27 Karnaugh map.

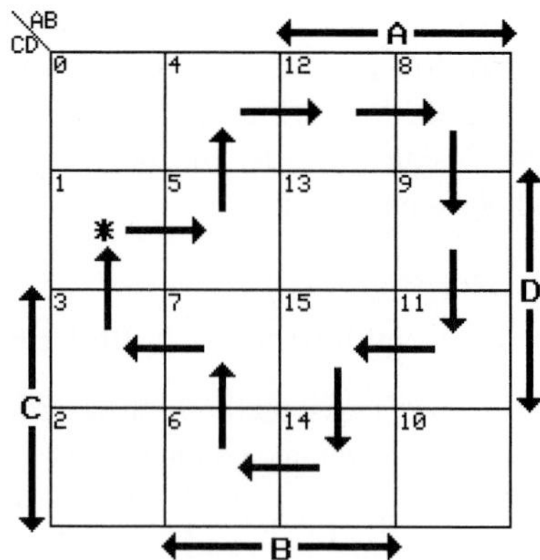

FIGURE 2.28 Next state map.

Coming up with glitch-free count sequences is accomplished by simply moving around the Karnaugh map one cell at a time. Reconsider the count sequence described previously:

0001 - 0101 - 0100 - 1100 - 1000 - 1001 - 1011 - 1111 - 1110 - 0110 - 0111 - 0011 - 0001 . . .

This same sequence can be represented in decimal as:

1 - 5 - 4 - 12 - 8 - 9 - 11 - 15 - 14 - 6 - 7 - 3 - 1

If we plot the transitions from one state to another on a Karnaugh map as a set of vectors, the map of Figure 2.28 is generated.

Note that the count sequence started in state 1. An asterisk (*) has been placed in the map at this point to designate the starting point of the sequence. State numbers were then chosen by moving in a relatively random fashion from one cell to another, moving up, down, right, or left until a count sequence was generated. This method can be used to assign state numbers to a state diagram and assure a glitch-free counter in the process.

Practical Example 2.3: Glitch-Free Counter Design

Consider the definition for a synchronous binary counter shown in Figure 2.29. The counter has six unique states, and the first is assigned a value of 0.

Your task is to use a next state map to assign the remaining states to numeric values in a manner that will be glitch-free.

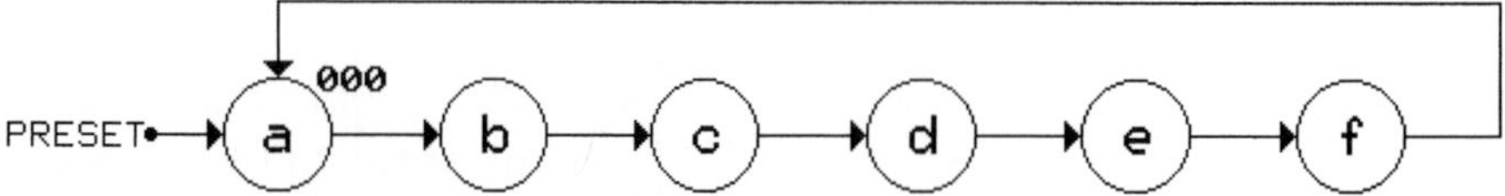

FIGURE 2.29 Practical Example 2.3, counter definition.

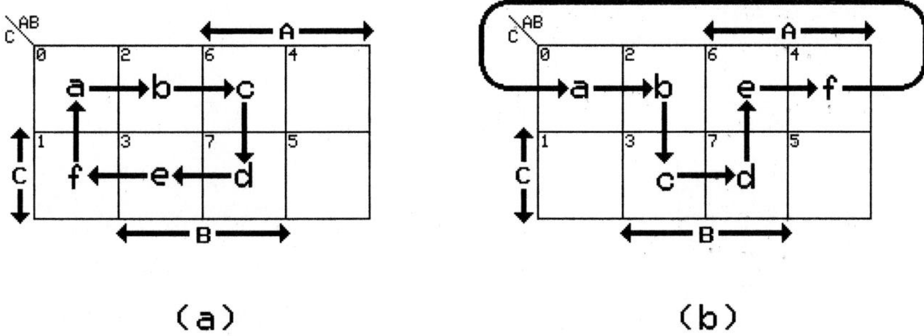

FIGURE 2.30 Practical Example 2.3, two solutions.

This is a three-bit counter by virtue of the facts that there are only six states, and the first state is revealed to be represented by a three-bit number. Therefore, a next state map must be drawn which uses a three-variable Karnaugh map. A count sequence can be generated by finding a path from state 0 back to itself after passing through five other states. Figure 2.30 illustrates two different solutions to this problem. The first solution (a) is the simplest. The second solution (b) appears more complex, but it is just as effective as the first. The count sequence defined by the first solution is:

000 - 010 - 110 - 111 - 011 - 001 - (repeat)

The second solution produces the following count sequence:

000 - 010 - 011 - 111 - 110 - 100 - (repeat)

Note that both solutions produce a six-count sequence where only one bit changes at at time. It is also reasonable to note that there are many different solutions to the problem, as there are many different permutations of pathways that can be taken on the map while staying within the constraints of the original definition.

Practical Exercise 2.3: Glitch-Free Counter Design

Considering the solution of Practical Example 2.3, use the same method to produce four other sequences that are glitch-free. Start each sequence at state 0.

Practical Example 2.4: Decision Making in Glitch-Free Counters

Consider the synchronous binary counter described by Figure 2.31. This problem includes input boxes in the state diagram. Though we have not yet looked at designing counters that make decisions based on external input, we can assign state numbers to the state diagram. When a state has an input box, the count sequence will branch to more than one other state

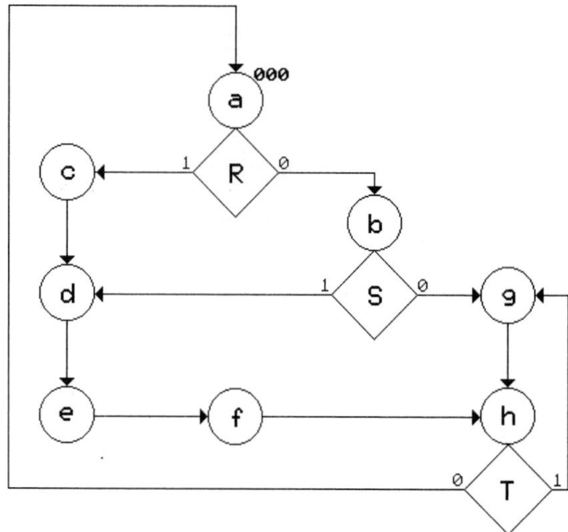

FIGURE 2.31 Practical Example 2.4, state diagram.

depending on the input value. When assigning states to the counter, we must still orient them so that each transition from one state to another requires moving only one cell in the Karnaugh map. As with the previous example, there are many correct solutions. This is a three-bit counter with all eight states defined in the sequence. The solution to this one will be a little harder, requiring some trial and error in the placement of the states on the map. Figure 2.32 illustrates one solution to the problem.

Carefully compare each transition (movement from one state to another) in the state diagram to the same transition in the next state map. Though this is a difficult example, it is evident that what we are doing is actually "drawing" the state diagram on the next state map. We are drawing only the state identifiers (the letters a, b, . . .) and the vectors (the lines with arrows). One benefit of this method is that the vectors on the next state map seldom cross one another. Often a complex state diagram, once entered on the next state map, can be redrawn so that the state boxes are in a similar orientation to the map. This can result in a much simpler layout that is easier to read and understand.

Once the vectors have been laid out on the next state map, the numbers from the cells can be used as the state assignments. These assignments are illustrated in Figure 2.33. Observe carefully that as the count moves from any state to another, regardless of the direction of the decisions involved, only one bit in the number changes at a time. Our counter is glitch-free.

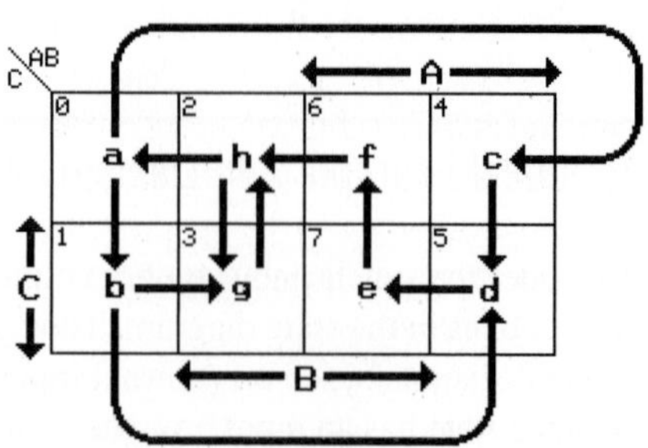

FIGURE 2.32 Practical Example 2.4, next state map.

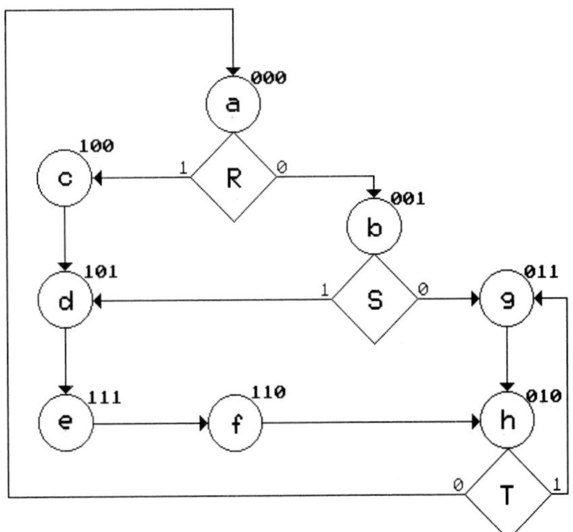

FIGURE 2.33 Practical Example 2.4, solution.

The two examples shown here are ideal. It is commonly impossible to create a counter from an original problem definition that is glitch-free. For example, if there is an odd number of states in a repeating sequence, it is impossible for the method to result in a closed loop on the next state map. In these cases, where it is impossible to get from one state to another on the map because of distance between them, a simple solution is to add states to the counter. This alternative is often an effective one. For example, if our state machine is being used to control a candy machine and it is clocked every millisecond, a difference in a few milliseconds between two states is trivial, since the operator cannot respond that quickly anyway.

Practical Exercise 2.4: Decision Making in Glitch-Free Counters

Using the same problem definition as Practical Example 2.4, determine at least two other correct solutions.

2.8.2 Delayed-State Timing

If we find ourselves in an environment where we cannot design a circuit to be glitch-free, an alternative is to allow the glitch to take place, but to delay the use of the signal until the glitches have had an opportunity to settle. This is effected by placing a delay between the glitch-laden controlling signal and the device under control. **Delayed-state timing (DST)** refers to a method of hazard protection that eliminates the effect of glitches by delaying the use of a signal until any glitches in it have had a chance to settle.

Consider Figure 2.34. The synchronous binary counter we have designed is illustrated in Figure 2.34 as two objects, the binary register and the input-forming logic. Let's assume that we have not used glitch-free design. There is a possibility that the outputs of

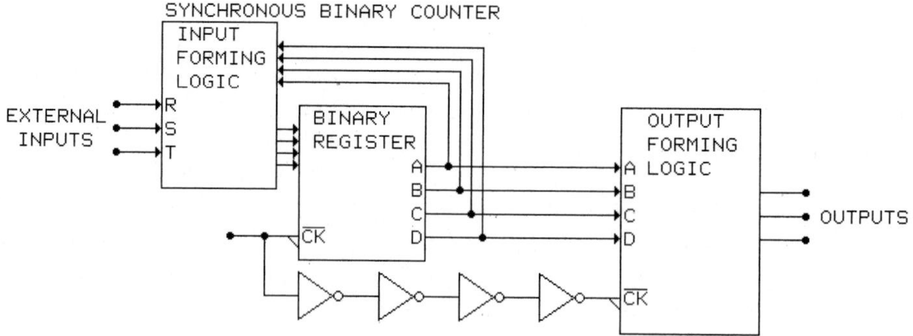

FIGURE 2.34 Delayed-state timing example.

the binary register may not all change state at exactly the same time, so output of the binary counter may glitch. Note that the output logic that is driven by the binary counter is enabled by the same clock signal. However, that signal has been delayed by the insertion of four inverters. If these inverters have an average propagation delay of four nanoseconds each, the output logic will not look at the output of the counter until 16 nanoseconds following the clock signal. If the average propagation delay of the counter is in the neighborhood of 10 nanoseconds, we are left with a safety margin of six nanoseconds.

Consider this concept: Delay the use of a glitch-laden signal until the glitches have had a chance to settle out. One characteristic of this method is that it is essentially asynchronous. The determination of the delay period must be calculated, and the insertion of such delays in several parts of a circuit can be cumbersome at best and impractical at worst. DST is an acceptable solution in many simple problems.

2.8.3 Alternate-State Timing

Another method of hazard protection we will consider is called **alternate-state timing (AST),** which refers to a hazard protection scheme that delays the use of a glitch-laden signal for one-half period of the clock. Consider Figure 2.35, which illustrates a situation

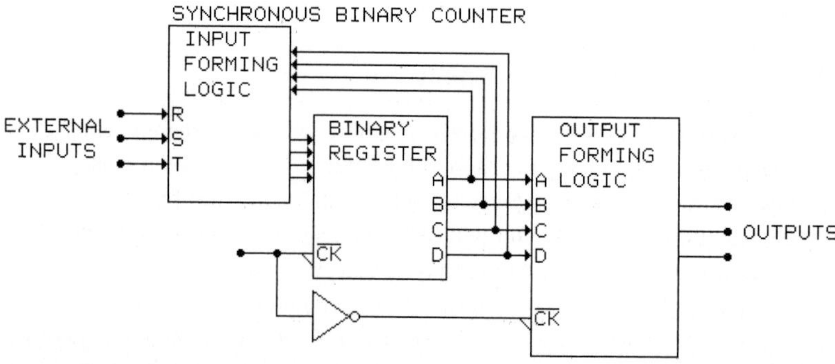

FIGURE 2.35 Alternate-state timing.

where the clock that drives the binary register is inverted prior to enabling the output logic. The falling edge of the clock enables the counter to change state. This same edge is seen by the output logic as a rising edge and has no effect on it. Half-way through the clock period, the original clock signal rises. Since the signal is inverted prior to presentation to the output logic, it is seen by the output logic as a falling edge, and the output logic is enabled.

Alternate-state timing has advantages over other methods. It is synchronous with the clock and quite predictable. We can enable circuitry that generates glitch-laden data on the falling edge of the clock and enable circuitry that uses that data on the rising edge of the clock. This provides the needed delay without requiring a lot of additional delay circuitry. In many microprocessor-based systems we see the use of alternate-state timing when devices attached to the microprocessor are enabled by an inverted clock.

QUESTIONS AND PROBLEMS

Define the terms in 1–17.

1. Synchronous binary counter
2. Binary register
3. Edge-triggered
4. Input-forming logic
5. Flip-flop excitation table
6. State diagram
7. State box
8. Input box
9. Output box
10. State table
11. Glitch
12. Hazard protection
13. Glitch-tolerant circuits
14. Glitch-sensitive circuits
15. Glitch-free design
16. Delayed-state timing (DST)
17. Alternate-state timing (AST)

18. Draw the excitation table for a positive SN7476-type JK flip-flop.
19. Draw the excitation table for a positive SN74107-type JK flip-flop.
20. Draw the excitation table for a positive D flip-flop.
21. Draw the excitation table for a positive SR flip-flop.
22. Draw the excitation table for an SR flip-flop with active-low SET and RESET inputs.
23. What is the purpose of a state diagram?
24. Describe the three major components of a state diagram.
25. Draw a state diagram of a counter that executes a repeating sequence of seven states.
26. How is an input box in a state diagram used to control the count sequence?
27. What is the purpose of a state table?

Consider the design of a three-bit synchronous binary counter that resets to the value 101_2 and then repeats the following sequence:

101 - 111 - 000 - 010 - 001

28. Draw a state diagram of this counter.
29. Construct a state table for the counter of Problem 28 using D-type flip-flops.
30. Determine the SOP expressions for the input-forming logic of Problem 29.
31. Draw the detailed schematic diagram of the counter of Problem 29.
32. Construct a state table for the counter of Problem 28 using JK-type flip-flops.
33. Determine the SOP expressions for the input-forming logic of Problem 32.
34. Draw the detailed schematic diagram of the counter of Problem 32.
35. Construct a state table for the counter of Problem 28 using SR-type flip-flops.
36. Determine the SOP expressions for the input-forming logic of Problem 35.
37. Draw the detailed schematic diagram of the counter of problem 35.

Consider the design of a three-bit synchronous binary counter that resets to the value 010_2 and then repeats the following sequence:

010 - 101 - 011 - 110 - 111 - 000

38. Draw a state diagram of this counter.
39. Construct a state table for the counter of Problem 38 using D-type flip-flops.
40. Determine the SOP expressions for the input-forming logic of Problem 39.
41. Draw the detailed schematic diagram of the counter of Problem 39.
42. Construct a state table for the counter of Problem 38 using JK-type flip-flops.
43. Determine the SOP expressions for the input-forming logic of Problem 42.
44. Draw the detailed schematic diagram of the counter of Problem 42.
45. Construct a state table for the counter of Problem 38 using SR-type flip-flops.
46. Determine the SOP expressions for the input-forming logic of Problem 45.
47. Draw the detailed schematic diagram of the counter of Problem 45.

Consider the design of a four-bit synchronous binary counter that resets to the value 0101_2 and then repeats the following sequence:

0101 - 0011 - 0110 - 0111 - 1010 - 1101 - 1011 - 1110 - 1111 - 1000

48. Construct a state diagram for this counter.
49. Construct a state table for the counter of Problem 48 using D-type flip-flops.
50. Determine the SOP expressions for the input forming logic of Problem 49.
51. Draw the detailed schematic diagram of the counter of Problem 49.
52. Construct a state table for the counter of Problem 48 using JK-type flip-flops.
53. Determine the SOP expressions for the input-forming logic of Problem 52.
54. Draw the detailed schematic diagram of the counter of Problem 52.
55. Construct a state table for the counter of Problem 48 using SR-type flip-flops.
56. Determine the SOP expressions for the input-forming logic of Problem 55.
57. Draw the detailed schematic diagram of the counter of Problem 55.

FIGURE 2.36

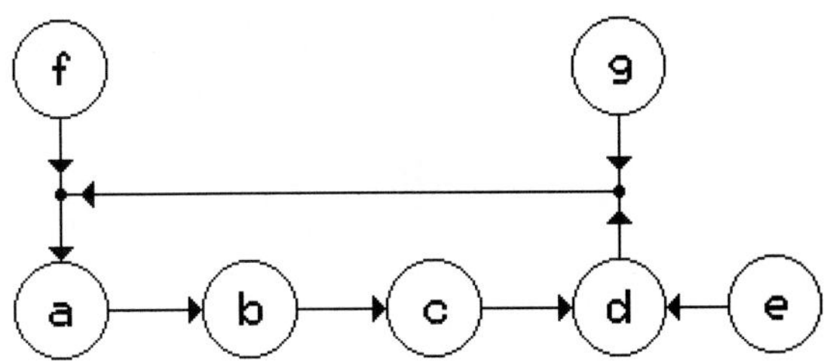

Refer to Figures 2.36 and 2.37 in order to answer Problems 58–65.

58. Assign state values to the state diagram shown in Figure 2.36 to assure glitch-free operation.
59. Assign state values to the state diagram shown in Figure 2.37 to assure glitch-free operation.
60. Develop the counter for Figure 2.36, showing the state table, development of the input-forming logic, and the schematic diagram using D-type flip flops.
61. Develop the counter for Figure 2.36, showing the state table, development of the input-forming logic, and the schematic diagram using JK-type flip flops.
62. Develop the counter for Figure 2.36, showing the state table, development of the input-forming logic, and the schematic diagram using SR-type flip flops.
63. Develop the counter for Figure 2.37, showing the state table, development of the input-forming logic, and the schematic diagram using D-type flip flops.
64. Develop the counter for Figure 2.37, showing the state table, development of the input-forming logic, and the schematic diagram using JK-type flip flops.
65. Develop the counter for Figure 2.37, showing the state table, development of the input-forming logic, and the schematic diagram using SR-type flip flops.
66. Describe the concept of delayed-state timing.
67. Describe the concept of alternate-state timing.

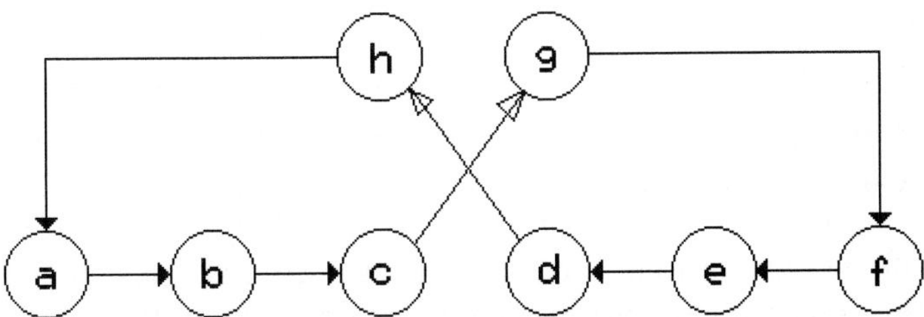

FIGURE 2.37

CHAPTER 3

Decision Making with Binary Counters

OBJECTIVES

After completing this chapter you should be able to:

- Use a programming language to provide a state machine problem definition.
- Draw a state diagram from a problem definition that includes input stimuli.
- Assign optimum state values to a state diagram providing for either of the following:
  - Glitch-free operation
  - Minimization of Input Forming Logic
- Develop state tables for a synchronous binary counter that will allow external input.
- Extract input-forming logic from those state tables.
- Construct synchronous binary counters from either D-type, JK-type, or SR-type flip-flops that exhibit the property of accepting input stimuli.

3.1 OVERVIEW

When one thinks of a synchronous binary counter, the first image that comes to mind is probably of a binary register that is configured to count from 0 to 10 or from 0 to 16 and repeat the sequence. It is only after a little bit of experience with digital counter design and application that we start to realize the count sequence can be much more flexible, meeting specific needs of a variety of applications. For example, we found in Chapter 2 that the count does not necessarily have to be in a numeric sequence, or in either ascending or descending order. Any desired sequence in a counter can be defined as long as the numeric value of the count at each state is unique. When we are designing state machines using synchronous binary counters, it is this property of the uniqueness of each state, rather than the numeric sequence of the state values, that will be of most importance in the design.

This chapter will add another feature to our synchronous binary counter: the ability to change the count direction based on the logic level of an external input. It is this feature that enables the binary counter to fulfill one of the two primary tasks of a state machine: monitor and control. The state machine must be able to change from one state to another, accepting inputs from some external source and driving outputs at each state according to a predetermined definition.

We will review the entire process of designing synchronous binary counters that was presented in Chapter 2. As the review is taking place we will be adding the monitoring of input stimuli to the procedure. When we are done we will be able to design synchronous binary counters that can monitor the state of external stimuli and provide the resource needed to control other external devices.

3.2 STATE MACHINE PROGRAMMING LANGUAGE

Communication requires common language, whether it be spoken, written, electronic, or by any other means. In order to describe the operation of a state machine in written form we will be using a textual language form. This language will define all of the activity generated within our state machine. A typical program can be stated as a state diagram, and a typical state diagram can also be stated as a program. The state machine programming language we will be using herein is similar to the CUPL language discussed in more detail in Chapter 5.

The binary counter proceeds through a sequence of states, each unique and stable. During each state there are at least three properties to define:

First, what state are we in? Each state must be unique and must have some unique identifier associated with it.

Second, what is the identifier of the next state in the sequence? In the case of binary counters that sense no external stimuli, the next state is always the same. If the count sequence is 1-2-3-4, we can always depend that if we are in state 2, the next state will be state 3. When we are sensing external stimuli, this will no longer be the case. We will need to specify the identifier of the next state as a function of inputs.

Third, what output from the state machine will be defined during this specific state? As we are monitoring external inputs, we are using those inputs to make decisions pertaining to the control of certain outputs.

Our state machine programming language will be able to define all three of these properties. This chapter will be concerned with only the first two. Output control will be added to our design in Chapter 4. Also, it may be useful to note at this point that the syntax used for these statements is the same as the syntax we will be using to program PALs later in the text. This makes learning this syntax worthwhile.

Program statements are case-sensitive. That is, the letter "a" is considered different from the letter "A." Upper- and lowercase letters will be used in a consistent fashion to improve readability. Variables will typically be written in lowercase; constants and program statement tokens are written in uppercase. Program commands and verbs are understood regardless of case.

3.2.1 The Semicolon (;)

The semicolon (;) is used to identify the end of a logical program line. A program statement that fits on a single line can be broken up into one or more lines using this convention. Also, some complex statements will not fit on a single line, making this free-form mechanism useful. The carriage return is ignored by the compiler we will be using to program PAL devices.

3.2.2 Comment Statement

The programming language also uses a free-form comment structure. It will use the delimiter "/*" to start a comment and the delimiter "*/" to close a comment. The following are examples of the format of a typical comment:

```
/* This is a comment */
/* This is also
a comment */
```

The free-form nature of the comment delimiters allows program segments to be "commented out" rather than deleted while program development and debugging takes place. Note that the comment can begin and end on different lines.

3.2.3 Numeric Constants and Logical Functions

Numeric constants can be represented in any of several number bases including decimal, hexadecimal, octal, and binary. The following are some examples of numeric constants in different bases:

Decimal:	'd'0	'd'21	'd'51	'd'7
Hexadecimal:	'h'0	'h'15	'h'33	'h'7
Octal:	'o'0	'o'25	'o'63	'o'7
Binary:	'b'0	'b'10101	'b'110011	'b'111

We will use the same logical function symbols as those defined in Chapter 2 as typographic symbols. These are repeated in Table 3.1 for convenience.

TABLE 3.1 Typographical logical operators.

Operator	*Character*	*Precedence*
NOT	!	1
AND	&	2
OR	#	3
XOR	$	4

3.2.4 Sequence Statement

The sequence statement declares that the statements within the brackets represent a complete counter sequence. The following is an example of a sequence statement:

```
SEQUENCE [<counter_output_bits>] {
...           /* transition statements */
}
```

Placed between the brackets "{" and "}" will be each of the transition statements that actually defines what is taking place in the sequence. The <counter_output_bits> will be a list of the labels of the counter output bits. This can be done in a variety of ways. For example:

```
[A,B,C,D]      [Q3,Q2,Q1,Q0]      [Q3..Q0]
```

All of these examples represent the outputs of a four-bit counter. The following is an example of the sequence statement used to define a three-bit counter that has outputs labeled A, B, and C:

```
SEQUENCE [A,B,C] {
...        /* transition statements */
}
```

The preceding example represents the general format of the sequences that we will be defining. We now need to add the transition statements that define the state machine activity.

3.2.5 Unconditional Transition Statement

Consider the following unconditional transition statement:

```
PRESENT <state0> NEXT <state1>;
```

The symbols "<" and ">" are used to indicate that these objects will be defined when the statement is written. The semicolon at the end signifies the end of the statement definition. This statement can be used to describe an unconditional count sequence. For example, if a count sequence is defined as 1, 2, 3, ... state 3 always follows state 2. The state machine programming language statement for state 2 would be:

```
PRESENT 2 NEXT 3;
```

Note that there is no output activity defined for this state. We will concern ourselves with output activity in Chapter 4. Figure 3.1 illustrates the state diagram for this statement.

Usually during the design stage, the numeric values of the states are not yet assigned. In that case, variables are often used. For example,

```
PRESENT a NEXT b;
```

Figure 3.2 illustrates the state diagram for this statement.

FIGURE 3.1 Unconditional numeric count sequence.

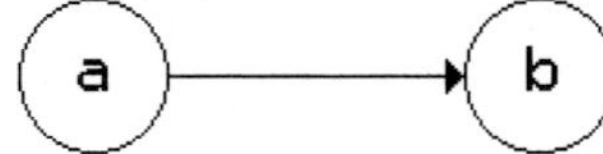

FIGURE 3.2 Unconditional variable count sequence.

Practical Example 3.1: Sequential Counter Program

Consider a three-bit counter that generates the repeating sequence 0, 4, 2, 1. Figure 3.3 illustrates the schematic and state diagram of this counter. Write a program that defines this state machine.

The solution is as follows:

```
/* Practical Example 3.1 */
SEQUENCE [X,Y,Z] {
              PRESENT 0 NEXT 4;
              PRESENT 4 NEXT 2;
              PRESENT 2 NEXT 1;
              PRESENT 1 NEXT 0;
              }
```

This program can be a model for describing any synchronous binary counter that has no external inputs or outputs. However, our state machines will be using counters that use input signals to make decisions. The next state following any given state may vary depending on input signals. For these situations we need the conditional transition statement.

3.2.6 Conditional Transition Statement

The conditional transition statement is used to describe a step in the count sequence that continues to one of two or more different next states as a function of one or more external inputs.

The following is an example of a conditional transition statement:

```
IF input_condition NEXT <state1>;
```

Again, we will not be concerned with the statement format that takes advantage of output arguments until Chapter 4.

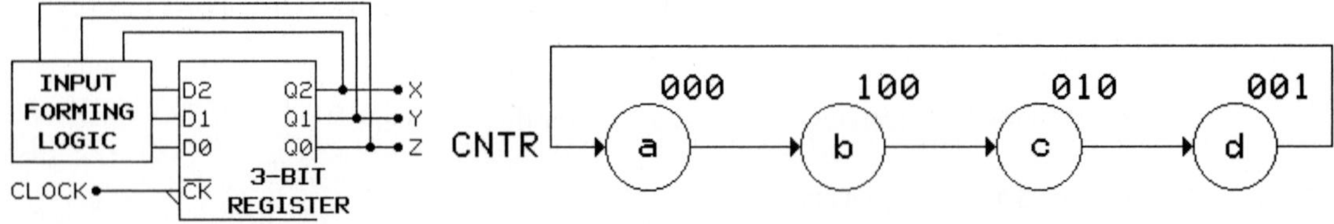

FIGURE 3.3 Practical Example 3.1, schematic and state diagram.

Practical Example 3.2: Conditional States in a State Machine Program

Consider the state machine defined by Figure 3.4. Note that when the counter is in state a, the next state of the state machine depends on the value of the logic level on input line R. If R = 1, the next state is to be state b. If R = 0, the next state is to be state c. Write the state machine programming language program for this counter.

The solution is as follows:

```
/* Practical Example 3.2 */
SEQUENCE [X,Y,Z] {
            PRESENT 2 IF R NEXT 5;
                      IF !R NEXT 7;
            PRESENT 1 NEXT 2;
            PRESENT 5 NEXT 1;
            PRESENT 7 NEXT 1;
            }
```

Note that the order of the transition statements within the sequence brackets is of no significance, but it is probably best to list them in an ascending or descending sequence of the present state in order to simplify the understanding of the overall counter activity.

3.2.7 Default Statement

The default statement is used to close a single IF argument, defining the destination for any arguments not previously stated. For example, the previous program can also be written:

```
/* Practical Example 3.2 */
SEQUENCE [X,Y,Z] {
            PRESENT 2 IF R=1 NEXT 5;
                      DEFAULT NEXT 7;
            PRESENT 1 NEXT 2;
            PRESENT 5 NEXT 1;
            PRESENT 7 NEXT 1;
            }
```

The default state will be the next state if all previous arguments fail.

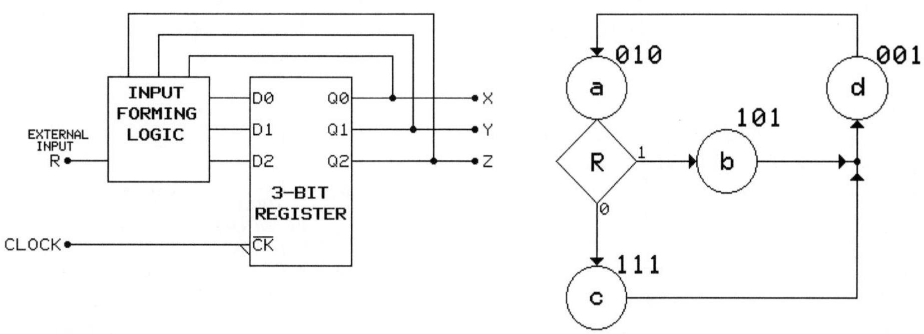

FIGURE 3.4 Practical Example 3.2, schematic and state diagram.

Practical Example 3.3: Creating a State Diagram from a State Machine Program

Consider the synchronous counter of Figure 3.5. The number of inputs to the counter has been increased from one to three when compared with the previous example. All other constraints are the same.

Often the design of a state machine starts with the problem that is stated as a program, not with the state machine diagram. In this example, we will start with the program and, using the information in it, generate the state diagram. Consider the following program:

```
/* Practical Example 3.3 Initial State Machine Program */
SEQUENCE [Q2..Q0] {
    PRESENT a IF R NEXT c;
              IF !R NEXT b;
    PRESENT b IF S NEXT a;
              IF !S NEXT d;
    PRESENT c IF S NEXT f;
              IF !S NEXT a;
    PRESENT d IF T NEXT e;
              IF !T NEXT b;
    PRESENT e NEXT a;
    PRESENT f IF T NEXT c;
              IF !T NEXT e;
              }
```

Your tasks are to use this program to (1) generate a state diagram, (2) assign numeric values to each state in the counter disallowing glitches, and (3) rewrite the program and state diagram using the assigned numeric values for the states.

State Diagram

The program is converted to a state machine by drawing each state, one at a time as defined by each transition statement in the program. Figure 3.6 illustrates a solution.

Assigning Values to Disallow Glitches

An effort was made to avoid crossing lines in the state machine diagram. When this is done, it is a much simpler task to observe the state machine operation. It also makes it easier to assign state values for the binary counter. Recall from Chapter 2 that we can use a

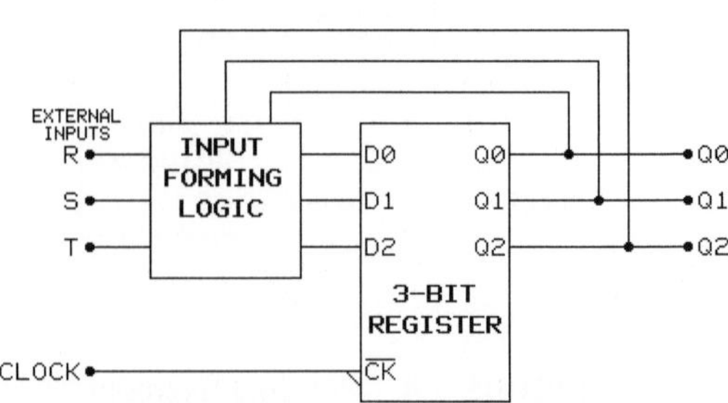

FIGURE 3.5 Practical Example 3.3, schematic diagram.

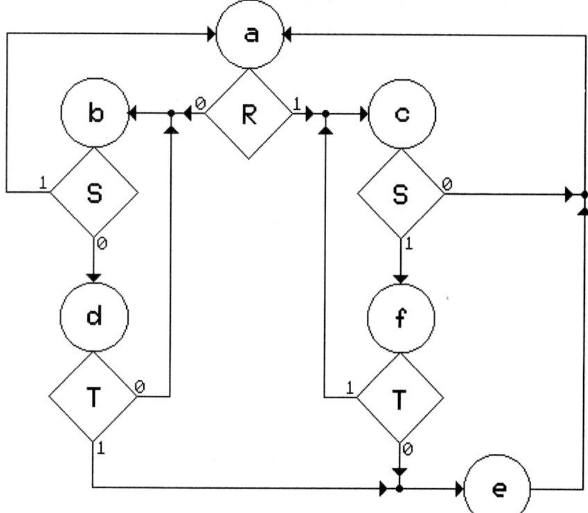

FIGURE 3.6 Practical Example 3.3, initial state diagram.

Karnaugh map as a next state map, arranging the states so that each transition on the diagram moves to an adjacent cell on the map. If the entire state machine can be placed on a map in this fashion, with all transitions moving between adjacent cells, the count sequence is glitch-free. Figure 3.7 illustrates the next state map of one solution for this problem.

Observe the results obtained in Figure 3.8 when the cell numbers from the next state map are drawn on the state diagram. Specifically, observe that as transitions take place from one state to another no more than one bit changes in the number at any point. This is the most effective way of obtaining glitch-free operation of the counter.

Rewriting the Program and State Diagram

Writing the final form of the state machine program, like the final form of the state diagram, is just a matter of substituting the cell numbers in the next state map for the variables used in the original program. The following is the solution:

```
/* Practical Example 3.3 State Machine Program, Second Solution */
SEQUENCE [Q2..Q0] {
    PRESENT 2 IF R NEXT 6;       /* State-a */
              IF !R NEXT 0;
    PRESENT 0 IF S NEXT 2;       /* State-b */
              IF !S NEXT 1;
    PRESENT 6 IF S NEXT 7;       /* State-c */
              IF !S NEXT 2;
    PRESENT 1 IF T NEXT 3;       /* State-d */
              IF !T NEXT 0;
    PRESENT 3 NEXT 2;            /* State-e */
    PRESENT 7 IF T NEXT 6;       /* State-f */
              IF !T NEXT 3;
              }
```

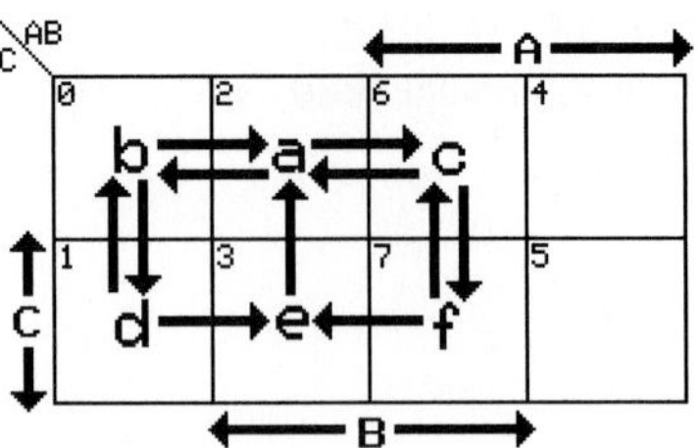

FIGURE 3.7 Practical Example 3.3, next state map.

An alternative way to draw this program is to organize the transition statements into a numeric sequence. The order of the transition statements within the sequence statement is not required to be in any sequence, but placing them in a sequence makes it a little easier to understand. Also, it may be convenient to represent the state values in binary, as shown in the following solution to the problem:

```
/* Practical Example 3.3 State Machine Program, Third Solution */
SEQUENCE [Q2..Q0]
{ PRESENT 'b'000 IF S NEXT 'b'010; DEFAULT NEXT 'b'001;
  PRESENT 'b'001 IF T NEXT 'b'011; DEFAULT NEXT 'b'000;
  PRESENT 'b'010 IF R NEXT 'b'110; DEFAULT NEXT 'b'000;
  PRESENT 'b'011 NEXT 'b'010;
  PRESENT 'b'110 IF S NEXT 'b'111; DEFAULT NEXT 'b'010;
  PRESENT 'b'111 IF T NEXT 'b'110; DEFAULT NEXT 'b'011; }
```

Note some of the changes in the format of this example. The state values are described in binary, and the conditional arguments are listed on a single line. Also, the command DEFAULT was used instead of the inverted argument to close the decision. This example shows that using the command is similar to using the word "ELSE" in similar logical arguments. The default state will be the next state if all previous arguments fail.

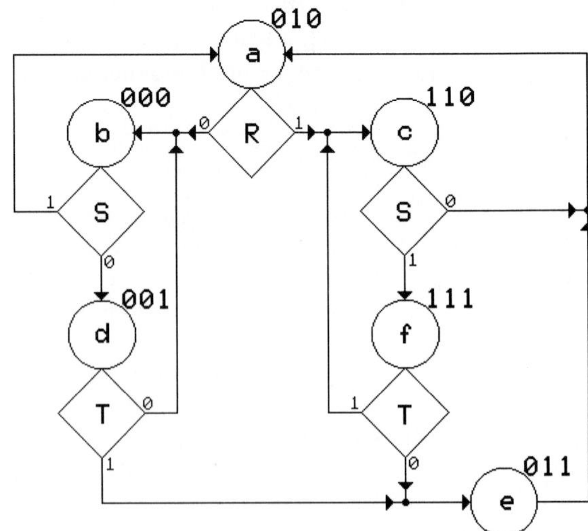

FIGURE 3.8 Practical Example 3.3, final state diagram.

Practical Exercise 3.1: Creating a State Diagram from a State Machine Program

Consider the following state machine program:

```
SEQUENCE [ ]  {
    PRESENT a NEXT b;
    PRESENT b IF R NEXT e; DEFAULT NEXT c;
    PRESENT c NEXT d;
    PRESENT d NEXT c;
    PRESENT e IF R NEXT f; DEFAULT NEXT k;
    PRESENT f NEXT g;
    PRESENT g NEXT h;
    PRESENT h IF R NEXT a; DEFAULT NEXT i;
    PRESENT i NEXT j;
    PRESENT j NEXT c;
    PRESENT k NEXT g;
    }
```

Note that the sequence field of the program is left blank since we have not yet determined the number of bits in the counter. Note also how the conditional transition statements were organized to fit on one line. Recall that the end of a logical line is determined by the semicolon, not by a carriage return.

Your tasks are to use this program to (1) generate a state diagram, (2) assign numeric values to each state in the counter disallowing glitches, (3) rewrite the program and state diagram using the assigned numeric values for the states, and (4) draw a block schematic diagram of the counter in a form similar to Figure 3.5.

3.3 DECISION MAKING IN STATE TABLES

Once a definition for a state machine has been determined and documented either as a state diagram or a state machine language program, we are ready to design the logic circuit to implement it. The next logical step in the design was shown in Chapter 2 to be the development of the state table. Recall from Chapter 2 that when no decisions are taking place, each line in the state table defines a single stable state of the counter. The state table shows the value of the counter for the present state, the value of the counter for the next state, and the stimulus needed to cause the counter to change from the present state to the next. When a decision is taking place during a state, there is more than one possible next state. The next state will depend on the logical value of one or more inputs to the state machine. Therefore, we will have to add these inputs to the arguments for determining the next state.

3.3.1 Present State, Input, and Next State Columns

Consider the following state machine definition shown in Figure 3.9. This is the same state machine as that presented earlier as Practical Example 3.2. As we build the state table, the

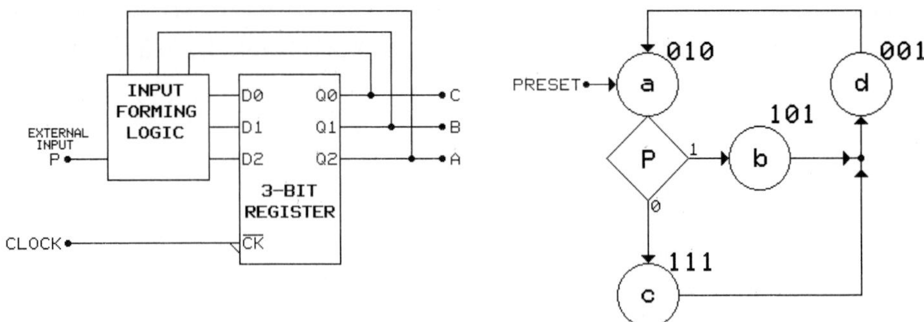

FIGURE 3.9 State machine with one decision.

method for writing entries for those unconditional transitions will be the same as was used in Chapter 2. However, observe that in state a there are two possible next states, state b if P = 1, state c if P = 0. Since there are two possible next states for state a, there will be two lines on the state table for this state.

The lines will be differentiated by adding a column for input P and declaring the permutations of that input and their effect on the decision regarding which state is the next one. Consider the state table illustrated in Figure 3.10.

Of the eight possible permutations of the counter, four are defined according to the state diagram of Figure 3.9. The transitions at state b, state c, and state d are unconditional. Their entries on the state table are in the same form as those designs previously investigated. The numeric value for each state is listed in the Present column, and the next state following in the state machine diagram is in the Next column. Note that since these entries are unconditional, it doesn't matter what the state of input P is, so a "don't care" is entered in the Input column for these bits. Also, as was done with previous designs, those states that are not defined in the sequence have no next state. "Don't cares" are entered for the next state for these. Also, since P has no effect on the next state for these unused states, a "don't care" is entered in the input column.

State a, (010_2) is conditional. The next state is dependent on the logical value on input P. Since P can take on the values 0 and 1, the two permutations of the input are entered as separate lines on the state table. If P = 0, the next state is 111_2. If P = 1, the next state is 101_2. Note how this is done on the table. One line is included for each permutation of the input

FIGURE 3.10 Single decision in a state table (state 010_2).

	PRES ABC	INPUT P	NEXT ABC
0	000	*	***
1	001	*	010
2	010	0	111
		1	101
3	011	*	***
4	100	*	***
5	101	*	001
6	110	*	***
7	111	*	001

* = Don't Care

FIGURE 3.11 Unconditional stimuli added to state table.

	PRES ABC	INPUT P	NEXT ABC	D_A	D_B	D_C
0	000	*	***	*	*	*
1	001	*	010	0	1	0
2	010	0	111			
		1	101			
3	011	*	***	*	*	*
4	100	*	***	*	*	*
5	101	*	001	0	0	1
6	110	*	***	*	*	*
7	111	*	001	0	0	1

* = Don't Care

that uniquely defines a next state. Therefore, on a line where P is defined as 0, a next state of 111 is entered. On a second line where P is defined as a 1, a next state of 101 is entered.

3.3.2 Stimulus Columns

It is now necessary for us to build the stimulus columns in the state table. Again, the method for unconditional transitions is identical to that presented in Chapter 2. However, when the state is conditional upon some input, a little more thought must go into what is being done. The principle is the same: Determine what logic will cause the flip-flop to go to the defined next state.

D-Type Flip-Flop Implementation. Consider Figure 3.11, assuming we will implement our state machine with D-type flip-flops.

Again, there should be no surprises here. The respective D inputs for states that are not in the sequence are undefined, and a "don't care" state is entered for each. These are in states 0, 3, 4, and 6. For the remaining unconditional transitions, the respective D inputs follow the data in the next state column. These are on lines 1, 5, and 7. Recall from Chapter 2 that a clocked D-type flip-flop will move to the state placed on its D input when it receives the clock edge. This principle is very important to consider when we observe what is necessary to define the D inputs when the next state is conditional, as illustrated on line 2. We will still be answering the question "What do I have to do to the D-type flip-flop to cause it to go to the next state as defined in the state diagram?"

Observe the transitions which take place in state 010_2, on line 2 in Figure 3.12:

- Bit A changes from a 0 to a 1, regardless of the value on input P. In order to get D_A to change to a 1, it is necessary to place a 1 on it.
- Bit C also changes from a 0 to a 1, regardless of the value on input P. In order to get D_C to change to a 1, it is necessary to place a 1 on it.
- Bit B remains at a 1 when input P = 0. Bit B changes from a 1 to a 0, when input P = 1. That is, $D_B = 1$ when P = 0, and $D_B = 0$ when P = 1. Therefore, $D_B = \overline{P}$.

Actually, observation of the state table reveals that the determination of D_B is simpler than that which has just been implied. Note that in the next state column, bit B is the complement of P. Therefore, D_B is the complement of P.

Certainly, of the flip-flops used in this text, the D-type, JK-type and SR-types, the D-type is the simplest to implement. Also, it will usually be the device of choice, since the

FIGURE 3.12 Conditional stimulus added to state table.

	PRES ABC	INPUT P	NEXT ABC	D_A	D_B	D_C
0	000	*	***	*	*	*
1	001	*	010	0	1	0
2	010	0 1	111 101	1	$\bar{P}$	1
3	011	*	***	*	*	*
4	100	*	***	*	*	*
5	101	*	001	0	0	1
6	110	*	***	*	*	*
7	111	*	001	0	0	1

* = Don't Care

most common PAL implementations will utilize the R-macrocell or V-macrocell types. We still need to look at how to use JK- type flip-flops and SR-type flip-flops, however.

JK-Type Flip-Flop Implementation. Consider the JK implementation illustrated in Figure 3.13. The determination of the J and K input stimulus for the "don't care" and unconditional states, defined by the excitation table, was discussed in Chapter 2. However, take a few moments to observe the stimulus entered on the state table for those unconditional states, 1, 5, and 7. Also, bit A and bit C in the conditional state 2 are unconditional since their transitions are not dependent upon input P. Only bit B in state 2 is dependent on the input and must be handled differently. Consider the following logic used to determine the J and K inputs of bit B in state 2:

- State 2 bit B goes from 1 to 1 when P = 0; so, J_B = * and K_B = 0.
- State 2 bit B goes from 1 to 0 when P = 1; so, J_B = * and K_B = 1.

The input K_B cannot be hardwired to a logic 0 and a logic 1 at the same time. However, it is the P input that is defining these two arguments. J_B is set to a "don't care" state, regardless of the logic level of input P. However, the K_B input is the same as input P, so we can place P on the K_B input for proper operation. Note how the variable P is placed in the state table.

SR-Type Flip-Flop Implementation. Figure 3.14 illustrates the SR-type flip-flop state table for this problem.

	PRES ABC	INPUT P	NEXT ABC	J_A	K_A	J_B	K_B	J_C	K_C
0	000	*	***	*	*	*	*	*	*
1	001	*	010	0	*	1	*	*	1
2	010	0 1	111 101	1	*	*	P	1	*
3	011	*	***	*	*	*	*	*	*
4	100	*	***	*	*	*	*	*	*
5	101	*	001	*	1	0	*	*	0
6	110	*	***	*	*	*	*	*	*
7	111	*	001	*	1	*	1	*	0

* = Don't Care

JK FF Excitation			
Q_t	Q_{t+1}	J_t	K_t
0	0	0	*
0	1	1	*
1	0	*	1
1	1	*	0

FIGURE 3.13 Conditional stimulus in a JK state table.

	PRES ABC	INPUT P	NEXT ABC	S_A	R_A	S_B	R_B	S_C	R_C
0	000	*	***	*	*	*	*	*	*
1	001	*	010	0	*	1	0	0	1
2	010	0	111	1	0	$\overline{P}$	P	1	0
		1	101						
3	011	*	***	*	*	*	*	*	*
4	100	*	***	*	*	*	*	*	*
5	101	*	001	0	1	0	*	*	0
6	110	*	***	*	*	*	*	*	*
7	111	*	001	0	1	0	1	*	0

* = Don't Care

SR FF Excitation			
Q_t	Q_{t+1}	S_t	R_t
0	0	0	*
0	1	1	0
1	0	0	1
1	1	*	0

FIGURE 3.14 Conditional stimulus in an SR state table.

As with the previous examples, the unconditional states follow the excitation table without modification. A "don't care" state is entered on the input column since it is not being observed by the circuitry during this state. Only bit B in state 2 is dependent on the input and must be handled differently. Consider the following logic used to determine the S_B and R_B inputs of bit B in state 2:

- State 2 bit B goes from 1 to 1 when P = 0; so, S_B = * and R_B = 0.
- State 2 bit B goes from 1 to 0 when P = 1; so, S_B = 0 and R_B = 1.

Note that the logic of the S_B input to the flip-flop is the complement of input P, and the logic of the R_B input to the flip-flop is the same as input P, so $S_B = \overline{P}$ and $R_B = P$.

3.4 KARNAUGH MAPPING OF INPUT ARGUMENTS

Our synchronous counter design method requires several design steps, including definition of the problem (illustrated as a state diagram or state machine program), creation of the state table, minimization of SOP functions taken from the state tables using Karnaugh maps, and drawing the schematic diagram. We have completed an overview of the first two steps of this method when applied to counters that are controlled from an external stimulus. The next step involves entering the data from the state tables onto the Karnaugh maps and uses an approach identical to that used when the counter had no conditional transitions. Figure 3.15 illustrates the mapping of the problem of Figure 3.9 using D-type flip-flops for implementation. There should be few surprises here.

The stimulus columns from Figure 3.12 are repeated here for convenience. Note that they are mapped directly onto the Karnaugh map. State 2, bit B is a variable, not a constant. This variable is also mapped directly. We have just entered a variable onto the D_B Karnaugh map at state 2. This defines the map as a **variable-entered map.** The cell where the variable is entered is "restricted" by the variable. In other words, this cell that defines the minterm $\overline{A}B\overline{C}$ becomes $\overline{A}B\overline{C}\overline{P}$. Figure 3.16 illustrates the schematic diagram of the counter.

This counter differs from our previous designs by virtue of the input P, which controls the count sequence. Recall that, according to our state table, its effect was on the second bit in the counter and affected it only during one of the states. This schematic diagram implies

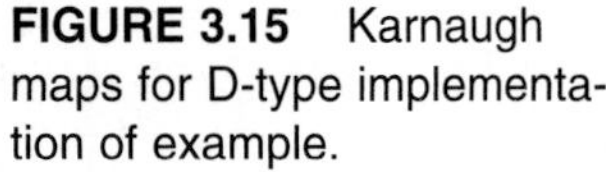

FIGURE 3.15 Karnaugh maps for D-type implementation of example.

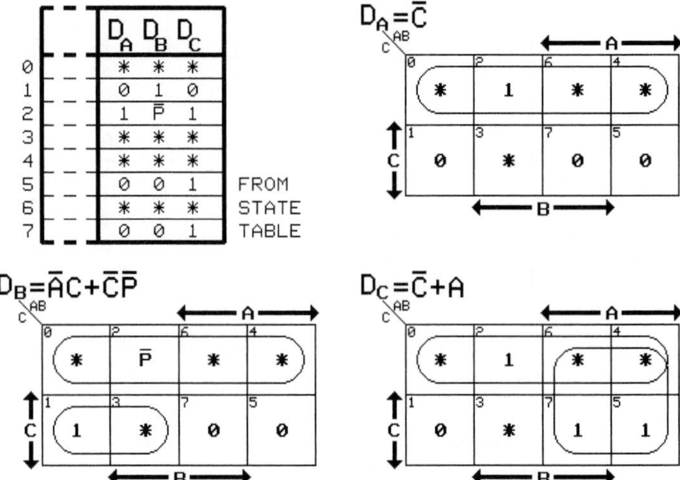

a similar argument. Note that the P input is applied to the second bit of the counter, and it is restricted by other logic.

Let's look at the results of placing the JK solution to this problem on a set of Karnaugh maps, as shown in Figure 3.17. Actually, there were multiple possibilities for several choices made in selecting input-forming logic expressions for this application. For example, $J_A = B$. J_A is also equal to $\overline{C}$. The former choice was made because it was a closer connection on the schematic. Note that when the input-forming logic for K_B was determined, there were similar choices. Also, when variables are entered on the map, the extracted terms may not be in simplest form. This is quite evident from this example. Figure 3.18 is an illustration of the schematic diagram of this circuit.

Finally, let's look at the results of placing the SR solution to this problem on a set of Karnaugh maps, as shown in Figure 3.19. Again, note how the variable P was entered on the map in the second bit of the counter. Figure 3.20 illustrates the final schematic diagram of this counter.

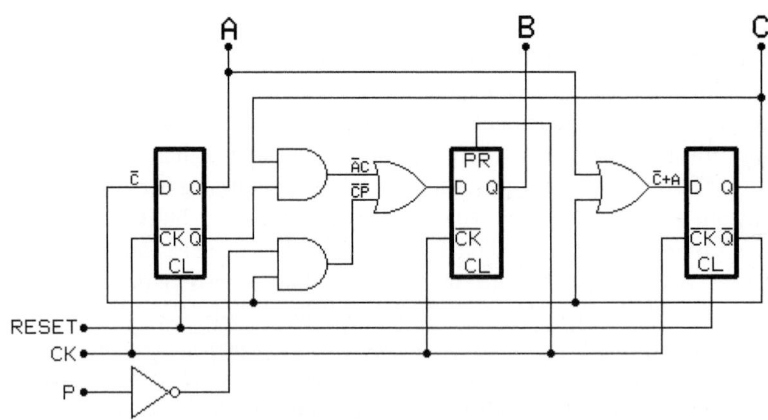

FIGURE 3.16 D-type implementation of controlled synchronous binary counter.

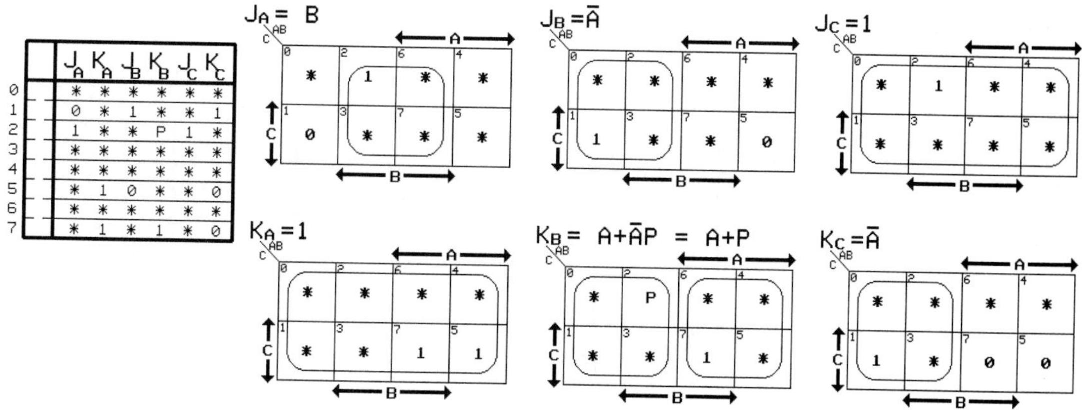

	J_A	K_A	J_B	K_B	J_C	K_C
0	*	*	*	*	*	*
1	0	*	1	*	*	1
2	1	*	*	P	1	*
3	*	*	*	*	*	*
4	*	*	*	*	*	*
5	*	1	0	*	*	0
6	*	*	*	*	*	*
7	*	1	*	1	*	0

FIGURE 3.17 Karnaugh maps for JK-type implementation of example.

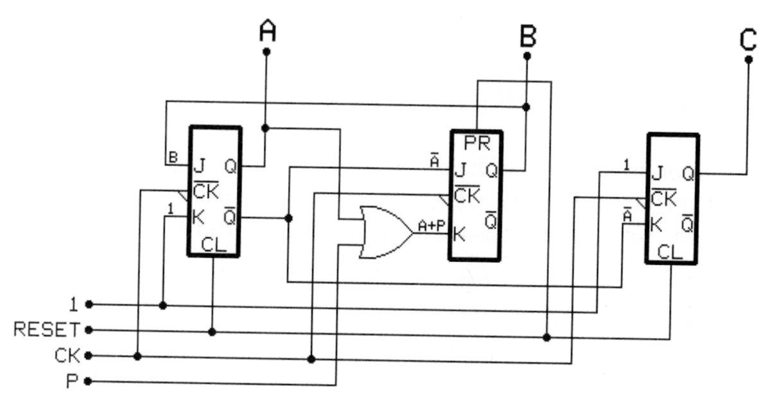

FIGURE 3.18 JK-type implementation of controlled synchronous binary counter.

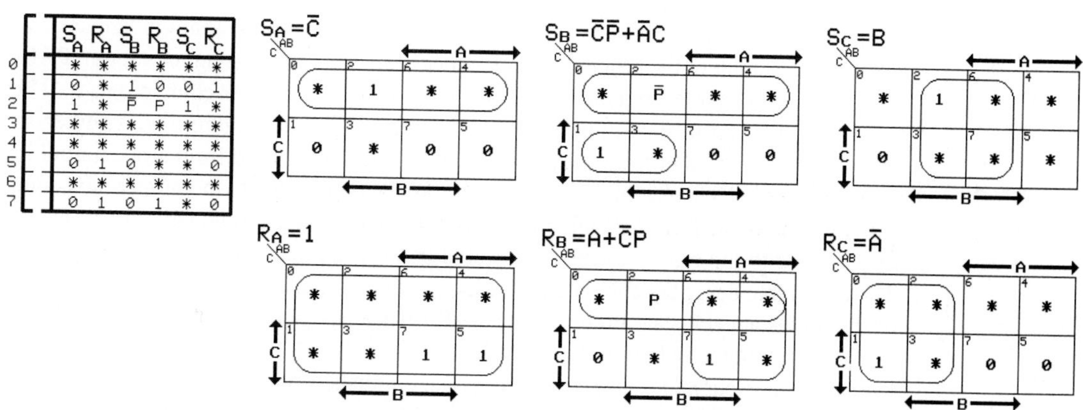

	S_A	R_A	S_B	R_B	S_C	R_C
0	*	*	*	*	*	*
1	0	*	1	0	0	1
2	1	*	$\bar{P}$	P	1	*
3	*	*	*	*	*	*
4	*	*	*	*	*	*
5	0	1	0	*	*	0
6	*	*	*	*	*	*
7	0	1	0	1	*	0

FIGURE 3.19 Karnaugh maps for SR-type implementation of example.

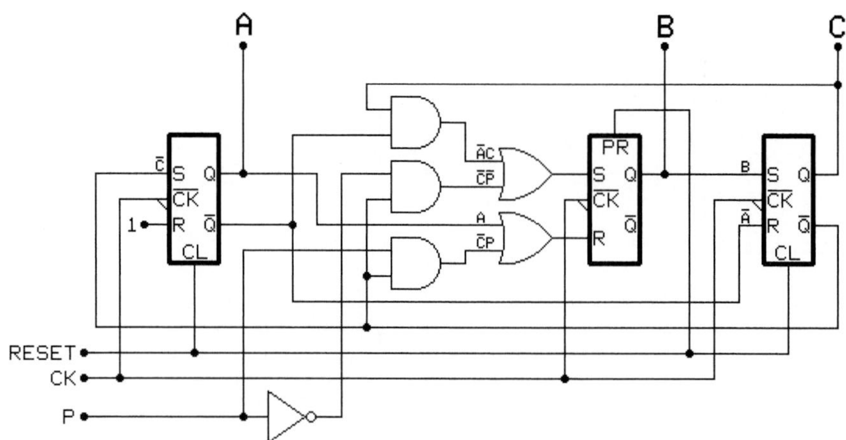

FIGURE 3.20 SR-type implementation of controlled synchronous binary counter.

3.5 DESIGN METHOD SUMMARY

This chapter is concerned with designing synchronous binary counters that are controlled from external logical stimuli. That is, the count sequence is modified by external inputs to the counter. We have investigated a specific design method that involves five steps:

1. Define the counter sequence. This step of the design process is determined by the nature and scope of the problem to be solved. We have yet to describe the solution to any specific problems. Instead, we have made up problems that essentially start at step two. We will be looking at implementing specific problems later in the text.
2. Document the counter sequence as a state diagram or as a state machine program or both. We will usually express the problem definition in both the state diagram and the state machine program forms. The state diagram gives an overall picture of the operation of the state machine. The state machine program will be useful in implementing the system using commercially available PAL chips. The state machine programming language we are using is similar to the programming language that we will be using to program PALs.
3. Construct a state table from either the state diagram or the state machine program. This state table is used to determine the input-forming logic that controls the counter sequence. It gets its input from the current state of the counter and from any external inputs. Its output drives the binary register to cause it to advance to the proper next state.
4. Simplify the input-forming logic functions extracted from the state table. Each D, JK, or SR input to the binary counter is created using combinational input forming logic. The state table describes the output of each combinational circuit needed as a function of the inputs at each state in the sequence. Karnaugh maps can be used to simplify the function for implementation.
5. Construct the schematic diagram. This is accomplished by creating a binary register from the type of flip-flop chosen for the problem and using the input-forming logic from step 4 to drive the inputs of the flip-flops.

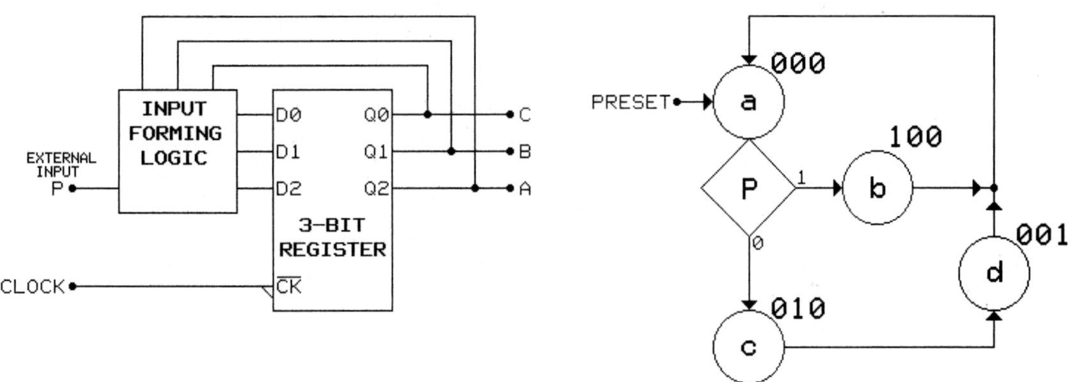

FIGURE 3.21 Practical Exercise 3.2, problem definition.

We have just observed an example of an implementation of a controlled synchronous binary counter using D-type, JK-type, and SR-type flip-flops. In many applications, the most effective solution will be to use PAL chips for implementation. Such chips are available with each type of flip-flop in its output macrocell, though the D-type is the most common. Because of this, most of the remaining examples will make use of the D-type flip-flop.

Practical Exercise 3.2: Decision Making in a Binary Counter

Consider the state machine described in Figure 3.21. Note that this counter is not glitch-free. It is possible for a glitch to take place between states c and d since two different bits are changing state.

Your tasks are to (1) write the state machine program, (2) create the state table, (3) determine the minimum input-forming logic equations, and (4) draw the schematic diagram of the counter that exhibits the count sequence shown.

Practical Example 3.4: Handling Multiple Inputs

The design method for handling controlled synchronous binary counters is the same, regardless of the number of control inputs. However, it would be reasonable at this point to observe such a design. Consider the problem statement implied by Figure 3.22. Note the manner in which multiple bits are used to make decisions at states a and c.

Your tasks are to (1) write the state machine program, (2) design the state table, (3) determine the input-forming logic functions, and (4) draw the schematic diagram of the counter defined by the state diagram of Figure 3.22.

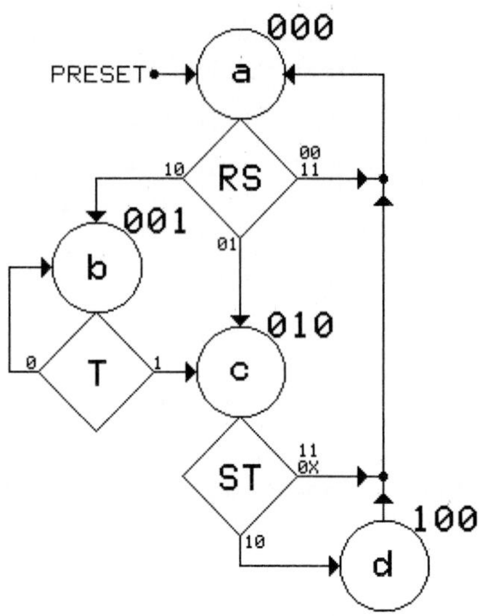

FIGURE 3.22 Practical Example 3.4, state diagram.

State Machine Program

Creating the state machine program is most easily accomplished by observing the state machine operation one state at a time, translating to the language as we go. Consider the following solution:

```
SEQUENCE [A,B,C] {
    PRESENT 'b'000   IF (R & !S) NEXT 'b'001;     /* STATE-a */
                     IF (!R & S) NEXT 'b'010;
                     DEFAULT NEXT 'b'000;
    PRESENT 'b'001   IF !T NEXT 'b'001;           /* STATE-b */
                     DEFAULT NEXT 'b'010;
    PRESENT 'b'010   IF (S & !T) NEXT 'b'100;     /* STATE-c */
                     DEFAULT NEXT 'b'000;
    PRESENT 'b'100   NEXT 'b'000;                 /* STATE-d */
    }
```

Note the use of the DEFAULT statement at state a. In state a there are two bits, R and S, which take part in the decision that determines the counter value of the next state. When RS = 01, the flow goes to state c. When RS = 10, the flow goes to state b. There are two more permutations of R and S, 00 and 11. Both of these are picked up by the DEFAULT statement.

State Table

Figure 3.23 illustrates the state table for this example. Observe how the Present, Input, and Next columns were generated. Specifically, observe state 000_2. This state is conditional and jumps to one of three destination states as a function of two bits, R and S. There are four permutations of these two bits, 00, 01, 10, and 11. Each of these is listed in the Input column, leaving the T input at a "don't care" state since it is not being used. The next state for each permutation is then listed. This is reviewed in Figure 3.24.

FIGURE 3.23 Practical Example 3.4, state table.

STATE	N	PRES ABC	INPUT RST	NEXT ABC	D_A	D_B	D_C
a	0	000	00*	000	0	$\overline{R}S$	$R\overline{S}$
			01*	010			
			10*	001			
			11*	000			
b	1	001	**0	001	0	T	$\overline{T}$
			**1	010			
c	2	010	*00	000	$S\overline{T}$	0	0
			*01	000			
			*10	100			
			*11	000			
	3	011	***	***	*	*	*
d	4	100	***	000	0	0	0
	5	101	***	***	*	*	*
	6	110	***	***	*	*	*
	7	111	***	***	*	*	*

* = Don't Care

Let's observe how the stimulus columns were determined for this state. Bit A in this state is at a logic 0 regardless of the values of the input bits. This is evident by the fact that it is always 0. Consequently, the D_A input should be set to 0 for this state. Observe Figure 3.25.

Bit B is conditional since it is not the same logic level for all permutations of the inputs. We need to determine the state of bit B as a function of the input bits R, S, and T. From observation we find that D_B is TRUE at $\overline{R}S$. Therefore, $D_B = \overline{R}S$.

Bit C is also conditional and its next state varies as a function of R, S, and T. From observation we find that D_C is TRUE when $R\overline{S}$. Therefore, $D_B = R\overline{S}$.

Consider state b. The next state is a function of input T. There are two permutations of input T, so there are two corresponding lines on the table for this state. Again, D_A is not conditional. Bit D_B is true when T, and bit D_C is true when $\overline{T}$. The function of each is annotated on the table shown in Figure 3.26.

Consider state c, as shown in Figure 3.27. The next state is a function of inputs S and T. There are four permutations of inputs S and T, so there are four corresponding lines on the table for this state. Input R is not involved in the decision, so its entries in the Input

FIGURE 3.24 Practical Example 3.4, state 0 Present, Input, and Next columns.

STATE	N	PRES ABC	INPUT RST	NEXT ABC
a	0	000	00*	000
			01*	010
			10*	001
			11*	000

FIGURE 3.25 Practical Example 3.4, state table, state a.

STATE	N	PRES ABC	INPUT RST	NEXT ABC	D_A	D_B	D_C
a	0	000	00*	000	0	$\overline{R}S$	$R\overline{S}$
			01*	010			
			10*	001			
			11*	000			

FIGURE 3.26 Practical Example 3.4, state table, state b.

STATE	N	PRES ABC	INPUT RST	NEXT ABC	D_A	D_B	D_C
b	1	001	**0	001	0	T	$\overline{T}$
			**1	010			

FIGURE 3.27 Practical Example 3.4, state table, state c.

STATE	N	PRES ABC	INPUT RST	NEXT ABC	D_A	D_B	D_C
c	2	010	*00	000	$S\overline{T}$	0	0
			*01	000			
			*10	100			
			*11	000			

column are "don't cares." This time D_A is conditional. It is true when S is TRUE and T is FALSE, or when $S\overline{T}$. The functions of D_B and D_C are each unconditional since they are 0 regardless of the inputs.

Karnaugh Maps

The Karnaugh maps are determined from the stimulus columns of the state table. The maps and the section of the state table are shown in Figure 3.28. Note how the variables, R, S, and T were entered on the map from the state table. Note also how they were used in extracting sum of product expressions from the maps. Recall that when variables are placed on a map, the expression extracted from the map may not be in simplest form. Some Boolean algebra may reveal simpler expressions. However, do not forget that if it is intended to implement these circuits using PAL devices, it is often easiest to leave expressions in their sum of products form, and minimization is not entirely necessary. Most PAL program compilers automatically simplify all expressions when possible.

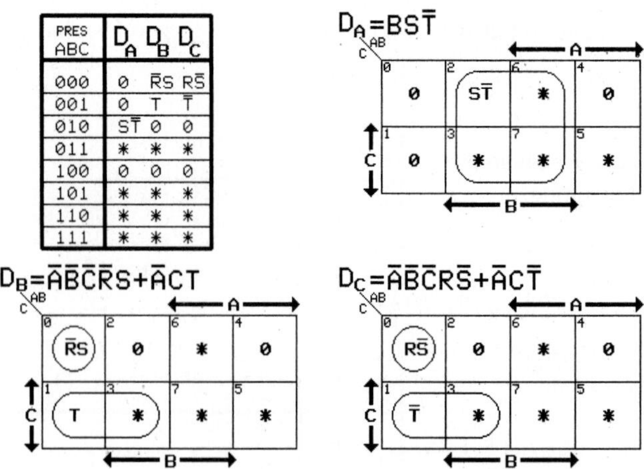

PRES ABC	D_A	D_B	D_C
000	0	$\overline{R}S$	$R\overline{S}$
001	0	T	$\overline{T}$
010	$S\overline{T}$	0	0
011	*	*	*
100	0	0	0
101	*	*	*
110	*	*	*
111	*	*	*

FIGURE 3.28 Practical Example 3.4, Karnaugh maps.

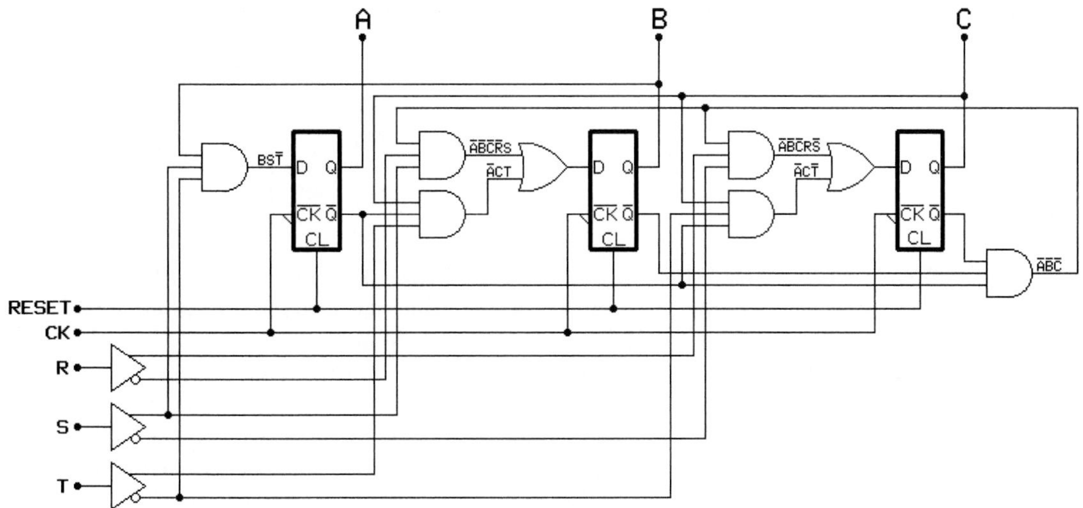

FIGURE 3.29 Practical Example 3.4, schematic diagram.

The schematic diagram of the controlled synchronous binary counter is illustrated in Figure 3.29. Note that the methodology has not changed. First, a binary register of D-type flip-flops is drawn with its clock and preset circuitry entered. Then, using the expressions from the Karnaugh maps of Figure 3.28, the input-forming logic is drawn. Note how the three inputs R, S, and T are shown on the schematic diagram.

3.6 INPUT-FORMING LOGIC SIMPLIFICATION

Practical Example 3.4 illustrated that the input-forming logic can get relatively complex pretty quickly as the complexity of the design increases. It is usually advantageous to consider circuit complexity during the design process in order to avoid having to redesign in order to provide such simplification. We have two considerations to make during the design process:

1. Do we need a glitch-free counter?
2. Do we need to minimize input-forming logic complexity?

Often these two arguments are mutually exclusive. If we need a glitch-free counter, we may not be able to minimize circuitry, since we will have little flexibility in choosing the count sequence of the state machine. The choice of designing with a glitch-free counter may create a design so complex as to make it an unfortunate one. There are other ways of dealing with glitches, as described in Chapter 2, including alternate-state timing and delayed-state timing methods.

If we can use alternate-state timing or delayed-state timing to remove counter glitches, there are no restrictions on the selected count sequence. When this is the case, we can make an effort to simplify the circuitry and the subsequent design complexity.

FIGURE 3.30 Practical Example 3.4, state table, state c.

STATE	N	PRES ABC	INPUT RST	NEXT ABC	D_A	D_B	D_C
c	2	010	*00	000	$S\bar{T}$	0	0
			*01	000			
			*10	100			
			*11	000			

Here's a simple rule: **Since sum of products expressions are extracted from the Karnaugh map by circling the TRUE terms, circuit complexity can be minimized by reducing the number of TRUE terms in a Karnaugh map.**

Consider Figure 3.30, which is a reprint of Figure 3.27. Note the choice of the next state values. For three of the four permutations, the next state is 000_2. For one permutation, the next state is 100_2. Of the four permutations, there is only a single 1 digit, and there is only one resulting term in the stimulus columns. Obviously, the fewer terms we find in the stimulus columns, the simpler the circuitry will be. Consider the difference that would be caused if the count sequence required more logic 1s in the next state. Figure 3.31 illustrates such a situation.

This example helps to remind us that we are looking for the expression that states D_A, D_B and D_C as a function of the inputs R, S, and T.

First note that the only difference in the design was in the choice of the next state counter values. Instead of using 000_2 and 100_2, values were chosen which had more logic 1s in them: 101_2, 011_2, and 110_2. This set of numbers had eight logic 1s in the four permutations of input bits. These four numbers generated eight respective terms in the stimulus columns. Also, because of their complexity, the expressions in the stimulus columns had to be manually simplified. These resulting expressions will each be placed into a single cell on a Karnaugh map. It is evident that the final circuit will be much more complex than the one of Practical Example 3.4.

To minimize the circuitry as much as possible, select counter state values such that those states which appear in the next state column most frequently have a *minimum number of logic 1s.* For example, if we are dealing with a three-bit counter, the most common next state should have a binary value of 000_2. The next most common states should have values of 001_2, 010_2, and 100_2. The next most common states should have values of 011_2, 110_2, and 101_2. The least common state should have a value of 111_2. When a counter does not use all of the possible binary permutations, the first numbers in this list should be chosen.

STATE	N	PRES ABC	INPUT RST	NEXT ABC	D_A	D_B	D_C
c	2	010	*00	101	$\bar{S}\bar{T}+S\bar{T}+ST$	$\bar{S}T+S\bar{T}+ST$	$\bar{S}\bar{T}+\bar{S}T$
			*01	011	$=S+\bar{T}$	$=S+T$	$=\bar{S}$
			*10	110			
			*11	110			

FIGURE 3.31 Practical Example 3.4, next state value selection with complex circuitry.

Often it is advantageous to utilize a counter that has more bits in order to minimize the number of logic 1s in the counter state values. For example, a five-bit counter has five states with a single 1 in the binary value, and ten states with two 1s in the binary value. If a five-bit counter is used, there can be up to 16 states without requiring a state value with more than two logic 1s in the binary number.

Practical Exercise 3.3: Circuit Complexity in Controlled Binary Counters

The purpose of this exercise is to observe the difference in circuit complexity realized when different choices of design criteria are made to apply to identical counters. Two designs will be made on a single counter with identical functions. The first example will minimize the number of logic 1s in the next state values. The second will not allow glitches. Consider the following controlled counter:

```
SEQUENCE [A,B,C,D] {
    PRESENT a IF (!R & S) NEXT b;      DEFAULT NEXT d;
    PRESENT b IF (R # !S) NEXT c;      DEFAULT NEXT a;
    PRESENT c IF (!R & !S) NEXT d;     DEFAULT NEXT e;
    PRESENT d IF S NEXT a;             DEFAULT NEXT d;
    PRESENT e NEXT f;
    PRESENT f NEXT g;
    PRESENT g NEXT h;
    PRESENT h IF !R NEXT e;            DEFAULT NEXT d; }
```

Your first task is to use the design methodology of this chapter to design a controlled binary synchronous counter that exhibits a minimized number of logic 1s in the next state values. Selecting state values a = 0, b = 2, c = 4, d = 8, e = 1, f = 3, g = 12, and h = 9, show (1) the state diagram, (2) the next state map, (3) the state table, (4) Karnaugh maps, and (5) the schematic diagram.

Your second task is to use the design methodology of this chapter to design a controlled binary synchronous counter that exhibits the characteristics of this program and generates no glitches. Selecting state values a = 0, b = 4, c = 12, d = 8, e = 13, f = 15, g = 11, and h = 9, show (1) the state diagram, (2) the next state map, (3) the state table, (4) Karnaugh maps, and (5) the schematic diagram. How many logic gate inputs are required to implement the circuit?

QUESTIONS AND PROBLEMS

Use Figure 3.32 to answer Problems 1–9.

1. Write the state machine program for Figure 3.32.
2. Assign state values to the state machine for Figure 3.32 for a glitch-free count sequence. Show the next state map.

FIGURE 3.32

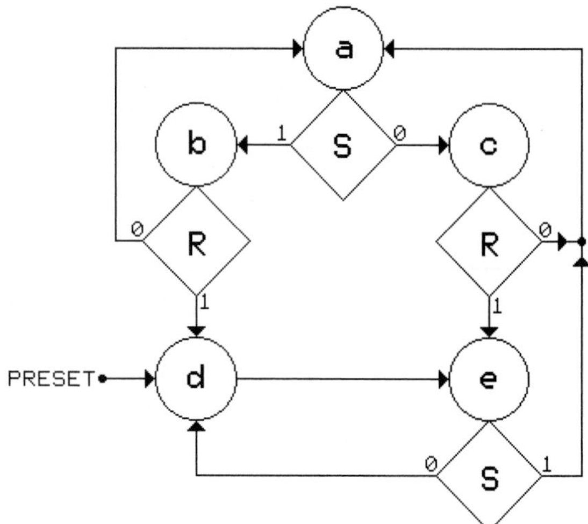

3. Draw the state table for implementing Problem 2 using D-type flip-flops. Determine the input-forming logic functions using Karnaugh map minimization. Draw the schematic diagram of the controlled counter.
4. Draw the state table for implementing Problem 2 using JK-type flip-flops. Determine the input-forming logic functions using Karnaugh map minimization. Draw the schematic diagram of the controlled counter.
5. Draw the state table for implementing Problem 2 using SR-type flip-flops. Determine the input-forming logic functions using Karnaugh map minimization. Draw the schematic diagram of the controlled counter.
6. Assign state values to the state machine for Figure 3.32 for a minimum number of logic 1s in the count sequence.
7. Draw the state table for implementing Problem 6 using D-type flip-flops. Determine the input-forming logic functions using Karnaugh map minimization. Draw the schematic diagram of the controlled counter.
8. Draw the state table for implementing Problem 6 using JK-type flip-flops. Determine the input-forming logic functions using Karnaugh map minimization. Draw the schematic diagram of the controlled counter.
9. Draw the state table for implementing Problem 6 using SR-type flip-flops. Determine the input-forming logic functions using Karnaugh map minimization. Draw the schematic diagram of the controlled counter.

Use Figure 3.33 to answer Problems 10–18.

10. Write the state machine program for Figure 3.33.
11. Assign state values to the state machine for Figure 3.33 for a glitch-free count sequence. Show the next state map. Note: It may be necessary to add states.
12. Draw the state table for implementing Problem 11 using D-type flip-flops. Determine the input-forming logic functions using Karnaugh map minimization. Draw the schematic diagram of the controlled counter.

FIGURE 3.33

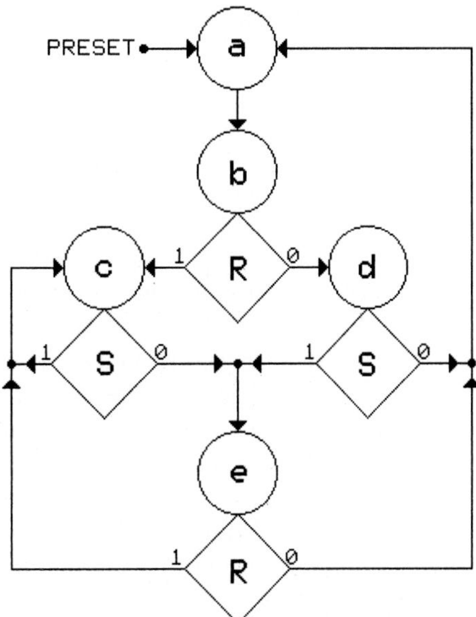

13. Draw the state table for implementing Problem 11 using JK-type flip-flops. Determine the input-forming logic functions using Karnaugh map minimization. Draw the schematic diagram of the controlled counter.
14. Draw the state table for implementing Problem 11 using SR-type flip-flops. Determine the input-forming logic functions using Karnaugh map minimization. Draw the schematic diagram of the controlled counter.
15. Assign state values to the state machine for Figure 3.33 for a minimum number of logic 1s in the count sequence.
16. Draw the state table for implementing Problem 15 using D-type flip-flops. Determine the input-forming logic functions using Karnaugh map minimization. Draw the schematic diagram of the controlled counter.
17. Draw the state table for implementing Problem 15 using JK-type flip-flops. Determine the input-forming logic functions using Karnaugh map minimization. Draw the schematic diagram of the controlled counter.
18. Draw the state table for implementing Problem 15 using SR-type flip-flops. Determine the input-forming logic functions using Karnaugh map minimization. Draw the schematic diagram of the controlled counter.

Use Figure 3.34 to answer Problems 19–27.

```
SEQUENCE[A,B,C] { PRESENT a NEXT b;
                  PRESENT b IF R NEXT c; DEFAULT d;
                  PRESENT c IF S NEXT d; DEFAULT a;
                  PRESENT d IF T NEXT a; DEFAULT b; }
```

FIGURE 3.34

19. Write the state machine diagram for Figure 3.34.
20. Assign state values to the state machine for Figure 3.34 for a glitch-free count sequence, adding states if necessary. Show the next state map.
21. Draw the state table for implementing Problem 19 using D-type flip-flops. Determine the input-forming logic functions using Karnaugh map minimization. Draw the schematic diagram of the controlled counter.
22. Draw the state table for implementing Problem 19 using JK-type flip-flops. Determine the input-forming logic functions using Karnaugh map minimization. Draw the schematic diagram of the controlled counter.
23. Draw the state table for implementing Problem 19 using SR-type flip-flops. Determine the input-forming logic functions using Karnaugh map minimization. Draw the schematic diagram of the controlled counter.
24. Assign state values to the state machine for Figure 3.34 for a minimum number of logic 1s in the count sequence.
25. Draw the state table for implementing Problem 24 using D-type flip-flops. Determine the input-forming logic functions using Karnaugh map minimization. Draw the schematic diagram of the controlled counter.
26. Draw the state table for implementing Problem 24 using JK-type flip-flops. Determine the input-forming logic functions using Karnaugh map minimization. Draw the schematic diagram of the controlled counter.
27. Draw the state table for implementing Problem 24 using SR-type flip-flops. Determine the input-forming logic functions using Karnaugh map minimization. Draw the schematic diagram of the controlled counter.

Use Figure 3.35 to answer Problems 28–36.

```
SEQUENCE[A,B,C,D] {
    PRESENT a  NEXT b;
    PRESENT b  NEXT c;
    PRESENT c  NEXT d;
    PRESENT d  NEXT e;
    PRESENT e  IF (!R&!S) NEXT f; DEFAULT NEXT b;
    PRESENT f  IF (!R#!S) NEXT g; DEFAULT NEXT f;
    PRESENT g  IF (R#S)   NEXT j; DEFAULT NEXT h;
    PRESENT h  NEXT a;
    PRESENT j  IF (R&S)#(!R&!S) NEXT e; DEFAULT NEXT a; }
```

FIGURE 3.35

28. Write the state machine diagram for Figure 3.35.
29. Assign state values to the state machine for Figure 3.35 for a glitch-free count sequence. Show the next state map.
30. Draw the state table for implementing Problem 28 using D-type flip-flops. Determine the input-forming logic functions using Karnaugh map minimization. Draw the schematic diagram of the controlled counter.
31. Draw the state table for implementing Problem 28 using JK-type flip-flops. Determine the input-forming logic functions using Karnaugh map minimization. Draw the schematic diagram of the controlled counter.

32. Draw the state table for implementing Problem 28 using SR-type flip-flops. Determine the input-forming logic functions using Karnaugh map minimization. Draw the schematic diagram of the controlled counter.
33. Assign state values to the state machine for Figure 3.35 for a minimum number of logic 1s in the count sequence.
34. Draw the state table for implementing Problem 33 using D-type flip-flops. Determine the input-forming logic functions using Karnaugh map minimization. Draw the schematic diagram of the controlled counter.
35. Draw the state table for implementing Problem 33 using JK-type flip-flops. Determine the input-forming logic functions using Karnaugh map minimization. Draw the schematic diagram of the controlled counter.
36. Draw the state table for implementing Problem 33 using SR-type flip-flops. Determine the input-forming logic functions using Karnaugh map minimization. Draw the schematic diagram of the controlled counter.
37. Use the design method of this chapter to create a three-bit synchronous binary counter that will count up when input $U = 1$ and count down when input $U = 0$.
38. Use the design method of this chapter to create a three-bit synchronous binary counter that will parallel load from an external source, shift right one bit if input $R = 1$, and shift left one bit if input $R = 0$.
39. Use the design method of this chapter to create a four-bit synchronous counter that will parallel load and shift to the right from one to three places depending on the binary value of inputs S_1 and S_0.
40. Use the design method of this chapter to create a synchronous counter that counts in a repeating cycle from 0 to 1 if inputs S_1 and $S_0 = 00$; from 0 to 3 if inputs S_1 and $S_0 = 01$; from 0 to 5 if inputs S_1 and $S_0 = 10$; and from 0 to 7 if inputs S_1 and $S_0 = 11$.

CHAPTER 4

Output-forming Logic Design

OBJECTIVES

After completing this chapter you should be able to:

- Draw state diagrams that include output-forming logic definitions.
- Write state machine programs that include output-forming logic definitions.
- Draw timing diagrams that illustrate output-forming logic signals.
- Describe unconditional output logic and define it using state diagrams and state machine programs.
- Describe conditional output logic and define it using state diagrams and state machine programs.
- Design output-forming logic circuitry that generates pulses of specific widths relative to the current state.
- Design output-forming logic circuitry that generates pulses of specific widths with beginning and ending points in different states.
- Design hazard protection circuitry for the variety of output-forming logic circuitry being used.
- Develop a complete design for a variety of simple state machines that use external inputs to control external outputs.

4.1 OVERVIEW

Adding output-forming logic to the externally controlled synchronous binary counter of the previous chapters is the last step in completing the design for the counter-type state machine presented in this text. Recall that the state machine we are to design is comprised of three major parts: (1) a binary register that provides the stable state, (2) input-forming logic that includes logic from the register output and external stimuli to create a controlled synchronous binary counter, and (3) output-forming logic. Output-forming logic is the circuitry that accepts input from the controlled binary counter and uses it to control external circuitry. When we put these three parts together, we have a simple yet complete state machine, a circuit that can provide monitoring and control in a digital environment more complex than is possible with discrete logic gates, yet simpler than one that may require a microcontroller or microprocessor.

The simplest form of output-forming logic would be that which is used to provide a defined logic level for the duration of a specific state of the state machine. The output of the synchronous binary counter is a binary number that represents the current state. Each numeric output from the counter represents a unique state. No two states have the same counter number. Generating a logical level during a specific state would be possible simply by decoding the counter bits using AND logic that is asserted by each of the counter output bits. Often this approach is sufficient for an application. However, we will take this concept a step further and determine the type of output-forming logic needed to produce pulses of various widths with respect to the clock pulse. We will also consider the type of circuitry needed to set and reset external devices at different points in the count sequence.

As we've done with previous examples in this text, we will use a method of documenting output-forming logic definitions in the state diagram, the state machine programming language, and the state table. We will introduce a new object in the state diagram, a new statement in the state machine programming language, and a new set of columns in the state table.

Following this chapter we should be prepared to start designing state machines that can accomplish real tasks.

4.2 SIMPLE OUTPUT-FORMING LOGIC

Output-forming logic (OFL) is the electronic circuitry that generates control signals that are a function of the output of the controlled binary counter in the state machine. Figure 4.1 illustrates the part that the output-forming logic plays in state machine design.

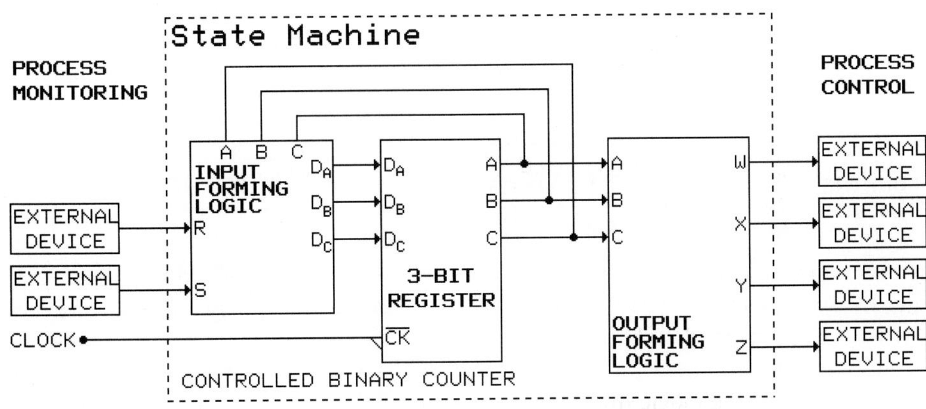

FIGURE 4.1 State machine overview.

As the controlled binary counter changes state, the outputs A, B, and C form the binary number of the counter for each state. Because of the design methodology used, the binary value assigned to each state is unique. This allows us to add the output-forming logic, so that external devices can be driven with signals that are generated by the state machine. The process monitoring is done by controlling the synchronous binary counter with signals from external devices. The design principles for this portion of the state machine were described in Chapters 2 and 3. Process control is accomplished by decoding the output of the binary counter using output-forming logic. This concept is introduced in this chapter.

There are two primary types of signals generated by the output-forming logic, pulses and sustained levels. An **output pulse** is generated by decoding the binary value of a single state. The pulse width is a function of the clock period and the design of the output-forming logic. Generally, pulses of a width equal to the clock period are commonly generated. Also, pulses of one-half clock period are often needed. When we generate half-period pulses that start halfway through the state period, there has been an opportunity for all glitches that occurred at the beginning of the state period to settle, thus cleaning up the signal. This method of hazard protection was referred to in Chapter 2 as alternate-state timing. **Sustained output levels** are generated by setting and resetting a flip-flop with pulses generated at different points in the count sequence. This allows for the generation of control signals that are longer in duration than a single clock period.

4.3 OUTPUT-FORMING LOGIC DESIGN

Up to this point in our designs, we have not been concerned with the output of the state machine. We have been using only two primary symbols in the state diagram, the state and the input boxes. We are now going to add the output box. Consider Figure 4.2.

Note that the state diagram shows a repeating two-bit count sequence that is glitch-free. It is preset to the binary value 01_2 and then continues through the cycle 11_2, 10_2, 00_2, 01_2, 11_2, ... without a change in the sequence. This example has no external stimuli. As the counter goes through the sequence we find that there is a new symbol at states 00_2 and 11_2.

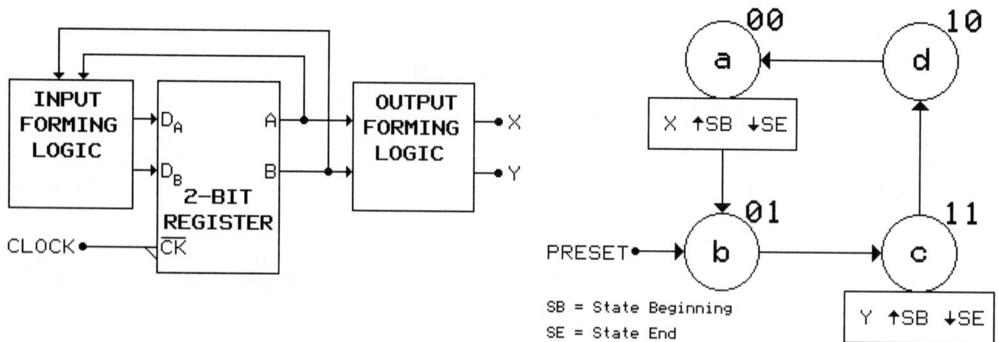

FIGURE 4.2 Two-bit state machine with output.

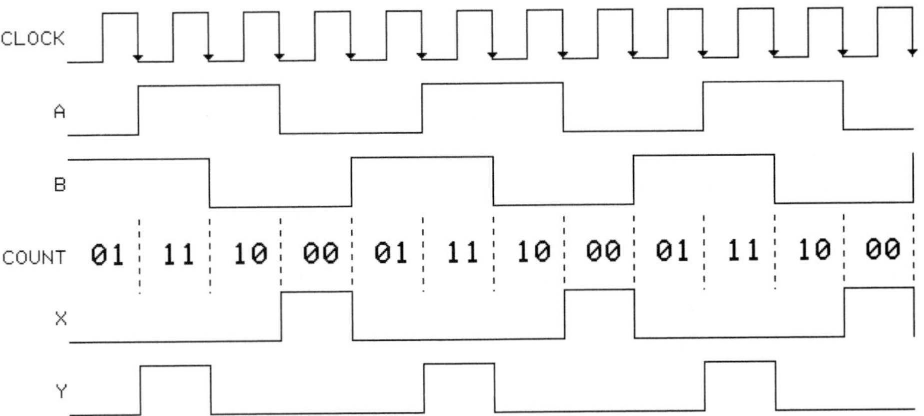

FIGURE 4.3 Example timing diagram.

This symbol, formed of a rectangle, is an output box. It refers to output activity that is to take place during the indicated state and is documented in two parts, the output object and its transition definition. The **output object** is the first variable presented in the box, and refers to the name given to the output control line to be asserted. In the above example, the object to be asserted during state 00_2 is X, and the object to be asserted during state 11_2 is Y. Following the output object is the transition definition. This transition definition describes the point in the state period when the object is to be asserted. It also indicates the logic level of that assertion. The output box at state-00_2 is read, "X goes high at the state beginning, goes low at the state end." The output box at state 11_2 is read, "Y goes high at the state beginning, goes low at the state end." Figure 4.3 describes the output that can be expected from the state machine of Figure 4.2.

Three complete cycles of the count sequence are illustrated here. Observe that the count sequence starts at state 01_2. At the next falling clock edge, the counter goes to state 11_2. According to our state diagram, output Y is to go high at the beginning of the state, and to return low at the end of the state. This is verified by observing the timing diagram. At the next falling clock edge, the counter goes to state 10_2. Note that the output lines are not asserted at this time. At the next falling clock edge, the counter goes to state 00_2. According to our state diagram, output X is to go high at the beginning of the state, and return low at the end of the state. This is verified by observing the timing diagram. Figure 4.4 illustrates a simple method for implementing the output-forming logic of this state machine.

According to the timing diagram and state diagram, anytime the counter is in state 00_2, the X output should be true. Note that this is accomplished by AND logic connected to the complement of the two bits. Using this method, the X output will be active during state 00_2. In the same manner, according to the timing diagram and state diagram, anytime the counter is in state 11_2, the Y output should be true. Note that this is accomplished by AND logic connected to the two bits. Using this method, the Y output will be active during state 11_2.

Another point of note: The count sequence was identified to be glitch-free. Recall from Chapter 2 that in order to be glitch-free, the count sequence cannot have any transitions where more than one counter bit changes state at the same time. This is the case for this example. Review the timing diagram of Figure 4.3. Observe that as the counter goes

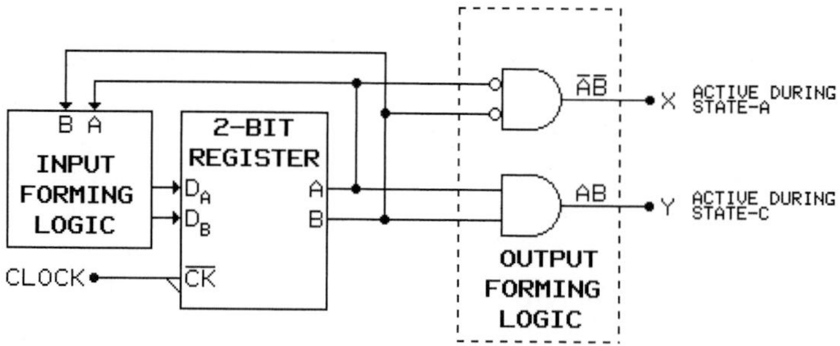

FIGURE 4.4 Simple output-forming logic.

through the sequence, there is no instance where the two counter output bits A or B change state at the same time. This way, there is no possibility that unequal propagation delays in the counter circuitry can create an output glitch, provided that the clock period is greater than the propagation delay of the counter register.

4.4 STATE TRANSITION DEFINITIONS

The output box in the state diagram defines the activity on the output lines on the state machine. In order to provide an accurate and clear definition of this activity, a shorthand method of describing the points in the state period is used. For example, Figure 4.2 stated that "X goes high at the state beginning, goes low at the state end." This was identified by the letters **SB** for "state beginning" and **SE** for "state end." Also, arrows were used to identify the direction of the transition on the output line, either high or low. Consider Figure 4.5.

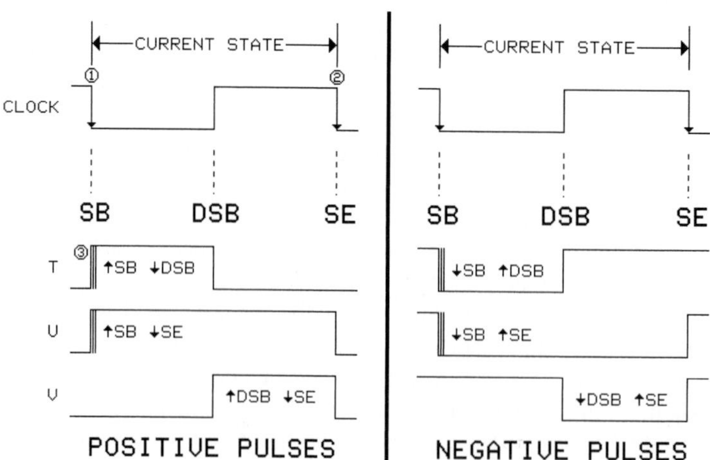

FIGURE 4.5 Non-delayed state machine output pulses.

This example, and all those used in this text, operate on an active-low clock signal. That is, the state machine counter changes state at a point on the low-going edge of the supplied clock signal. Therefore, a state is defined to begin at a low-going clock edge and to end on the next low-going clock edge, when the counter goes to the next state. These two points are illustrated in Figure 4.5. The encircled number 1 shows the point of the beginning of the state, and the encircled number 2 shows the point of the end of the state, which is also the beginning of the next state.

When referring to points in the state period, we always describe those points relative to the current state. Those shown in Figure 4.5 are described as **SB, DSB,** and **SE,** referring to "state beginning," "delayed state beginning," and "state end," respectively.

The beginning and ending points of an output pulse are dependent on the needs of the device being driven.

Recall that it is possible for glitches to occur in the counter output if the count sequence is not glitch-free and there is a difference in the propagation delay of the individual flip-flops in the counter. These glitches would appear for a short period of time following the low-going transition of the clock. This is illustrated in Figure 4.5 at the point labeled by the encircled number 3. It should be noted that any transition that is generated by the output-forming logic that coincides with the SB point has the potential for glitches. If glitches in the signal cannot be tolerated, this edge should be avoided. Otherwise, alternate-state timing or delayed-state timing techniques can be used to eliminate the glitches from the signal after they have left the state machine. Three possible types of pulses are shown in Figure 4.5 and are labeled T, U, and V. Let's examine the generation of each of these pulses one at a time.

4.4.1 SB to SE Pulse

The SB to SE pulse is the simplest to generate. This pulse is to remain true throughout the duration of the specified state. Since the state is defined by the placement of a unique binary value on the output of the state machine counter, it is necessary only to decode that state using AND logic, as illustrated in Figure 4.4. The AND logic will be used to provide an active output only when the selected state bits are true. This is illustrated in Figure 4.6. Note that this pulse has the potential of glitches at the SB point.

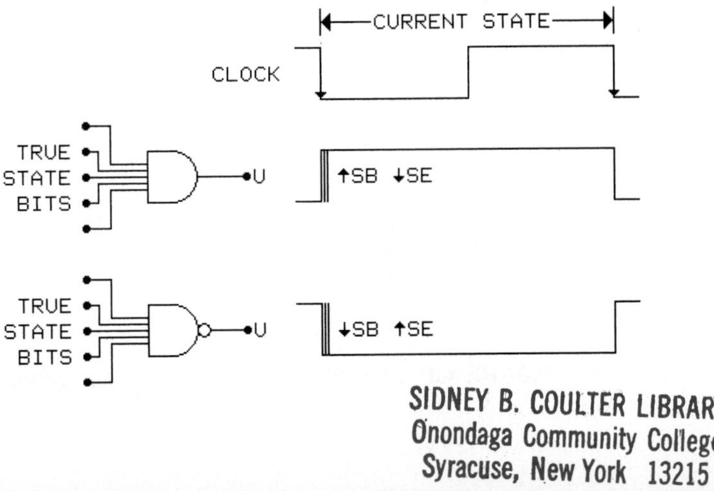

FIGURE 4.6 Unconditional SB to SE pulse generation.

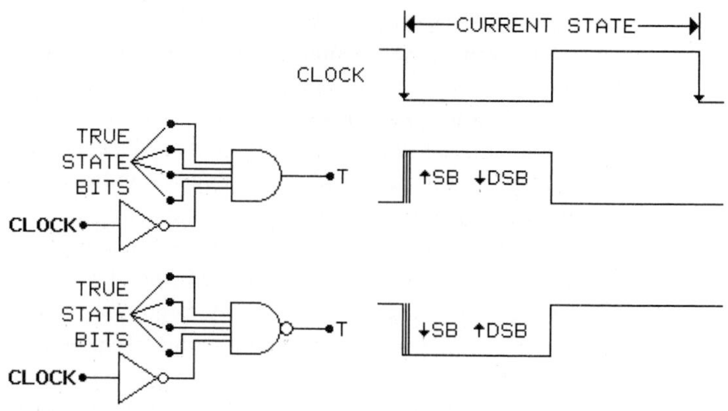

FIGURE 4.7 Unconditional SB to DSB pulse generation.

4.4.2 SB to DSB Pulse

The SB to DSB pulse is very similar to the SB to SE pulse. Casual observation of Figure 4.5 reveals that the SB to DSB pulse is true during the first half of the clock cycle. This is also the period of time when the clock signal is low. If we invert the clock signal and add it to the inputs to the AND logic of Figure 4.6, Figure 4.7 is realized. By inverting the clock and adding the resulting logic to the input of the AND gate, the signal generated by the true state bits is enabled only during the first half of the clock cycle. Again, since this pulse is started at the SB point, there is a possibility of glitches at its beginning.

4.4.3 DSB to SE Pulse

The DSB to SE pulse is probably the most useful of the pulses generated by the state machine. The true state bits and the clock are ANDed together to generate the pulse. The result is shown in Figure 4.8. Since the pulse does not start until DSB, halfway through the clock period, any glitches present in the system have had time to settle. That makes this signal glitch-free regardless of the state machine count sequence. In effect, this circuit

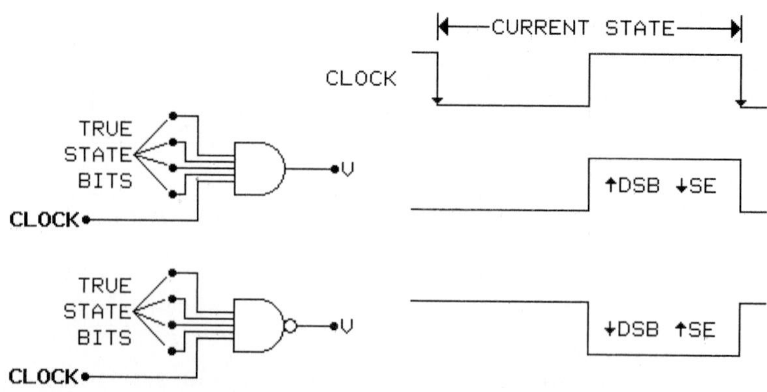

FIGURE 4.8 Unconditional DSB to SE pulse generation.

delays the start of the pulse for one-half clock period. This method of hazard protection was referred to in Chapter 2 as alternate-state timing.

The SB to SE, SB to DSB, and DSB to SE pulses are examples of output signals that can be easily generated by the state machine for the purpose of driving external circuits. Only one of the signals was guaranteed glitch-free. Often the external device may require pulses of different duration and starting points. Figure 4.9 illustrates a few more examples of possible state machine output signals that can be generated using relatively simple output-forming logic. These outputs, labeled W, X, Y, and Z, are generated using the clock signal and the true state bits as were the previous signals, T, U, and V.

These signals are delayed by the use of a D-type flip-flop so that they occur at the output of the state machine from one-half to one full clock period following the SB point. This means we need to define more points in the state period. These are illustrated in Figure 4.9 and are referred to as **DSE** and **DDSE,** "delayed-state end" and "double-delayed state end," respectively.

4.4.4 DSB to DSE Pulse

Figure 4.10 illustrates the schematic and timing diagrams of the circuitry that will generate the DSB to DSE pulse. This circuit uses a D-type flip-flop to delay the SB to SE pulse by one-half clock cycle. It also cleans up any glitches in the process. This is another form of alternate-state timing. At the SB point, the output of the AND logic changes to its asserted state. However, the input to the flip-flop is not enabled until halfway through the clock period. This delay is accomplished by inverting the clock signal so that its high-going edge can be used to enable the flip-flop. The flip-flop output remains stable until one full clock period later, when the inverted clock signal generates another trigger for the flip-flop. At this point, the input to the flip-flop has been deasserted for a half clock period, so the output of the flip-flop deasserts. This method of generating an output pulse has the advantage of generating a pulse that is the same width as the clock period but, due to the half-period delay, is free of glitches.

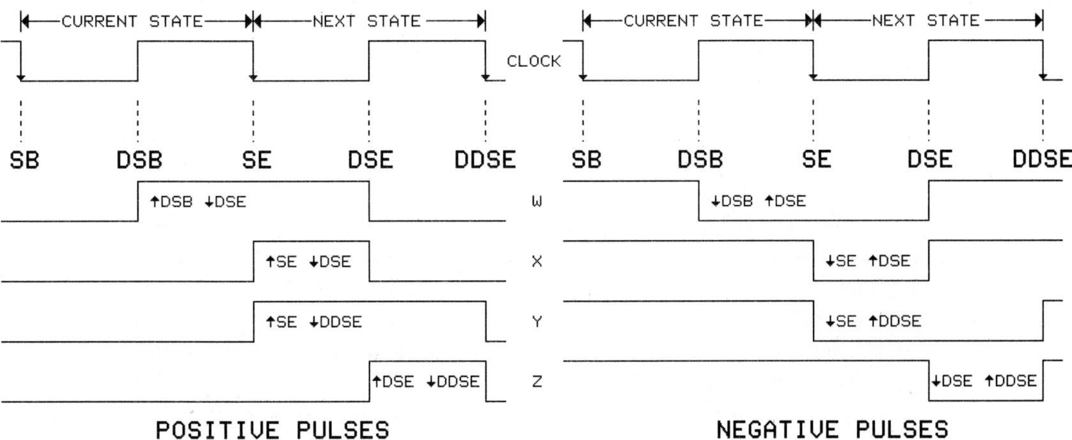

FIGURE 4.9 Delayed state machine output pulses.

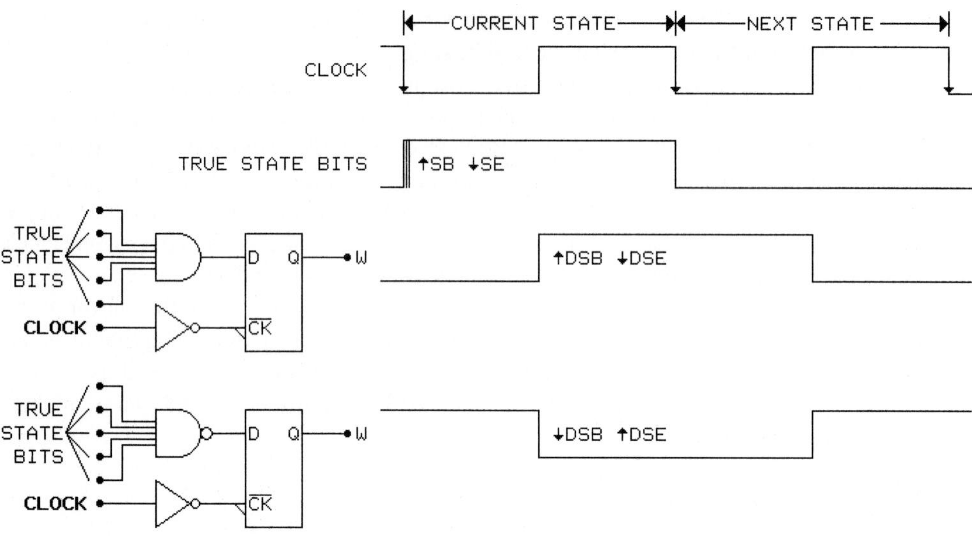

FIGURE 4.10 Unconditional DSB to DSE pulse generation.

4.4.5 SE to DSE Pulse

The SE to DSE pulse is a half-cycle width pulse that starts at the end of the state. Due to its principle of generation, it is glitch-free. Casual observation of this circuitry, as shown in Figure 4.11, reveals that it is very similar to the previous pulse. The D-type flip-flop output is actually a DSB to DSE pulse, as previously described. The output is further inhibited by ANDing it with the inverted clock. This has the effect of enabling the output signal only during the first half of the clock signal.

It should be evident by this time that these output-forming logic circuits are not unique solutions to the problem. For example, the difference between the two circuits in Figure 4.11 is in the use of an AND gate versus using a NAND gate. A similar set of different circuits can be produced simply by selecting the noninverting or inverting outputs of the flip-flop.

4.4.6 SE to DDSE Pulse

The SE to DDSE pulse is a glitch-free pulse of the same width as the clock signal and delayed by a single clock period. It is another form of alternate-state timing. Consider Figure 4.12. The D-type flip-flop clock input is enabled by the provided clock signal without inversion. At SB, when the clock goes low, it instructs the counter to change from the previous state to the current one. By the time the change takes place, the clock edge is long gone. Consequently, the flip-flop loads with the deasserted level that was on its inputs in the previous state. At SE, when the low-going clock instructs the flip-flop to load, its input is asserted by the output of the AND logic gate, which itself is asserted during the current state. The flip-flop will be asserted at this point. When the next active clock edge at DDSE comes, the flip-flop will be returned to its deasserted state, since its input is deasserted immediately prior to the clock edge.

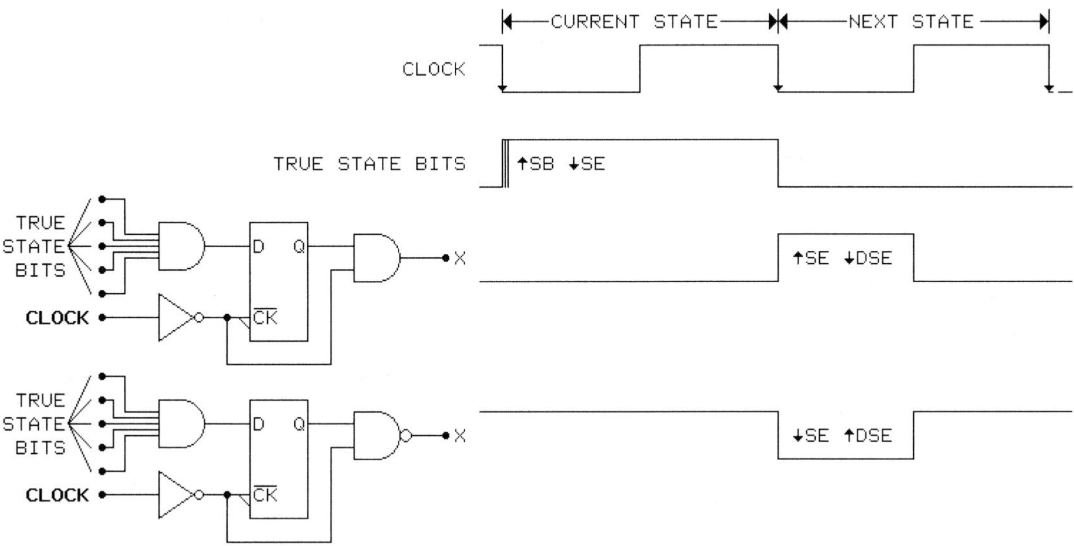

FIGURE 4.11 Unconditional SE to DSE pulse generation.

4.4.7 DSE to DDSE Pulse

The last output pulse to be considered for this study is one that lasts from DSE to DDSE. Consider Figure 4.13. The generation of this pulse is similar to the previously described SE to DDSE pulse, except that it is allowed to assert only during the last half of the clock period. This is accomplished by ANDing the SE to DDSE pulse with the clock signal.

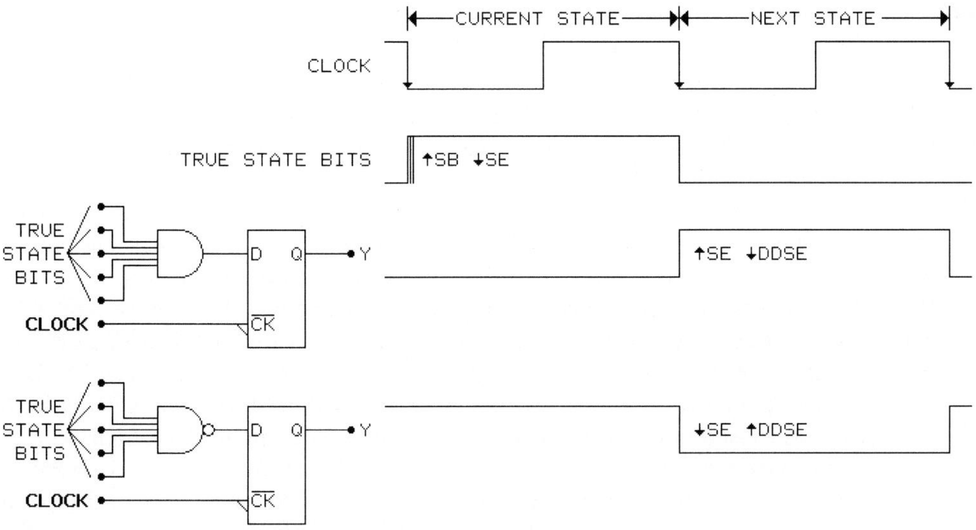

FIGURE 4.12 Unconditional SE to DDSE pulse generation.

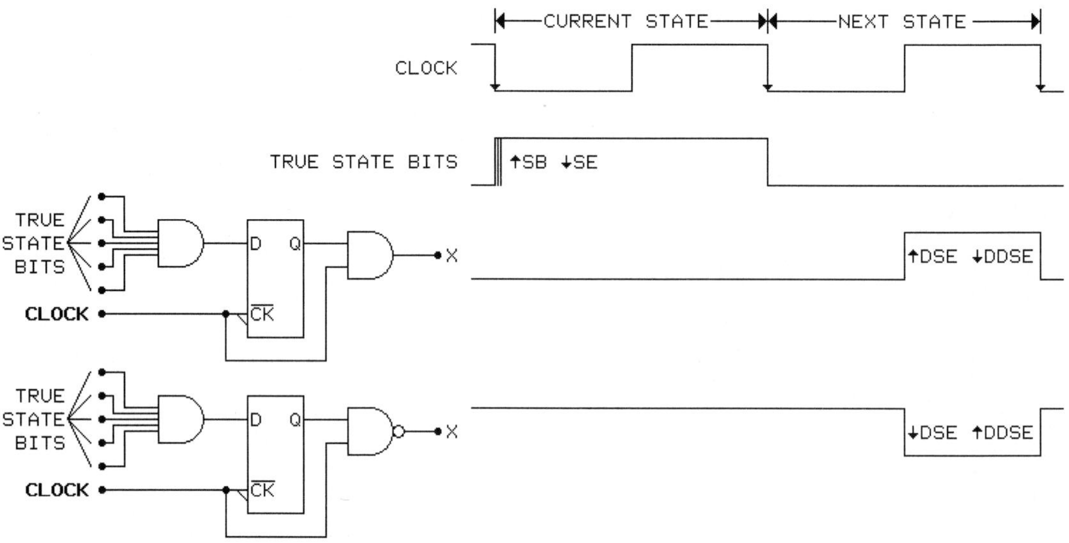

FIGURE 4.13 Unconditional DSE to DDSE pulse generation.

4.5 A COMPLETE STATE MACHINE DESIGN

By this point in the text we have covered all the major components of a state machine: the input-forming logic, the synchronous binary counter, and the output-forming logic. We are now ready to do a complete state machine design.

Practical Example 4.1: Counter-Type State Machine Design

A counter-type state machine consists of a binary register, input-forming logic to that register that defines and controls the count sequence, and output-forming logic that drives external circuitry. Now that we have been introduced to output-forming logic design, we are ready to observe a complete yet simple state machine design. Consider the state machine sequence defined by the state diagram of Figure 4.14. This state machine presets to state a. While in this state, it is considering input R in order to determine the next state. If R = 1, the next state is state b and the Y output is asserted high from DSB through DSE. This is followed by a subsequent state e, which asserts Z low from DSB to SE. If R = 0, the state following state a is state c, where W is asserted high from SB through SE. This is followed by a subsequent state d, where X is asserted high from DSB through SE. Both paths taken as a result of the decision at state a return to state a upon completion.

Based on the definition of the problem, state values were assigned as shown in Figure 4.14. This was not a glitch-free count sequence. State values were selected to (1) minimize logic by using the fewest number of logic 1s in the next state column of the state table, and (2) minimize logic by keeping as close to a numeric sequence as possible.

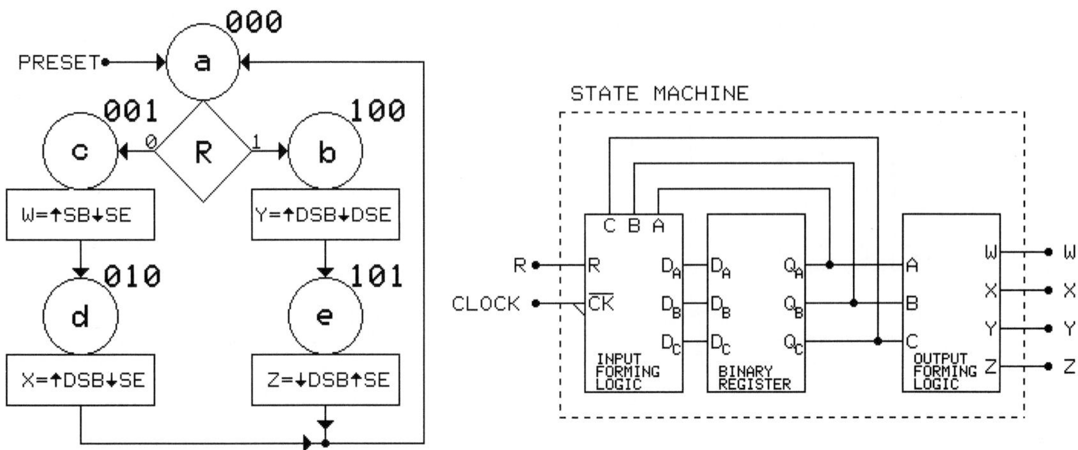

FIGURE 4.14 Practical Example 4.1, state diagram and block diagram.

This may not produce the optimum solution, but it will be a simpler solution than one that assigns the state values in a random fashion. Note also that the only output that could create a glitch is at state c, since all others assert at the DSB point, after glitches have settled. Anytime the output of the state machine counter is 001_2, the W line will be active. If any of the other transitions go through this state as it changes from one intended state to another, the glitch will appear on the W line.

Once the state diagram is established, a block diagram of the state machine can be drawn, as illustrated in Figure 4.14. It is evident that we will be using a three-bit counter to drive the output-forming logic that will generate the outputs W, X, Y, and Z as required. Using the information from the state diagram, the state table of Figure 4.15 is drawn. From this, the input-forming logic functions are defined. These are also shown in Figure 4.15.

Armed with the input-forming logic functions, we may draw the schematic diagram using the same concepts of the previous chapters. We are now in a position to add the output-forming logic W, X, Y, and Z using the circuit fragments discussed in the previous text of this chapter. Consider Figure 4.16. The input-forming logic and counter components of the schematic diagram are drawn in the same orientation as previous examples. Its design should be rather straightforward.

The definition of the output-forming logic circuitry is added to the state table as shown in Figure 4.15. This circuitry is drawn across the top half of the schematic. To draw this, the noninverted and inverted outputs of each bit of the counter are brought up and across the schematic in a bus architecture between the counter and the output-forming logic. This way, we can select the necessary counter output bits and assertion levels as are needed by the output-forming logic. Take a very close look at the output-forming logic circuits and how their architectures were derived. Since the functions $\overline{A}\overline{B}\overline{C}$ and $A\overline{C}$ are used twice, they are generated with discrete logic once and the result is sent to the two destinations as required. This is done instead of generating the data twice, which would require additional logic circuitry.

Review this example carefully. The development of this circuitry is a fundamental skill of our study.

	PRES ABC	INPUT R	NEXT ABC	D_A	D_B	D_C	W	X	Y	Z
							OUTPUT FORMING LOGIC			
a	000	0	001	R	0	$\overline{R}$				
		1	100							
c	001	*	010	0	1	0	↑SB↓SE			
d	010	*	000	0	0	0		↑DSB↓SE		
	011	*	***	*	*	*				
b	100	*	101	1	0	1			↑DSB↓DSE	
e	101	*	000	0	0	0				↓DSB↑SE
	110	*	***	*	*	*				
	111	*	***	*	*	*				

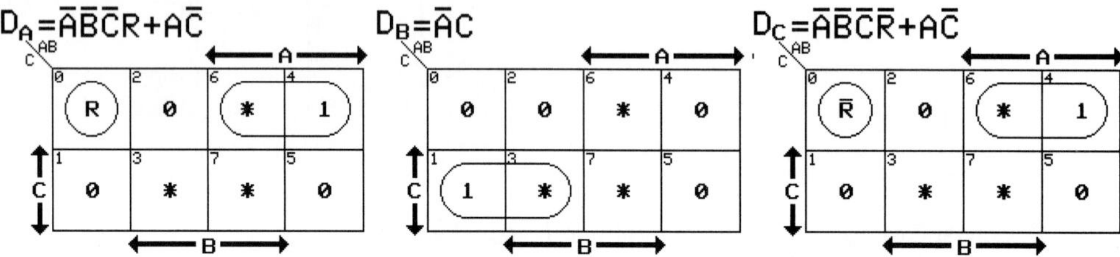

FIGURE 4.15 Practical Example 4.1, state table and Karnaugh maps.

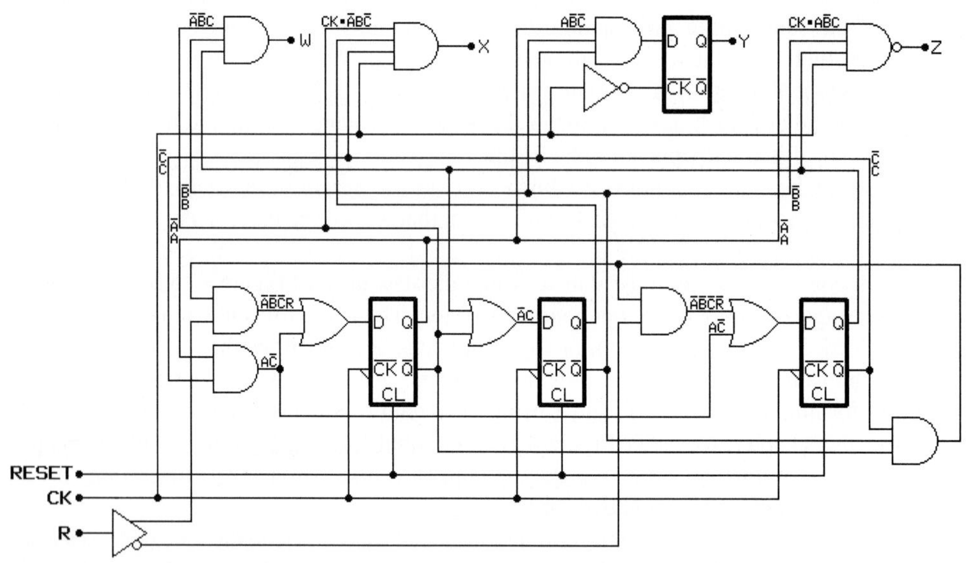

FIGURE 4.16 Practical Example 4.1, schematic diagram.

Practical Example 4.2: Output-Forming Logic in State Machine Programs

Consider the problem completed in Practical Example 4.1 and its definition in the form of a state machine program. Chapter 3 included a discussion on the format of the state machine program statements. The output-forming logic information is placed in the transition statement using the OUT command. Consider the following example:

```
NEXT <state1> OUT <output_var >;
```

The output-forming logic statement is placed between the <state1> argument and the semicolon that closes the line. The state machine program for Practical Example 4.1 can be generated by observation of the state diagram of Figure 4.14. The following is a solution:

```
/* Practical Example 4.2 */
SEQUENCE [A,B,C] {
    PRESENT 'b'000
        IF !R           NEXT 'b'001;
        DEFAULT         NEXT 'b'100;
    PRESENT 'b'001      NEXT 'b'010 OUT W;
    PRESENT 'b'010      NEXT 'b'000 OUT X;
    PRESENT 'b'100      NEXT 'b'101 OUT Y;
    PRESENT 'b'101      NEXT 'b'000 OUT Z; }
```

Take a few moments to compare this program with the state diagram of Figure 4.14. It is important to note that the output-forming logic pulse generated by the OUT statement is always asserted during the entire state. That is, it asserts at SB and deasserts at SE. If the state machine definition requires a different output-forming argument, additional circuitry must be added. However, the OUT argument provides the logic created by the true state bits for each state. Also, the NOT operator (!) may be used to identify that an output argument asserts low.

Practical Exercise 4.1: Unconditional Output-Forming Logic

Consider the state machine defined by Figure 4.17.

Your tasks are to (1) draw the state table and state machine program, (2) determine the input-forming logic (IFL) by using Karnaugh maps, and (3) draw a schematic diagram, including the IFL, the binary register, and the OFL.

4.6 TIMING DIAGRAMS

Timing diagrams are useful parts of the documentation of a problem. Often a problem statement starts with a timing diagram. Just as a state machine can be defined using a state diagram or a state table, it can also be documented using a timing diagram. Such a diagram gives a clear illustration of the state machine operation, as it gives a "snapshot" of the status of each of the selected state machine resources at each clock cycle.

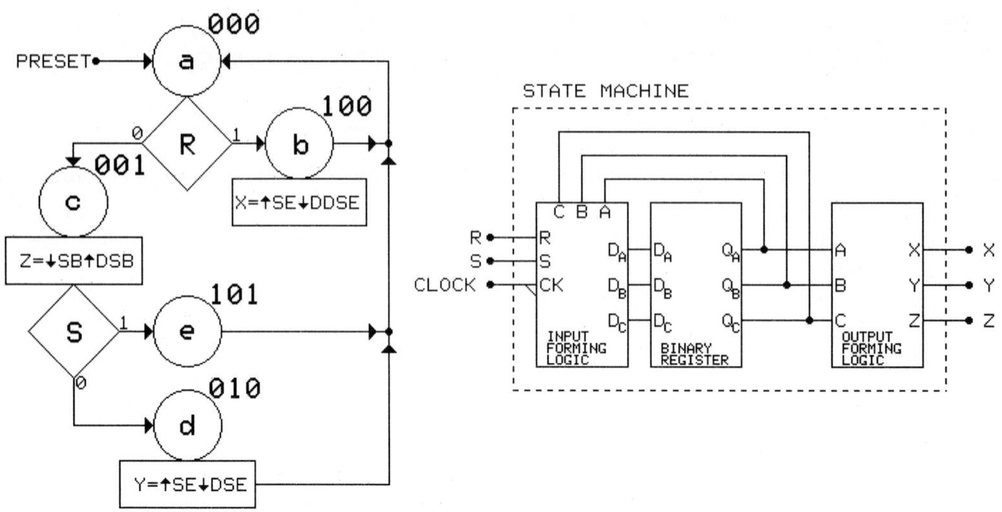

FIGURE 4.17 Practical Exercise 4.1, state diagram and block diagram.

Creating a timing diagram from a state diagram is a relatively trivial matter. In order to generate the timing diagram we are going to parse the diagram, recording the logic levels of all of the desired variables as we do so. To **parse** means to analyze using a step-by-step approach. In this case, it means to look at each state one at a time in the defined sequence, making sure to "touch" each state at least once. As we touch each state, we will record the values of the counter and output bits as they change on entering that state.

Practical Example 4.3: State Machine Timing Diagram

Consider Figure 4.18. This is similar to the state diagram of Practical Example 4.1. Superimposed over the state diagram is a dotted line that starts at state a (at the encircled 1). Note that the state following this one is conditional, that is, the line arbitrarily takes one route, leaving the other untouched. Continuing through the sequence brings us back to state a. Since we have missed part of the diagram, the line continues, taking the other route. This process continues until all states in the diagram have been touched.

The purpose of this parsing algorithm is to record the logic levels of the important variables in the system at each state. As we can see from Figure 4.18, the sequence will be to start at state a and continue through the following sequence:

a - b - e - a - c - d - a

The variables we will record will be the counter bits A, B, and C; the input bit, R; and the output bits W, X, Y, and Z. When we have recorded the state of each of these variables through the above indicated sequence, the timing diagram of Figure 4.19 is realized.

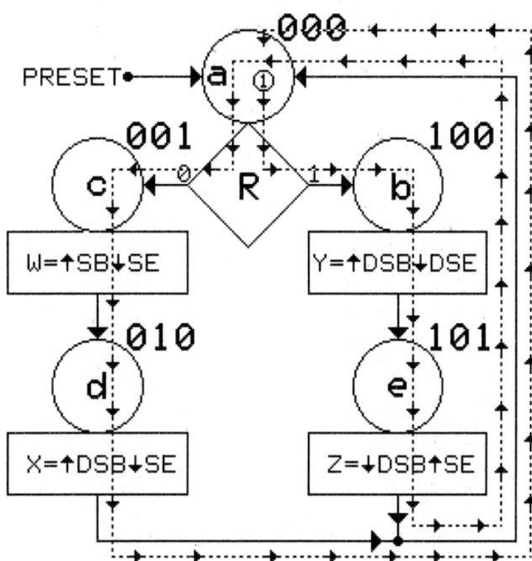

FIGURE 4.18 Practical Example 4.3, parsing algorithm.

Note how the output bits follow the definitions in each output box of the state diagram. Recall that it is the function of the output-forming logic to generate these signals.

Take a moment to observe each transition on each line of the timing diagram and relate it to the state diagram. Pay particular interest to the output signals and their starting and ending points.

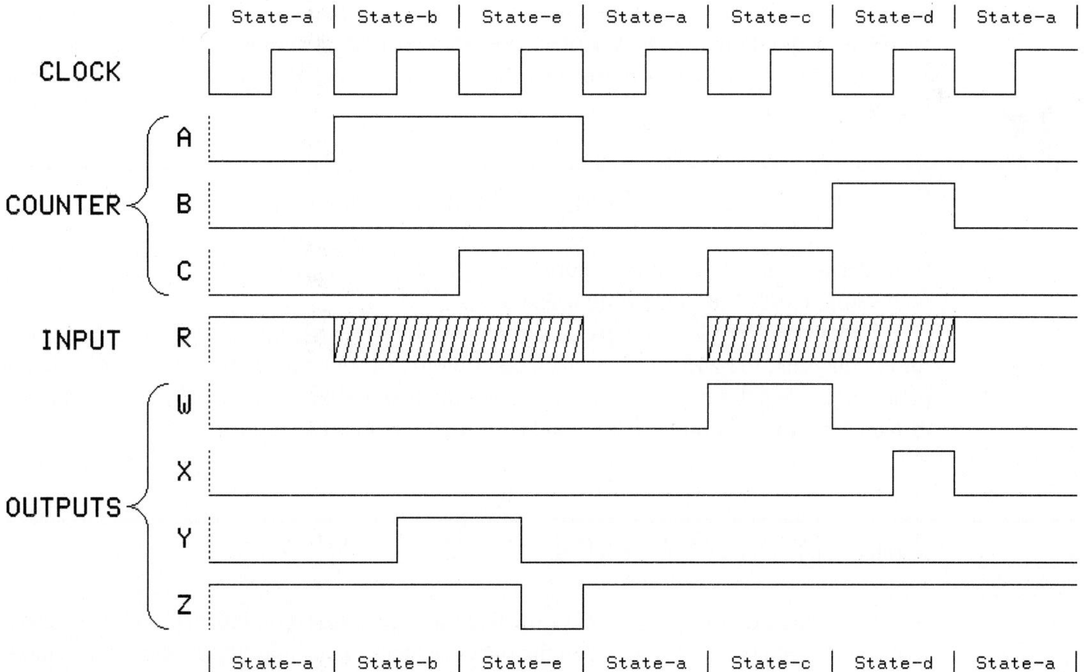

FIGURE 4.19 Practical Example 4.3, timing diagram.

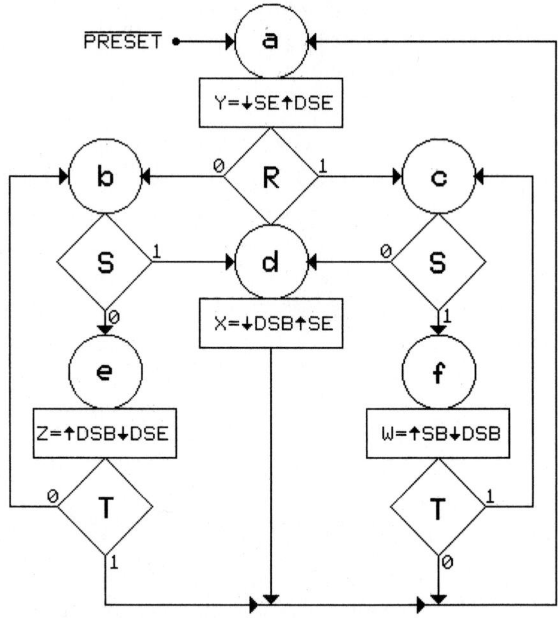

FIGURE 4.20 Practical Exercise 4.3, state diagram.

Practical Exercise 4.2: State Machine Timing Diagram

Using the method previously described, construct a timing diagram that describes the logic levels of all of the variables defined by the state diagram of Practical Exercise 4.1, which is illustrated in Figure 4.17.

Practical Exercise 4.3: A Complete State Machine Design

Consider the state diagram of Figure 4.20. Develop a complete state machine design by completing each of the following tasks:

Your tasks are to (1) assign state values for simplest implementation, (2) draw a timing diagram, (3) write a state machine program, (4) develop a state table that includes present state, input stimulus, next state, and output signal information, (5) develop the set of Karnaugh maps in order to (6) draw a detailed schematic diagram.

4.7 HANDLING MULTIPLE INPUTS

All of the examples to this point have involved decisions based on input stimuli that were a function of a single bit. That is, when the state machine is in a conditional state, only a single bit is being observed, and there are two possible next states as a result of that observation. Actually, it is very common when designing state machines that the next state will be decided based on a function of more than one bit. This means that there may be more than two possible next states to such a decision. In order to overview the design process, the following Practical Example will go through all of the design processes we have observed up to this point.

Practical Example 4.4: Handling Multiple Inputs

Consider the state diagram of Figure 4.21. Notice that there is another change in the design constraints, compared to previous examples. Instead of pulses, the outputs are set to specific levels and left there for action in subsequent states. For example, output W is set to a logic 1 in state a and reset to a logic 0 in state c. This will be accomplished by generating pulses that set and reset a flip-flop that generates the indicated output. Develop a complete state machine design by completing each of the following tasks:

Your tasks are to (1) assign states for glitch-free operation, (2) draw a timing diagram, (3) write a state machine program, (4) develop a state table that includes present state, input stimulus, next state, and output signal information, (5) use Karnaugh maps to determine the input-forming logic functions, and (6) determine the output-forming logic.

Assigning States for Glitch-Free Operation

Recall the method. We will place the identifier for each state on a Karnaugh map in such a manner that each state is immediately adjacent to another on the map in the same orientation as the state diagram. That is, each transition between states will be from one cell in the Karnaugh map to an adjacent cell. Figure 4.22 illustrates one solution. Note that the variables were placed on the map in the same orientation that they appear in the state diagram.

With the state values assigned from the Karnaugh cell numbers, we are ready to continue with the design. It is necessary to do this part of the design first, since subsequent design steps require the state numbers.

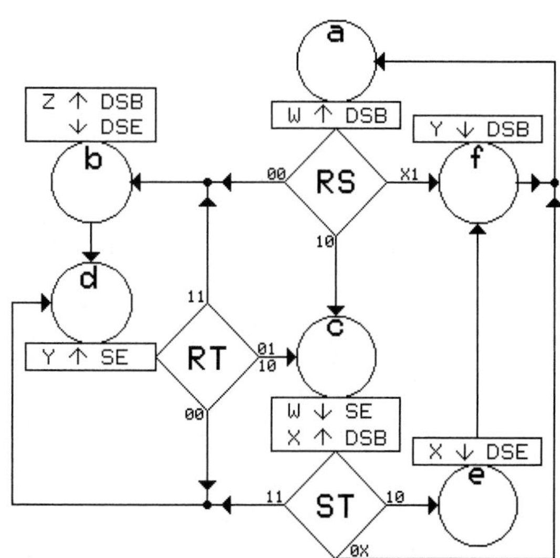

FIGURE 4.21 Practical Example 4.4, state diagram.

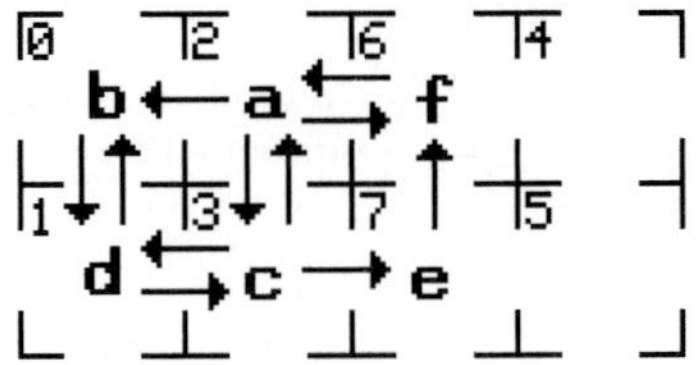

FIGURE 4.22 Practical Example 4.4, next state map.

Timing Diagram

Parsing this state diagram will be a little more difficult than the one previously described. Since there is more than one exit point from some of the decisions, several cycles through the sequence will be required in order to hit every state. Figure 4.23 illustrates the parsing algorithm used for this solution. Note that it starts with the encircled 1 and ends with the encircled 2. It winds around the state diagram several times in order to establish at least one pass through every transition identified in the table. The timing diagram is not complete unless this is done. If a transition is not shown in the timing diagram, the next state that follows that particular decision is not defined. Obviously, the timing diagram can get rather long for a complex state machine.

Note how each of the decisions in this example are a function of the state of two of the three input bits. This is also indicated on the timing diagram. If we follow the parsing algorithm we find that the counter sequence to be illustrated is:

a - f - a - c - e - f - a - b - d - b - d - c - a - b - d - c - d - d

Again, the parsing algorithm is designed to follow each transition at least once.

When logic levels of each variable are recorded at each state, the timing diagram is realized, as shown in Figure 4.24. Obviously, this timing diagram is rather lengthy and complex when compared with the previous example. Take a few moments to examine the

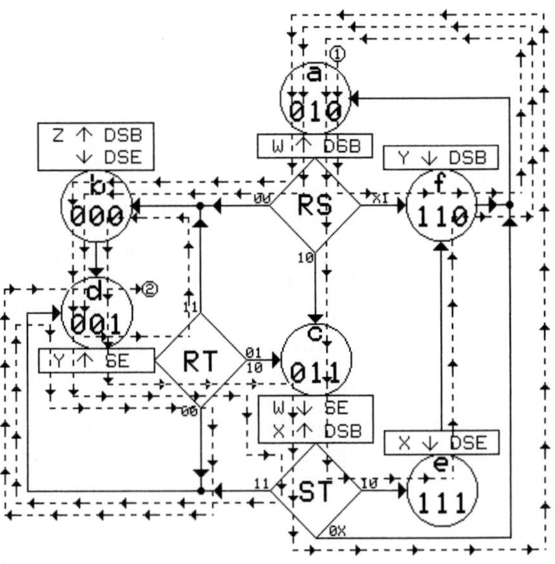

FIGURE 4.23 Practical Example 4.4, parsing algorithm.

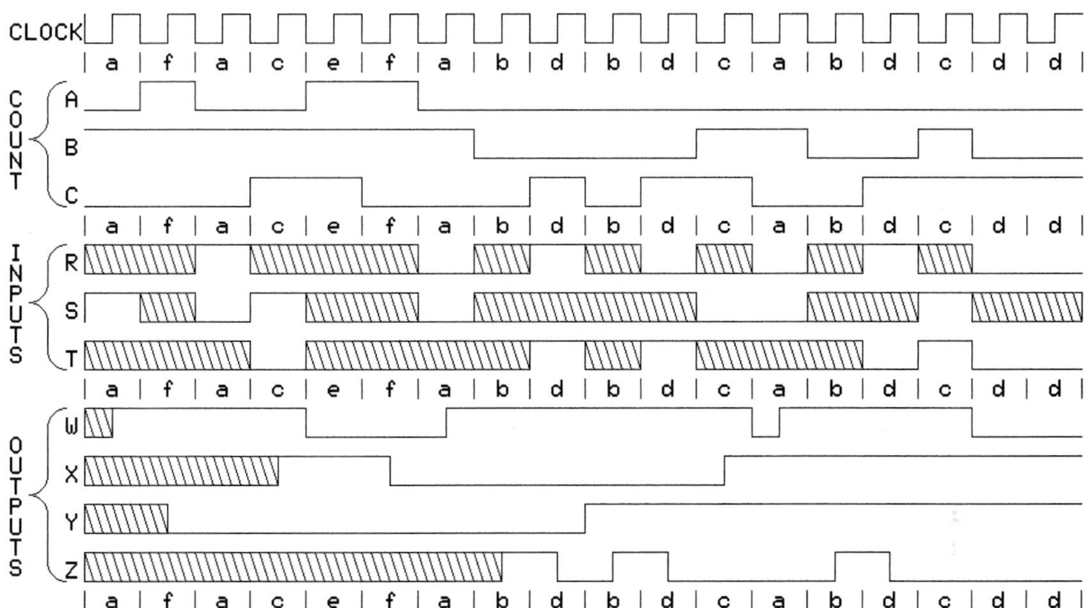

FIGURE 4.24 Practical Example 4.4, timing diagram.

levels one state at a time and determine how they were obtained from the state diagram. Note also that the logic levels of the outputs are not defined until the first time they are asserted, so they are shown as "don't care" states. Also, the logic levels of the inputs are identified as "don't care" states when they are not being used.

Recall that it was necessary to parse the state diagram so that the parsing algorithm passes along each transition at least once. When this is done, every possible permutation of decision is indicated in the timing diagram. Most of these combinations are rather straightforward, except the transition from state d to state c. There are two specific permutations of inputs R and T that follow this direction: $\overline{R}T$ and $R\overline{T}$. It is necessary to follow this transition twice, once for each permutation of the arguments.

Using this approach for generating the timing diagram produces a comprehensive illustration. Each possible permutation of state machine response is described. A particular advantage of this tool is for troubleshooting. With the timing diagram in hand, it is a relatively simple task to drive the constructed state machine with the input stimulus in the proper sequence. When this is done, the state machine variables can be monitored and recorded with a digital logic analyzer. The timing diagram output from the logic analyzer should match the drawn timing diagram. Any differences would identify errors in the design.

State Machine Program

In this situation, the problem was initially defined in the form of a state diagram. When this is the case, the state machine program uses it as its source of data. When the problem starts with a state machine program, the state diagram is created from it. In either event, each resource, whether it be the program or state diagram, completely defines the functions of the state machine.

We will create the state machine program by observation of the state diagram of Figure 4.21 one state at a time in numerical sequence:

```
SEQUENCE (A,B,C)    {
/* State-b */ PRESENT 'b'000  NEXT 'b'001 OUT Z;
/* State-d */ PRESENT 'b'001  OUT Y
              IF !R & !T      NEXT 'b'001;
              IF R & T        NEXT 'b'000;
              DEFAULT         NEXT 'b'011;
/* State-a */ PRESENT 'b'010  OUT W
              IF !R & !S      NEXT 'b'000;
              IF R & !S       NEXT 'b'011;
              DEFAULT         NEXT 'b'110;
/* State-c */ PRESENT 'b'011  OUT X
              IF S & !T       NEXT 'b'111;
              IF S & T        NEXT 'b'001;
              DEFAULT         NEXT 'b'010;
/* State-f */ PRESENT 'b'110  NEXT 'b'010  OUT Y;
/* State-e */ PRESENT 'b'111  NEXT 'b'110  OUT X; }
```

Note a couple of features of this program as it relates to the problem. State B is pretty straightforward. The choice of next state is unconditional, and output Z is pulsed. State D shows a slightly new statement architecture since the next state is conditional, and the output Y activity is unconditional. Since Y is set to a logic 1 regardless of the decision, it is listed prior to the decision statements. The alternative would have been to include it on each of the three decision lines. Also, of the four permutations of arguments of the two inputs, two are unique. These decisions are listed first so that the DEFAULT statement can be used to pick up the two remaining permutations. Note that the outputs W, X, Y, and Z are true for the entire state period. Additional output-forming logic circuitry must be added to condition the pulse as required by the state machine definition.

A hint of things to come: The remainder of the design is the most complex and error-prone part of the overall process. Most PAL compilers will take the above program and generate the PAL program from it alone. If we implement the state machine with PAL chips, the remainder of the design process is unnecessary. (Note that the compiler language will support only SB and SE transitions.)

State Table

The method for development of the state table is the same regardless of the number of inputs to the state machine or the combinations of inputs used to create arguments. We are still looking to find the function that is needed to stimulate the inputs to the binary register. First, observe the state machine for the problem of Practical Example 4.4.

Consider Figure 4.25. Consider the generation of the logical arguments for inputs to the binary register at state b. Since this is an unconditional state, the assignment of bits to D_A, D_B, and D_C are straightforward. However, when we get to state d, a little more analysis is needed. Let's look at that state in detail. Consider Figure 4.26.

First, observe how the input column is formatted. The next state is a function of inputs R and T, which can take on one of four permutations of combinations. Input S is not considered during this state and is left as a "don't care" state. Each of the four permutations of

	ABC RST ABC	DA	DB	DC	W	X	Y	Z
b	000 XXX 001	0	0	1				Z↑DSB↓DSE
d	001 0X0 001 001 0X1 011 001 1X0 011 001 1X1 000	0	$\bar{R}T+R\bar{T}$	$\bar{R}+\bar{T}$			Y↑SE	
a	010 00X 000 010 X1X 110 010 10X 011	S	R+S	$R\bar{S}$	W↑DSB			
c	011 X0X 010 011 X10 111 011 X11 001	$S\bar{T}$	$\bar{S}+\bar{T}$	S	W↓SE	X↑DSB		
f	110 XXX 010	0	1	0			Y↓DSB	
e	111 XXX 110	1	1	0		X↓DSE		

FIGURE 4.25 Practical Example 4.4, state table.

FIGURE 4.26 Practical Example 4.4, state table segment for state D.

	ABC RST ABC	DA	DB	DC
d	001 0X0 **001** 001 0X1 **011** 001 1X0 **011** 001 1X1 **000**	0	$\bar{R}T+R\bar{T}$	$\bar{R}+\bar{T}$

the arguments of R and T are then listed in numeric order so that the output function can be determined. Determination of that output function is now possible by observation. We find, for example, that counter bit B is true when $\bar{R}T$ or when $R\bar{T}$. Therefore:

$$D_B = \bar{R}T + R\bar{T}$$

Also, counter bit C is true when $\bar{R}\bar{T}$ or when $\bar{R}T$ or when $R\bar{T}$.

$$D_C = \bar{R}\bar{T} + \bar{R}T + R\bar{T}$$

$$D_C = \bar{R}(\bar{T} + T) + R\bar{T}$$

Therefore, $$D_C = \bar{R} + R\bar{T}$$

$$D_C = \bar{R} + \bar{T}$$

The same principle is applied in order to come up with the remainder of the functions.

Minimizing the IFL Functions

Now we are ready to minimize the functions using Karnaugh maps, as shown in Figures 4.27, 4.28, and 4.29.

Extracting terms from the Karnaugh map of Figure 4.27, we obtain the function:

$$D_A = AC + \bar{A}B\bar{C}S + \bar{A}BCS\bar{T}$$

Extracting terms from the Karnaugh map of Figure 4.28, we obtain the function:

$$D_B = A + \bar{A}B\bar{C}(R + S) + \bar{A}BC(\bar{S} + \bar{T}) + \bar{B}C(\bar{R}T + R\bar{T})$$

$$D_B = A + \bar{A}B\bar{C}R + \bar{A}B\bar{C}S + \bar{A}BC\bar{S} + \bar{A}BC\bar{T} + \bar{B}C\bar{R}T + \bar{B}CR\bar{T}$$

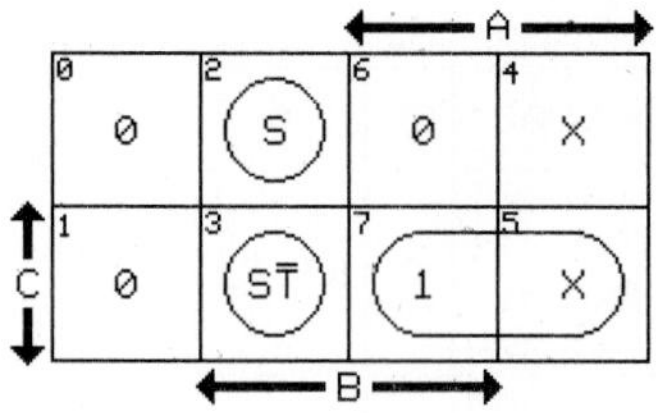

FIGURE 4.27 Practical Example 4.4, Karnaugh map for function D_A.

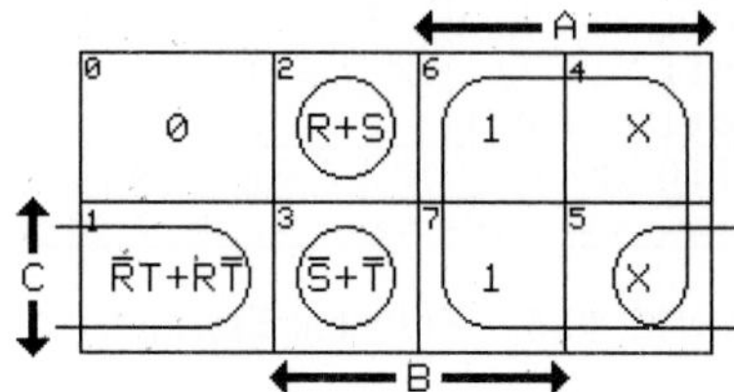

FIGURE 4.28 Practical Example 4.4, Karnaugh map for function D_B.

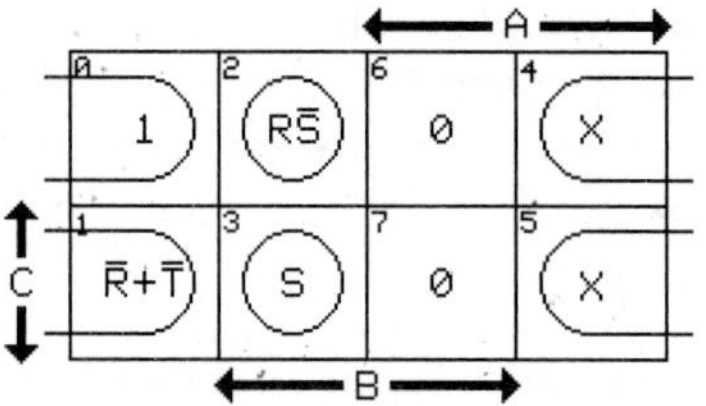

FIGURE 4.29 Practical Example 4.4, Karnaugh map for function D_C.

Extracting terms from the Karnaugh map of Figure 4.29, we obtain the function:

$$D_C = \overline{B}\,\overline{C} + \overline{A}B\overline{C}R\overline{S} + \overline{A}BCS + \overline{B}C(\overline{R} + \overline{T})$$

$$D_C = \overline{B}\,\overline{C} + \overline{A}B\overline{C}R\overline{S} + \overline{A}BCS + \overline{B}C\overline{R} + \overline{B}C\overline{T}$$

As we can see from these examples, when the state machine includes external control of the count sequence, the functions can get complex rather quickly. We are now to the point where implementation of this circuitry using discrete logic is inappropriate. However, note that a PAL with registered output macrocells would fit the situation quite nicely. These functions have a maximum of seven minterms, four counter variables and three external inputs. A 16R4 PAL chip would do quite nicely.

Also, now that we have come up with such a complex solution to the functions, it is no longer appropriate to draw the discrete gates when drawing the schematic diagram. We will use a shortcut when this happens.

Output-Forming Logic Design

Since the input-forming logic functions have been determined by this time in the design process, we are left only with the determination of the output-forming logic. If we go back to the state table of Figure 4.25, we find the information we need to determine the output-forming logic. Consider Figure 4.30.

The manner in which we are using the output-forming logic for this example is a little different from previous examples. Up to this point we have looked at generating

	ABC	W	X	Y	Z
b	000				Z↑DSB↓DSE
d	001 001 001 001			Y↑SE	
a	010 010 010	W↑DSB			
c	011 011 011	W↓SE	X↑DSB		
f	110			Y↓DSB	
e	111		X↓DSE		

FIGURE 4.30 Practical Example 4.4, OFL portion of state table.

pulses at each state. Actually, there is nothing significantly different here. The outputs W, X, and Y are set and reset in different states. Therefore, a flip-flop will have to be used in order to store the values of these variables between states. We will set and reset the respective flip-flops using pulses generated in the indicated states.

First consider the W output, which is supposed to be set at the DSB point of state a. That is, we need to set the flip-flop on the rising edge of the clock in the center of the state when the counter is at a 010_2. A pulse that lasts from DSB to SE will do the trick. The generation of this pulse was described in Section 4.4. Output W is supposed to RESET at the SE point of state c. This means we need to generate a pulse that starts at the end of the state, when the counter is at a 011_2. Either a pulse from SE to DSE, or a pulse from SE to DDSE will do the trick. From Section 4.4 we find the SE to DDSE pulse to be the simplest to implement. The result of this argument is the schematic diagram of Figure 4.31. Note that by observation we can see that the W flip-flop is set by state 010_2 and reset by state 011_2.

Next, consider the X output. We need to generate a pulse at the DSB point of state c to set the X flip-flop and to generate a pulse at the DSE point of state e to reset the X flip-flop. Of the pulses we can generate that start at DSB, the DSB to SE pulse is the simplest. The only pulse that starts at DSE is the DSE to DDSE pulse. The result of this argument is illustrated in the circuitry of Figure 4.32.

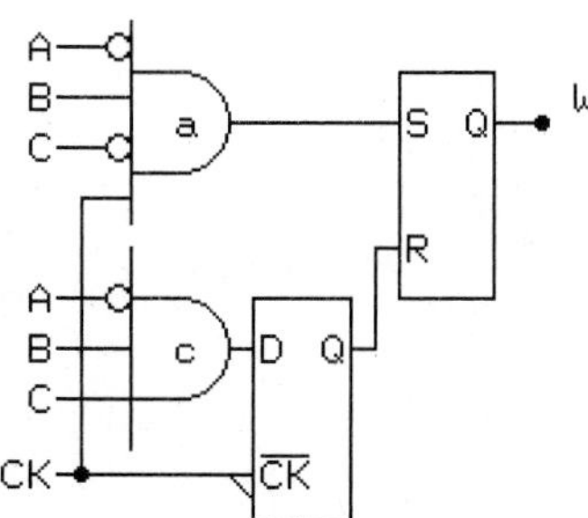

FIGURE 4.31 Practical Example 4.4, W output circuit.

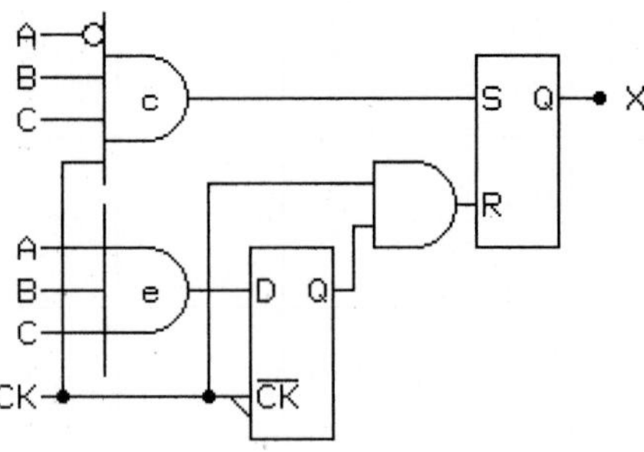

FIGURE 4.32 Practical Example 4.4, X output circuit.

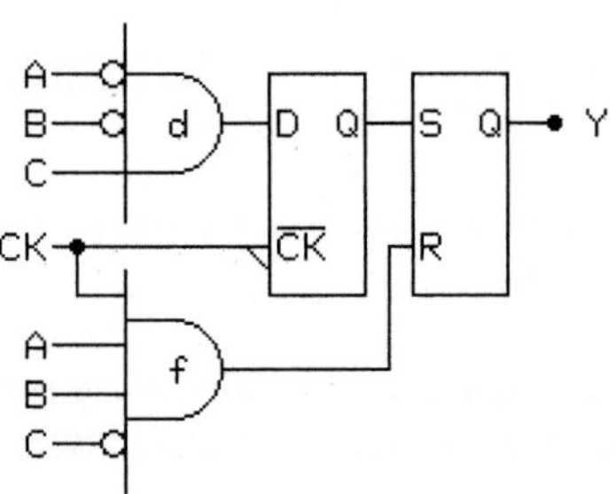

FIGURE 4.33 Practical Example 4.4, Y output circuit.

Then, consider the Y output. We need to set the Y flip-flop at the end of state d. The SE to DDSE pulse can be used for this. We need to reset the flip-flop in the center of state f. The DSB to SE pulse can be used. The schematic diagram of Figure 4.33 is then realized.

Finally, consider the Z output. It is simply an active-high pulse from DSB to DSE. From Section 4.4 we find we can utilize the circuit of Figure 4.34.

Practical Exercise 4.4: Output Pulses

Consider the state diagram of Figure 4.34. Note that all outputs are pulses. Develop a complete design for this example. Hint: Similar outputs can be pulsed at different states as shown in this example. Since no two states are ever true at the same time, the OFL from each state can be logically ORed together to produce the final output.

Given the state diagram of Figure 4.35, your tasks are to (1) develop a detailed state machine program, (2) the timing diagram, (3) the state table, and (4) minimum IFL and OFL circuitry.

FIGURE 4.34 Practical Example 4.4, Z output circuit.

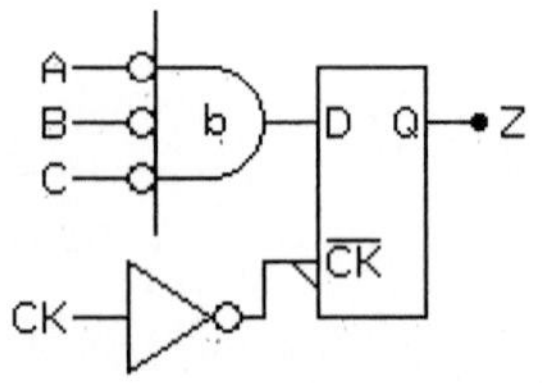

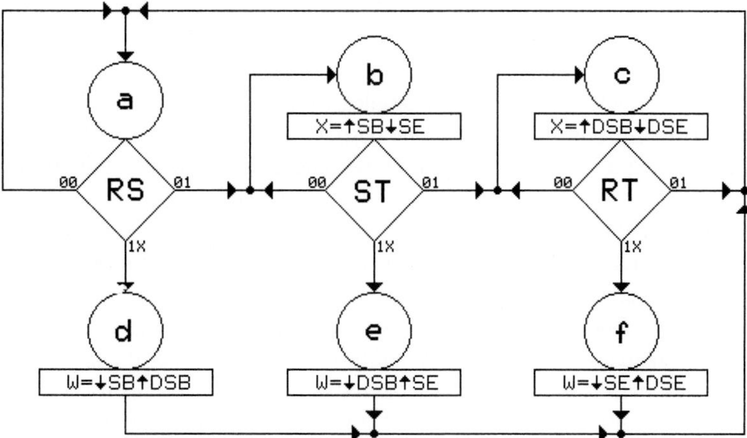

FIGURE 4.35 Practical Example 4.4, state diagram.

Practical Exercise 4.5: Output Levels

Consider Figure 4.36. In this example, two outputs are either set or reset by pulses at different states. Note that all outputs are pulses that set final output logic levels. Develop a complete design for this example. Hint: Similar outputs can be pulsed at different states as shown in this example. Since no two states are ever true at the same time, the OFL from each state can be logically ORed together to produce the final output to an SR flip-flop.

Given the state diagram of Figure 4.36, your tasks are to develop (1) a detailed state machine program, (2) the timing diagram, (3) the state table, and (4) minimum IFL and OFL circuitry.

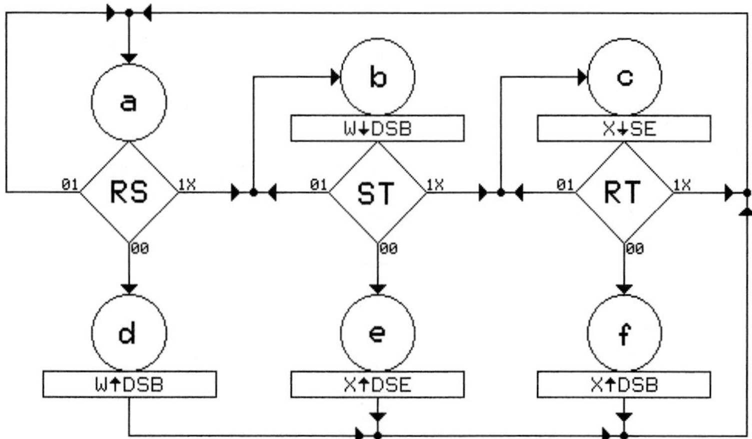

FIGURE 4.36 Practical Example 4.5, state diagram.

4.8 UNCONDITIONAL OUTPUT-FORMING LOGIC

All of the output-forming logic examples we have seen to this point have been unconditional. **Unconditional OFL** generate output signals that are asserted everytime a specific state is encountered. This is shown in the state diagram by attaching the output box to the state box. Figure 4.37 illustrates varying examples of unconditional OFL.

Note that though the state diagrams to this point have consistently placed the output box immediately following the state box, such a convention is probably not necessary. Unconditional output can be as effectively documented simply by attaching the output box to the state box. Sometimes this improves the overall appearance of a state diagram and enhances its readability. When an output box is unconditional, the logic level it describes is asserted everytime the state is encountered in the sequence. The following are some examples of unconditional OFL statements in state machine program form:

```
PRESENT 'b'0000 NEXT 'b'0001 OUT X;
PRESENT 'b'0001 OUT W;
  IF !R NEXT 'b'0010;
  DEFAULT NEXT 'b'0011;
```

The first example sets output X high during state 0000_2. The second example sets output W high during state 0001_2, regardless of the decision that determines the next state. When unconditional OFL is desired, the output logic is always asserted without regard to the determination of the next state.

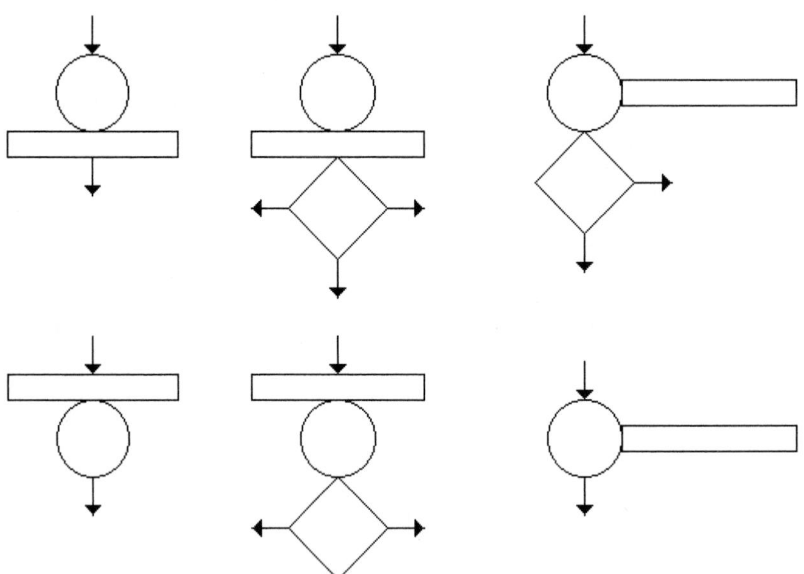

FIGURE 4.37 Unconditional output-forming logic.

4.9 CONDITIONAL OUTPUT-FORMING LOGIC

Often in the process of state machine design we find that the assertion of outputs from the state machine are dependent on inputs to the state machine. For example, if a state machine is used to control a candy machine, the solenoid that releases a specific candy should activate only when that candy is selected. At a single state in the sequence, one of several products may be selected by user input. This is a very common situation that should be addressed. Its implementation is rather simple once the design of unconditional OFL is fully understood.

Conditional OFL are output signals that are asserted in a specific state only when a specific pattern of input arguments is true. This is implemented by adding the input arguments to the ANDed true state bits that enable the output-forming logic circuitry.

Consider Figure 4.38. Note the obvious difference in the placement of the output boxes in the state diagrams. Rather than being attached to the state box, they are attached to the exit points of the decision boxes. The implication is rather straightforward: The OFL is still a function of the indicated state, but its assertion is further restricted by the input arguments described at the exit point of the decision box.

Consider the following example. Output W is to be asserted during state 100_2. If this were implemented in an unconditional fashion, the three counter bits A, B, and C would be logically ANDed together to produce an unconditional W output. However, if W is only to be asserted when input R is true, we would have to add the logic of input R to the AND gate. Figure 4.39 illustrates the difference between these two applications. Note that the only difference between the OFL is the addition of the input argument to the AND logic.

When applied to the state machine program, the unconditional OFL statements are moved to the conditional transition statements. Consider the following examples:

```
PRESENT 'b'100
  IF R NEXT 'b'011 OUT W;
  DEFAULT NEXT 'b'100;
PRESENT 'b'001
  IF !R NEXT 'b'010 OUT X;
  DEFAULT NEXT 'b'011;
```

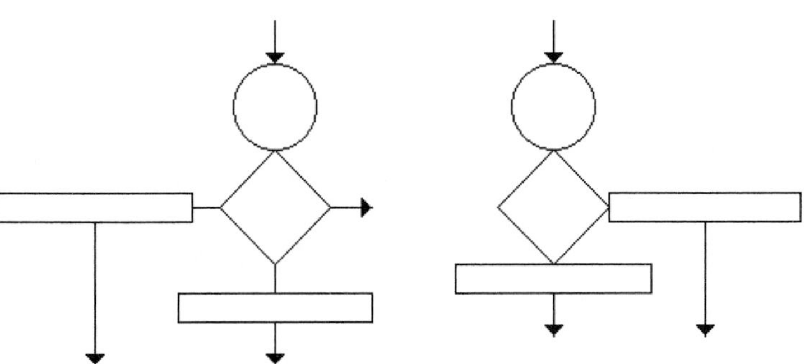

FIGURE 4.38 Conditional output-forming logic.

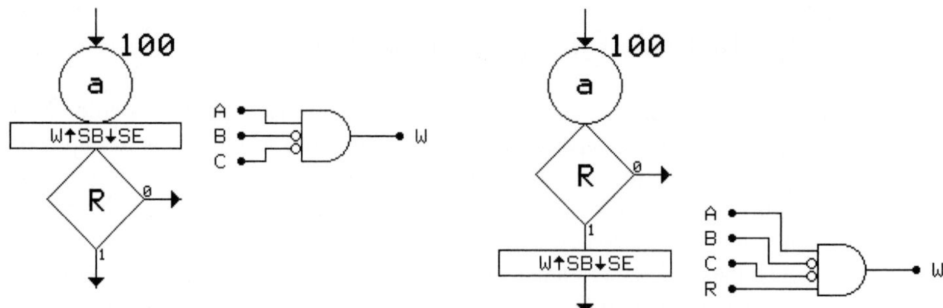

FIGURE 4.39 Unconditional/conditional OFL comparison.

The first case is an implementation of the conditional OFL example of Figure 4.39. Output W is asserted during state 100_2 only if input R is true. The second case asserts X during state 001_2 only if input R is false. It is also reasonable to mix conditional and unconditional output arguments in the same state.

Consider the example illustrated in Figure 4.40. In this case, two outputs are considered for assertion during state 101_2. The Y output is unconditional, as it is not restricted by input R. However, it is restricted by the clock signal and is therefore asserted from DSB to SE. The Z output is conditional on the logic level of input R. Again, asserted only during state 101_2, the assertion is further restricted by input R, so Z is true only when the sequence is in state 101_2 and only when R is also true. OFL arguments in a single state can become quite complex in real applications.

Practical Example 4.5: A Simple Soda Machine Controller

At last we can design something real! We should now be able to define a practical application of a simple state machine and follow through its design to a successful culmination.

FIGURE 4.40 Mixed OFL.

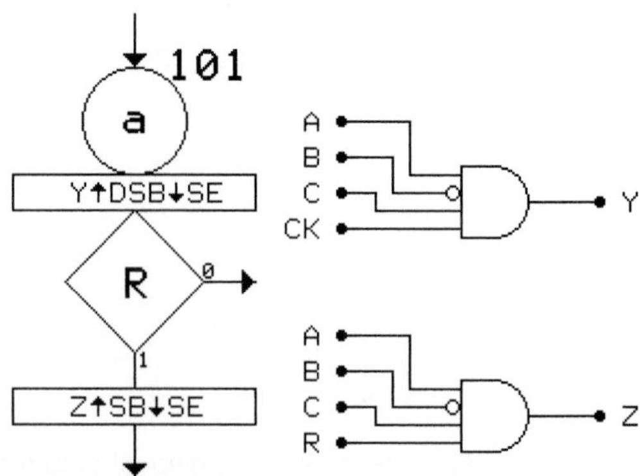

The material already presented is sufficient that, with practice, relatively complex designs can be completed. For a first design, we will do a simple problem involving a soda machine controller that dispenses only two types of soda, accepting only quarters from the buyer.

Problem: A buyer walks up to a simple soda machine that dispenses only two types of soda: cola, and diet cola. The machine accepts only quarters. Also, anytime there is money in the machine, the coin return may be depressed. Finally, the user is human and may spend considerable time in front of the machine trying to make a decision. The price of the soda is 50 cents.

The problem must be described as a series of states. That is, break down the problem into a list of all of the possible user actions as a sequence of events.

State a: Initial State. The buyer may either (1) wait or (2) insert a quarter. If the quarter is inserted, go to state b. Anytime we are waiting, remain in the same state.

State b: There is a quarter in the machine. The buyer may either (1) wait, (2) press the coin return, or (3) insert a quarter. If the coin return is pressed, go to state d, where a quarter is dispensed. If a quarter is inserted, go to state c.

State c: There are two quarters in the machine. The buyer may either (1) wait, (2) press the coin return, (3) request cola, or (4) request diet cola. If the coin return is pressed, output a quarter and go to state d, where another quarter will be outputted. If cola is dispensed, output a cola and return to state a. If diet cola is dispensed, output a diet cola and return to state a.

State d: Dispense the remaining quarter. Output a quarter and return to state a.

Figure 4.41 illustrates a block diagram of this state machine with the inputs and output defined as described in the problem statement. Observe each of the input and output variables used and how they relate to the problem statement.

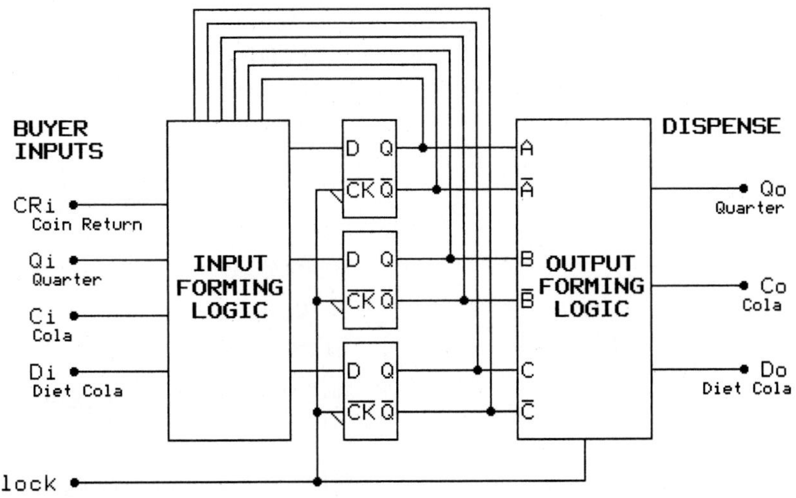

FIGURE 4.41 Practical Example 4.5, block diagram.

State Diagram

Figure 4.42 illustrates the state diagram of this problem. Observe how each state was drawn from the problem statement, considering the following conventions:

- "NA" refers to "no activity." Since this is a clocked counter, it will be continually changing from one state to another. If the buyer is waiting, the counter must remain in the same state until an input is asserted. Three of the four states are waiting on buyer input.
- The decisions are labeled a little differently than previous examples. Exit paths are taken when the indicated input is true. No two inputs can be recognized at the same time.
- The output pulses are glitch-free by virtue of the fact that each is of a half clock period duration starting at the rising edge of the clock. Therefore, the count sequence is not designed to be glitch-free.
- The counter uses three bits to minimize IFL complexity.

Observe how each state is derived from the problem statement.

State a. Only a single input is observed by the state machine. No outputs are defined.

State b. Two inputs are observed by the state machine. If a quarter is inserted, go to state c. If the coin return is pressed, go to state d.

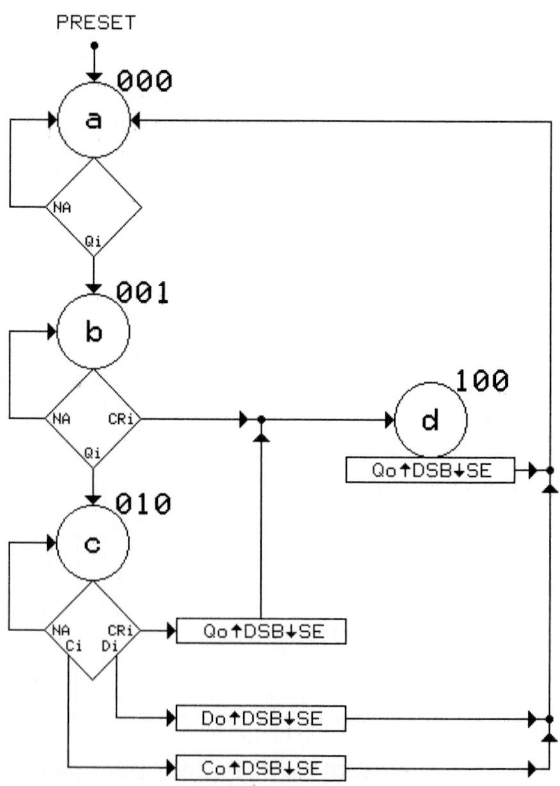

FIGURE 4.42 Practical Example 4.5, state diagram.

State c. Three inputs are observed by the state machine. Two will conditionally dispense products and go to state a. One will conditionally dispense a quarter and go to state d.

State d. A quarter is unconditionally dispensed; go to state a.

State Table

Take a good look at the state table illustrated in Figure 4.43.

One of the features of the design is that only one input can be recognized at a time. This limitation greatly simplifies the input column in the state table. Each conditional state must have only one condition identified for each permutation of buyer entry. Consider state c. The buyer has four options, (1) wait, (2) coin return, (3) cola, and (4) diet cola. If the first option is exercised, then all three buyer inputs are at a logical 0. In this permutation the next state is state 010_2, and the counter stays in the current state. If the second option is exercised, the coin return input is a logic 1. This is the only time this input will be a 1 in this state and when it is asserted, all other inputs are disabled. Therefore, all other inputs are in a "don't care" state. This same concept applies to the other two inputs in this state.

The method for determination of the input-forming logic has not changed, though as inputs are added to the design, it gets progressively more complex.

The method for determining the output-forming logic is the same for the unconditional state, state d. Note, however, the conditional outputs of state c. The output-forming logic is indicated as being asserted only in those permutations defined by the state diagram. This is important to note. This implies that as the output logic is asserted as a function of the true state bits AND the identified input arguments.

Input-Forming Logic

Figures 4.44, 4.45, and 4.46 illustrate the three Karnaugh maps used to minimize the input-forming logic of this example.

The input function for D_A is quite straightforward. Note how the placement of the "don't care" states worked out to advantage. If this counter were implemented with a two-bit counter, the maps would have been significantly more complex. The dots were inserted in the expression to identify the AND argument so that the expression would be more easily read.

	PRES ABC	INPUTS Qi	Cri	Ci	Di	NEXT ABC	D_A	D_B	D_C	Qo	Co	Do
a	000	0	X	X	X	000	0	0	Q	0	0	0
		1	X	X	X	001						
b	001	0	0	X	X	001	Cr	Q	$\overline{Q}\bullet\overline{Cr}$	0	0	0
		1	X	X	X	010						
		X	1	X	X	100						
c	010	X	0	0	0	010	Cr	$\overline{Cr}\bullet\overline{C}\bullet\overline{D}$	0	0	0	0
		X	1	X	X	100				↑DSB↓SE	0	0
		X	X	1	X	000				0	↑DSB↓SE	0
		X	X	X	1	000				0	0	↑DSB↓SE
	011	X	X	X	X	XXX	X	X	X	X	X	X
d	100	X	X	X	X	000	0	0	0	↑DSB↓SE	0	0
	101	X	X	X	X	XXX	X	X	X	X	X	X
	110	X	X	X	X	XXX	X	X	X	X	X	X
	111	X	X	X	X	XXX	X	X	X	X	X	X

FIGURE 4.43 Practical Example 4.5, state table.

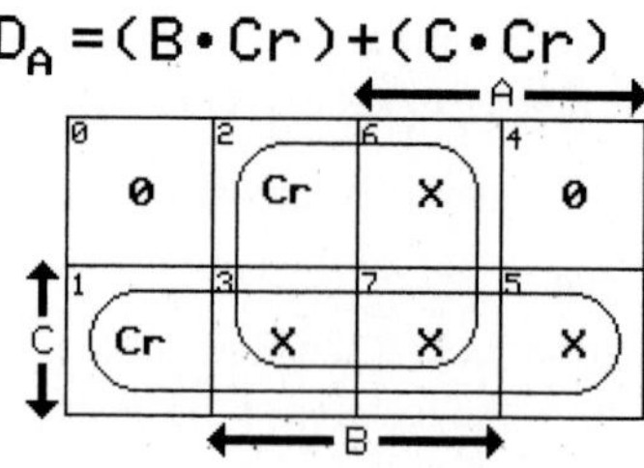

FIGURE 4.44 Practical Example 4.5, input function D_A.

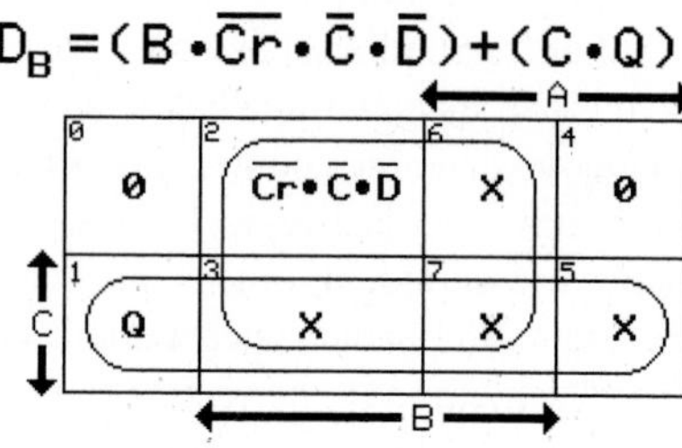

FIGURE 4.45 Practical Example 4.5, input function D_B.

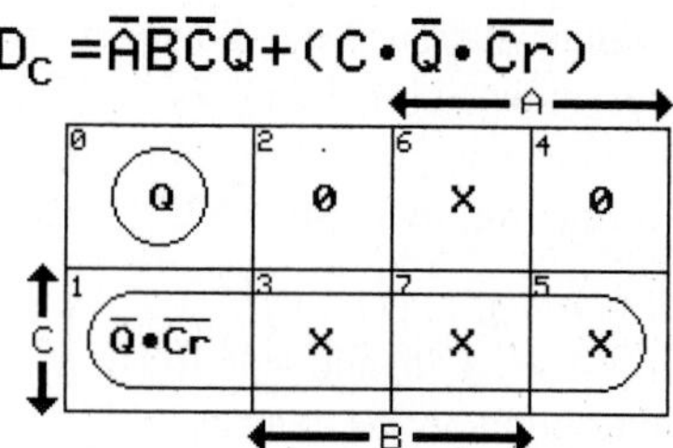

FIGURE 4.46 Practical Example 4.5, input function D_C.

The logic for state 2 of input function D_B shows the logical sum of three different arguments. Again, the resulting function extraction is relatively straightforward.

From these Karnaugh maps and the output-forming logic functions described in the state table, the schematic diagram of Figure 4.47 is realized. Review this example in detail and establish an understanding of each of the connections made in Figure 4.47.

Practical Exercise 4.6: A Simple Candy Machine Controller

Using the same approach as Practical Example 4.5, design a candy machine controller with the following constraints. Show all development steps in detail.

Inputs: No activity
Press coin return
Input quarter
Select chewing gum (25¢ cost)

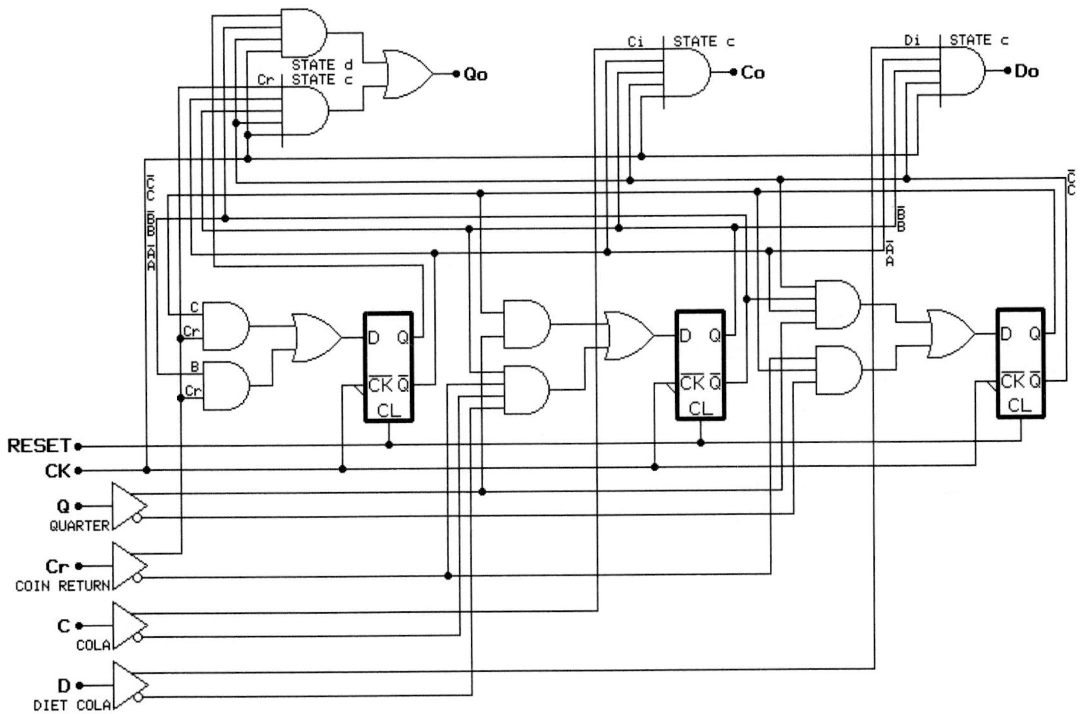

FIGURE 4.47 Practical Example 4.5, schematic diagram.

	Select potato chips	(50¢ cost)
	Select chocolate bar	(50¢ cost)
Outputs:	Dispense quarter	
	Dispense chewing gum	
	Dispense potato chips	
	Dispense chocolate bar	

Practical Exercise 4.7: A Moderate Candy Machine Controller

Using the same approach as Practical Example 4.5, design a candy machine controller that has the constraints of Practical Exercise 4.6. In addition, implement the following inputs and outputs:

Inputs:	Input dime
	Select cheese crackers (40¢ cost)
Outputs:	Dispense dime
	Dispense nickel
	Dispense cheese crackers

Assume that the machine never runs out of product. Once 50¢ or more has been inserted into the machine, assume additional coins are automatically sent to the coin return. Show all development steps in detail.

QUESTIONS AND PROBLEMS

Define the terms in 1–10.

1. Output-forming logic
2. Output pulse
3. Sustained output level
4. Output object
5. Transition statement
6. SB, DSB
7. SE, DSE, DDSE
8. State diagram parsing
9. Unconditional OFL
10. Conditional OFL
11. What is the purpose of output-forming logic as implemented in a state machine?

The following are instructions for Problems 12–25: Draw the schematic diagram of the logic circuitry that will generate the following pulses as a function of a state referred to as state a.

12. Positive pulse from SB to DSB
13. Negative pulse from SB to DSB
14. Positive pulse from SB to SE
15. Negative pulse from SB to SE
16. Positive pulse from DSB to SE
17. Negative pulse from DSB to SE
18. Positive pulse from DSB to DSE
19. Negative pulse from DSB to DSE
20. Positive pulse from SE to DSE
21. Negative pulse from SE to DSE
22. Positive pulse from SE to DDSE
23. Negative pulse from SE to DDSE
24. Positive pulse from DSE to DDSE
25. Negative pulse from DSE to DDSE
26. Draw the schematic of OFL that sets at SB of state a, and resets at SB of state b.
27. Draw the schematic of OFL that sets at DSB of state a, and resets at DSB of state b.
28. Draw the schematic of OFL that sets at SE of state a, and resets at SE of state b.
29. Draw the schematic of OFL that sets at DSE of state a, and resets at DSE of state b.
30. Draw the schematic of OFL that sets at SB of state a, and resets at DSB of state b.
31. Draw the schematic of OFL that sets at SB of state a, and resets at DSE of state b.
32. Draw the schematic of OFL that sets at DSB of state a, and resets at DSE of state b.
33. Draw the schematic of OFL that sets at DSE of state a, and resets at SB of state b.
34. Draw the schematic of OFL that sets at DSE of state a, and resets at DSB of state b.
35. Draw the schematic of OFL that sets at SE of state a, and resets at SB of state b.

36. Draw the schematic of OFL that sets at SE of state a, and resets at DSB of state b.
37. Draw the schematic of OFL that sets at SE of state a, and resets at DSB of state b.
38. Describe a general state machine program statement for an unconditional positive output pulse from SB to DSB.
39. Describe a general state machine program statement for an unconditional negative output pulse from SB to SE.
40. Describe a general state machine program statement for an unconditional positive output pulse from DSB to SE.
41. Describe a general state machine program statement for an unconditional negative output pulse from DSB to DSE.
42. Describe a general state machine program statement for an unconditional positive output pulse from SE to DSE.
43. Describe a general state machine program statement for a conditional positive output pulse from SB to DSB.
44. Describe a general state machine program statement for a conditional negative output pulse from SB to SE.
45. Describe a general state machine program statement for a conditional positive output pulse from DSB to SE.
46. Describe a general state machine program statement for a conditional negative output pulse from DSB to DSE.
47. Describe a general state machine program statement for a conditional positive output pulse from SE to DSE.
48. Draw a detailed timing diagram of the state machine defined in Figure 4.17.
49. Draw a detailed timing diagram of the state machine defined in Figure 4.35.
50. Draw a detailed timing diagram of the state machine defined in Figure 4.36.
51. Draw a detailed state table of the state machine defined in Figure 4.36.

Use Figure 4.48 to answer Problems 52–63.

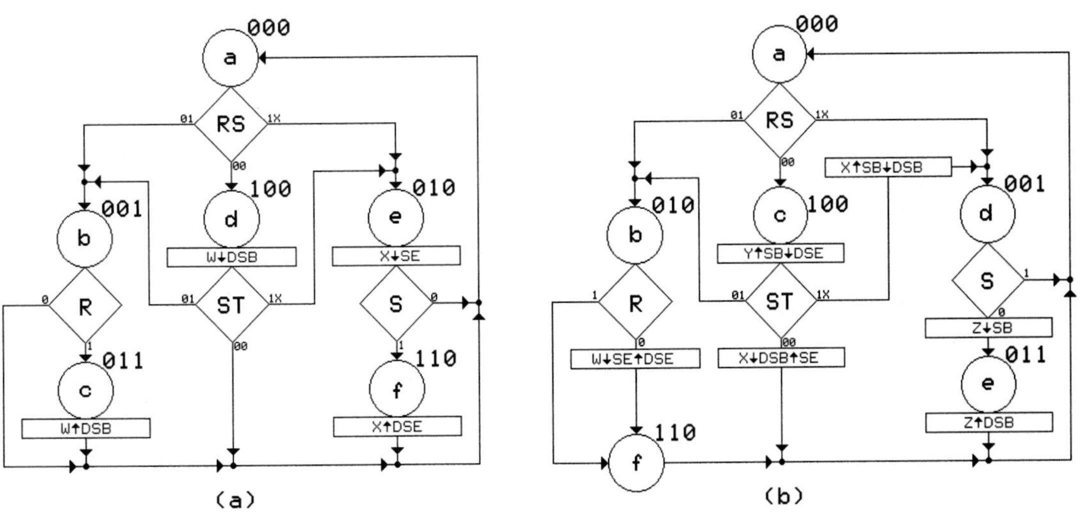

FIGURE 4.48

52. Draw the timing diagram of the state machine of Figure 4.48(a).
53. Draw the state machine program of the state machine of Figure 4.48(a).
54. Draw the state table of the state machine of Figure 4.48(a).
55. Determine the minimum IFL of the state machine of Figure 4.48(a).
56. Determine the OFL of the state machine of Figure 4.48(a).
57. Which outputs in the state machine of Figure 4.48(b) are unconditional?
58. Which outputs in the state machine of Figure 4.48(b) are conditional?
59. Draw the timing diagram of the state machine of Figure 4.48(b).
60. Draw the state machine program of the state machine of Figure 4.48(b).
61. Draw the state table of the state machine of Figure 4.48(b).
62. Determine the minimum IFL of the state machine of Figure 4.48(b).
63. Determine the OFL of the state machine of Figure 4.48(b).

CHAPTER 5

Programming PLD Circuits with CUPL

OBJECTIVES

After completing this chapter you should be able to:

- Describe the development sequence used to design with programmable logic.
- Describe a significant subset of the CUPL language elements that can be used to implement counter-type state machines.
- Convert a digital schematic diagram of a counter-type state machine to CUPL language.

5.1 OVERVIEW

In the state machine design examples described in previous chapters, their implementation with programmable logic devices (PLDs), such as PALs, PLAs, and ROMs has been mentioned; however, the discussions used discrete logic devices for example solutions. As the design problems become more complex, we find that using discrete devices is impractical, particularly when programmable devices are available. This chapter will introduce the methodology of implementing solutions using programmable devices.

The methodology for programmable device application is, as most concepts in this text, a multistep process. Once the desired circuitry for the state machine is determined, a software program must be written that will generate the binary program that can then be loaded into a programmable logic device. This necessitates the application of a specific programming language that will translate the information from a schematic diagram to the binary pattern that must be "burned" into the programmable device in order to emulate the original logic. There are several languages used in the field, including PALASM®, Pro-Logic®, CUPL®, and others. These programs are available from and registered trademarks of Texas Instruments, Intel Corporation, and Logical Devices, Inc., respectively.

This text will use CUPL, Universal Compiler for Programmable Logic, to create programs for PALs, PLAs, and ROMs. This language was chosen because of its intuitive syntax, its suitable power, and its broad range of application in the industry. This compiler can be obtained by writing to: Logical Devices, Inc., 692 South Military Trail, Deerfield Beach, FL 33442. A free starter kit is available as a student version. It contains all of the significant features of the CUPL language and can be used to implement most of the applications in this text.

The use of the language, its elements, and examples are provided here by permission of Logical Devices, Inc. This chapter presents only a subset of this powerful language. For more information, refer to the data book published by Logical Devices, Inc. entitled *CUPL, Universal Compiler for Programmable Logic.*

Once a binary program has been generated by the compiler, it must be loaded into the programmable logic device by using a device programmer. Most device programmers designed to "burn" EPROMs, EEPROMs, PALs, and PLAs are suitable for this task. One should refer to a list of supported devices and program formats in order to determine which device programmer is most suitable for a given task.

Once the language has been overviewed, this chapter will present several examples of programs for counter-type state machines similar to those described in the previous three chapters.

5.2 PROGRAMMABLE LOGIC DEVELOPMENT SEQUENCE

The complete design of a system that uses programmable logic to implement a state machine requires several design steps. Essentially, each step converts the problem to a different form, with the final form being the programmed chip. These steps can be illustrated as shown in Figure 5.1.

5.2.1 Problem Definition

The first step in the development process is to create a problem definition that is in a form suitable for implementation using the procedures outlined in the previous chapters. The final circuitry is determined entirely by the problem to be solved, so the problem definition must be accomplished first. The problem definition must be complete enough to describe the scope of the problem, all of its requirements, and all of its constraints. The most common form of the problem definition is one that describes all of the inputs to the state machine, all of the outputs of the state machine, and both internal and external activity at each state.

A simple form of a problem definition is a simple written algorithm: a step-by-step description of what takes place at each state in the sequence. For example, we could define a state machine using the following information:

```
State-a: If input R=0 then goto State b else goto State c;
State-b: Output Y=1, goto State a;
... etc.
```

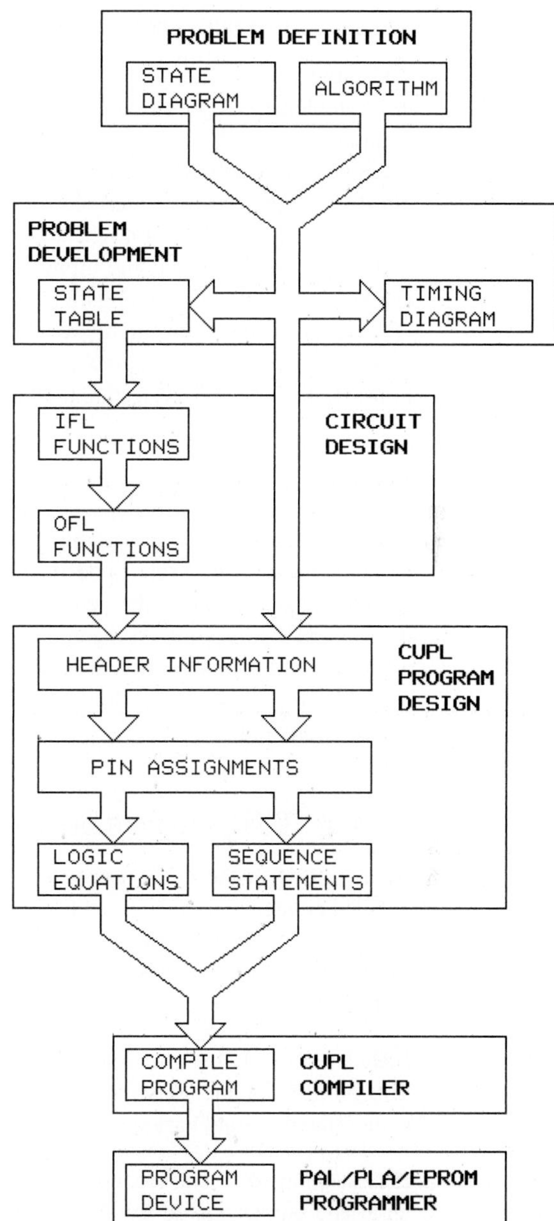

FIGURE 5.1 Programmable logic development sequence.

A second form of a problem definition is the state diagram. We have made significant use of the state diagram in previous chapters.

Regardless of the form of the problem definition, it must be complete enough that a state machine can be designed from it. When it comes to digital electronics, there is absolutely no mercy when it comes to problem definitions. Any vagueness in the problem definition will result in an ineffective solution.

5.2.2 Problem Development

The activity at this step in the process varies with the application. If the state machine is to be implemented with PALs, it is possible that the program to be written for the PAL compiler can be written directly from the problem definition. However, if the state machine is to be implemented with discrete logic, with ROMs, or is too complex for a simple PAL program, some intermediate problem development may be required. Figure 5.1 implies that this is the point where the timing diagram can be produced. This timing diagram will reveal conflicts in output definitions. It also provides a troubleshooting tool. It can be compared with actual data taken from the circuit using a digital logic analyzer.

If the design is unsuitable for direct application in the PAL programming language, it may be necessary to complete the development necessary to define the input logic functions. If this is necessary, a state table must be generated.

5.2.3 Circuit Design

Again, if the design is unsuitable for direct application in the PAL programming language, it may be necessary to define the input logic functions. If that is the case, it would be done at this point in the program. In addition, the output logic functions would be determined at this point. At the conclusion of this step in the design, it is possible to draw a detailed and complete logical schematic diagram. If it is intended to develop the circuit using discrete logic rather than programmable logic, the development sequence can be exited at this point.

5.2.4 CUPL Program Design

At this point in the development sequence we have either a problem statement that is detailed enough and in the proper form that the CUPL language can be used to implement it or we have a detailed description of the digital logic to be implemented. In either case, we are ready to write a CUPL program. The program is typically written using a simple text editor that maintains its text in ASCII form. That program is made up of several parts.

Header Information. The first part of the CUPL program includes the header information. This information includes a name given to the program, the device to be programmed, the author, the date, and various other items.

Pin Assignments. The second part of the CUPL program identifies the assignments of the input and output pins of the device to the input and output variables used in the design. Care should be taken at this point to determine that the selected device, the pin I/O directions, and the macrocell requirements are consistent.

Logic Equations, Sequence Statements. Following the pin assignments is the portion of the CUPL program that defines the relationships among the assigned pins. Implementation of the logic for those relationships may take the form of logic equations for the input-forming logic and output-forming logic. It may also take the form of a set of sequence statements, as described in previous chapters. It may also take the form of a combination of these.

5.2.5 CUPL Compiler

Once the CUPL program has been designed using a text editor, it must be converted to binary pattern to be loaded into the programmable logic device. This is accomplished using a software compiler designed to translate the CUPL statements into the binary pattern. This text uses the CUPL compiler available from Logical Devices, Inc. The compiler generates another file that is formatted in a manner that can be loaded into a hardware device that programs the PALs, PLAs, or ROMs.

5.2.6 PAL/PLA/EPROM Programmer

Finally, a hardware device is used to program the PALs, PLAs, or ROMs. This device takes the binary pattern generated by the CUPL program and programs it into the programmable logic device. The device is now ready for use.

The previous chapters have prepared us for the first three steps just described. We will at this point look at a method for completing the CUPL program design, operating the CUPL compiler, and using a programmable device programmer to "burn" the binary pattern into a target device.

5.3 CUPL PROGRAM STRUCTURE

A CUPL program makes use of a free-form structure that includes three primary sections: (1) header information, (2) pin assignments, and (3) logic expressions. Those logic expressions may be described in many different forms. Figure 5.2 illustrates a simple CUPL program. This example is modified slightly from the form distributed by Logical Devices, Inc. in order to better illustrate the three sections of the program.

5.4 CUPL LANGUAGE ELEMENTS

The following is a subset of the CUPL language, taken from *CUPL, Universal Compiler for Programmable Logic,* published by Logical Devices, Inc.

5.4.1 Variables

Variables are strings of alphanumeric characters that specify device pins, internal nodes, constants, input signals, output signals, intermediate signals, or sets of signals. Variables can start with a numeric digit, alphabet character, or underscore, but must contain at least one alphabetic character. They are case sensitive; that is, variables distinguish between uppercase and lowercase letters.

Variables can contain up to 31 characters. Longer variables are truncated to 31 characters. Also, variables cannot contain any of the CUPL reserved symbols or reserved keywords illustrated in Tables 5.1 and 5.2.

```
/***********************************************************************/
/* Header Information */
/***********************************************************************/
Name          Gates;
Partno        CA0001;
Revision      04;
Date          9/12/89;
Designer      G. Woolhiser;
Company       Logical Devices, Inc.;
Location      None;
Assembly      None;
Device        P16L8;

/***********************************************************************/
/*      This is a example to demonstrate how CUPL compiles            */
/*      simple gates.                                                  */
/*      Target Devices: P16L8, P16LD8, P16P8, EP300, and 82S153        */
/***********************************************************************/

/***********************************************************************/
/* Pin Assignments                                                     */
/***********************************************************************/
/* Inputs: define inputs to build simple gates from */

Pin 1 = a;
Pin 2 = b;

/* Outputs: define outputs as active HI levels
 * Note: For PAL16L8 and PAL16LD8, DeMorgan's Theorem is applied to
 * invert all outputs due to fixed inverter in the device. */

Pin 12 = inva;
Pin 13 = invb;
Pin 14 = and;
Pin 15 = nand;
Pin 16 = or;
Pin 17 = nor;
Pin 18 = xor;
Pin 19 = xnor;

/***********************************************************************/
/* Logic Expressions                                                   */
/***********************************************************************/
inva = !a;                /* inverters */
invb = !b;
and  = a & b;             /* and gate */
nand = !(a & b);          /* nand gate */
or   = a # b;             /* or gate */
nor  = !(a # b);          /* nor gate */
xor  = a $ b;             /* exclusive or gate */
xnor = !(a $ b);          /* exclusive nor gate */
```

FIGURE 5.2 CUPL program, GATES.PLD.

TABLE 5.1 CUPL reserved keywords.

APPEND	DEVICE	IF	OUT	SEQUENCED
ASSEMBLY	ELSE	JUMP	PARTNO	SEQUENCEJK
ASSY	FIELD	LOC	PIN	SEQUENCERS
COMPANY	FLD	LOCATION	PINNODE	SEQUENCET
CONDITION	FORMAT	MACRO	PRESENT	TABLE
DATE	FUNCTION	MIN	REV	
DEFAULT	FUSE	NAME	REVISION	
DESIGNER	GROUP	NODE	SEQUENCE	

Indexed Variables. Variable names can be used to represent a group of variables by ending them with an index number from 0 through 31. Index values must be decimal. For example, the following are indexed variables that might be used to describe an eight-bit data bus:

D0 D1 D2 D3 D4 D5 D6 D7

5.4.2 Numerical Constants and Logical Functions

These are the same as the numerical constants and logical functions described in Chapter 3. Their description is repeated here for convenience. The default number base for all numbers used in the CUPL source file is hexadecimal, except for device pin numbers and indexed variables, which are always decimal. Numbers for a different base may be used by preceding them with the prefix shown below. Once a base change has occurred, that new base is the default base. All operations involving numbers in the CUPL compiler are calculated to 32-bit accuracy; therefore, numbers in the range of 2^{-31} to $2^{+31}-1$ are supported.

Numeric constants can be represented in any of several number bases, including decimal, hexadecimal, octal, and binary. The following are some examples of numeric constants in different bases:

Decimal:	'd'0	'd'21	'd'51	'd'7
Hexadecimal:	'h'0	'h'15	'h'33	'h'7
Octal:	'o'0	'o'25	'o'63	'o'7
Binary:	'b'0	'b'10101	'b'110011	'b'111

The letter "X" may be substituted for any binary, octal, or hexadecimal digit to identify a "don't care" value.

TABLE 5.2 CUPL reserved symbols.

&	#	(	)	-
*	+	[	]	/
:	.	..	/*	*/
;	,	!	‘	=
@	$	^		

We will use the same logical function symbols as those defined in Chapter 1 as typographic symbols. These are illustrated in Table 5.3.

5.4.3 Comments

The CUPL language uses the same form of comment structure as described in Chapter 3. Its description is repeated here for convenience. It will use the delimiter "/*" to start a comment, and the delimiter "*/" to close a comment. The following are examples of the format of a typical comment:

```
/* This is a comment */
```

or

```
/*
This is also a comment
*/
```

The free-form nature of the comment delimiters allows program segments to be "commented out" rather than deleted while program development and debugging take place. Note that the comment can begin and end on different lines.

5.4.4 List Notation

It is often convenient to refer to lists of numbers in a shorthand manner. This shorthand notation is commonly used in pin and node declarations, bit field declarations, logic equations, and set operations. It is simply a list of the variables separated by commas and enclosed in brackets. For example, the following are examples of list notation:

```
[D0, D1, D2, D3, D4, D5, D6, D7]
[ON, OFF]
[ONE, TWO, THREE, FOUR]
```

When all of the variable names in the list are sequentially numbered, either in ascending or descending sequence, the following format may be used:

```
[D0..D7]      [A00..A15]     [INPUT3..INPUT0]
[D0..7]       [A7..0]        [I2..1]
```

Note that leading zeros in the index of an indexed variable are ignored.

TABLE 5.3 CUPL logical operators.

Operator	*Character*	*Precedence*
NOT	!	1
AND	&	2
OR	#	3
XOR	$	4

TABLE 5.4 CUPL header information keywords.

NAME	The program filename. When using MS-DOS systems, use a valid filename without the extension. That is, limit the filename to eight alphanumeric characters. Do not include a filename extension, as the extension for the different files associated with the program vary.
PARTNO	The company's proprietary part number (not the device number).
REVISION	Start with 01 and increment each time the program is altered.
DATE	Change to the current date each time the program is altered.
DESIGNER	The designer's name.
COMPANY	The company name.
ASSEMBLY	Identify the assembly on which the PLD will be used.
LOCATION	Identify the PC board reference or board location of the device.
DEVICE	Set the default device type for the compilation. The device type set in the compiler command line overrides this line.
FORMAT	Identify the download output format used by the compiler. Valid values are: h ASCII-hex output i Signetics HL output j JEDEC output This line overrides any output format option included in the compiler command line.

5.4.5 Header Information

The header information section of the source file describes the file and many of the parameters needed for identification, application, and compilation. It is best to keep the header information at the top of the source program. Though not all of the header keywords are required, if a required header line is missing, the compiler will issue a warning and continue compilation. Table 5.4 includes a short description of the header information keywords.

5.4.6 Pin Declaration Statements

Pin declaration statements declare the pin numbers and assign them symbolic variable names. The general format for a pin declaration is:

```
PIN pin_number = [!]var;
```

where "PIN" is a keyword to declare the pin numbers, "pin_number" is a decimal pin number or a list of pin numbers, "[!]" refers to an optional inversion of the variable, and "var" refers to a single variable name or a list of variables. The statement must end with a semicolon. Consider the following examples:

Pin 1 = Clock;

Pin 2 = S;

Pin 11 = !OE; Note the use of "!" to assert low.

Pin [3..6] = [D0..D3]; Note the use of list notation.

5.4.7 Node Declaration Statements

Many devices contain functions that are not available on external pins. For example, Figure 1.23 illustrates the internal schematic diagram of a Philips PLS105 FPLS device. It contains six registered functions and one logical function that are not pinned out; instead, their outputs return to the gate array. They have been assigned pin node numbers that are numerically greater than the number of pins on the device. Assigning internal nodes to these additional pin numbers allows those nodes to be used in the same way that other externally accessible nodes are. The general format for a pin node declaration is:

```
PINNODE pin_number = [!]var;
```

Its use is identical to the application of the pin declaration statement.

5.4.8 Preprocessor Commands

CUPL supports a variety of preprocessor commands. These are commands that operate on the source file before the compilation process. Though several commands are available in the language, we will review only three, $DEFINE, $INCLUDE, and $REPEAT.

$DEFINE. This command replaces one character string with another. Its purpose is primarily to replace a CUPL operator, number, or symbol with a character string that is more consistent with a specific application. The format for the statement is as follows:

```
$DEFINE argument1 argument2
```

where "argument1" is a character or character string and "argument2" is a valid operator, number, or variable name. Once the statement appears in the source program, the new string is substituted for the one it replaces from that point on. Consider the following examples:

```
$DEFINE ON 'b'1
$DEFINE OFF 'b'0
$DEFINE PORTA 'h'C000
$DEFINE + #
$DEFINE * &
```

The first example shows how the words "ON" and "OFF" can be used for the binary values of 0 and 1, respectively. The second example allows the use of the word "PORTA" to represent the hexadecimal address C000. The last example replaces the CUPL OR and AND logical symbols with "+" and "*", respectively. Consider the following additional example:

```
$DEFINE STATE-a    'b'000
$DEFINE STATE-b    'b'001
$DEFINE STATE-c    'b'010
$DEFINE STATE-d    'b'100
```

This set of $DEFINE statements replaces binary state numbers with a mnemonic label. If this is done, the statement

```
PRESENT 'b'100 NEXT 'b'000;
```

becomes

```
PRESENT STATE-d NEXT STATE-a
```

This can make the CUPL program much more readable and self-explanatory. Also, by grouping the $DEFINE statements at the beginning of the program, changes can be made to any value, affecting the entire program. For example, if we chose to change "STATE-B" from the binary value 'b'001 to the binary value 'b'111, we need only to change the $DEFINE statement. Then all occurrences of the object in the program will be suitably interpreted. It is usually best to place the $DEFINE statements at the beginning of the program section, following the heading information.

$INCLUDE. A second preprocessor command we will consider is the $INCLUDE statement. Its purpose is to include the code from another CUPL program at the point of declaration. Its form is:

```
$INCLUDE filename
```

where "filename" is the name of the file to be included at the declaration point. The file to be included must be a legitimate set of CUPL instructions. The entire file is included at the point of declaration at compile time. Consider the following application: Assume there is a block of CUPL statements that perform a common, complex operation and that block of statements is used in several programs. Instead of repeating the code in each program, the code can be stored as a separate file and included as needed using the $INCLUDE command. Note that the included code is inserted at the point of declaration, and must be compatible with the remainder of the program just as if it were typed into the source program at that point. Program code segments that are included using this command may themselves contain $INCLUDE statements, allowing nesting. However, be careful not to create a circular definition when nesting $INCLUDE statements.

$REPEAT, $REPEND. These two preprocessor commands provide the mechanism to repeat a section of CUPL source code using an index variable. Consider the following form:

```
$REPEAT index = [number1,number2,...,numbern]
    body of loop
    $REPEND
```

In preprocessing, the program statements making up the body of the loop will be repeated for each value from number_1 through and including number_n.

5.4.9 Variable Extensions

Extensions can be added to variable names to indicate specific functions associated with the major nodes within a programmable device. These nodes are defined by the architecture of the target device and are indicated in the specification sheets for that device. Care must be taken to be sure that the devices support such variable extensions and that the extensions are used correctly. Table 5.5 is not a complete list, but it includes extensions needed for designs applicable to this text.

Consider the example circuit of Figure 5.3. The output pin from the macrocell of Figure 5.3 is labeled OUT. The mechanism for referencing those other nodes within that macrocell are indicated by the extensions shown. For example, in order to indicate the D input of the D-type flip-flop, the variable name OUT.D is used. If the clock input to the flip-flop comes from the gate array, that input can be referenced by the variable name OUT.CK.

TABLE 5.5 CUPL variable extensions.

Extension	*Description*
.AP	Asynchronous preset of a flip-flop
.AR	Asynchronous reset of a flip-flop
.CK	Programmable clock of a flip-flop
.D	D input of a D-type flip-flop
.DQ	Q output of a D-type flip-flop
.INT	Internal feedback path for registered macrocell
.IO	Pin feedback path selection
.J	J input of a JK-type flip-flop
.K	K input of a JK-type flip-flop
.L	D input of a transparent latch
.LE	Programmable latch enable
.LQ	Q output of a transparent latch
.OE	Programmable output enable
.R	R input of a Set-Reset type flip-flop
.S	S input of a Set-Reset type flip-flop
.SP	Synchronous preset of a flip-flop
.SR	Synchronous reset of a flip-flop
.T	T input of a toggle-type flip-flop

Consider the following CUPL segment:

```
PIN 1 = CLOCK;
PIN 17 = OUT;
OUT.CK = !CLOCK;
```

The first line declares the variable CLOCK to refer to an input pin 1. The second line declares the variable OUT to refer to an output pin 17. The third line declares the function to be connected to the programmable clock input of the flip-flop in the macrocell. In this case, the logic on pin 1, CLOCK, is to be inverted and then connected to the clock input of

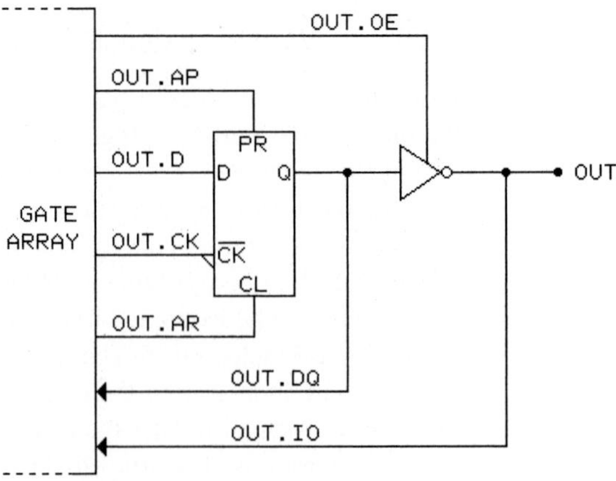

FIGURE 5.3 Typical variable extension application.

the flip-flop. Often, the D input of the flip-flop is some function of the input and output pins, such as the external stimulus and counter bits of a counter-type state machine. The following CUPL program segment provides such an example:

```
/* Practical Example 4.4 Alternate Solution */
PIN 2 = R;   /* Input Stimulus */
PIN 3 = S;   /* Input Stimulus */
PIN 16 = A; /* Counter MSBit */
PIN 15 = B; /* Counter */
PIN 14 = C; /* Counter LSBit */
A.D = B&S&!T;
B.D = (!A&!B&!C&!R&S) # (!A&C&T);
C.D = (!A&!B&!C&R&!S) # (!A&C&!T);
```

Recall that the SEQUENCE statements previously used to generate state machine programs were, for the most part, legitimate CUPL code. This example is an actual CUPL program segment that implements the counter design of Practical Example 4.4. Up to this point in our discussions, we have used SEQUENCE statements to define the counter functions in the CUPL program. It is also possible to declare the counter input function logic directly as shown here. Spend a few moments and compare this program with the design in Practical Example 4.4. The last three lines of this program segment replace the SEQUENCE statements of that example.

Pins 2 and 3 are assigned to input lines on a PAL device, possibly a 16R4. Pins 14, 15, and 16 are assigned to three registered output pins on the PAL. The last three lines identify that the three outputs A, B, and C are the outputs from macrocells containing D-type flip-flops. They also assign the indicated logical functions to the D inputs of those flip-flops, creating the correct input-forming logic.

In order to utilize this method of writing the CUPL program for a counter-type state machine, it is necessary to know the input function logic definitions. Consequently, it is necessary to go through the steps of creating the state table and minimization of the input-forming logic. However, if this is done, the CUPL program is simpler, since the SEQUENCE statements produce a longer source program and a more complex compilation. Note also that input-forming logic minimization in this case may not be necessary, as the CUPL compiler always minimizes logic functions during the compilation process.

The other variable extensions indicated in Table 5.5 are used in the same manner with macrocells that are consistent with their definitions. For example, the .J and .K extensions refer to the J and K inputs of JK-type flip-flops. There are several other extensions used with a variety of macrocell types not covered in this text. (For a complete list and complete documentation, refer to the CUPL text described in the chapter overview.)

5.4.10 The APPEND Statement

The APPEND statement allows multiple expressions to be assigned to a single variable. The effect of appending an expression to a variable is identical to a logical OR function. Consider the following example:

```
A.D = !A & !B & R;
APPEND A.D = A & !C & S;
APPEND A.D = !A & B & R & !S;
```

These three statements are equivalent to the following statement:

```
A.D = (!A & !B & R) # (A & !C & S) # (!A & B & R & !S);
```

The APPEND statement may be useful in a variety of situations. Consider a situation where a logical function is generated in a program segment and is later used in several places. Instead of repeating the logic in each place, the logic can be assigned to a single variable, and that variable can be appended where used. Making a change to the initial logic function will effect a correct change throughout the program.

5.4.11 The SEQUENCE Statement

Previous chapters have made a lot of use of the SEQUENCE statement in state machine programs. The syntax for those statements is consistent with the CUPL language, with the exception of the form of the OUT statement. Some of the description from Chapter 3 is repeated here for convenience. The sequence statement declares that the statements within the brackets represent a complete counter sequence. The following is an example of a sequence statement:

```
SEQUENCE [<counter_output_bits>] {
  ...         /* transition statements */
  }
```

Placed between the brackets "{" and "}" will be each of the transition statements that actually defines what is taking place in the sequence. The <counter_output_bits> will be a list of the labels of the counter output bits. This can be done in a variety of ways. For example:

```
[A,B,C,D]     [Q3,Q2,Q1,Q0]     [Q3..Q0]
```

All of these examples represent the outputs of a four-bit counter. The following is an example of the sequence statement used to define a three-bit counter that has outputs labeled A, B, and C:

```
SEQUENCE [A,B,C] {
...          /* transition statements */
}
```

The above example represents the general format of the sequences that we will be defining. We now need to add the transition statements that define the state machine activity.

Unconditional Transition Statement. Consider the following unconditional transition statement:

```
PRESENT <state0> NEXT <state1>;
```

or

```
PRESENT <state0> NEXT <state1> OUT <output_var>;
```

When the sequence is in <state0>, the next state following the clock edge is <state1>. If the OUT <output_var> statement is included, the output variable defined as <output_var> will be asserted during the entire current state. It is important to note that there is no facility to define the starting point and ending point of an output transition in the CUPL language. The OUT statement asserts the output line at the beginning of the state, and deasserts it at the end of the state. That is, the only output form the OUT statement generates is the SB to SE pulse. This is a significant variation from the state machine language of Chapter 3.

This statement can be used to describe an unconditional count sequence. For example, if a count sequence is defined as 1, 2, 3, . . . state 3 always follows state 2. The state machine programming language statement for state 2 would be:

```
PRESENT 2 NEXT 3;
```

Usually during the design stage, the numeric values of the states are not yet assigned. In that case, variables are often used. For example,

```
PRESENT a NEXT b;
```

uses variables instead of constants. An alternative method to describing the states is to make use of the $DEFINE command described previously in this chapter. Consider the following code example:

```
$DEFINE STATE-a     'b'000
$DEFINE STATE-b     'b'001
$DEFINE STATE-c     'b'010
$DEFINE STATE-d     'b'100
SEQUENCE[A,B,C,D]
  { PRESENT STATE-a NEXT STATE-b;
    PRESENT STATE-b NEXT STATE-c;
    PRESENT STATE-c NEXT STATE-d;
    PRESENT STATE-d NEXT STATE-a OUT X; }
```

This code defines a counter that repeats the three-bit values 0, 1, 2, 4, 0, 1, 2, 4, ... The output variable X will be asserted during state D.

Conditional Transition Statement. The conditional transition statement is used to describe a step in the count sequence that continues to one of two or more different next states as a function of one or more external inputs. The following is an example of a conditional transition statement:

```
IF input_condition NEXT <state1>;
```

or

```
IF input_condition NEXT <state1> OUT <output_var>;
```

This concept is similar to that discussed in Chapter 3.

Default Statement. The default statement is used to close a single IF argument, defining the destination for any arguments not previously stated. The following example is from Chapter 3:

```
/* Practical Example 3.2 */
SEQUENCE [X,Y,Z] {
               PRESENT 2 IF R=1 THEN NEXT 5;
                         DEFAULT NEXT 7;
               PRESENT 1 NEXT 2;
               PRESENT 5 NEXT 1;
               PRESENT 7 NEXT 1; }
```

As we can see, the CUPL form of the SEQUENCE, PRESENT, NEXT, IF, and DEFAULT statements is the same as that used in our state machine programs used in Chapter 3.

Output Assertion. The only difference is that the OUT statement does not support transition definitions. The OUT statement asserts the indicated output for the duration of the state. This is the equivalent of a transition of SB to SE. If the output logic is to be further processed, additional logic will be necessary. This can be done by treating the output as an "intermediate variable" by selecting a macrocell where the output is fed back to the gate array. It can then be used as an operand in subsequent logical expressions. We will deal with intermediate variables shortly. If the output variable is indicated as noninverted, the output will go to a logic 1 during the state. The output can be asserted low by attaching the "!" symbol prior to the variable. For example,

```
OUT !dtack;
```

will assert an output line labeled dtack to a logic 0 during the state.

5.4.12 The FIELD Statement

Many times it is convenient to refer to a group of bits by a single name. The bit FIELD declaration provides such a capability. The format is as follows:

```
FIELD var = [var,var,...,var];
```

where "FIELD" is the keyword, "var" is any valid variable name and "[var,var,...,var]" is the list of variables in list notation.

A convenient application for the FIELD statement is in describing the bits in a single device, such as a counter. If a binary counter can be represented by variables A, B, C, D, a FIELD statement such as:

```
FIELD COUNTER = [A,B,C,D];
```

allows us to use the name COUNTER to represent the list of variables. For example, consider how the following statements are equivalent once the FIELD statement has been declared.

```
SEQUENCE [A,B,C,D] {
```

and

```
SEQUENCE COUNTER {
```

5.4.13 Truth Tables

Sometimes the clearest way to express logical relationships is with a state table or truth table. The TABLE keyword may be used to relate two logical arguments. The format for using the keyword is as follows:

```
TABLE var_list_1 => var_list_2 {
    input_1 => output_1;
    input_2 => output_2;
    ...
    input_n => output_n;
    }
```

For example, consider a code converter that converts a four-bit binary nibble to seven-bit ASCII. This circuit converts the four bits defined as I3, I2, I1, I0 (assigned to four input pins of the PLD) to the seven-bit character O6, O5, I4, O3, O2, O1, O0 (assigned to seven output pins of the PLD.) The following CUPL segment would provide such a conversion:

```
FIELD binput = [I3..0];
FIELD aoutput = [O6..0];
TABLE binput => aoutput {
    'h'0 => 'h'30; 'h'1 => 'h'31; 'h'2 => 'h'32; 'h'3 => 'h'33;
    'h'4 => 'h'34; 'h'5 => 'h'35; 'h'6 => 'h'36; 'h'7 => 'h'37;
    'h'8 => 'h'38; 'h'9 => 'h'39; 'h'A => 'h'41; 'h'B => 'h'42;
    'h'C => 'h'43; 'h'D => 'h'44; 'h'E => 'h'45; 'h'F => 'h'46;
    }
```

5.4.14 Set Operations

Any operation that can be performed on a single bit can be performed on a set of bits by enclosing that set in brackets, using a comma as a delimiter between operands. For example,

```
[I0, I1, I2, I3] & mask
```

is the same as

```
[I0 & mask, I1 & mask, I2 & mask, I3 & mask]
```

Operations can be performed between two sets. In this event, the sets must be the same size. Consider the following example:

```
[A3, A2, A1, A0] & [B3, B2, B1, B0]
```

is the same as

```
[A3 & B3, A2  & B2, A1 & B1, A0 & B0]
```

Sets of variables can be combined with the FIELD statement in order to reference the set using a single variable name. For example:

```
FIELD inputs = [I3, I2, I1, I0, OE];
```

5.4.15 Equality Operations

The equality operation checks for bit equality between a set of variables and a constant. The equality operator is the colon, ":". The format of the equality operation is as follows:

```
[var, var, ..., var] : constant;
```

or

```
bit_field_var : constant;
```

Bit positions of the constant number are checked against the corresponding bit positions in the set. Where the constant bit position is a binary 1, the set element is unchanged. Where the constant bit position is a binary 0, the set element is complemented. Where the bit is declared as a "don't care" (X), the set element is eliminated. For example, consider the following equality operation where the hexadecimal value D (binary 1101) is checked against the set A3, A2, A1, A0.

```
select = [A3..0] : 'h'D;
```

The result is the equivalent of the expression:

```
select = A3 & A2 & !A1 & A0;
```

The following example shows the use of the "don't care" state:

```
select = [A3..0] : 'b'1X0X;
```

The result is the equivalent of the expression:

```
select = A3 & !A1;
```

Note that the A2 and A0 variables were eliminated by the "don't care" state, and variable A1 was complemented.

5.5 COMBINATIONAL CIRCUIT DESIGN

Though our discussion in this text has focused on using PLDs to implement state machines, it would be useful to look first at the use of PLDs to implement the functions of common digital logic circuits. State machines, though a very useful subset of digital logic circuits, are only a subset. Programmable logic devices can be programmed to execute the functions of a very wide range of digital circuitry. Usually simple logic circuits can be divided into two categories, combinational and sequential. We will look first at the implementation of some combinational logic circuits. The basic design strategy is rather straightforward:

- Design the logic circuitry.
- Select an applicable PLD.
- Write the PLD source program.
- Program the PLD.
- Test the PLD.

The first step of the strategy is considered prerequisite to this study. The logic circuit to be designed has already been determined. This study is concerned with implementation of the circuitry on the PLD. Therefore, assuming circuits have been designed, we will continue with the design process. The first example to consider is the implementation of simple combinational logic circuits.

5.5.1 Design the Logic Circuitry

Consider Figure 5.4. This figure illustrates the implementation of eight simple and common combinational logic circuits. Note that each can be represented as a summation expression and as a typographic expression. It is important that we realize that combinational arguments can always be expressed in a summation or sum of products form since we will usually be using programmable logic devices that implement SOP functions. Also, it is important to

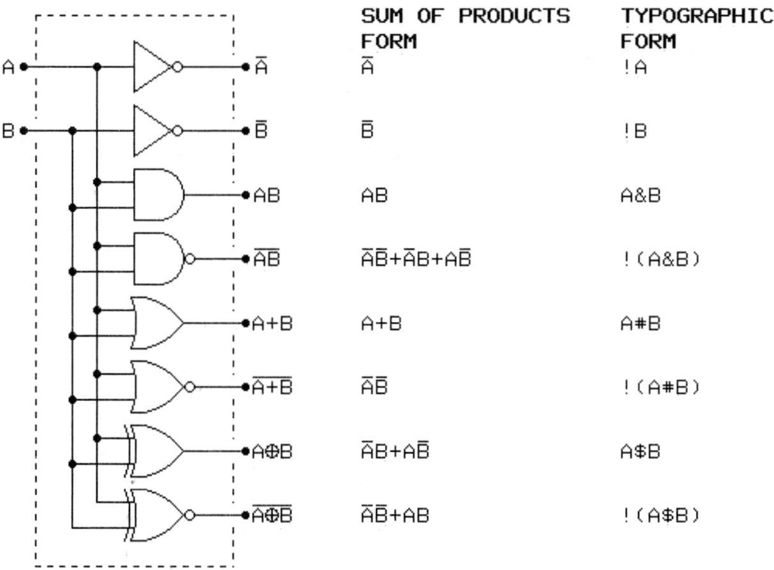

FIGURE 5.4 Combinational logic circuits.

note that these arguments can be placed in typographic form, since the source programs for the PLDs are written in CUPL, which uses this form.

5.5.2 Select the Applicable PLD

Note that this example circuit has two inputs and eight combinational outputs. The most complex combinational output has three minterms. We need at this point to locate a PLD that has enough capability to handle this task. That is, we need a PLD that has at least two inputs and eight combinational outputs. The device must also be able to provide enough minterms to implement the expressions. Recall from Chapter 1 that an L-type macrocell in a typical PAL device can be programmed to implement sum of products expressions. A PAL device with eight L-type macrocells will be able to implement the circuit of Figure 5.4.

Observing the specifications of a variety of common PAL devices, we find that the 16L8 device has eight combinational outputs and ten inputs. Consider Figure 5.5. Of the eight outputs, six can also be configured to be used as inputs. These outputs support up to seven minterms each. This device will be able to support the design we are considering.

The selection of the device for implementation requires that the circuit design be complete and its application in a programmable device be understood so that a device can be selected. A simple way of determining applicability is to determine if the target PLD has sufficient circuitry to implement the circuit design. The PALs we will observe utilize AND-OR logic to implement sum of products expressions. Combinational expressions that are not in this form may be converted to this form to ascertain the number of minterms required. However, for many applications, common programmable logic devices contain sufficient minterm generation capacity to make this step unnecessary.

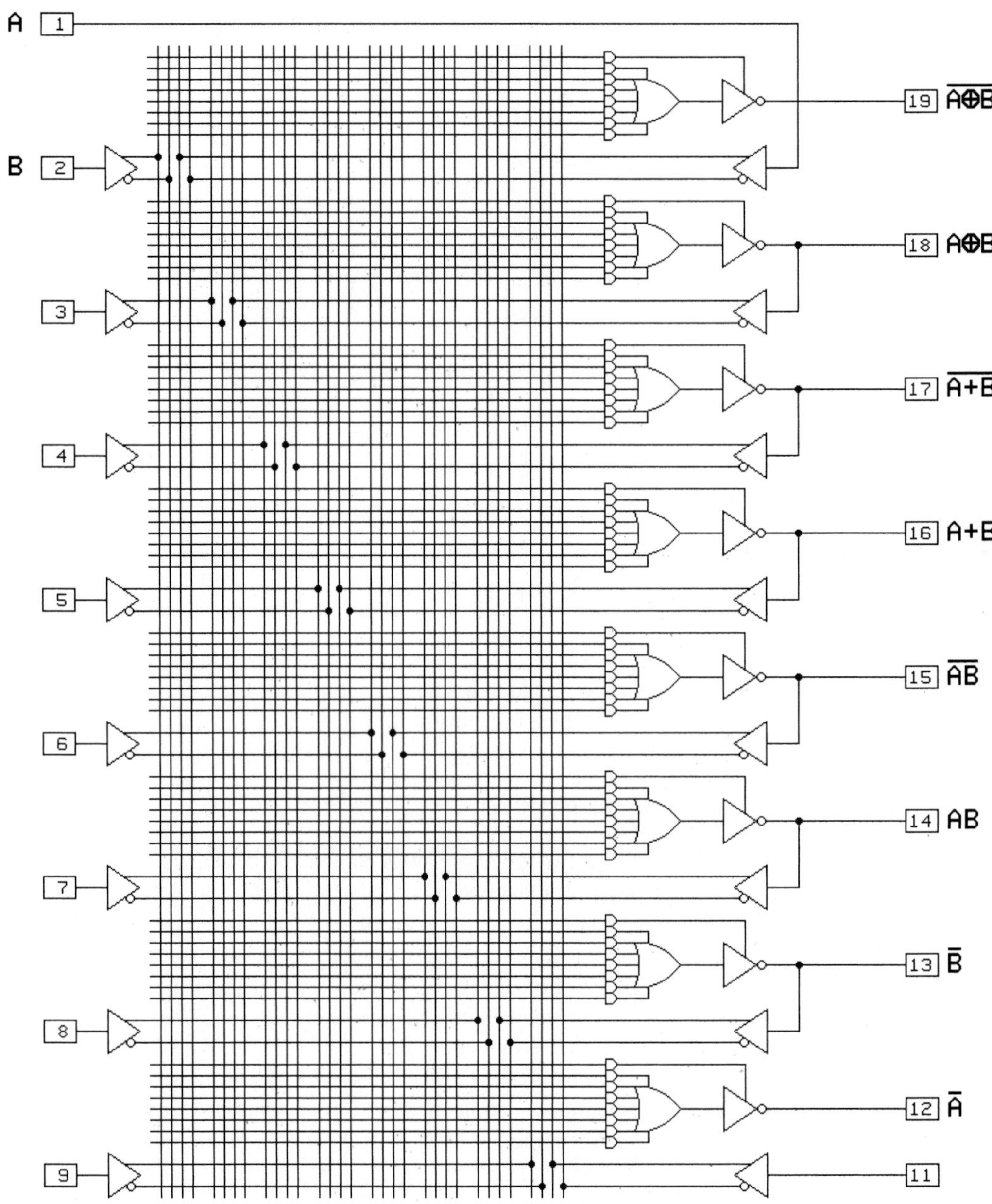

FIGURE 5.5 16L8 PAL circuit diagram with pin assignments.

When circuit designs become complex, it may be necessary to break the circuit down into parts and program multiple PLDs for implementation. In situations where a very large number of minterms is needed, we may find that we can avoid breaking down the circuit by using a ROM instead of a PAL or PLA.

Once a suitable programmable logic device has been found, the source program needed to program the device can be written. As previously stated, the program is made of three basic components, the header information, the pin assignments, and the logic equations that define the relationships among the pins. The header information is described in Section 5.4.

5.5.3 Write the PLD Source Program

Pin Declaration Statements. Following the header information, the use of the PLD pins must be declared. This step will associate the pins with the input and output variables used in the circuit. When assigned in this manner, the variable usage is defined in terms of the resources of the selected device. For example, if a 16L8 PAL is used, any variables assigned to pins 1 through 9 and pin 11 are understood by the compiler to be inputs to the circuit. Pins 12 through 19 are understood to be outputs. The assignment of pins is simply a matter of identifying those pins on the PLD that correlate to the functions of the circuit to be implemented. The inputs and outputs of the example circuit of Figure 5.4 are shown as assigned to the pins of the 16L8 device in Figure 5.5. When applied to a CUPL source program, the statements appear as follows:

```
/* Inputs: */
       Pin 1 = a;
       Pin 2 = b;
/* Outputs: */
       Pin 12 = inva;
       Pin 13 = invb;
       Pin 14 = and;
       Pin 15 = nand;
       Pin 16 = or;
       Pin 17 = nor;
       Pin 18 = xor;
       Pin 19 = xnor;
```

Note that to the right of each equals sign is the name of a variable, not a logical function.

CUPL Logic Expressions. The final section of the program is the assignment of the logical expressions that define the relationships among the assigned pins and the consequent function of the device. The current example defines eight outputs as combinational logic functions of the two inputs. This implies that there will be eight logical expressions to define them. Consider the following CUPL logic expressions:

```
inva = !a;         /* inverters */
invb = !b;
and = a & b;       /* AND gate */
nand = !(a & b);   /* NAND gate */
or = a # b;        /* OR gate */
nor = !(a # b);    /* NOR gate */
xor = a $ b;       /* EXCLUSIVE OR gate */
xnor = !(a $ b);   /* EXCLUSIVE NOR gate */
```

Each expression defines an output in terms of the respective combinational logic function of the indicated inputs. The completed program for this example is illustrated in Figure 5.2.

5.6 INTERMEDIATE VARIABLES

We find from the previous example that implementation of combinational logic is quite straightforward when the expressions are written in sum of products form. This implies that there are only two levels of combinational logic between input and output, the first being product circuitry and the second being summation circuitry. Often, the expression cannot easily be placed in this form. For a variety of reasons, there may be multiple levels of logic. When multiple levels of logic are encountered, it is necessary to use intermediate variables. These are formed by breaking down the layers of logic into sections that each can be generated by a PAL output. Recall that the logic output of a PAL or PLA is often fed back to the gate array. Using this principle, a variable generated by a logical combination of inputs can be fed back to the gate array and used as an operand in yet another layer of circuitry. This variable is referred to as an intermediate variable and requires the assignment of an output pin for implementation.

Consider Figure 5.6. Though this circuit can be implemented in a single layer of logic by performing a little Boolean algebra on it, we will use this as a simple example of an intermediate variable. To create the first layer of logic, the variable R is generated by an output of a PAL. This output is fed back to the gate array, and used in combination with variable C to create the output X. Consider the pin assignments on a 16L8 PAL device:

```
/* Inputs: */
      Pin 1 = a;
      Pin 2 = b;
      Pin 3 = c;
/* Outputs: */
      Pin 13 = r;      /* Intermediate Variable */
      Pin 12 = x;
```

The circuit diagram of the 16L8 must again be considered when the pins are assigned. It is necessary that the intermediate variable R be fed back to the gate array so that it can be used as an operand in subsequent logic circuits. Observe in Figure 5.5 that this is the case: Pin 13 of the 16L8 is a logical output that is also fed back to the gate array.

The CUPL program will now require two logical expressions in order to describe the use of the intermediate variable:

```
r = !(a $ b);
x = r & c;
```

We can see from this example that an intermediate variable is used to create multiple levels of logic. The levels are linked by using an output pin on the PLD.

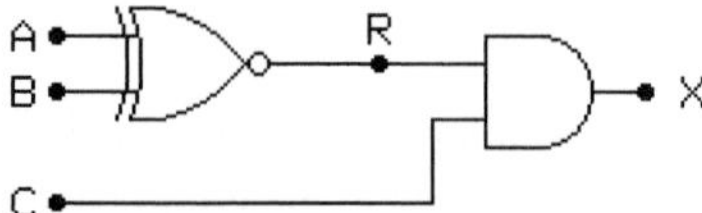

FIGURE 5.6 Circuit with intermediate variable R.

Practical Example 5.1: Parity Generator

Consider a circuit to generate a parity bit based on an eight-bit input word. A parity bit identifies whether the number of logic 1 bits in a binary word is even or odd. A binary word is considered to have even parity when there is an even number of logic 1 bits in it. A binary word is considered to have odd parity when there is an odd number of logic 1 bits in it.

Consider Figure 5.7. This circuit is most easily applied by using intermediate variables between the layers of logic. To do this, we will consider each term generated by the first layer of logic as outputs and feed them back to the gate array as operands for the second layer. The terms generated by the second layer can be fed back to the gate array to form the operands for the third layer.

Consider the assignment of intermediate variables illustrated in Figure 5.8. Variables named V0 through V5 have been assigned as intermediate variables. The 16L8 PAL will support this logic, and pins can be assigned as follows:

```
/* Inputs: */
        Pin [1..8] = [D0..D7];
/* Outputs: */
        Pin [13..18] = [V0..V5];      /* Intermediate Variables */
        Pin 12 = P;
```

Note the use of list notation in order to simplify the program. This implies that D0 is assigned to pin 1, D1 is assigned to pin 2, etc. We are now able to write the CUPL logic expressions:

```
V0 = D0 $ D1;
V1 = D2 $ D3;
V2 = D4 $ D5;
V3 = D6 $ D7;
V4 = V0 $ V1;
V5 = V2 $ V3;
P  = V4 $ V5;
```

The complete program requires header information that identifies, in addition to other things, the device to be programmed, either a P16L8 PAL or a G16L8 GAL. Recall that V-type macrocells will implement the functions of either combinational or registered outputs, so the program illustrated also applies to a P16V8 PAL or a G16V8 GAL.

FIGURE 5.7 Practical Example 5.1, eight-bit parity generator.

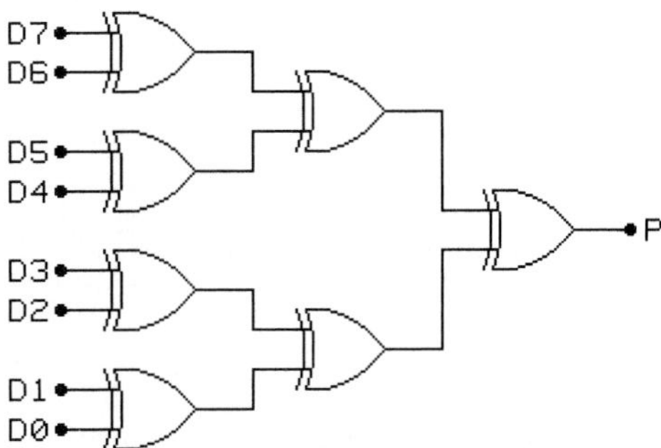

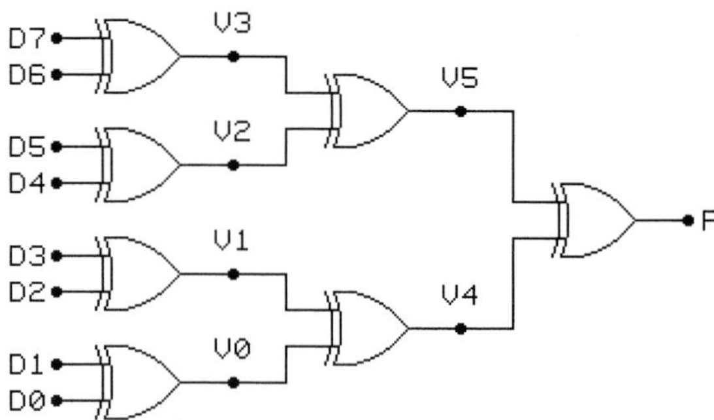

FIGURE 5.8 Practical Example 5.1, intermediate variable assignments.

5.7 SEQUENTIAL CIRCUIT DESIGN

As with combinational circuit design, sequential circuit design with PLDs is a matter of applying known circuitry to devices that provide sufficient resources. Consider the R-type macrocell that was introduced in Chapter 1.

Consider in detail the capabilities of a typical R-type macrocell as illustrated in Figure 5.9. This macrocell was taken from a 16R4 PAL device. It is essentially a D-type flip-flop with its D input driven by the gate array through a sum of products circuit. The output of the flip-flop is also fed back to the gate array to be used as operands in other macrocells. The clock input to the flip-flop is provided via an external pin and is often common to each of the macrocells. Often, the clock input comes from the gate array, making it programmable. Also, the output-enable function of the macrocell comes from an external pin and is common with other macrocells. This output-enable function is also commonly programmable, coming from the gate array instead of an external pin.

Many sequential circuits are made up of D-type flip-flops driven by a combinational logic argument. This is the case for the synchronous binary counter and the counter-type state machines we have been observing so far in this text. These circuits have been typically three- and four-bit counters with each bit stimulated by combinational logic that defines the next state stimulus. Also, output-forming logic was needed for the state machine designs, and this logic is simply combinational functions of the counter bits, with an occasional D-type flip-flop thrown in. A simple synchronous binary counter or state machine may implement well with a PAL device such as a 16R4, which is discussed and illustrated in this section.

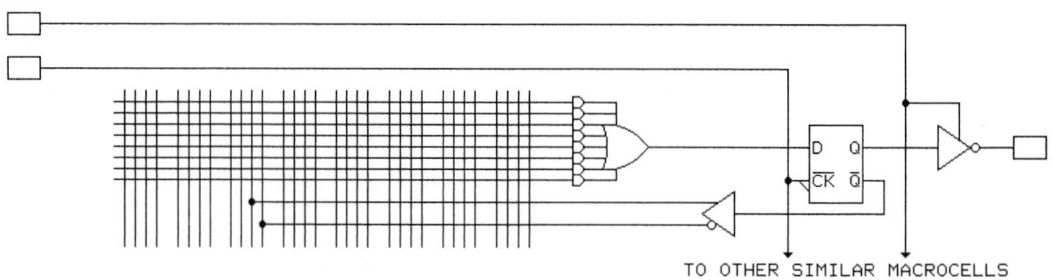

FIGURE 5.9 Typical R-type macrocell.

Consider Figure 5.10. This circuit is a three-bit binary synchronous counter. We can see from observation that its basic structure is a sequence of three D-type flip-flops, each stimulated by a combinational logic function. These functions are:

$$D_A = BC$$
$$D_B = \overline{A}B + \overline{A}C$$
$$D_C = \overline{A}B + A\overline{B} + \overline{C}$$

Note that these expressions are in sum of products form, ideal for implementation with a PAL. In order to use a PAL device, we will need one that has at least three macrocells that contain D-type flip-flops. One must support up to three minterms, another two, and another one. Each of the Q outputs of the flip-flops must be fed back to the gate array since they are used as operands in the combinational functions. The 16R4 device includes these resources. Figure 5.11 illustrates the schematic diagram of this device with the pins assigned for this problem. This device has four R-type macrocells tied to pins 14, 15, 16, and 17. Notice that the outputs of these macrocells are enabled by pin 11 and asserted low. Pins 12, 13, 18, and 19 are L-type macrocells, suitable for implementing additional combinational logic as needed, including output-forming logic if possible. They can also be used as inputs when not configured as outputs. Pins 1 through 9 are external inputs to the gate array.

Note that there are no inputs defined on the synchronous counter. When the counter is to be externally controlled, as with a typical counter-type state machine, external inputs will also be used. We'll examine this shortly. At this point we can complete the CUPL source program by assigning the pins and writing the logical expressions. Note how the variable extension .D is used to identify the D input to the flip-flops on the indicated pins.

```
/* Outputs: */
Pin 16 = A        /* Intermediate Variables */
Pin 15 = B;
Pin 14 = C;
/* Logical Expressions */
A.D = B&C;
B.D = (!A&B) # (!A&C);
C.D = (!A&B) # (A&!B) # !C;
```

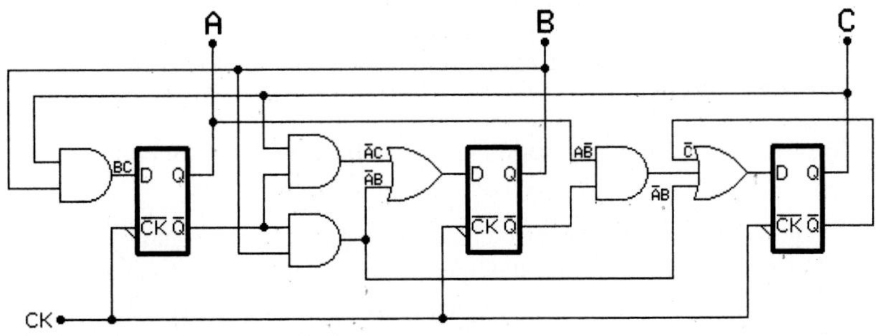

FIGURE 5.10 Example sequential circuit.

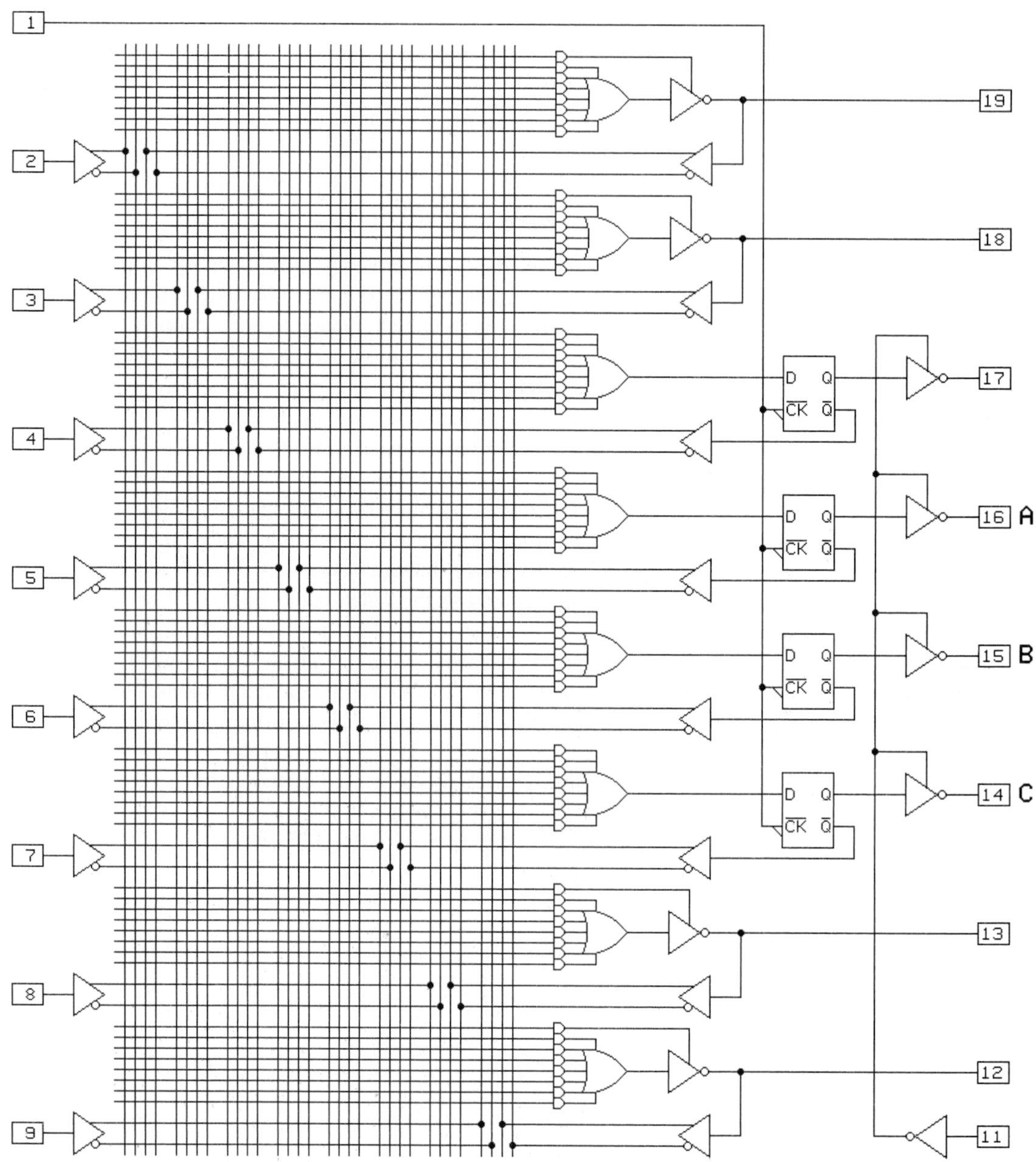

FIGURE 5.11 16R4 PAL with pin assignments for a synchronous three-bit counter.

Another way to write a CUPL program for a counter is through the use of a SEQUENCE statement. This statement can be used in lieu of the logical expressions when those expressions are defining a synchronous binary counter. SEQUENCE statements include the syntax to handle external stimulus to the counter as well as simple output-forming logic. This makes this application ideal for simple counter-type state machines.

This counter was designed to execute a repeated sequence: 1 - 2 - 3 - 7 - 4 - 1. The other three states, 0, 5, and 6 are to go to state 1 if present. Instead of calculating the logic expressions, sequence statements can be used. Consider the following CUPL program:

```
Pin 16 = A; Pin 17 = B; Pin 18 = C;
SEQUENCE [A,B,C] {
     PRESENT 'b'000 NEXT 'b'001;
     PRESENT 'b'001 NEXT 'b'010;
     PRESENT 'b'010 NEXT 'b'011;
     PRESENT 'b'011 NEXT 'b'111;
     PRESENT 'b'100 NEXT 'b'001;
     PRESENT 'b'101 NEXT 'b'001;
     PRESENT 'b'110 NEXT 'b'001;
     PRESENT 'b'111 NEXT 'b'100;
     }
```

Though the program is longer, it is simpler to write when given the counter states.

Practical Example 5.2: State Machine Design in CUPL

A complete state machine design was illustrated in Practical Example 5.1 when a problem similar to that shown in Figure 5.12 was completed. The output-forming logic has been simplified.

The circuit design produced a three-bit synchronous counter-type state machine with outputs W, X, Y, and Z, controlled by input R. The schematic diagram shown in Figure 5.13 is obtained when the method discussed in Chapter 4 is used. Study of the schematic

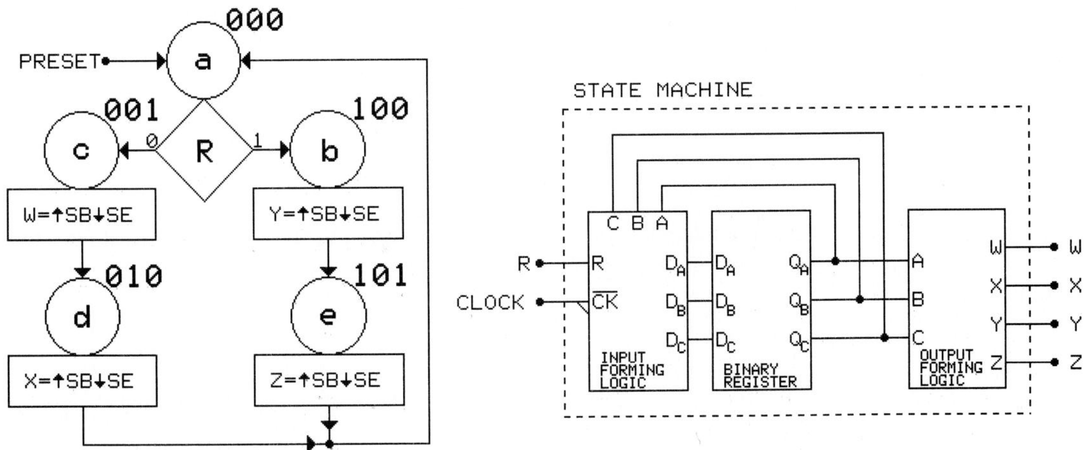

FIGURE 5.12 Practical Example 5.2, state diagram and block diagram.

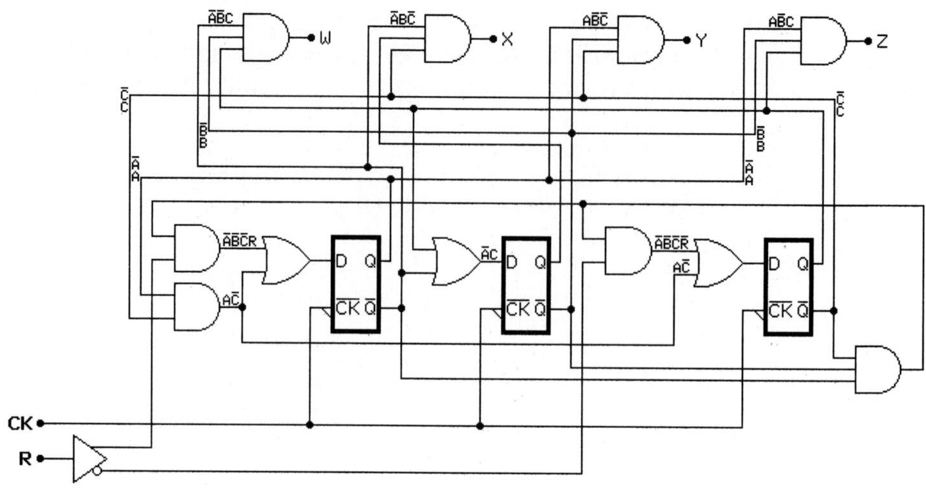

FIGURE 5.13 Practical Example 5.2, schematic diagram.

diagram of Figure 5.13 reveals that the 16R4 PAL device can be used to implement this state machine. Consider the following pin assignments:

```
Pin 1 = R;       /* Input R */
Pin 16 = A;      /* Counter Bits A, B and C */
Pin 15 = B;
Pin 14 = C;
Pin 19 = W;      /* Output Forming Logic Bits W, X, Y and Z */
Pin 18 = X;
Pin 13 = Y;
Pin 12 = Z;
```

When the logic circuitry was designed for this example, the following logic equations were realized:

$$D_A = \overline{A}\,\overline{B}\,\overline{C}R + A\overline{C}$$

$$D_B = \overline{A}C$$

$$D_C = \overline{A}\,\overline{B}\,\overline{C}\,\overline{R} + A\overline{C}$$

If it is desired to complete the logic expression portion of the CUPL program with discrete logic expressions, the following result is obtained:

```
A.D = (!A & !B & !C & R) # (A & !C);
B.D = !A & C;
C.D = (!A & !B & !C & !R) # (A & !C);
W = !A & !B & C; X = !A & B & !C; Y = A & !B & !C; Z = A & !B & C;
```

The expressions for this portion of the CUPL program were obtained using the manual method of input-forming logic and output-forming logic design discussed in

Chapter 4. As with the previous example, it is simpler to use the SEQUENCE statement in the CUPL language to define the state machine program. These statements can be used to react to the R input, and to drive the W, X, Y, and Z outputs. Consider the following complete CUPL program:

```
Name       EXP0502;
Revision   01;
Date       02/14/96;
Designer   J.W. Carter;
Company    University of North Carolina, Charlotte;
Location   None;
Assembly   None;
Device     P16R4;

/*****************************************************************/
/* Practical Example No. 5.2 Complete State Machine            */
/*****************************************************************/

Pin 1  = R;   /* Input R */
Pin 16 = A;   /* Counter Bits A, B and C */
Pin 15 = B;
Pin 14 = C;
Pin 19 = W;   /* Output Forming Logic Bits W, X, Y and Z */
Pin 18 = X;
Pin 13 = Y;
Pin 12 = Z;

/*****************************************************************/
/* State Machine Program */
/*****************************************************************/

SEQUENCE [A,B,C] {
      PRESENT 'b'000
             IF !R NEXT 'b'001;
           DEFAULT NEXT 'b'100;
    PRESENT 'b'001 NEXT 'b'010    OUT W;
    PRESENT 'b'010 NEXT 'b'000    OUT X;
    PRESENT 'b'100 NEXT 'b'101    OUT Y;
    PRESENT 'b'101 NEXT 'b'000    OUT Z; }
```

It is important to note that when the SEQUENCE statement is used to specify output-forming logic for the state machine, the only form supported enables the output during the entire clock cycle. However, this represents what is described as the product of the true state bits used to determine output-forming logic. Therefore, in order to implement the more complex transition positions for output-forming logic, the output signals from the SEQUENCE statements become the intermediate logic arguments for the discrete circuitry used to generate that output-forming logic. This example has used all of the available combinational outputs for a 16R4 device, so to implement more complex output forming logic would require a larger chip.

Practical Exercise 5.1: CUPL Program of a Combinational Circuit

Write a CUPL program that will program a 16L8 PAL device to act as a code converter that will accept a three-bit number and complement any numbers that are odd.

Practical Exercise 5.2: CUPL Program of a Sequential Circuit

Consider a synchronous binary counter that will count in the repeating sequence 0 - 2 - 4 - 5 - 7 - 3 - ... If the counter is in a state outside the sequence, cause it to go to count 0.

1. Write a CUPL program that uses discrete logic expressions to implement the counter design on a 16R4 PAL device.
2. Write a CUPL program that uses SEQUENCE statements to implement the counter design on a 16R4 PAL device.

Practical Exercise 5.3: CUPL Program of a State Machine

Consider the state machine definition illustrated in Figure 5.14.

1. Write a CUPL program to implement this state machine on a 16R4 PAL device using discrete logic expressions.
2. Write a CUPL program to implement this state machine on a 16R4 PAL device using SEQUENCE statements.

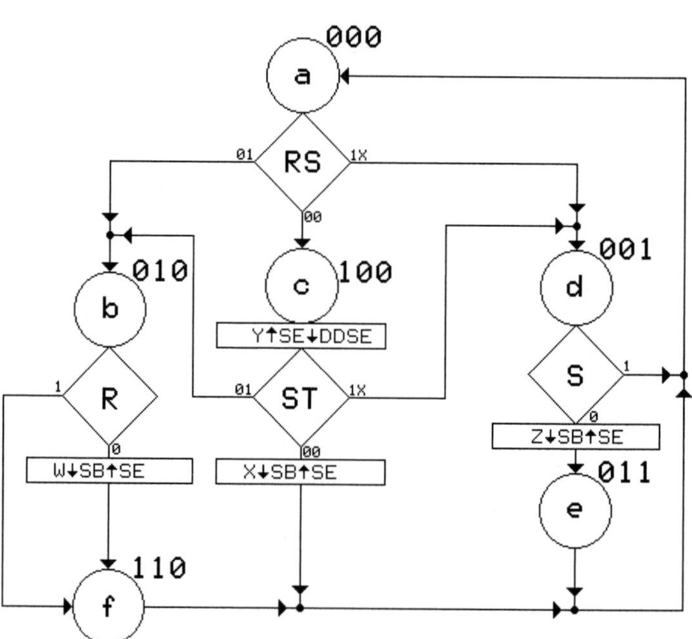

FIGURE 5.14 Practical Exercise 5.3, state diagram.

Hint: There are two new twists to this design. First, outputs W, X, and Z are active-low. To declare outputs to be asserted low, use a sequence statement similar to OUT !W, instead of OUT W. Also, output Y is delayed by a full clock cycle. You will need to use the remaining D-type flip-flop on the chip to provide the delay.

5.8 THE CUPL PROGRAMMING ENVIRONMENT

The CUPL compiler software is provided on floppy diskettes that are intended to be downloaded and run off of a hard disk using an operating system such as Microsoft Corporation's MS-DOS® or UNIX. The version used in this text runs under MS-DOS®. Some of the important files that are required in the disk directory include:

CUPL.DL	Compiler Device Library
CUPL.EXE	Compiler Main Executable (command line compiler)
CUPLA.EXE	Compiler Library
CUPLB.EXE	Compiler Library
CUPLC.EXE	Compiler Library
CUPLLOGO.OVR	Compiler Library
CUPLM.EXE	Compiler Library
CUPLX.EXE	Compiler Library
EZEDIT.EXE	Text Editor
MCONFIG.OVR	Compiler Configuration Library
MCUPL.CFG	Compiler Configuration File (provides file locations)
MCUPL.EXE	Compiler Shell Executable (menu-driven compiler)

File MCUPL.CFG can be edited to change the text editor used with the menu shell. If it is modified, the file EZEDIT.EXE is not needed. Several other files have been omitted from this list, as they are not immediately needed for our programming environment.

The software package, when purchased from Logical Devices, Inc. also includes a PLD simulator. This set of programs is referred to as the PALexpert® Package. The software can be run from the command prompt with options, or it can be run from the menu shell, MCUPL.EXE. This text will use the menu shell for the programming environment. Detailed information concerning running from the command prompt is available in the CUPL data book described in the introduction to this chapter. The command shell can be executed by entering the following at the MS-DOS command prompt:

```
C>mcupl
```

When installed and executing properly, the menu shown in Figure 5.15 should appear.

Again, we are not going to be concerned in this text with all of the functions of the CUPL environment. For example, a very useful and powerful part of the environment provided by Logical Devices, Inc. is the CUPL simulator, which allows computer-based simulation of a device after the program has been compiled. This simulator can be used to

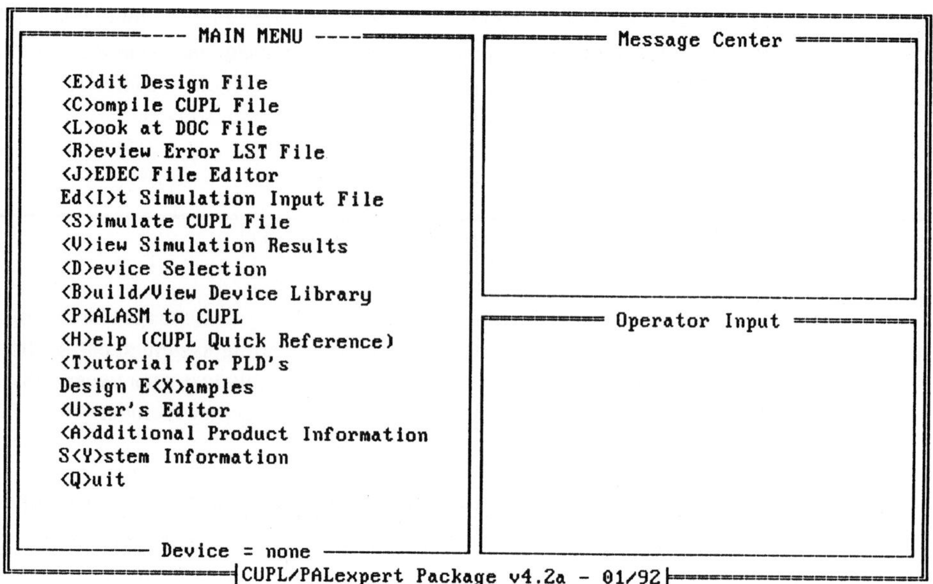

FIGURE 5.15 CUPL/PALexpert® menu shell display.

test and debug a design before committing it to production. This feature will not be addressed in this text.

Those menu functions that are common and useful to the compiling of CUPL programs into a downloadable format will be considered here. Our purpose is to take the CUPL source program and produce another program that is a text form of the binary pattern that is actually "burned" into the programmable logic device.

The menu shell illustrated in Figure 5.15 is divided into three parts: the menu command list, the message center, and the position for user input. The menu command list provides menu-oriented access to all of the functions needed to write and compile PLD source programs. Following are some of the commands that are useful for this sequence of activity:

<E>dit Design File	Calls the text editor to create or edit the PLD source file.
<C>ompile CUPL File	Compiles the PLD source file into a form the GAL programmer can use to load the PLD.
<L>ook at DOC File	Calls up the text editor and loads the DOC file generated by the compilation.
<R>eview Error LST File	Calls up the text editor and loads the LST file generated by the compilation.
<D>evice Selection	Used to select the PLD device. Overrides the device indicated in the PLD source file.
<H>elp	CUPL Quick Reference
<Q>uit	Exit to DOS.

The other commands in the menu are also useful in the overall task of developing PLD programs, but they are not necessary for simple compilation.

Once a program has been written, it is necessary to compile that program into a form that a PLD programmer (any typical EPROM, PAL, or PLA programming device) can use for the purpose of downloading from the computer to the chip. That is, the preparation of PLD programming takes two steps: text editing and compiling.

5.8.1 Text Editing

To invoke the text editor for the purposes of creating or editing a PLD source program, select the menu item entitled <E>dit Design File. This command will first prompt the user for the name of the file to be edited. Instructions are displayed in the message center portion of the screen, and the prompt is displayed in the operator input portion of the screen. A list of available files is provided in the menu portion of the screen, and one can be selected by use of the cursor keys and the enter key.

The text editor software available with the PALexpert software is a very simple program, EZEDIT.EXE. If the use of another editor is desired, the file MCUPL.CFG can be modified to provide this capability. The line in the file that contains the path and program name for EZEDIT.EXE must be modified to show the path and name of the desired editor. Once this change is made, the menu command will call the desired editor program.

The editing process is necessary to produce the source program in the CUPL language. This program must adhere to the rules and syntax of that language so that the compiler can generate the binary pattern that is to be downloaded to the PLD.

5.8.2 Compilation

The compilation step converts the ASCII source file to a form the PLD programming software can use to program the device. The PLD programmer requires the binary program to be in a specified format, typically ASCII hex, Signetics HL, or JEDEC. Consider Figure 5.16. This is the compiler menu. Some of the important commands include:

`<J>EDEC download format`	Creates a JEDEC formatted download file.
`<H>EX download format`	Creates an ASCII hex formatted download file.
`<L>isting file`	Creates an .LST file that displays compilation errors.
`<F>use plot/chip diagram`	Creates a .DOC file that displays a plot of the programmed fuses and a chip diagram

The compilation process requires two steps. First, the compiler options are set by selecting items from the CUPL compiler section of the menu of Figure 5.16. These typically include the download format and a selection of listing file. When this selection is complete, the compilation process is initiated by pressing softkey F5 on the computer keyboard.

5.8.3 Downloading

The download process depends on the hardware being used to program the PLD. The PLD programming environment will dictate the format of the file as well as the procedure for device programming. Refer to the subsequent documentation for the PLD programmer for information pertaining to the download process.

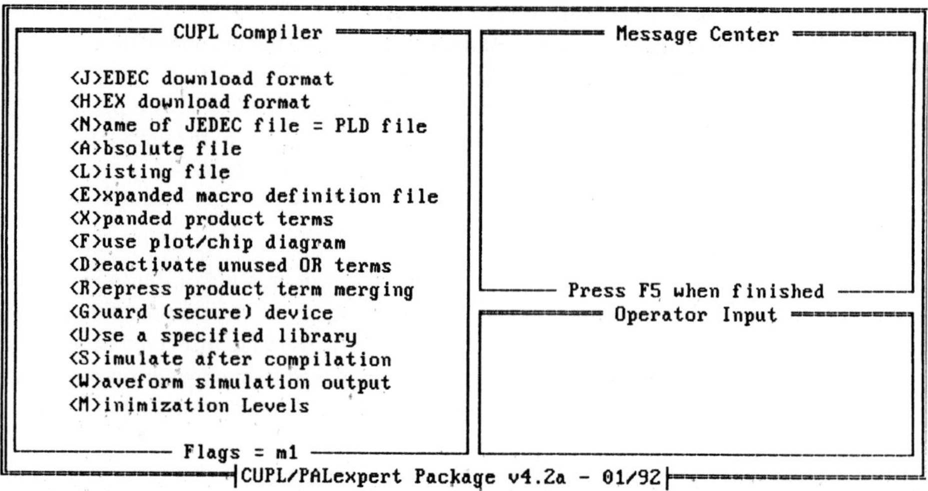

FIGURE 5.16 CUPL/PALexpert® compiler menu display.

Practical Exercise 5.4: Simple Traffic Intersection Controller

Consider the simple vehicle traffic intersection illustrated in Figure 5.17. Traffic entering the intersection from each of the four directions sees one set of traffic lights. A clock pulse is generated by the system clock every five seconds. Traffic is to flow for 15 seconds before a yellow light clears an intersection for a period of five seconds. There is no input stimulus to this design. The output signals are those logic levels necessary to drive the six

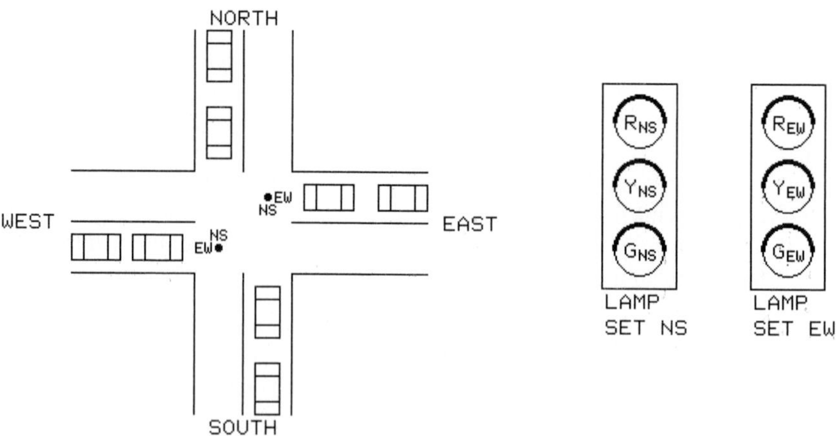

FIGURE 5.17 Practical Exercise 5.4, simple traffic intersection.

lamps indicated in Figure 5.17. Design a PAL-based controller that will provide the logic signals needed to drive the lamps.

Consider the following design approach:

- Consider each of the possible states in the system. List them, describing the active lamps in each state.
- Repeat states in order to extend the time to periods longer than five seconds.
- Design a state diagram that organizes those states into a reasonable sequence.
- Write a CUPL program for a selected PAL device.

Practical Exercise 5.5: Moderate Traffic Intersection Controller

Consider the vehicle traffic intersection illustrated in Figure 5.18. This problem differs from the previous one by adding an advanced turn lane to the northbound and southbound traffic. There is a transducer in each of the turn lanes that indicates that there is traffic waiting to turn. When this logic level is TRUE, an advanced green should be provided. Assume that the time delay between clock pulses is determined outside of the state machine.

The following is a possible scenario: Traffic is flowing in the north-south direction. When the intersection is being cleared, check the input stimulus coming from the two transducers on the east-west lanes to determine one of four possible next states. One of

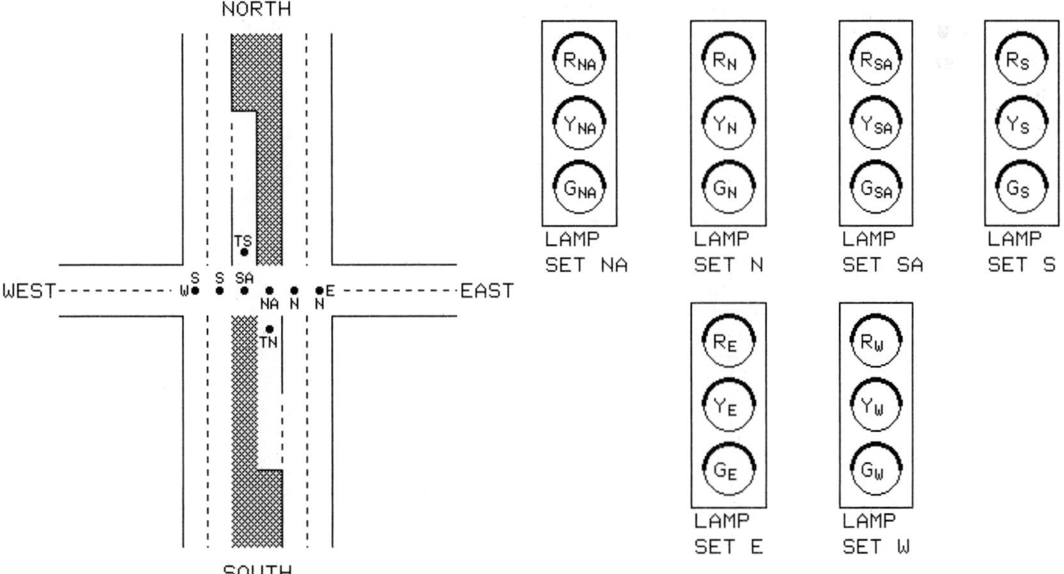

FIGURE 5.18 Practical Exercise 5.5, moderate traffic intersection.

those states, where no one is waiting, will go to the state where east-west traffic is allowed to move. If both transducers indicate waiting traffic, allow the east and west turn lanes to move. If only one of the transducers indicates waiting traffic, allow the advance movement and also enable the adjacent lane traffic to move.

Design a PAL-based state machine that accepts the two inputs and a clock and generates the logic signals needed to drive the lamps. Note that the output logic can be significantly reduced if common patterns in lamp arrangement are observed. (For example, whenever one direction has a green or yellow lamp burning, the red lamp on the opposite lane is burning.)

Include in the design:

- Problem statement
- List of possible states
- State diagram
- CUPL program

Practical Exercise 5.6: Complete Traffic Intersection Controller

Consider the intersection of Figure 5.19. This intersection is very similar to that described in Practical Example 5.5, except turn lanes have been added for the east-west traffic. Using the discussion from that example, provide a similar PAL-based state machine design.

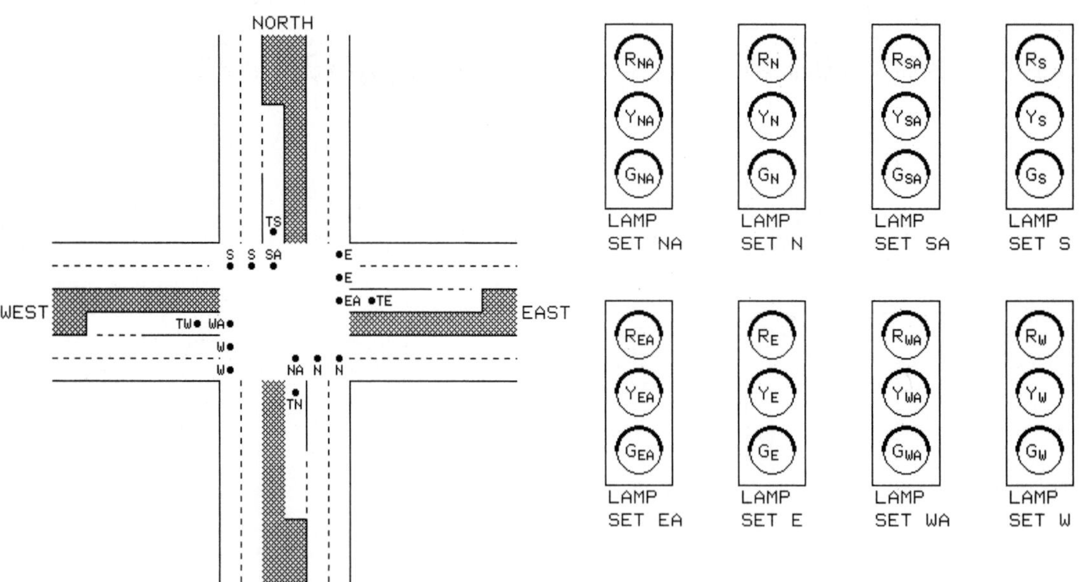

FIGURE 5.19 Practical Exercise 5.6, complete traffic intersection.

QUESTIONS AND PROBLEMS

1. Describe the sequence of steps used to program a PLD.
2. Describe the three major portions of a CUPL program.
3. Describe the header information included in a CUPL program.
4. Describe the format of a pin declaration in a CUPL program.
5. What is the purpose of the CUPL compiler?
6. How is a PLD actually programmed with a bit pattern?
7. Describe what limitations are placed on CUPL variables.
8. How are indexed variables defined in the CUPL language?
9. How are different number bases identified in the CUPL language?
10. What is CUPL list notation? Give some examples.
11. What header information is required in a CUPL program?
12. Describe the format and function of the $DEFINE preprocessor command.
13. Describe the format and function of the $INCLUDE preprocessor command.
14. What are CUPL variable extensions? What are they used for?
15. What is the function of the APPEND statement in the CUPL language?
16. Describe the format of the sequence statement in the CUPL language.
17. Describe the format of an unconditional transition statement in the CUPL language.
18. Describe the format of a conditional transition statement in the CUPL language.
19. What is the purpose and use of a DEFAULT statement in the CUPL language?
20. List the steps needed to complete a design using a programmable logic device.
21. How is an applicable PLD selected?
22. Describe the logical operators in the CUPL language.
23. What is an intermediate variable in the CUPL language?
24. How is a CUPL intermediate variable implemented on a PAL device?
25. What types of PAL macrocells will effectively implement sequential logic circuits that use D-type flip-flops?
26. How is a CUPL source program written and prepared?
27. What is the purpose of a CUPL source program?
28. What are the steps needed to prepare a CUPL source program?
29. What is the purpose of the CUPL compilation process?
30. How is a CUPL source program compiled?

CHAPTER 6

Design Examples Using PLDs

OBJECTIVES

After completing this chapter you should be able to:

- Be more familiar with the variety of applications of programmable logic devices.
- Be more familiar with the use of the CUPL language as applied in a variety of PLD applications.

6.1 OVERVIEW

Up to this point in the text, we have observed a mixture of theory and application. At times the discussion may have seemed dry, leaving the reader longing for more examples. With a good part of that theory out of the way, we are now prepared to apply the methods to a few more design examples.

6.2 SELECTED PRACTICAL EXAMPLES

These examples are selected to further illustrate the use of these design methods and the wide scope of the technology wherein these methods can be used.

It is the thesis of this author that there is a design gap between digital control circuits that require discrete digital devices and those that require microcontrollers. All too often, designers with an ignorance of PLD-based design jump to the use of a microcontroller in a digital controller design, when the less expensive programmable logic device is a better solution.

Practical Example 6.1: Pattern Recognition in a Serial Data Stream

Consider the following digital signal processing problem. A serial data stream of binary bits is transmitted from one point to another. Imbedded in that stream are specific and unique eight-bit binary patterns that provide commands to the receiving device. For this example, let's assume the receiving device operates in two modes: command mode and data mode. When in a command mode, the information received on the line is treated by the receiver as command information. When in a data mode, the information received on the line is treated by the receiver as data. The receiver toggles back and forth between these modes.

Now let's make the problem a little more complex by adding a third unique pattern in the binary stream, which is used to generate a pulse whenever it is encountered. The purpose of that pulse is to synchronize the receiver with the data stream. Those familiar with serial data communications using an RS-232 or similar protocol are familiar with the method of synchronization of the receiver using start and stop bits imbedded in the data stream. The protocol used in this example allows those bits to be eliminated, increasing the throughput of the system. Once synchronized, many characters of information can be sent without any additional data between.

Note that this protocol is not important to this problem. We simply want to detect three unique eight-bit binary patterns and use them to drive two control lines in a defined protocol.

Figure 6.1 illustrates a block diagram of this control circuit. In summary, the control circuit is to place a pulse on the SYNCH line when the serial input pattern is 11111111_2, the $C/\overline{D}$ line is to latch at a logic 1 when the serial input pattern is 11111101_2, and latch at a logic 0 when the serial input pattern is 11111110_2.

Solution Algorithm. This is an excellent candidate for a counter-type state machine that can be implemented with programmable logic using the design approaches of this text. Consider this solution algorithm: A counter will be controlled by the SI input line, testing each bit against the prescribed patterns. If at any time the serial data pattern varies from one of the three target patterns, the process starts over again.

Figure 6.2 is a state diagram of this algorithm. Observe how, in this algorithm, the first logic 1 is tested in state a. If the bit is a logic 0, the algorithm restarts. If a logic 1 is detected, the next bit received is tested in state b. This sequence continues until a pattern of 111111_2 has been detected at state f. At this point additional decision making takes place.

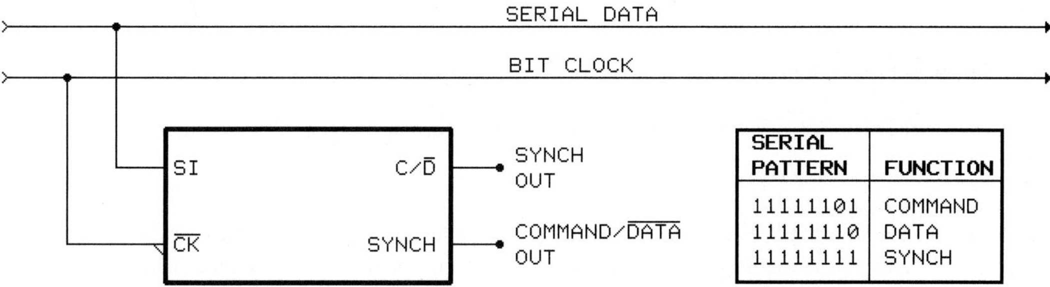

SERIAL PATTERN	FUNCTION
11111101	COMMAND
11111110	DATA
11111111	SYNCH

FIGURE 6.1 Practical Example 6.1, serial pattern recognition circuit block diagram.

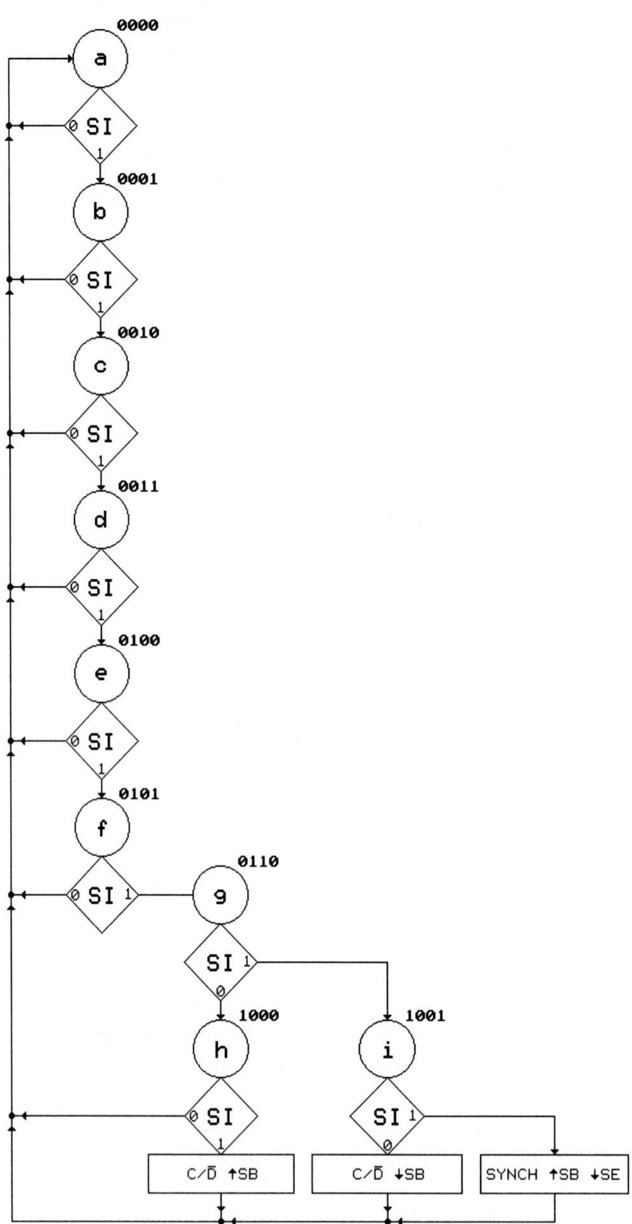

FIGURE 6.2 Practical Example 6.1, state diagram, serial pattern recognition.

If a logic 0 is detected at state g, the pattern is either 11111100_2 or 11111101_2. The condition is identified by the eighth bit tested in the sequence, in state h, where a logic 1 identifies a COMMAND state. If this is the case, the output line $C/\overline{D}$ is set to a logic 1. If state h finds a logic 0, control reverts to the beginning of the algorithm.

If a logic 1 is detected at state g, the pattern is either 11111110_2 or 11111111_2. The condition is identified by the eighth bit tested in the sequence, in state i, where a logic 1 identifies a SYNCH pulse. If this is the case, the output line SYNCH is driven high for one clock period. If state i finds a logic 0, the binary pattern for the DATA state has been found. If this is the case, the output line C/$\overline{D}$ is set to a logic 0.

Count Sequence Selection. The count sequence identified in Figure 6.2 was selected to minimize the number of logic 1 states, keeping the remainder of the count as close to a binary count as possible. For this reason, the count value of 111_2 was skipped.

Device Selection. This circuit can be implemented using a controlled binary counter with the SI input providing count sequence control. If a PAL device is used, this would be the only input to that device other than the clock. Two combinational outputs are defined. The SYNCH output is pulsed during its active state. However, the C/$\overline{D}$ output uses a Set-Reset type flip-flop. We will use two combinational outputs configured as a Set-Reset flip-flop circuit to implement this. Added to these three outputs is the necessity of having four registered outputs assigned to the four-bit binary counter. Therefore, we need a device that has at least one input and seven outputs, where three outputs are combinational and four outputs are registered. A 16R4 device will fill the bill nicely. Figure 6.3 illustrates a 16R4 PAL configured in this manner.

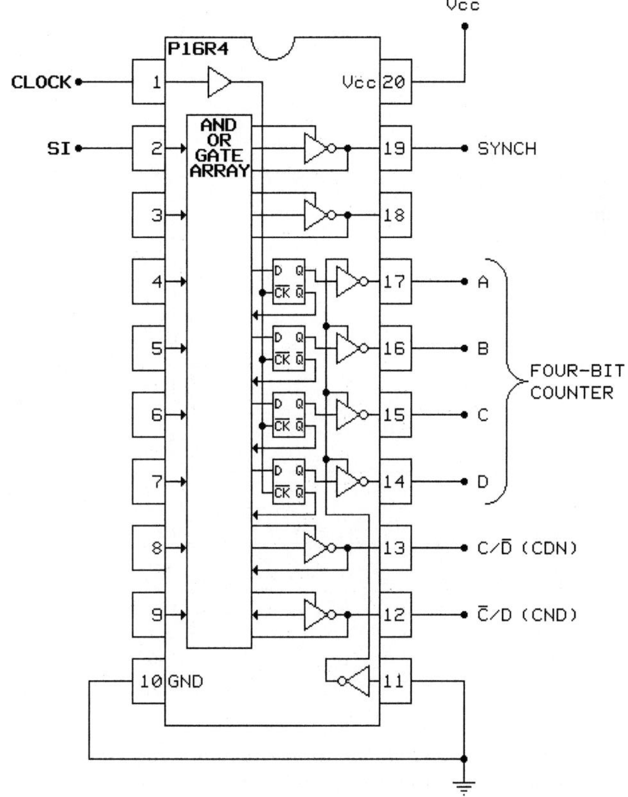

FIGURE 6.3 Practical Example 6.1, 16R4 PAL implementation of serial pattern recognition circuit.

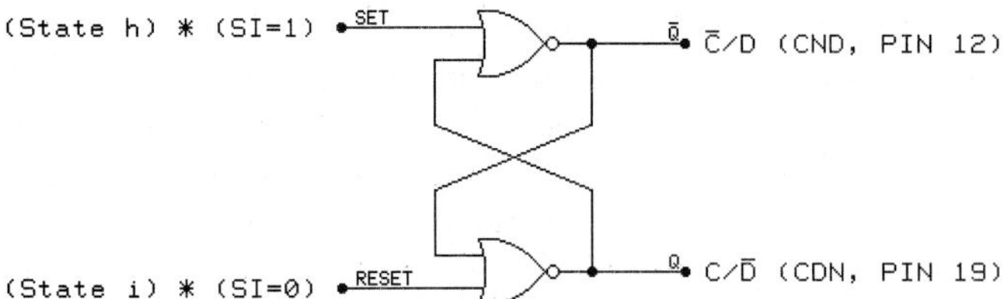

FIGURE 6.4 Practical Example 6.1, two 16R4 combinational outputs configured as a Set-Reset flip-flop.

Since the two combinational outputs are being configured as a Set-Reset flip-flop, they are labeled differently in order to distinguish their function. Figure 6.4 illustrates the use of these two outputs in a Set-Reset flip-flop configuration.

Observation of the state diagram of Figure 6.2 reveals that the C/$\overline{D}$ line is set in state h when SI = 1. The C/$\overline{D}$ line is reset in state i when SI = 0. Note how this is shown in Figure 6.4. The labels CND and CDN will be used to distinguish the two output pins.

Recall the two ways we have been solving this type of counter-type state machine problem. One method is to use a truth table based on the state diagram. From this truth table, the input-forming logic can be determined using Karnaugh map minimization methods. Figure 6.5 illustrates the truth table, Karnaugh maps, and input-forming logic expressions realized from the state diagram of Figure 6.2.

As a review of the previous chapters, observe how the information in Figure 6.5 was developed. The sequence identified by the state diagram is first used to define the input-forming logic. The unused states, 7, 10, 11, 12, 13, 14, and 15, each have a next assigned state of 0.

	ABCD	SI	ABCD	DA	DB	DC	DD	CDN	SYNCH
a	0000	0	0000	0	0	0	SI		
		1	0001						
b	0001	0	0000	0	0	SI	0		
		1	0010						
c	0010	0	0000	0	0	SI	SI		
		1	0011						
d	0011	0	0000	0	SI	0	0		
		1	0100						
e	0100	0	0000	0	SI	0	SI		
		1	0101						
f	0101	0	0000	0	SI	SI	0		
		1	0110						
g	0110	0	1000	1	0	0	SI		
		1	1001						
	0111	X	0000	0	0	0	0		
h	1000	0	0000	0	0	0	0		
		1	0000					↑SB	
i	1001	0	0000	0	0	0	0	↓SB	
		1	0000						↑SB ↓SE
	1010	X	0000	0	0	0	0	Note: Unused states go to State a to avoid hanging the counter.	
	1011	X	0000	0	0	0	0		
	1100	X	0000	0	0	0	0		
	1101	X	0000	0	0	0	0		
	1110	X	0000	0	0	0	0		
	1111	X	0000	0	0	0	0		

DA

0: 0	4: 0	12: 0	8: 0
1: 0	5: 0	13: 0	9: 0
3: 0	7: 0	15: 0	11: 0
2: 0	6: 1	14: 0	10: 0

DB

0: 0	4: SI	12: 0	8: 0
1: 0	5: SI	13: 0	9: 0
3: SI	7: 0	15: 0	11: 0
2: 0	6: 0	14: 0	10: 0

DC

0: 0	4: 0	12: 0	8: 0
1: SI	5: SI	13: 0	9: 0
3: 0	7: 0	15: 0	11: 0
2: SI	6: 0	14: 0	10: 0

DD

0: SI	4: SI	12: 0	8: 0
1: 0	5: 0	13: 0	9: 0
3: 0	7: 0	15: 0	11: 0
2: SI	6: 0	14: 0	10: 0

DA = $\overline{A}BC\overline{D}$ DB = $\overline{A}B\overline{C}\bullet SI + \overline{A}\overline{B}CD\bullet SI$ DC = $\overline{A}\overline{C}D\bullet SI + \overline{A}\overline{B}C\overline{D}\bullet SI$ DD = $\overline{A}\overline{C}\overline{D}\bullet SI + \overline{A}\overline{B}\overline{D}\bullet SI$

FIGURE 6.5 Practical Example 6.1, state table solution of serial pattern recognition circuit.

This is done so that if the counter ever falls into one of these states, it will not hang up, but rather it will return to the first state. Karnaugh maps are then used to minimize each of the input-forming logic functions. Finally, the output-forming logic is identified in the last two columns. Armed with this information, we can now write the Boolean functions for both the input- and output-forming logic. The Set-Reset circuit of Figure 6.4 is used to implement the C/$\overline{\text{D}}$ output. When this is done, the following CUPL program can be written. (Note that some of these programs may not fall on a single page.)

```
Name      EXP0601A;
Date      01/01/96;
Revision  01;
Designer  J. Carter;
Device    P16R4;
Format    j;
/*****************************************************************/
/* SERIAL PATTERN RECOGNITION CIRCUIT                             */
/* Example 6.1A  Using counter input logic functions.             */
/*****************************************************************/
Pin 2       = SI        ;       /* Serial Input, SI               */
Pin 19      = SYNCH     ;       /* SYNCH output                   */
Pin 17      = A         ;       /* Four-Bit Counter               */
Pin 16      = B         ;       /* "                              */
Pin 15      = C         ;       /* "                              */
Pin 14      = D         ;       /* "                              */
Pin 13      = CDN       ;       /* Command/Data Output            */
Pin 12      = CND       ;       /* Complement of Pin 13           */
/* Input Forming Logic */
 A.D = (!A &  B &  C & !D);
 B.D = (!A &  B & !C & SI)  # (!A & !B &  C &  D & SI);
 C.D = (!A & !C &  D & SI)  # (!A & !B &  C & !D & SI);
 D.D = (!A & !C & !D & SI)  # (!A & !B & !D & SI);
/* Output Forming Logic */
 CDN = !((!SI & A & !B & !C &  D) # CND);
 CND = !(( SI & A & !B & !C & !D) # CDN);
 SYNCH =   SI & A & !B & !C & D;
```

The second development method uses the SEQUENCE statement in the CUPL language to implement the input-forming logic. This simpler approach eliminates the need for developing the state table and Karnaugh maps, since the program can be written from the information provided in the state diagram. The C/$\overline{\text{D}}$ output line still uses the Set-Reset flip-flop circuit of Figure 6.4. The SYNCH output is true during state i when the SI input is TRUE. Consequently, the OUT statement in the CUPL language can be used to define this during state i. When the state diagram is translated to a CUPL program, the result is as follows:

```
Name      EXP0601B;
Date      01/01/96;
Revision  01;
Designer  J. Carter;
Device    P16R4;
Format    j;
```

```
/*****************************************************************/
/* SERIAL PATTERN RECOGNITION CIRCUIT                            */
/* Example 6.1B  Using counter sequence statements.              */
/*****************************************************************/
Pin 2       = SI      ;     /* Serial Input, SI                  */
Pin 19      = SYNCH   ;     /* SYNCH output                      */
Pin 17      = A       ;     /* Four-Bit Counter                  */
Pin 16      = B       ;     /* "                                 */
Pin 15      = C       ;     /* "                                 */
Pin 14      = D       ;     /* "                                 */
Pin 13      = CDN     ;     /* Command/Data Output               */
Pin 12      = CND     ;     /* Complement of Pin 13              */
sequence[A,B,C,D] {
   present 'b'0000
             if SI next 'b'0001;   default next 'b'0000;
   present 'b'0001
             if SI next 'b'0010;   default next 'b'0000;
   present 'b'0010
             if SI next 'b'0011;   default next 'b'0000;
   present 'b'0011
             if SI next 'b'0100;   default next 'b'0000;
   present 'b'0100
             if SI next 'b'0101;   default next 'b'0000;
   present 'b'0101
             if SI next 'b'0110;   default next 'b'0000;
   present 'b'0110
             if SI next 'b'1001;   default next 'b'1000;
   present 'b'0111 next 'b'0000;
   present 'b'1000 next 'b'0000;
   present 'b'1001
             if SI next 'b'0000 OUT SYNCH; default next 'b'0000 ;
   present 'b'1010 next 'b'0000;
   present 'b'1011 next 'b'0000;
   present 'b'1100 next 'b'0000;
   present 'b'1101 next 'b'0000;
   present 'b'1110 next 'b'0000;
   present 'b'1111 next 'b'0000;
   }
 CDN = !((!SI & A & !B & !C &  D) # CND);
 CND = !(( SI & A & !B & !C & !D) # CDN);
```

Practical Example 6.2: Sorting Objects in an Assembly Line

This example is similar to Practical Example 6.1 in concept, but explores an entirely different application. Consider a situation where objects are being manufactured on an assembly line. Each object has a bar code that identifies it. At some point in the assembly line, these objects must be sent to different destinations according to the bar codes.

This sorting operation could take place anywhere along an assembly line, but to make our example more practical, let's assume it happens at the end of the assembly line and determines the customer to whom the product will be shipped. Our assembly line looks like a single-line queue, much like motor vehicles in line at an intersection. In this case, as each product enters the intersection, a rotary carousel upon which the product rests rotates to deliver any of six different products to one of three destinations. Figure 6.6 is a sketch of the layout of this system.

The table sensor provides a falling edge on the CLOCK line when it is time to turn the carousel. The bar code reader input, S, is a logic 0 in its inactive state. Each of six products has a unique four-bit bar code that identifies it. The code is indicated in the table in Figure 6.6. The first digit of the bar code is always a logic 1. Therefore, based on the bar code, the states of output lines X and Y are set to the binary value of the destination number.

We can observe from Figure 6.6 that there is a physical distance between the bar code reader and the carousel. Ideally, the bar code reader would be on the carousel, but in this example it is placed on the conveyor so that we can explore an additional complication. We must delay the state of the X and Y control lines by a clock cycle. A one-cycle delay can be generated by running each signal through a D-type flip-flop with its clock tied to the same clock as the counter. We will simply add this feature at the end of the controller design sequence.

The design of the state machine is centered around the data coming from the bar code reader. Figure 6.7 illustrates the state diagram of this controlled counter.

To observe how this state diagram was developed, recall the data values coming from the bar code reader. When there is no object in front of the reader, it outputs a logic 0, causing the state machine to remain in state a. When an object passes in front of the bar code reader, the first bit of the product code, a logic 1, moves the state machine to state b. The next three bits of the product code move the state machine to the state that will provide the proper carousel position, state e, state f, or state g. In each of these states, the output-forming logic lines X and Y are defined, turning the carousel to the proper position as indicated by the table in Figure 6.6.

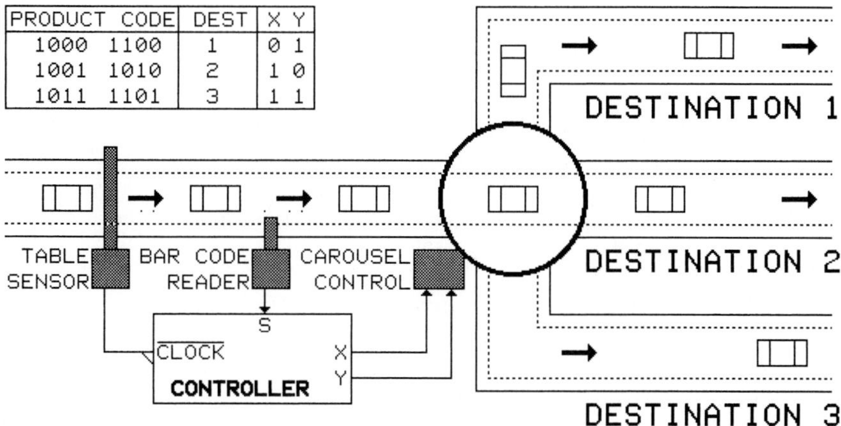

FIGURE 6.6 Practical Example 6.2, assembly line.

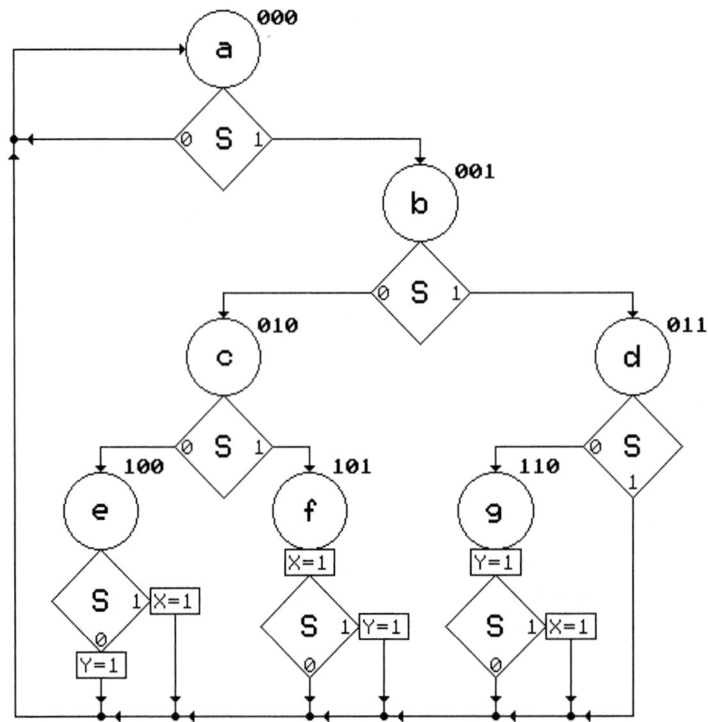

FIGURE 6.7 Practical Example 6.2, state diagram, assembly line carousel controller.

Again, we can investigate two design methods, the use of a truth table or the application of the SEQUENCE statement in the CUPL language. Let's first observe the truth table solution illustrated in Figure 6.8.

Device Selection. This controller will have one input, S, two combinational outputs, X and Y, and two delayed outputs, X′ and Y′. It will also require a three-bit counter with bits A, B, and C. Therefore, we need a device that has two combinational and five registered outputs. A 16R6 PAL will do the job. Figure 6.9 illustrates a 16R6 configured to be used as this assembly line carousel controller.

We now have sufficient information to write the CUPL program using the logic equations from Figure 6.8 for the counter input-forming logic and the output-forming logic that drives the carousel. Note that we have added a single clock cycle delay to the X and Y outputs by driving them into two D-type flip-flops and creating a second delayed output set, XP and YP.

```
Name      EXP0602A;
Date      01/01/96;
Revision  01;
Designer  J. Carter;
Device    P16R6;
Format    j;
```

	ABC	S	ABC	DA	DB	DC	X	Y
a	000	0	000	0	0	S	0	0
		1	001					
b	001	0	010	0	1	S	0	0
		1	011					
c	010	0	100	1	0	S	0	0
		1	101					
d	011	0	110	$\bar{S}$	$\bar{S}$	0	0	0
		1	000					
e	100	0	000	0	0	0	0 =S	1 =$\bar{S}$
		1	000				1	0
f	101	0	000	0	0	0	1	0 =S
		1	000				1	1
g	110	0	000	0	0	0	0 =S	1
		1	000				1	1
	111	X	000	0	0	0	0	0

DA

	←	A	→	
	0	2	6	4
↑	0	1	0	0
C	1	3	7	5
↓	0	$\bar{S}$	0	0

← B →

DB

	←	A	→	
	0	2	6	4
↑	0	0	0	0
C	1	3	7	5
↓	1	$\bar{S}$	0	0

← B →

DC

	←	A	→	
	0	2	6	4
↑	S	S	0	0
C	1	3	7	5
↓	S	0	0	0

← B →

X

	←	A	→	
	0	2	6	4
↑	0	0	S	S
C	1	3	7	5
↓	0	0	0	1

← B →

Y

	←	A	→	
	0	2	6	4
↑	0	0	1	$\bar{S}$
C	1	3	7	5
↓	0	0	0	S

← B →

$$DA = \bar{A}B\bar{C} + \bar{A}BC\bar{S}$$
$$DB = \bar{A}BC\bar{S} + \bar{A}\bar{B}C$$
$$DC = \bar{A}\bar{C}S + \bar{A}\bar{B}S$$
$$X = A\bar{C}S + A\bar{B}C$$
$$Y = AB\bar{C} + A\bar{B}\bar{C}\bar{S} + A\bar{B}CS$$

FIGURE 6.8 Practical Example 6.2, truth table solution, assembly line carousel controller.

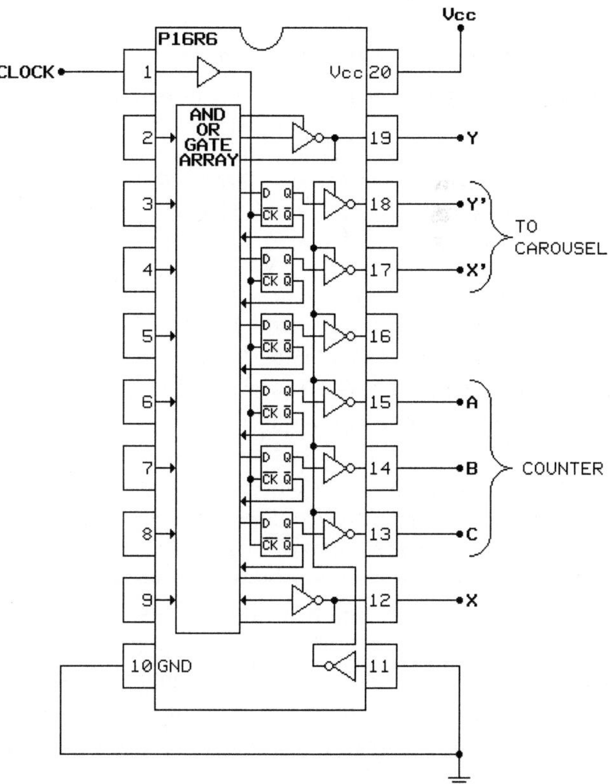

FIGURE 6.9 Practical Example 6.2, P16R6 PAL configured for the assembly line carousel controller.

```
/*************************************************************/
/* ASSEMBLY LINE CAROUSEL CONTROLLER                         */
/* Example 6.2A Using counter input logic functions.         */
/*************************************************************/
Pin 2      = S     ;      /* Serial Input from Bar Code     */
Pin 18     = YP    ;      /* X Output to Carousel           */
Pin 17     = XP    ;      /* Y Output to Carousel           */
Pin 19     = Y     ;      /* Intermediate Y Calculation     */
Pin 12     = X     ;      /* Intermediate X Calculation     */
Pin 15     = A     ;      /* Three-Bit Counter              */
Pin 14     = B     ;      /* "                              */
Pin 13     = C     ;      /* "                              */
/* Input Forming Logic */
 A.D = (!A &  B & !C) # (!A &  B & C & !S);
 B.D = (!A & !B & C)  # (!A &  B & C & !S);
 C.D = (!A & !C & S)  # (!A & !B & S);
/* Output Forming Logic */
 X = (A & !C &  S) # (!A & !B &  C);
 Y = (A &  B & !C) # ( A & !B & !C & !S) # (A & !B & C & S);
 XP.D = X;
 YP.D = Y;
```

The second form of problem solution uses the SEQUENCE statement in CUPL to eliminate the need for the state table and Karnaugh maps. The CUPL header and pin information on the CUPL program is the same as above. The remainder of the program is replaced with:

```
sequence[A,B,C] {
 present 'b'000 if S next 'b'001; default next 'b'000;
 present 'b'001 if S next 'b'011; default next 'b'010;
 present 'b'010 if S next 'b'101; default next 'b'100;
 present 'b'011 if S next 'b'000; default next 'b'110;
 present 'b'100 if S next 'b'101 OUT X; default next 'b'000 OUT Y;
 present 'b'101 next 'b'000 OUT X; if S OUT Y;
 present 'b'110 next 'b'000 OUT Y; if S OUT X;
 present 'b'111 next 'b'000;
 }
XP.D = X;
YP.D = Y;
```

Practical Example 6.3: Glitch-Free Three-Bit Up-Down Binary Counter

Consider a three-bit glitch-free count sequence. Such a sequence can be determined by linking adjacent cells on a "next state" map. Figure 6.10 illustrates such a map.

This figure illustrates that the count-up sequence repeats the binary values:

0, 1, 3, 2, 6, 7, 5, 4, ...

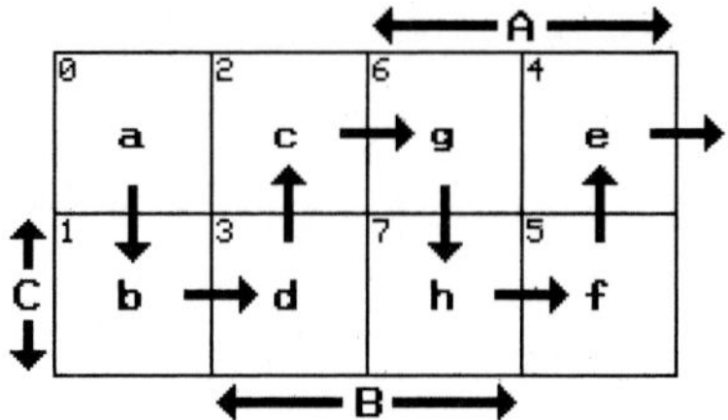

FIGURE 6.10 Practical Example 6.3, next state map for glitch-free counter.

It is evident by inspection that this is a glitch-free sequence. That is, as the sequence goes from one binary pattern to another, only one bit changes state at each transition. The count-down sequence would repeat the binary values:

4, 5, 7, 6, 2, 3, 1, 0

Figure 6.11 illustrates a functional diagram of the counter to be designed.

As with the previous examples, we will observe two methods of solution. The first uses a state table and Karnaugh maps to reduce the input-forming logic for the counter. Note that this circuit has no defined output-forming logic since it is itself only a counter. Figure 6.12 illustrates this solution.

Device Selection. This counter requires only three registered outputs from the selected device. Because it has four registered outputs, we will use the 16R4 PAL chip. Figure 6.13 illustrates a configuration of the 16R4 PAL for this purpose.

Using the information from this development process, the following CUPL program is generated:

```
Name       EXP0603A;
Date       01/01/96;
Revision   01;
Designer   J. Carter;
Device     P16R4;
Format     j;
/*****************************************************************/
/* GLITCH-FREE THREE-BIT UP/DOWN BINARY COUNTER                  */
/* Example 6.3A  Using counter input logic functions.            */
/*****************************************************************/
Pin 2       = UDN       ;     /* UP/DOWN' (1=UP, 0=DOWN)        */
Pin 16      = A         ;     /* Three-Bit Counter              */
```

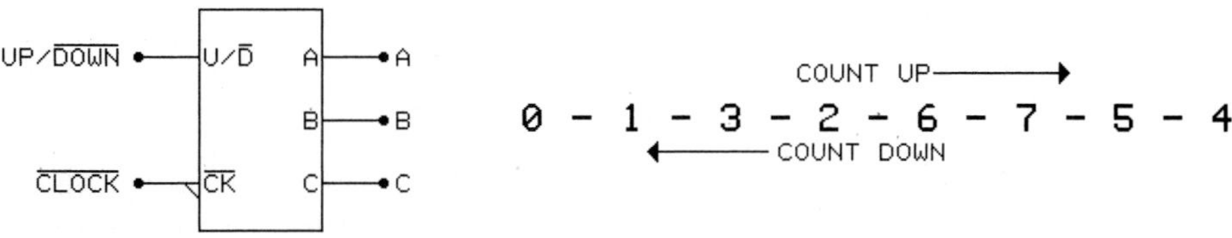

FIGURE 6.11 Practical Example 6.3, functional diagram of a glitch-free up-down binary counter.

	ABC	U	ABC	DA	DB	DC
a	000	0	100	$\bar{U}$	0	U
		1	001			
b	001	0	000	0	U	U
		1	011			
c	010	0	011	U	1	$\bar{U}$
		1	110			
d	011	0	001	0	U	$\bar{U}$
		1	010			
e	100	0	101	$\bar{U}$	0	$\bar{U}$
		1	000			
f	101	0	111	1	$\bar{U}$	$\bar{U}$
		1	100			
g	110	0	010	U	1	U
		1	111			
h	111	0	110	1	$\bar{U}$	U
		1	101			

DA

	0	2	6	4
C=0	$\bar{U}$	U	U	$\bar{U}$
	1	3	7	5
C=1	0	0	1	1

DC

	0	2	6	4
C=0	U	$\bar{U}$	$\bar{U}$	U
	1	3	7	5
C=1	U	$\bar{U}$	$\bar{U}$	U

DB

	0	2	6	4
C=0	0	1	1	0
	1	3	7	5
C=1	U	U	$\bar{U}$	$\bar{U}$

$$DA = \bar{B}\bar{C}\bar{U} + B\bar{C}U + AC$$

$$DB = \bar{A}CU + AC\bar{U} + B\bar{C}$$

$$DC = B\bar{U} + \bar{B}U$$

FIGURE 6.12 Practical Example 6.3, state table solution for glitch-free up-down counter.

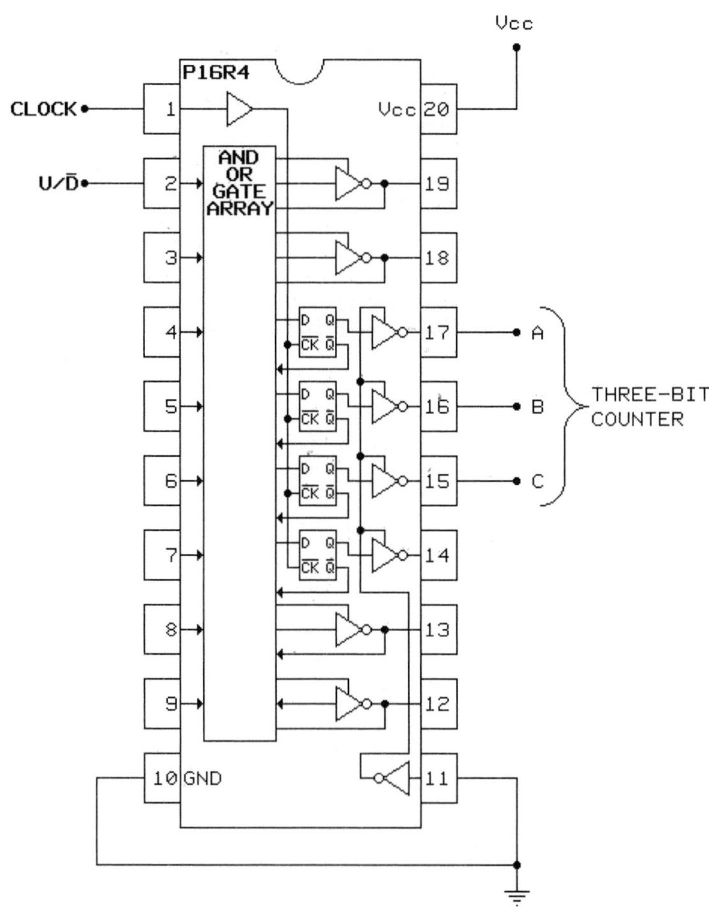

FIGURE 6.13 Practical Example 6.3, 16R4 PAL configured as a three-bit up-down counter.

```
Pin 15      = B          ;     /* "                                   */
Pin 14      = C          ;     /* "                                   */
/* Input Forming Logic */
 A.D = (!B & !C & !U) # (B & !C &  U) # (A &  C);
 B.D = (!A &  C &  U) # (A &  C & !U) # (B & !C);
 C.D = ( B & !U) # (!B & U);
```

If the SEQUENCE statement in CUPL is used instead of the state table, the following code will replace the input-forming logic portion of the previous program:

```
sequence[A,B,C] {
   present 'b'000 if UDN next 'b'001; default next 'b'100;
   present 'b'001 if UDN next 'b'011; default next 'b'000;
   present 'b'010 if UDN next 'b'110; default next 'b'011;
   present 'b'011 if UDN next 'b'010; default next 'b'001;
   present 'b'100 if UDN next 'b'101; default next 'b'101;
   present 'b'101 if UDN next 'b'100; default next 'b'111;
   present 'b'110 if UDN next 'b'111; default next 'b'010;
   present 'b'111 if UDN next 'b'101; default next 'b'110;
   }
```

Note that no state diagram was drawn in the development of this problem. Figure 6.11 illustrates sufficient information to take the place of the state diagram since the count sequence is shown so simply.

Practical Example 6.4: PM-Type Stepper Motor Controller

Another excellent application of a state machine is a stepper motor controller. Figure 6.14 illustrates the schematic diagram and timing diagram of an example stepper motor control circuit.

A stepper motor shaft rotates in a discrete number of stable steps for each rotation. For example, if the specifications for a given stepper motor indicate that it has 48 stable steps per revolution, then each time its coils receive a proper step stimulus, the shaft will rotate 7.5°. The timing diagram of Figure 6.14 relates the stimulus requirements to the schematic diagram. This diagram illustrates the turning of the stepper motor in both full- and half-step forms. The coil stimulus must be initialized in one of the illustrated stable states. To rotate the shaft one full step clockwise, the stimulus must move two states to the right. To rotate the shaft one full step counterclockwise, the stimulus must move two states to the left. Half-step movements require similar movements to adjacent states. The timing diagram is continuous, such that to the right of state h is state a and to the left of state a is state h. It is interesting to note that the torque available from the stepper motor is greater when rotation is done in full steps.

Note that the controller that is to provide this timing activity has three control inputs, STEP, CW, and FULL. The step input is the motor clock. A falling edge on this input will step the controller to the next state. The CW input identifies the direction of rotation. If CW = 1, the rotation is clockwise. If CW = 0, the rotation is counterclockwise. The FULL input determines if the rotation is to be a full- or a half-step movement. If FULL = 1, the shaft is to rotate a full step. If FULL = 0, the shaft is to rotate a half step.

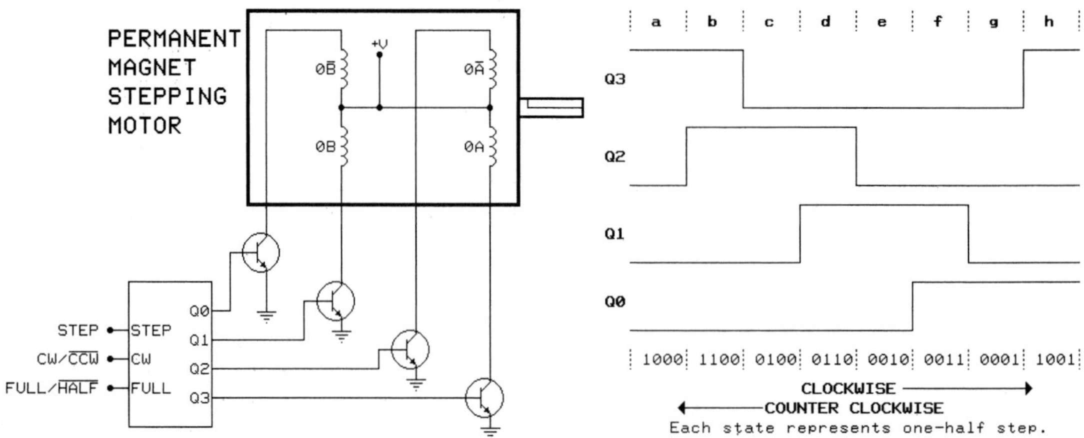

FIGURE 6.14 Practical Example 6.4, stepper motor controller schematic and timing diagram.

Development Method. The first step in the development is to draw a state diagram that illustrates the relationship between the states and the input stimuli. Figure 6.15 is this state diagram. This is an example of an instance where it may be simpler to use the timing diagram to create a state table or CUPL program than it is to use the state diagram.

Again, we will use the two approaches to problem solution. This is the first example where the different methods produce different results. The first method, using a state table, requires the development of the state table from either the timing diagram or the state diagram. When this is done, the solution of Figure 6.16 is realized. Note that the column labels for the binary state numbers are labeled A, B, C, and E. The label D is skipped, since the name D is reserved in the CUPL language. It is used as an extension for identifying the D input of a D-type flip-flop.

Note also, that the unused states 0101_2, 0111_2, 1010_2, 1011_2, 1101_2, 1110_2, and 1111_2 each go to state 0000_2. State 0000_2 goes to state 0001_2. Therefore, if an illegal state is entered, it will take two step pulses to get back on track. The decision to use this method was made to simplify the logic. If each illegal state went to state 0001_2, the DE expression would be too complex.

Device Selection. From the state table solution of Figure 6.16, we find that we will be using a four-bit feedback counter to create the timing waveforms of Figure 6.14. Since there are four bits in the counter, we can use a PAL device that has at least four registered outputs. Also, we may want to consider the number of minterms that each output expression may require. In this example, there are five minterms in each expression, so the registered outputs must be able to sum at least five minterms. It would appear that the 16R4 PAL chip will work fine for this solution. However, note that the expressions are quite complex. If the following CUPL program is compiled for the 16R4 device, the error "Excessive product terms" is encountered. It is necessary to use a device that has more product terms, such as the G20V8 GAL. Figure 6.17 illustrates the configuration of the G20V8 GAL device for this use.

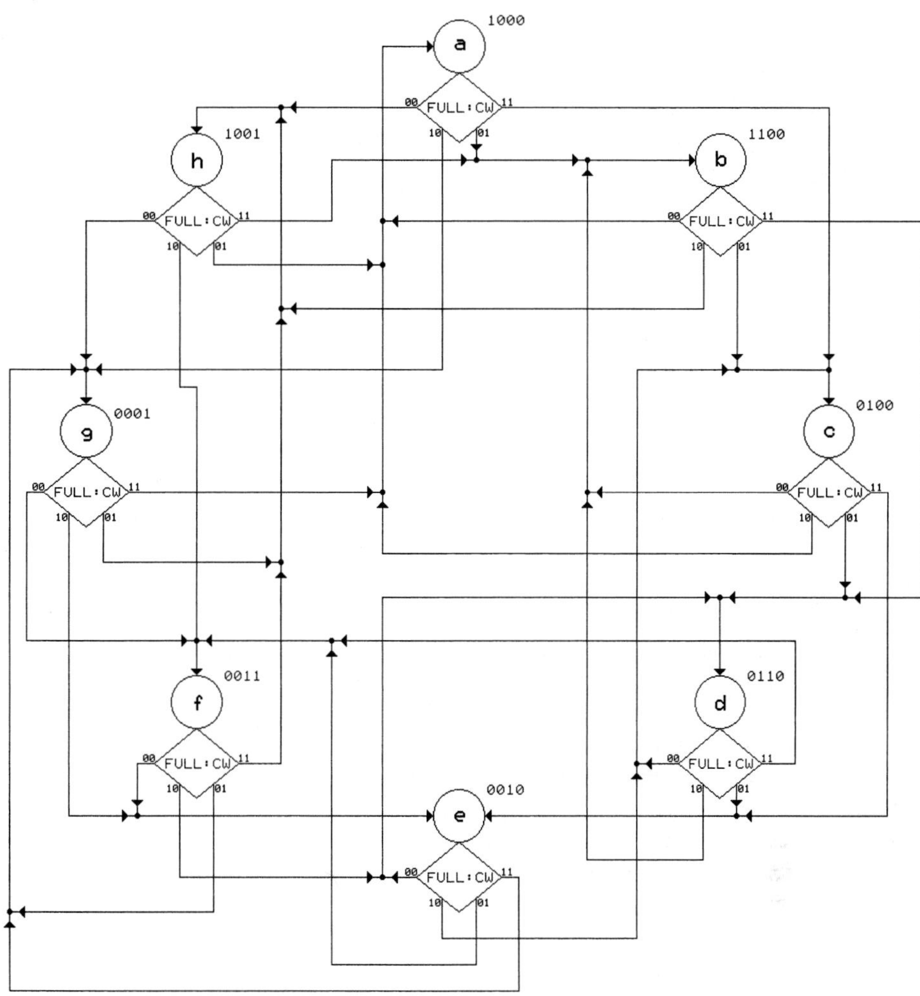

FIGURE 6.15 Practical Example 6.4, stepper motor state diagram.

With this material in hand, we are ready to write our CUPL program for this form of the solution that uses the logical expressions developed in Figure 6.16 to define the input-forming logic for the four-bit counter. The following is the complete CUPL program solution.

```
Name      EXP0604A;
Date      01/01/96;
Revision  01;
Designer  J. Carter;
Device    G20V8;
Format    J;
```

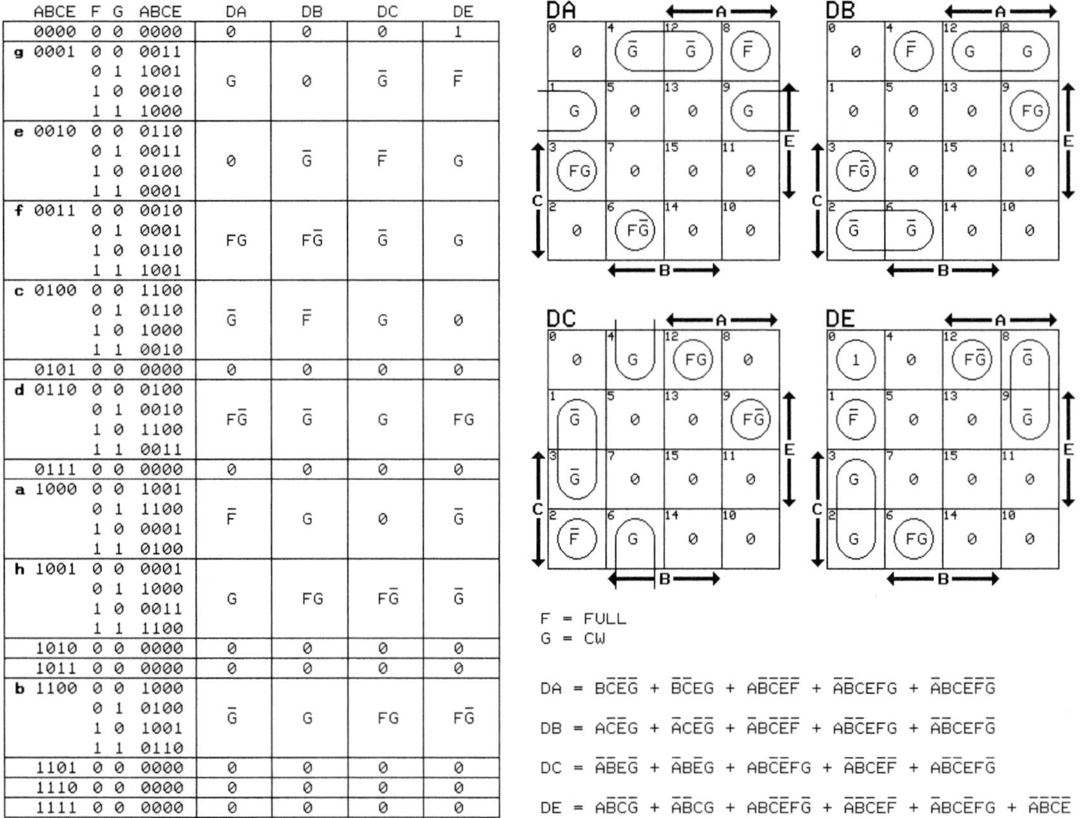

	ABCE	F G	ABCE	DA	DB	DC	DE
	0000	0 0	0000	0	0	0	1
g	0001	0 0	0011	G	0	$\overline{G}$	$\overline{F}$
		0 1	1001				
		1 0	0010				
		1 1	1000				
e	0010	0 0	0110	0	$\overline{G}$	$\overline{F}$	G
		0 1	0011				
		1 0	0100				
		1 1	0001				
f	0011	0 0	0010	FG	$F\overline{G}$	$\overline{G}$	G
		0 1	0001				
		1 0	0110				
		1 1	1001				
c	0100	0 0	1100	$\overline{G}$	$\overline{F}$	G	0
		0 1	0110				
		1 0	1000				
		1 1	0010				
	0101	0 0	0000	0	0	0	0
d	0110	0 0	0100	$F\overline{G}$	$\overline{G}$	G	FG
		0 1	0010				
		1 0	1100				
		1 1	0011				
	0111	0 0	0000	0	0	0	0
a	1000	0 0	1001	$\overline{F}$	G	0	$\overline{G}$
		0 1	1100				
		1 0	0001				
		1 1	0100				
h	1001	0 0	0001	G	FG	$F\overline{G}$	$\overline{G}$
		0 1	1000				
		1 0	0011				
		1 1	1100				
	1010	0 0	0000	0	0	0	0
	1011	0 0	0000	0	0	0	0
b	1100	0 0	1000	$\overline{G}$	G	FG	$F\overline{G}$
		0 1	0100				
		1 0	1001				
		1 1	0110				
	1101	0 0	0000	0	0	0	0
	1110	0 0	0000	0	0	0	0
	1111	0 0	0000	0	0	0	0

FIGURE 6.16 Practical Example 6.4, state table solution for stepper motor controller.

```
/****************************************************************/
/* PM TYPE STEPPER MOTOR CONTROLLER WITH HALF STEP CONTROL      */
/* Example 6.4A  Using counter input logic functions.           */
/****************************************************************/
/* Inputs */
Pin 2 = F;                /* 1=Full Step 0=Half Step             */
PIN 3 = G;                /* 1=Clockwise 0=Counterclockwise      */
/* Outputs */
Pin 20 = A;               /* Motor Coil Control Lines Q3         */
Pin 19 = B;               /*                          Q2         */
Pin 18 = C;               /*                          Q1         */
Pin 17 = E;               /*                          Q0         */
/* Input Forming Logic */
A.D = (      B & !C & !E &      !G)
    # (     !B & !C &  E &       G)
    # ( A & !B & !C & !E & !F)
    # (!A & !B &  C &  E &  F & G)
    # (!A &  B &  C & !E & !F & !G);
```

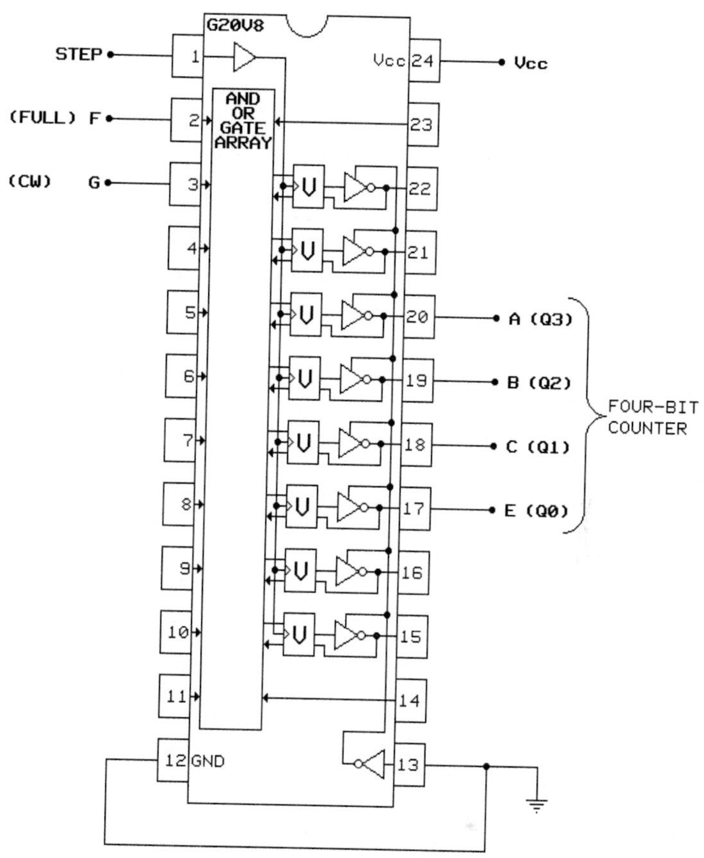

FIGURE 6.17 Practical Example 6.4, 20V8 GAL configuration for stepper motor controller.

```
B.D = ( A &          !C & !E &          G)
    # (!A &           C & !E &         !G)
    # (!A &  B & !C & !E & !F)
    # ( A & !B & !C &  E &  F &  G)
    # (!A & !B &  C &  E &  F & !G);
C.D = (!A & !B &           E &         !G)
    # (!A &  B &         !E &           G)
    # ( A &  B & !C & !E &  F &  G)
    # (!A & !B &  C & !E & !F)
    # ( A & !B & !C &  E &  F & !G);
E.D = ( A & !B & !C &                  !G)
    # (!A & !B &  C &                   G)
    # ( A &  B & !C & !E &  F & !G)
    # (!A & !B & !C &  E & !F)
    # (!A &  B &  C & !E &  F &  G)
    # (!A & !B & !C & !E);
```

The second solution method makes use of the SEQUENCE statement in the CUPL language. When the state diagram (or timing diagram) is translated to CUPL, the following statements replace the input-forming logic statements in the preceding program:

```
Sequence [A,B,C,E] {
        Present 'b'0000    next 'b'0001;
        Present 'b'0001                                  /* STATE-g */
                if !F & !G next 'b'0011;
                if !F &  G next 'b'1001;
                if  F & !G next 'b'0010;
                if  F &  G next 'b'1000;
        Present 'b'0010                                  /* STATE-e */
                if !F & !G next 'b'0110;
                if !F &  G next 'b'0011;
                if  F & !G next 'b'0100;
                if  F &  G next 'b'0001;
        Present 'b'0011                                  /* STATE-f */
                if !F & !G next 'b'0010;
                if !F &  G next 'b'0001;
                if  F & !G next 'b'0110;
                if  F &  G next 'b'1001;
        Present 'b'0100                                  /* STATE-c */
                if !F & !G next 'b'1100;
                if !F &  G next 'b'0110;
                if  F & !G next 'b'1000;
                if  F &  G next 'b'0010;
        Present 'b'0101    next 'b'0000;
        Present 'b'0110                                  /* STATE-d */
                if !F & !G next 'b'0100;
                if !F &  G next 'b'0010;
                if  F & !G next 'b'1100;
                if  F &  G next 'b'0011;
        Present 'b'0111    next 'b'0000;
        Present 'b'1000                                  /* STATE-a */
                if !F & !G next 'b'1001;
                if !F &  G next 'b'1100;
                if  F & !G next 'b'0001;
                if  F &  G next 'b'0100;
        Present 'b'1001                                  /* STATE-h */
                if !F & !G next 'b'0001;
                if !F &  G next 'b'1000;
                if  F & !G next 'b'0011;
                if  F &  G next 'b'1100;
        Present 'b'1010    next 'b'0000;
        Present 'b'1011    next 'b'0000;
        Present 'b'1100                                  /* STATE-b */
                if !F & !G next 'b'1000;
                if !F &  G next 'b'0100;
                if  F & !G next 'b'1001;
                if  F &  G next 'b'0110;
        Present 'b'1101    next 'b'0000;
        Present 'b'1110    next 'b'0000;
        Present 'b'1111    next 'b'0000; }
```

Practical Example 6.5: Wait State Generation in an Embedded Processor Design

Consider the placement of a microprocessor into a typical application such as instrumentation, process control, or monitoring. In such an application, the processor is used to manipulate data among several forms of input/output (I/O) devices and memory devices such as RAM, ROM, and EPROM. Often the clock speed of the processor is faster than the access times of the memory and I/O devices. Data would be lost on READ and WRITE operations if the processor is not instructed to wait for the slower device. When the processor is instructed to wait, that waiting period is referred to as a wait state. The circuitry that generates the control signal to accomplish this task is referred to as a wait state generator.

Consider the application of a Motorola M68000 microprocessor in an environment that has some RAM, EPROM, an A/D converter, a serial interface such as an MC68681 DUART, and a parallel interface such as an MC68230 PIT. The signal line used to hold the M68000 processor is referred to as the DTACK line. When this active-low line is asserted by an external open-collector buffer, the processor does not respond to the incoming system clock, placing it in a wait state until the line is deasserted. It is the responsibility of the external circuitry to control this line. Some devices, such as the DUART and the PIT, are designed to operate in this environment, and a DTACK output is provided that may be connected directly to the processor. When a memory reference is made to one of these two devices, the DTACK line is held low until the data transfer is complete.

Other devices that are typically placed in the memory map of the embedded processor system, such as RAM, EPROM, and A/D converters, do not have the capability of holding the processor in a wait state. Therefore, a circuit must be designed that will generate the DTACK signal whenever the devices are accessed by the processor. Consider the following example: An M68000 processor is driven by an 8 MHz clock. Therefore, the clock period is 125 nanoseconds. The environment includes some RAM, ROM, a PIT, DUART, and an A/D converter.

Since we are concerned with an embedded processor application rather than a microcomputer application, we are not obligated to allocate a lot of memory space. Let's assume we will apply the following memory map:

Address Range	*Device*	*Access Speed*	*Added Wait States*
\$00000 – \$0FFFF	ROM	150 nanoseconds	1
\$10000 – \$103FF	RAM	300 nanoseconds	2
\$20000 – \$2003F	PIT	(Contains its own DTACK driver)	
\$30000 – \$3000F	DUART	(Contains its own DTACK driver)	
\$40000	A/D	400 nanoseconds	3

To accomplish the task at hand, we need a circuit that will determine, based on the address, the amount of time the DTACK line must be held low and then hold it for that amount of time.

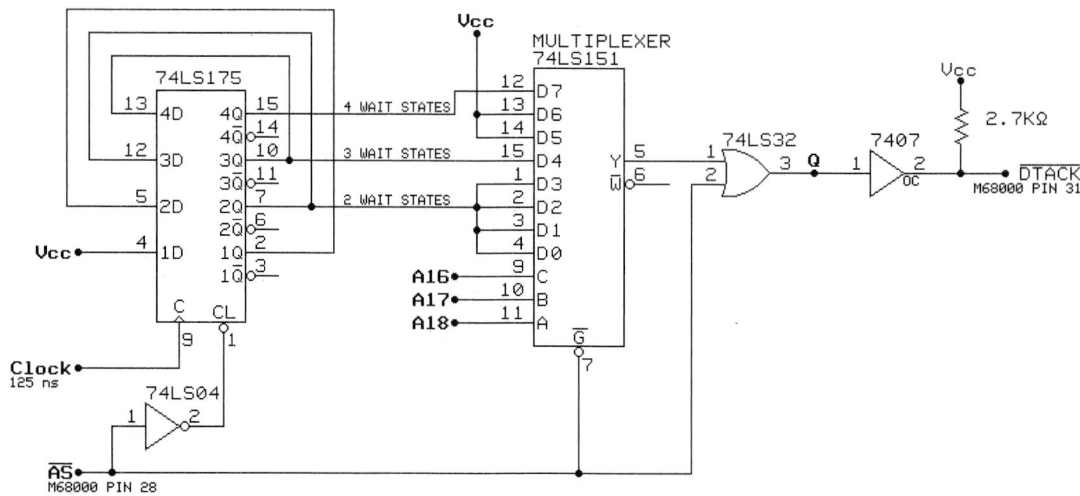

FIGURE 6.18 Practical Example 6.5, wait state generation circuitry.

A digital multiplexer chip can be used to determine the first of these two decisions. Note that the first nibble of the address determines the device and, consequently, the number of wait states to be inserted. Consider the circuit of Figure 6.18.

While the address strobe from the M68000 is deasserted, the 74LS175 is maintained in a cleared state. When the address strobe is asserted, the first flip-flop is set after one clock cycle by the application of Vcc to the 1D input. Since the output 1Q is fed to the 2D input, the 2Q output will go high after two clock cycles. This same pattern repeats, causing the outputs 1Q, 2Q, 3Q, and 4Q each to go to a logic 1 state after a delay of 1, 2, 3, and 4 respective clock cycles. Figure 6.19 illustrates this timing.

The multiplexer shown in Figure 6.18 is used to generate the DTACK signal of Figure 6.19 as a function of the memory address and the shift register output. When the address strobe is deasserted, the output of the circuit is a logic 1, the quiescent state for the DTACK line. The address strobe is asserted by the prccessor when a reference is being made to a valid memory address. The multiplexer consequently selects, based on that address, one of the lines coming off of the shift register of D-type flip-flops. The DTACK line will be pulled low when the selected input to the multiplexer is being held low for about 190 nanoseconds for inputs D0–D3, about 315 nanoseconds for input D4, and about 440 nanoseconds for input D7. Since inputs D5 and D6 are held high, they will not pull the DTACK line low when addressed. This is done because the devices at these addresses, the PIT and DUART, handle the DTACK line themselves.

The example shown in Figure 6.19 illustrates an address of $4XXXX on the bus and generates two extra wait states, for a DTACK delay period of about 315 nanoseconds. This timing diagram assumes the use of 74LSXX type chips and takes into account the propagation delays of the devices.

Design Method. If this circuit is to be implemented using the discrete logic of Figure 6.18, the system is rather bound by the design. The chips must be placed on a circuit board and

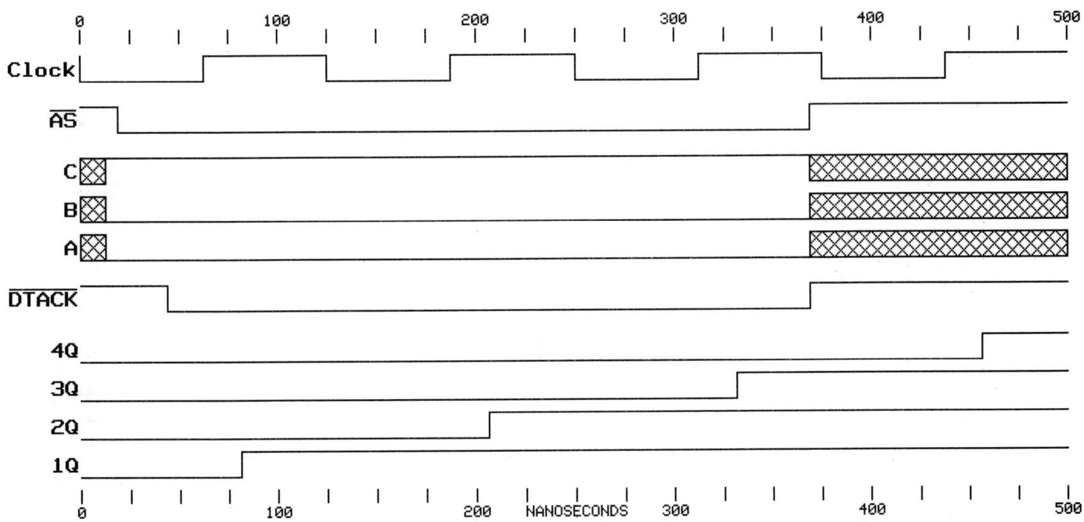

FIGURE 6.19 Practical Example 6.5, wait state generator timing.

runs "burned" into the board, generating an effective but unalterable solution. If the circuitry is replaced with a programmable logic device, the design is programmed into the device rather than into the circuit board. This way, not only are fewer devices used, but also the circuitry can be modified in the field without having to create a new circuit board. For this circuit replacement to take place, we must simply select a device that has the capabilities of the circuit. Then the circuitry can be applied, device-by-device, within the programmable logic.

Device Selection. This circuitry utilizes four D-type flip-flops and a single multiplexer output. That output is simply a single logic expression. Therefore, a 16R4 PAL may be used. Figure 6.20 illustrates the device configuration for this application.

CUPL Program. In this example, the program is simply a mirror of the schematic diagram. That is, each device in the DTACK generator is implemented with one of the resources in the programmable logic device. The following is the CUPL program for this application.

```
Name       EXP0605;
Date       01/01/96;
Revision   01;
Designer   J. Carter;
Device     P16R4;
Format     J;
/*******************************************************************/
/* Example 6.5  DTACK Generator Circuit                            */
/*******************************************************************/
/* Inputs */
Pin 2 = Vcc;              /* Input to D-Type Shift Register        */
Pin 3 = AS;               /* Address Strobe                        */
```

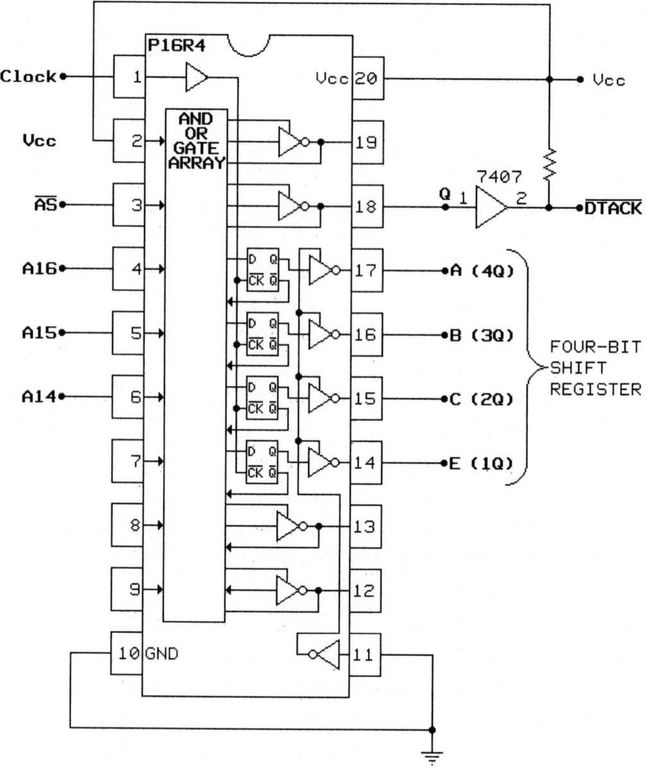

FIGURE 6.20 Practical Example 6.5, 16R4 PAL implementation of the wait state generator.

```
Pin 4 = A16;                /* Address Bus                         */
Pin 5 = A15;
Pin 6 = A14;
/* Outputs */
Pin 18 = Q;                 /* Generator Output to Buffer          */
Pin 17 = A;                 /* Shift Register Bit 1Q               */
Pin 16 = B;                 /*                    2Q               */
Pin 15 = C;                 /*                    3Q               */
Pin 14 = E;                 /*                    4Q               */
/* Shift Register */
A.D = Vcc;
B.D = A;
C.D = B;
E.D = C;
/* Multiplexer */
Q = ( (!A16 & B)                /* D3, D2, D1, D0                  */
    # ( A16 & !A15 & !A14 & C)  /* D4                              */
    # ( A16 & !A15 &  A14)      /* D5                              */
    # ( A16 &  A15 & !A14)      /* D6                              */
    # ( A16 &  A15 &  A14 & E) ) /* D7                             */
    # AS;
```

Practical Example 6.6: Priority Interrupt Encoder with Interrupt Mask

Consider an environment where it is necessary to respond to interrupts from a minimum of eight sources and to establish a priority system for those requests. A simple priority encoder will provide a three-bit binary value that corresponds to the request of the most significant interrupt source. For example, if eight sources are provided to the processor and a simple priority encoder is used, the interrupting device with the lowest priority would be device 0 and the interrupting device with the highest priority would be device 7. For further complexity, we will add the use of a three-bit interrupt mask register. We will allow only those interrupts with a numeric value greater than or equal to the contents of the mask register to be sent on to the processor. To do this, we will first need a three-bit binary comparitor circuit that will provide a logic level only when one three-bit input value is greater than or equal to another. Consider Figure 6.21, which illustrates the logic diagram of this three-bit binary magnitude comparitor circuit.

The Y output of the circuit is a logic 1 only when the binary value $A_2A_1A_0$ is greater than or equal to the binary value $B_2B_1B_0$. Obviously, this circuit is already a prime candidate for PLD application. The next circuit we will need is the actual priority encoder. This is another combinational circuit and is illustrated in Figure 6.22.

With these two tools in hand, we are ready to develop the overall structure of our circuit. The device contains a three-bit register, which can be loaded from external circuitry using a line labeled LOAD_MASK. When one of the eight inputs to the priority encoder is asserted, its three-bit encoded value is compared with the contents of the mask register. If the priority of the incoming signal is equal to or greater than the contents of the mask register, the IRQ output asserts low. We will also add two more features to our priority encoding circuit: the ability to clear the interrupt mask register (allowing all interrupts) and to set the interrupt mask register to all logic 1s (allowing only the highest level interrupt, priority 7, which is not maskable.) Figure 6.23 illustrates a block diagram of this completed design.

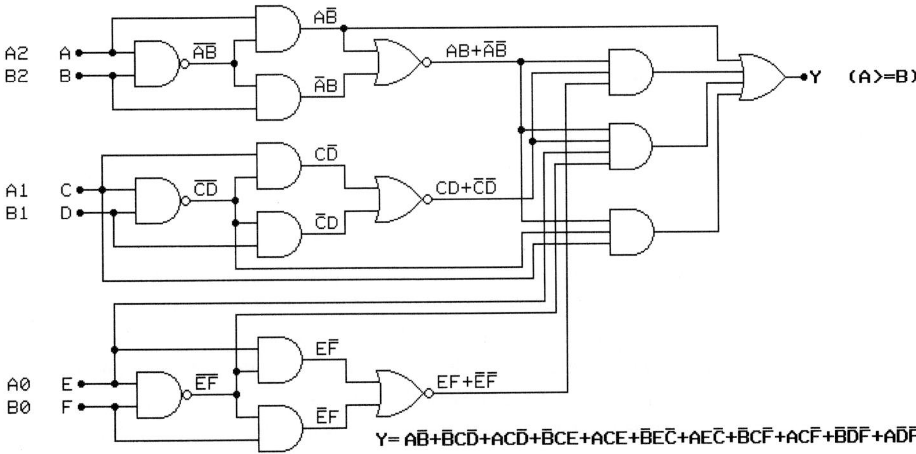

Figure 6.21 Practical Example 6.6, three-bit magnitude comparitor.

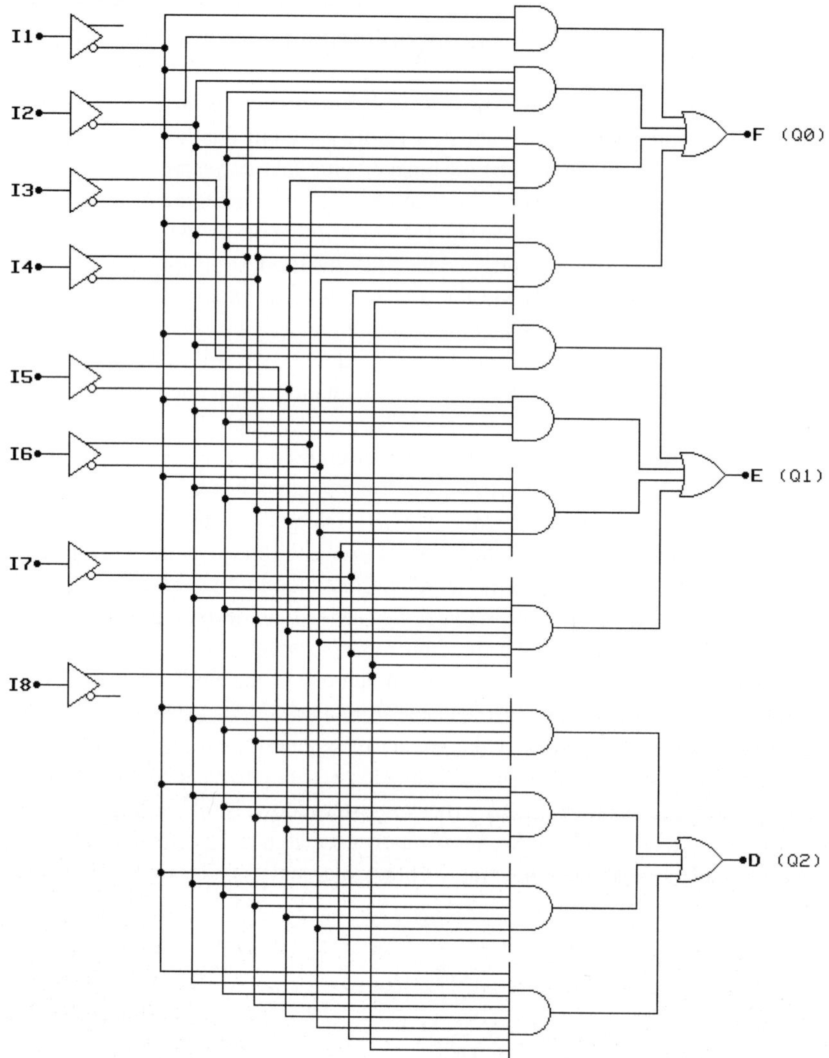

FIGURE 6.22 Practical Example 6.6, eight-bit to three-bit priority encoder circuit.

Let's briefly review the operation of this circuit, using Figure 6.23 as a guide. The eight inputs, I1 through I8 are interrupt request signals (active-high) from eight individual sources. The output of the eight-input OR gate asserts to a logic 1 if there is activity on any of the input request signals. The priority encoder will encode the active input to a binary number such that inputs I1 to I8 encode to binary numbers 000_2 to 111_2, respectively. The lowest priority is represented by 000_2, the highest by 111_2. In the event that two or more inputs are asserted at the same time, the input with the highest priority is encoded. The output of the priority encoder is provided to external circuitry for use, if needed, for interrupt acknowledgment.

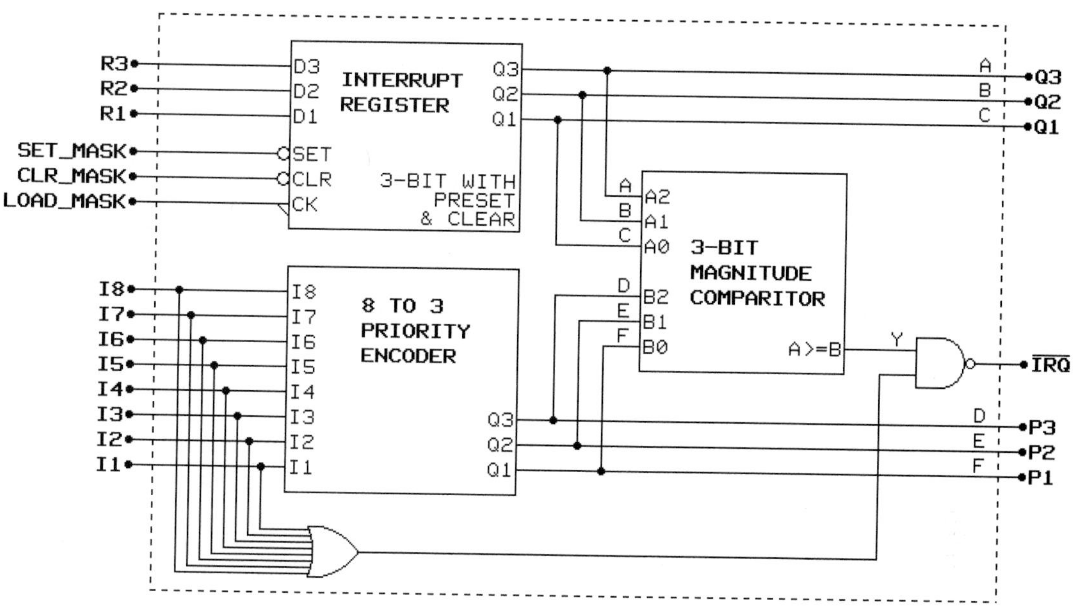

FIGURE 6.23 Practical Example 6.6, priority encoder with interrupt register, block diagram.

The interrupt mask register contains a three-bit number that can be loaded from inputs R3, R2, and R1. The register can be externally cleared to 000_2 or externally set to 111_2. In order to assert the IRQ output, the priority identified by the interrupt request signal must be greater than or equal to the number in the interrupt mask register. Therefore, if the interrupt mask register is cleared, all interrupts will be acknowledged. If the mask register is set to 111_2, only the highest priority, number 7, will be acknowledged. This priority level is, therefore, not maskable. The output of the interrupt mask register is provided to external circuitry for use, if needed.

Device Selection. This device requires 13 inputs and seven outputs, only three of which are registered. Also, the calculated Boolean functions for Y and IRQ contain more than seven minterms. Considering these requirements, a P22V10 may be selected and configured as illustrated in Figure 6.24. Note that there was no spare room in this design.

CUPL Program. In this example, the program is simply a mirror of the schematic diagrams previously described. Observe the following CUPL program. The inputs and outputs were assigned as shown in Figure 6.24. Note the use of the list structure to identify the individual interrupt request inputs. The three-bit comparitor circuit was implemented with a single logic function, Y. However, note that this function has a lot of minterms, so our choice of PAL device was rather limited. The priority encoder is another combinational circuit of three logical functions, D, E, and F. The interrupt mask register is loaded from the external inputs R3, R2, and R1 using the clock input to the chip. The preset and clear inputs are presented at pins 15 and 14, respectively. Finally, the IRQ output is enabled when there is an IRQ request and that request is of an equal or greater value to that calculated by the comparitor.

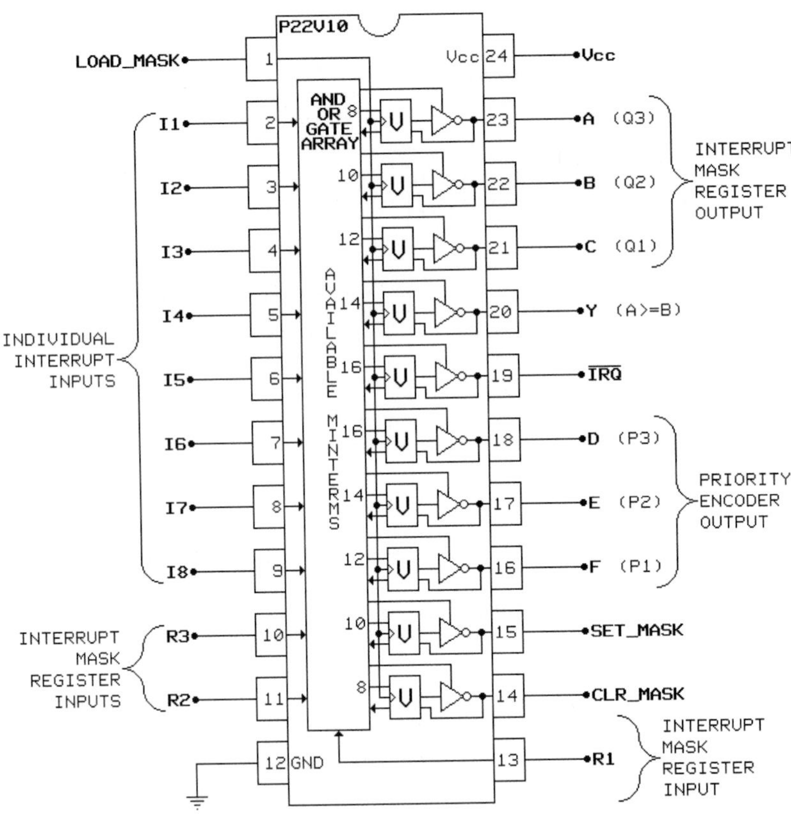

FIGURE 6.24 Practical Example 6.6, P22V10 configured as the priority encoder with mask register.

```
Name        EXP0606;
Date        01/01/96;
Revision    01;
Designer    J. Carter;
Device      G22V10;
Format      J;
/*****************************************************************/
/* Example 6.6  8 TO 3 Priority Encoder with Interrupt Masking   */
/*****************************************************************/
/* Inputs */
Pin 1 = LOAD_MASK;
Pin [2..9] = [I1..I8];      /* Individual IRQ inputs.             */
Pin 10 = R3;                /* Interrupt Mask Register Inputs     */
Pin 11 = R2;
Pin 13 = R1;
Pin 14 = CLR_MASK;          /* Set Interrupt Mask Register to 000 */
Pin 15 = SET_MASK;          /* Set Interrupt Mask Register to 111 */
```

```
/* Outputs */
Pin 23 = A;                /* Interrupt Mask Register R3        */
Pin 22 = B;                /*                         R2        */
Pin 21 = C;                /*                         R1        */
Pin 20 = Y;
Pin 19 = IRQ;              /* !IRQ Output to Processor          */
Pin 18 = D;                /* Priority Encoder Output Q3        */
Pin 17 = E;                /*                         Q2        */
Pin 16 = F;                /*                         Q1        */
/* Three-Bit Comparitor */
Y = (A & !B) # (!B & C & !D) # ( A & C & !D) #
 (!B &  C & E) # ( A & C &  E) # (!B & !C &  E) # ( A & !C &  E) #
 (!B & C & !F) # ( A & C & !F) # (!B & !D & !F) # ( A & !D & !F);
/* Priority Encoder */
D = (!I1 & !I2 & !I3 & !I4 & I5) #
    (!I1 & !I2 & !I3 & !I4 & !I5 & I6) #
    (!I1 & !I2 & !I3 & !I4 & !I5 & !I6 & I7) #
    (!I1 & !I2 & !I3 & !I4 & !I5 & !I6 & !I7 & I8);
E = (!I1 & !I2 & I3) # (!I1 & !I2 & !I3 & I4) #
    (!I1 & !I2 & !I3 & !I4 & !I5 & !I6 & I7) #
    (!I1 & !I2 & !I3 & !I4 & !I5 & !I6 & !I7 & I8);
F = (!I1 & I2) # (!I1 & !I2 & !I3 & I4) #
    (!I1 & !I2 & !I3 & !I4 & !I5 & I6) #
    (!I1 & !I2 & !I3 & !I4 & !I5 & !I6 & !I7 & I8);
/* Interrupt Mask Register */
A.D = R3; A.AR = !CLR_MASK; A.SP = !SET_MASK;
B.D = R2; B.AR = !CLR_MASK; B.SP = !SET_MASK;
C.D = R1; C.AR = !CLR_MASK; C.SP = !SET_MASK;
/* IRQ Output */
IRQ = !(I1 & I2 & I3 & I4 & I5 & I6 & I7 & I8) # !Y;
```

Practical Example 6.7: Programmable Retriggerable Monostable Multivibrator

A monostable multivibrator is a binary logic circuit with one stable state. When the circuit is triggered, it toggles to its unstable state for a predetermined amount of time. Once the time has expired, it returns to its stable state. Monostable multivibrators are often used as timers, since the length of time the circuit is in its unstable state is definable and repeatable. Simply, the monostable multivibrator provides a single pulse of a predetermined width when triggered.

The length of the pulse generated by the monostable multivibrator is usually determined by the value of certain circuit components, often resistors and capacitors, placed in a network. Such a multivibrator is considered retriggerable if, while in its unstable state, it can be triggered again, extending the period of the unstable state. Usually if a retriggerable monostable multivibrator is triggered during its unstable state, it will remain in that state for its predetermined period as if it were first triggered. Usually the length of the pulse generated by the monostable multivibrator is fixed by circuit components. If the length of the pulse is programmable, the monostable multivibrator is referred to as a programmable monostable multivibrator. There are a variety of ways of accomplishing this programmability.

The circuit we will develop for this example is both programmable and retriggerable. Rather than using a timing circuit to determine the length of the generated pulse, we are going to use a binary counter that counts down to 0 from a predetermined binary number. Once at 0, it will stay there. The logical states of each of the counter bits will be ORed together to generate the output pulse. Anytime the counter is in a non-zero state, the output pulse will be a logic 1. Therefore, the pulse width is determined by the binary value loaded into the counter. To exhibit this capability, we must be able to load the binary counter with a preset value. Consider Figure 6.25. This is a block diagram of the circuit we will be developing. Note that the four-bit binary counter is loaded from an external source by assertion of the $\overline{\text{PRESET}}$ input. When this is done, the counter is preset to the count on the input lines D3, D2, D1, and D0. Since the number is non-zero, the output Q asserts immediately. The counter now counts down to 0, at the end of the 1 count, the Q output goes back to its stable state, resulting in a pulse width equal to the length of the number of clock periods determined by the value originally loaded into the register. In this example, the number 12 was loaded into the register, and the output pulse was active for 12 clock cycles.

Design Methodology. The main portion of this circuit is the binary counter. This example uses a four-bit counter that will be able to generate output pulses from one to 15 clock periods in length. Because of the need to preset the register to a known state, the SEQUENCE statements in CUPL cannot be used to directly design the system. It will be necessary for us to go through the state table design of the counter input-forming logic. We can then insert the logic needed to preset the counter. Figure 6.26 illustrates the state diagram, state table, and Karnaugh map reduction of the input forming logic for the counter. Note that the count sequence is quite simple. The counter may be preset to a value as large as 15, and in all cases,

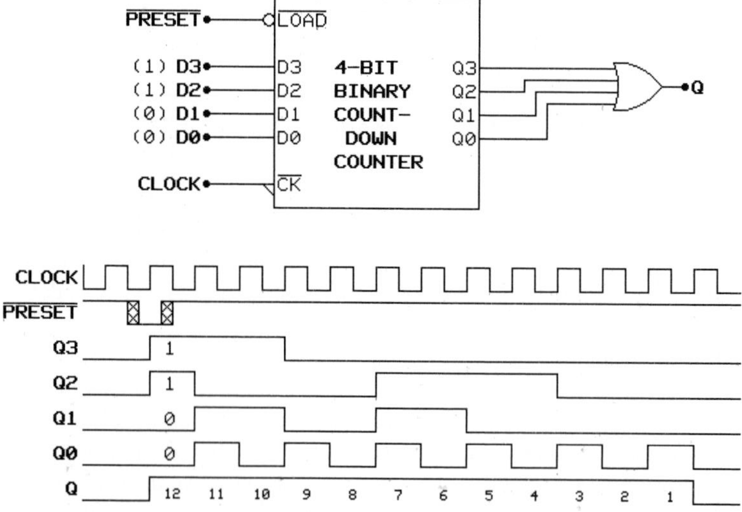

FIGURE 6.25 Practical Example 6.7, programmable retriggerable monostable block diagram and timing.

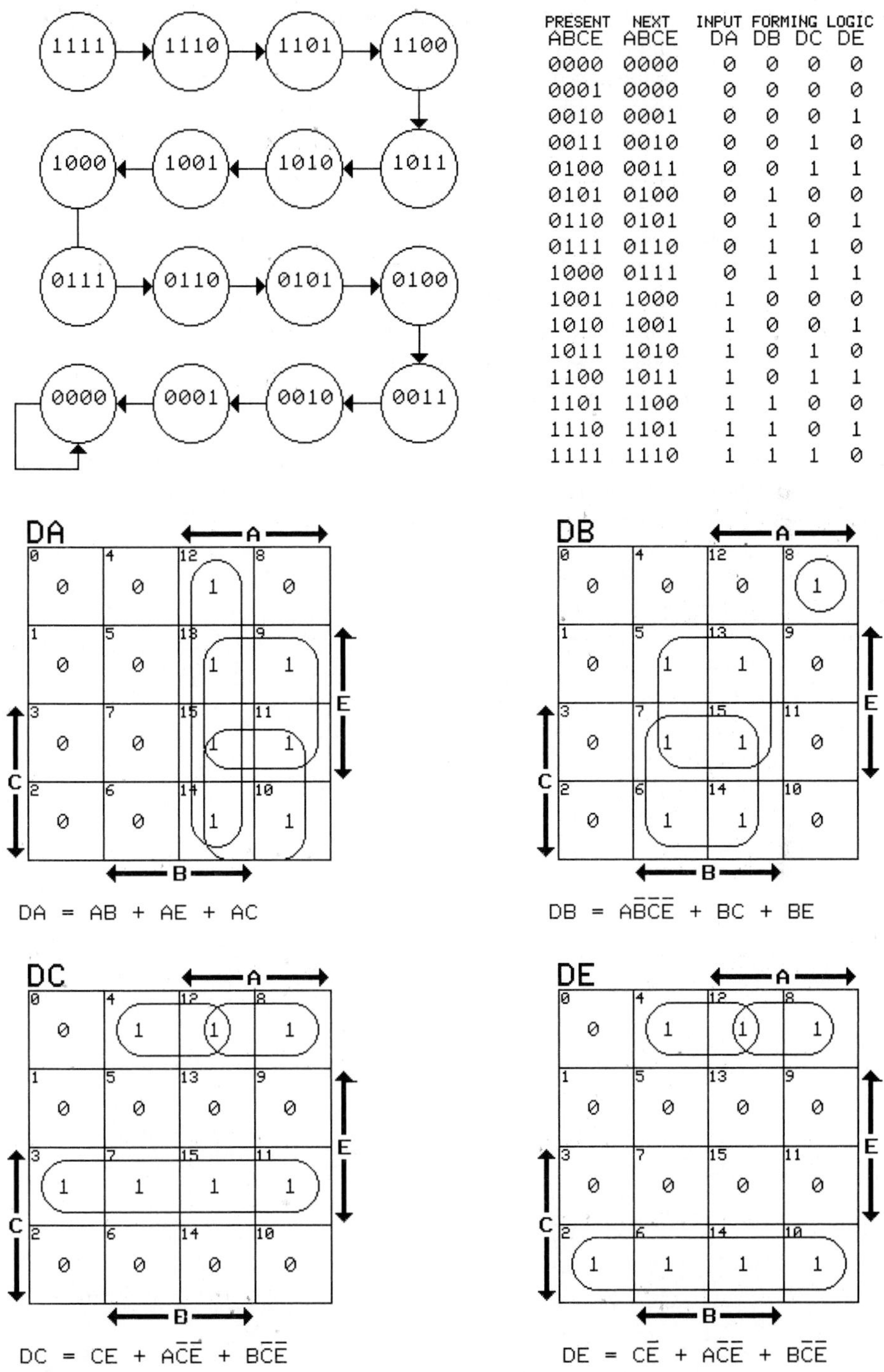

PRESENT ABCE	NEXT ABCE	DA	DB	DC	DE
0000	0000	0	0	0	0
0001	0000	0	0	0	0
0010	0001	0	0	0	1
0011	0010	0	0	1	0
0100	0011	0	0	1	1
0101	0100	0	1	0	0
0110	0101	0	1	0	1
0111	0110	0	1	1	0
1000	0111	0	1	1	1
1001	1000	1	0	0	0
1010	1001	1	0	0	1
1011	1010	1	0	1	0
1100	1011	1	0	1	1
1101	1100	1	1	0	0
1110	1101	1	1	0	1
1111	1110	1	1	1	0

FIGURE 6.26 Practical Example 6.7, state diagram, state table, and Karnaugh map reduction.

the count sequence is descending. When at state 0, the counter is to remain in that state until retriggered. From that point, the design of the counter is relatively straightforward.

With the information of Figure 6.26, we are ready to proceed. The Boolean equations will provide us with the input-forming logic necessary to generate the needed count sequence. However, let us not forget that we must be able to preset this counter to an externally applied binary value. Therefore, the flip-flop we use for the counter must support a large number of minterms. The 22V10 and similar PAL devices contain such flip-flops, so it is with this chip selection that we finish the design. Figure 6.27 is a detailed schematic diagram of this circuitry. It is drawn in detail because of the addition of the two-to-one multiplexers.

Note with care the method used to preset the flip-flops. In the absence of external individually controllable preset and clear inputs to the flip-flops, we must come up with another method to preset the register to an externally applied logic state. For this example, we have added a two-to-one multiplexer to the data input of each flip-flop. The multiplexer is used to select which of two sources provides the next state for the state machine. Those two sources are the input-forming logic and the externally applied data. While the input-forming logic is selected, the counter will count in its defined count sequence. When the input data is selected, that data is presented to the inputs of the register, so the register jumps to that state on the next clock edge. Note that this method is synchronized with the clock, so the pulse output will always be an accurate multiple of the clock period in length. For this to operate correctly, the $\overline{\text{PRESET}}$ signal must be defined prior to the clock edge when the external data is used. This is noted in the timing diagram of Figure 6.25. This architecture also implies that the input-forming logic will be treated as an intermediate variable in the CUPL program. The output of the multiplexers will drive the counter.

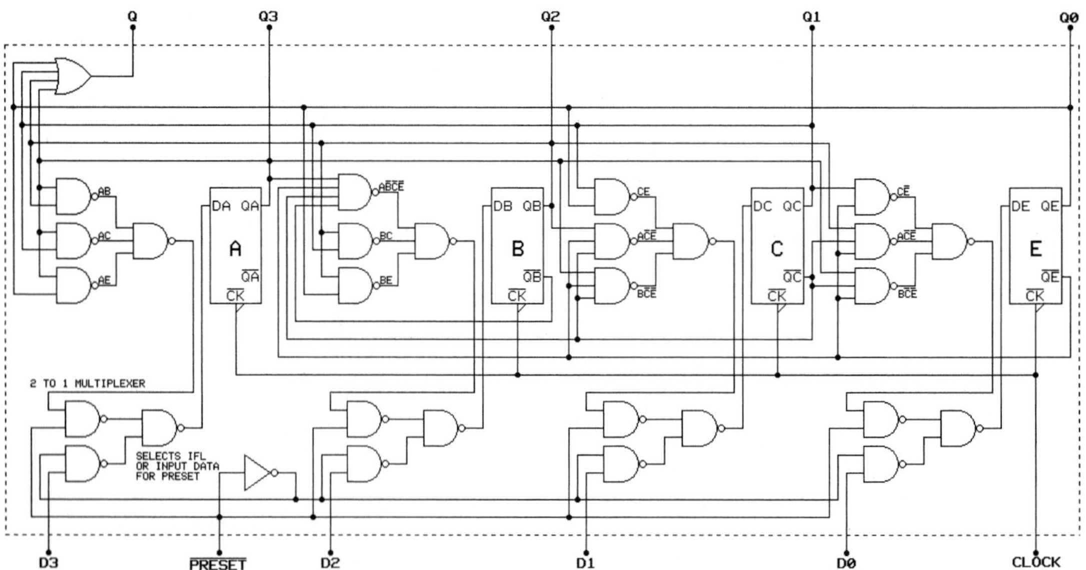

FIGURE 6.27 Practical Example 6.7, detailed schematic, programmable retriggerable monostable multivibrator.

As already stated, a 22V10 PAL device may be used for this application. It was chosen because of its programmable preset and clear inputs. Figure 6.28 illustrates this device.

The writing of the CUPL program is rather straightforward. The input-forming logic for the counter was developed in Figure 6.26, and the logic for the preset and clear functions is illustrated in Figure 6.27. The resulting CUPL program is shown below. Note that the only change from that of a simple design is the addition of a two-to-one multiplexer between the input-forming logic and the register input. Again, this was done so that the input to the register could be selected from two sources: the input-forming logic or the input data. It is this selection logic that is actually assigned to the register inputs. It would be possible to express this logic in two forms. Consider the following expressions:

```
DA = (A & B) # (A & E) # (A & C);
A.D = (DA & PRESET) # (D3 & !PRESET);
```

This expression set first calculates the input-forming logic as an intermediate variable. That variable is then used in the assignment of the register input. Consider combining the two expressions:

```
A.D = ((A & B) # (A & E) # (A & C) & PRESET) # (D3 & !PRESET);
```

FIGURE 6.28 Practical Example 6.7, 22V10 PAL configured as a monostable multivibrator.

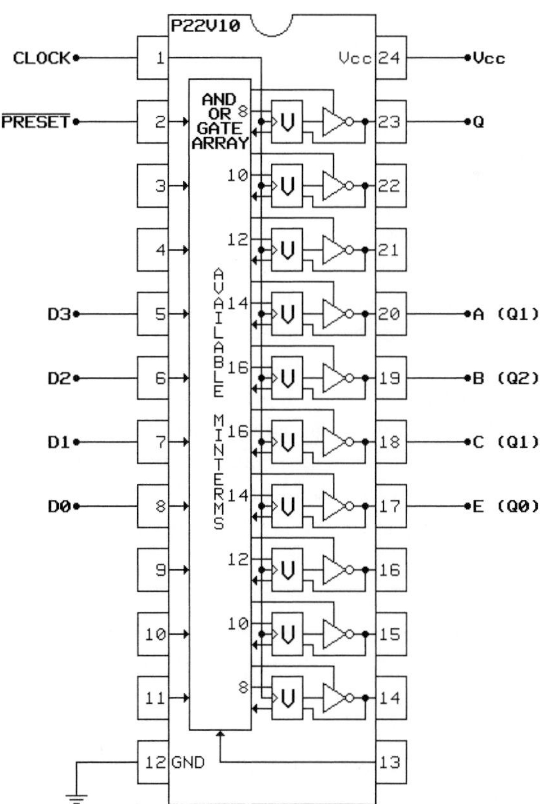

It is with this latter methodology that the following CUPL program was written:

```
Name      EXP0707;
Date      01/01/96;
Revision  01;
Designer  J. Carter;
Device    P22V10;
Format    J;
/**********************************************************************/
/* Example 6.7                                                        */
/* Programmable Retriggerable Monostable Multivibrator                */
/**********************************************************************/
/* Inputs */
Pin 1 = CLOCK;
Pin 2 = PRESET;                 /* Trigger input Loads Count Register */
Pin [5..8] = [D3..0];           /* Input Data for Count Register      */
/* Outputs */
Pin 16 = E;                     /* Four-Bit Binary Counter Q0         */
Pin 17 = C;                     /*                         Q1         */
Pin 18 = B;                     /*                         Q2         */
Pin 19 = A;                     /*                         Q3         */
Pin 23 = Q;                     /* Multivibrator output pulse         */
/* Four-Bit Binary Counter */
A.D = ((A & B) # (A & E) # (A & C)            & PRESET) # (D3 & !PRESET);
B.D = ((A & !B & !C & !E) # (B & C) # (B & E)  & PRESET) # (D2 & !PRESET);
C.D = ((C & E) # (A & !C & !E) # (B & !C & !E)  & PRESET) # (D1 & !PRESET);
E.D = ((C & !E) # (A & !C & !E) # (B & !C & !E) & PRESET) # (D0 & !PRESET);
/* Multivibrator Output */
Q = A # B # C # E;
```

That certainly was a lot of development for such a small CUPL program. This example illustrates how a relatively complex logic circuit such as that illustrated in Figure 6.27 can be implemented in a single PLD device with relative ease. The development process involved the calculation of the input-forming logic equations and the application of the two-to-one multiplexer on each register input so that the counter could be preset to an externally determined logic state.

Consider again the application of this circuit. By presetting the counter, the circuit outputs a pulse of a width equal to the clock period times the preset number, where that number can be any value from 1 through 15. If a five-bit counter were used, that number could be increased to 31. As the number of bits in the counter increases, the resolution of the pulse increases.

Practical Example 6.8: Eight-Bit Combinational Shifter

Up to this point in this chapter, all of our examples have been sequential in nature, requiring the use of clocked flip-flops. There are occasions in the technology where traditionally clocked circuits are replaced with combinational ones. This is becoming more prevalent

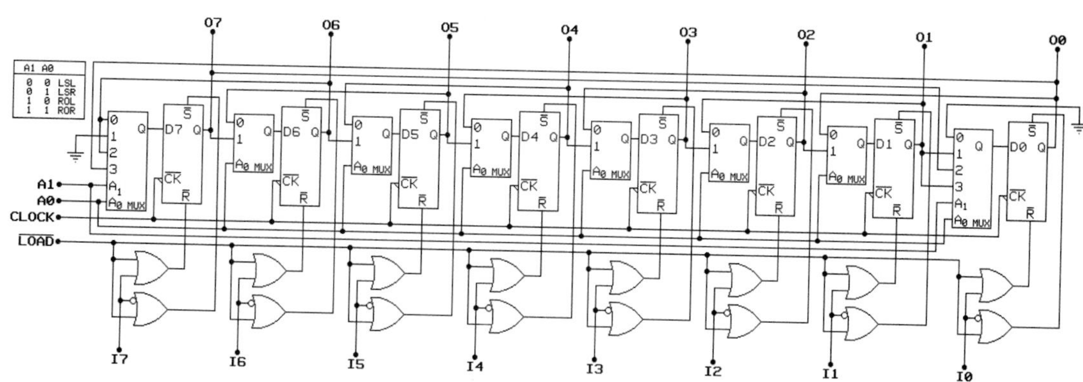

FIGURE 6.29 Practical Example 6.8, eight-bit shift register with logical shift and rotate linkage.

with the availability of large combinational logic circuits such as PLDs. When large combinational applications are being addressed, PLAs, PALs, and ROMs all play a major part in systems design.

A shifter is a circuit that takes a multibit data word and shifts that data right or left. Traditionally this application made use of a register of flip-flops. Once the register is loaded, the data may be shifted right or left using a clock signal. This circuit, known as a shift register, has provided the basis for this logic for a long time.

There is a second circuit that plays a part in this game of logic. When a multibit data word is shifted right or left, some decision must be made as to what data is shifted into the register and where the data that is shifted out is to go. Another circuit needs to be connected to the ends of the shift register to provide the data path for information being shifted into and out of its ends. This circuit may be referred to as the shift linkage multiplexer. Together these two circuits, the shift register and the shift linkage multiplexer, form the shifter circuitry. Figure 6.29 illustrates a shifter capable of handling byte-length data.

Careful study of this circuit will show how it is loaded with eight parallel bits and how the multiplexers that drive the inputs of the flip-flops select the four types of rotates and shifts this circuit supports.

Though this is a very common and useful circuit, it has one drawback that has become significant only in recent years: It requires a clock pulse to shift. Competition in the marketplace has demanded circuits that are more powerful and respond faster. This circuit can be designed to operate without a clock pulse if it is implemented entirely with combinational circuitry. The redesign is quite simple: Use a multiplexer on each output data line that selects the proper data source to generate the necessary shift. These arguments may be expressed using the table illustrated in Figure 6.30.

FIGURE 6.30 Practical Example 6.8, combinational shifter output functions.

A1	A0	MNEM	O7	O6	O5	O4	O3	O2	O1	O0
0	0	LSL	I6	I5	I4	I3	I2	I1	I0	0
0	1	LSR	0	I7	I6	I5	I4	I3	I2	I1
1	0	ROL	I6	I5	I4	I3	I2	I1	I0	I7
1	1	ROR	I0	I7	I6	I5	I4	I3	I2	I1

The logic expressions for each output of the shifter can be calculated by writing summations of the true minterms of each. Consider the following set of resulting expressions:

$$O7 = (\overline{A1} \cdot \overline{A0} \cdot I6) + (\overline{A1} \cdot A0 \cdot 0) + (A1 \cdot \overline{A0} \cdot I6) + (A1 \cdot A0 \cdot I0) = (\overline{A0} \cdot I6) + (A1 \cdot A0 \cdot I0)$$

$$O6 = (\overline{A1} \cdot \overline{A0} \cdot I5) + (\overline{A1} \cdot A0 \cdot I7) + (A1 \cdot \overline{A0} \cdot I5) + (A1 \cdot A0 \cdot I7) = (\overline{A0} \cdot I5) + (A0 \cdot I7)$$

$$O5 = (\overline{A1} \cdot \overline{A0} \cdot I4) + (\overline{A1} \cdot A0 \cdot I6) + (A1 \cdot \overline{A0} \cdot I4) + (A1 \cdot A0 \cdot I6) = (\overline{A0} \cdot I4) + (A0 \cdot I6)$$

$$O4 = (\overline{A1} \cdot \overline{A0} \cdot I3) + (\overline{A1} \cdot A0 \cdot I5) + (A1 \cdot \overline{A0} \cdot I3) + (A1 \cdot A0 \cdot I5) = (\overline{A0} \cdot I3) + (A0 \cdot I5)$$

$$O3 = (\overline{A1} \cdot \overline{A0} \cdot I2) + (\overline{A1} \cdot A0 \cdot I4) + (A1 \cdot \overline{A0} \cdot I2) + (A1 \cdot A0 \cdot I4) = (\overline{A0} \cdot I2) + (A0 \cdot I4)$$

$$O2 = (\overline{A1} \cdot \overline{A0} \cdot I1) + (\overline{A1} \cdot A0 \cdot I3) + (A1 \cdot \overline{A0} \cdot I1) + (A1 \cdot A0 \cdot I3) = (\overline{A0} \cdot I1) + (A0 \cdot I3)$$

$$O1 = (\overline{A1} \cdot \overline{A0} \cdot I0) + (\overline{A1} \cdot A0 \cdot I2) + (A1 \cdot \overline{A0} \cdot I0) + (A1 \cdot A0 \cdot I2) = (\overline{A0} \cdot I0) + (A0 \cdot I2)$$

$$O0 = (\overline{A1} \cdot \overline{A0} \cdot 0) + (\overline{A1} \cdot A0 \cdot I1) + (A1 \cdot \overline{A0} \cdot I7) + (A1 \cdot A0 \cdot I1) = (A0 \cdot I1) + (A1 \cdot \overline{A0} \cdot I7)$$

These expressions can be compared with the multiplexers applied in Figure 6.29. With these expressions we can write the CUPL program to use a 16L8 PAL to implement them.

```
Name        EXP0608;
Date        01/01/96;
Revision    01;
Designer    J. Carter;
Device      P16L8;
Format      J;
/*****************************************************************/
/*                                                               */
/* Example 6.8                                                   */
/* 8-Bit Combinational Shifter with Logical Shift & Rotate       */
/*****************************************************************/
/* Inputs */
Pin [1..2] = [A1..0];      /* Function Control Lines            */
                           /* A1,A0 = 0,0 = Logical Shift Left  */
                           /* A1,A0 = 0,1 = Logical Shift Right */
                           /* A1,A0 = 1,0 = Rotate Left         */
                           /* A1,A0 = 1,1 = Rotate Right        */
Pin [3..9] = [I7..I1];     /* Input Data Byte                   */
Pin     11 = I0;
/* Outputs */
Pin [19..12] = [O7..O0];   /* Output Data Byte                  */
/* Logical Output Functions */
O7 = (!A0 & I6) # (A1 & A0 & I0);       O7.OE = 'b'1;
O6 = (!A0 & I5) # (A0 & I7);            O6.OE = 'b'1;
O5 = (!A0 & I4) # (A0 & I6);            O5.OE = 'b'1;
O4 = (!A0 & I3) # (A0 & I5);            O4.OE = 'b'1;
O3 = (!A0 & I2) # (A0 & I4);            O3.OE = 'b'1;
O2 = (!A0 & I1) # (A0 & I3);            O2.OE = 'b'1;
O1 = (!A0 & I0) # (A0 & I2);            O1.OE = 'b'1;
O0 = ( A0 & I1) # (A1 & !A0 & I7);      O0.OE = 'b'1;
```

Because this application lacks the flip-flops of Figure 6.29, it can be placed in an eight-bit data stream shifting or rotating the data that passes through. This circuit exhibits the advantage of speed, as the shift process is delayed only by the propagation delay of the PAL device, instead of the introduction of a delay to the next clock pulse. It does require, however, that the input data is stable. Consequently, it would be typical that the data is stored in a register and is available on some type of data bus.

One application of this type of circuit is in the design of an RISC processor. A fundamental philosophy in RISC design is that all defined arithmetic and logical functions can be concluded within the period of a single clock cycle. This circuit can be placed in the data stream in the arithmetic logic unit and provide the shift of registered data without the need for an additional clock pulse.

Practical Example 6.9: Four-Bit Combinational Barrel Shifter

A digital logic circuit that provides applications in arithmetic processes and digital signal processing is a barrel shifter. Traditional shift registers use a sequence of flip-flops that transfer their data from one flip-flop to the next when given a clock pulse. In order to move data to a multiple of positions, a multiple of clock pulses must be provided. An alternate approach provides the same function much faster by eliminating the use of the sequential circuitry, using only combinational circuitry instead. This circuit, the barrel shifter, accepts multibit input data, a line that determines if the shift is to the right or to the left, and a set of inputs that determines the number of positions to shift. The output is simply a function of these inputs.

Figure 6.31 illustrates a simple four-bit barrel shifter. It accepts four bits of data on its inputs, along with the shifting commands, and provides four bits of shifted data on its outputs.

In addition to the four input bits I3, I2, I1, and I0, two additional input data bits are provided, SI3 and SI0. SI3 is used to shift data into the most significant position of the data word when executing a shift to the right. The SI0 is used to shift data into the least significant position of the data word when executing a shift to the left. Two additional outputs are

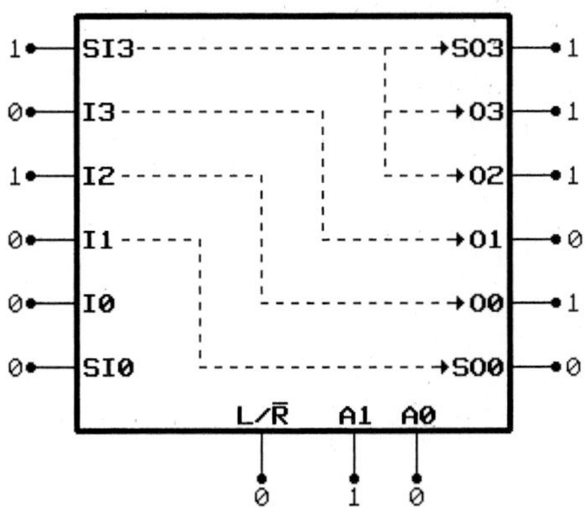

L/R̄	A1	A0	SO3	O3	O2	O1	O0	SO0
0	0	0	SI3	I3	I2	I1	I0	SI0
0	0	1	SI3	SI3	I3	I2	I1	I0
0	1	0	SI3	SI3	SI3	I3	I2	I1
0	1	1	SI3	SI3	SI3	SI3	I3	I2
1	0	0	SI3	I3	I2	I1	I0	SI0
1	0	1	I3	I2	I1	I0	SI0	SI0
1	1	0	I2	I1	I0	SI0	SI0	SI0
1	1	1	I1	I0	SI0	SI0	SI0	SI0

FIGURE 6.31 Practical Example 6.9, four-bit combinational barrel shifter, block diagram, and function table.

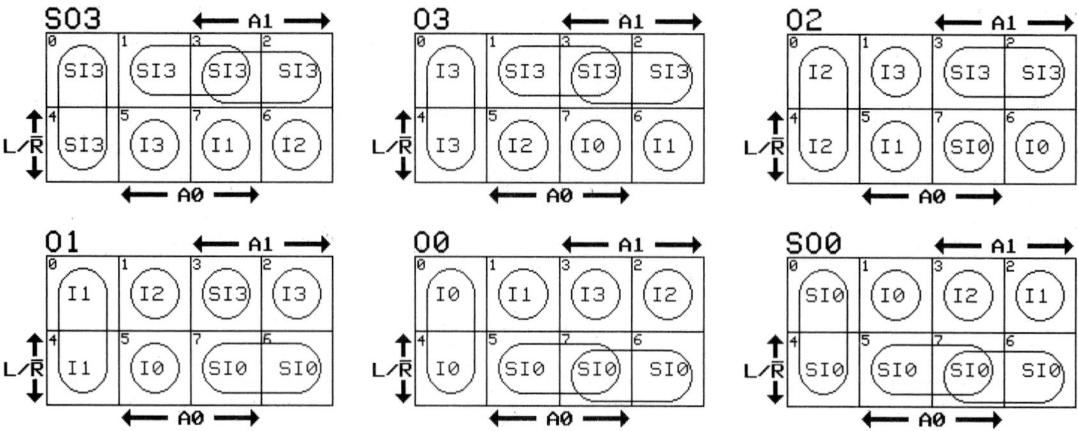

FIGURE 6.32 Practical Example 6.9, Karnaugh maps of barrel shifter functions.

provided which have a similar definition. SO3 is set to the logical value of the data bit shifted out of the most significant position of the data word when executing a shift to the left. SO0 is set to the logical value of the data bit shifted out of the least significant position of the data word when executing a shift to the right. The data for these bits are typically driven by a multiplexer that selects input data from a variety of sources. Often this multiplexer is referred to as a shift linkage multiplexer.

The state table illustrated in Figure 6.31 can be used to calculate each of the Boolean output functions. Figure 6.32 illustrates the Karnaugh map reduction of these functions.

From the Karnaugh maps of Figure 6.32, the following set of output expressions is realized:

$$\begin{aligned} SO3 = {} & (\overline{A0} \cdot \overline{A1} \cdot SI3) + (A0 \cdot \overline{LR} \cdot S13) + (A1 \cdot \overline{LR} \cdot SI3) \\ & + (A0 \cdot \overline{A1} \cdot LR \cdot I3) + (\overline{A0} \cdot A1 \cdot LR \cdot I2) + (A0 \cdot A1 \cdot LR \cdot I1) \\ O3 = {} & (\overline{A0} \cdot \overline{A1} \cdot I3) + (A0 \cdot \overline{LR} \cdot S13) + (A1 \cdot \overline{LR} \cdot SI3) \\ & + (A0 \cdot \overline{A1} \cdot LR \cdot I2) + (\overline{A0} \cdot A1 \cdot LR \cdot I1) + (A0 \cdot A1 \cdot LR \cdot I0) \\ O2 = {} & (\overline{A0} \cdot \overline{A1} \cdot I2) + (A0 \cdot \overline{A1} \cdot \overline{LR} \cdot I3) + (A1 \cdot \overline{LR} \cdot SI3) \\ & + (A0 \cdot \overline{A1} \cdot LR \cdot I0) + (\overline{A0} \cdot A1 \cdot LR \cdot SI0) + (A0 \cdot A1 \cdot LR \cdot I1) \\ O1 = {} & (\overline{A0} \cdot \overline{A1} \cdot I1) + (A0 \cdot \overline{A1} \cdot \overline{LR} \cdot I2) + (A0 \cdot \overline{A1} \cdot \overline{LR} \cdot I3) \\ & + (A0 \cdot A1 \cdot \overline{LR} \cdot SI3) + (A0 \cdot \overline{A1} \cdot LR \cdot I0) + (A1 \cdot LR \cdot SI0) \\ O0 = {} & (\overline{A0} \cdot \overline{A1} \cdot I0) + (A0 \cdot \overline{A1} \cdot \overline{LR} \cdot I1) + (A0 \cdot \overline{A1} \cdot \overline{LR} \cdot I2) \\ & + (A0 \cdot A1 \cdot \overline{LR} \cdot I3) + (A0 \cdot LR \cdot SI0) + (A1 \cdot LR \cdot SI0) \\ SO0 = {} & (\overline{A0} \cdot \overline{A1} \cdot SI0) + (A0 \cdot \overline{A1} \cdot \overline{LR} \cdot I0) + (A0 \cdot \overline{A1} \cdot \overline{LR} \cdot I1) \\ & + (A0 \cdot A1 \cdot \overline{LR} \cdot I2) + (A0 \cdot LR \cdot SI0) + (A1 \cdot LR \cdot SI0) \end{aligned}$$

With these logic expressions, the following CUPL program was developed.

```
Name      EXP0609;
Date      01/01/96;
Revision  01;
Designer  J. Carter;
```

```
Device     P16L8;
Format     J;
/*************************************************************/
/* Practical Example 6.9                                     */
/* 4-Bit Combinational Barrel Shifter                        */
/*************************************************************/
/* Inputs */
Pin 1 = SI3;             /* Shift into Most Significant Bit   */
Pin [2..5] = [I3..I0];   /* Input Data Lines                  */
Pin 6 = SI0;             /* Shift into Least Significant Bit  */
Pin 7 = LR;              /* Left/!Right 1=Left, 0=Right       */
Pin [8..9] = [A1..A0];   /* Number of Shifts: 3, 2, 1 or 0    */
/* Outputs */
Pin 19 = SO3;            /* Data from Most Significant Bit    */
Pin [18..15] = [O3..O0]; /* Output Data Lines                 */
Pin 14 = SO0;            /* Data from Least Significant Bit   */
/* Logical Output Functions */
SO3 = (!A0 & !A1 & SI3)      # ( A0 & !LR & SI3)
    # ( A1 & !LR & SI3)      # ( A0 & !A1 & LR & I3)
    # (!A0 &  A1 & LR & I2)  # ( A0 &  A1 & LR & I1);
    SO3.OE = 'b'1;
O3  = (!A0 & !A1 & I3)       # ( A0 & !LR & SI3)
    # ( A1 & !LR & SI3)      # ( A0 & !A1 & LR & I2)
    # (!A0 &  A1 & LR & I1)  # ( A0 &  A1 & LR & I0);
    O3.OE = 'b'1;
O2  = (!A0 & !A1 & I2)       # ( A0 & !A1 & !LR & I3)
    # ( A1 & !LR & SI3)      # ( A0 & !A1 &  LR & I0)
    # (!A0 &  A1 & LR & SI0) # ( A0 &  A1 &  LR & I1);
    O2.OE = 'b'1;
O1  = (!A0 & !A1 & I1)       # ( A0 & !A1 & !LR & I2)
    # ( A0 & !A1 & !LR & I3) # ( A0 &  A1 & !LR & SI3)
    # ( A0 & !A1 &  LR & I0) # ( A1 &  LR & SI0);
    O1.OE = 'b'1;
O0  = (!A0 & !A1 & I0)       # ( A0 & !A1 & !LR & I1)
    # ( A0 & !A1 & !LR & I2) # ( A0 &  A1 & !LR & I3)
    # ( A0 &  LR & SI0)      # ( A1 &  LR & SI0);
    O0.OE = 'b'1;
SO0 = (!A0 & !A1 & SI0)      # ( A0 & !A1 & !LR & I0)
    # ( A0 & !A1 & !LR & I1) # ( A0 &  A1 & !LR & I2)
    # ( A0 &  LR & SI0)      # ( A1 &  LR & SI0);
    SO0.OE = 'b'1;
```

Practical Example 6.10: Eight-Bit Combinational Barrel Shifter

Practical Example 6.9 discussed the concept of the barrel shifter. This example, included in the documentation published with the CUPL software, was developed to illustrate the use of the language. Observe closely how the shifting is accomplished using the set and equality.

```
Name       Barrel22;
Date       05/11/89;
Designer   Kahl;
Company    Logical Devices, Inc.;
Device     P22V10;
/**************************************************************/
/*                                                            */
/* 8-Bit Registered Barrel Shifter                            */
/*                                                            */
/* This 8-bit registered barrel shifter takes 8 data inputs   */
/* and cyclically rotates the data from 0 to 7 places under   */
/* control of the select ( S0, S1, S2 ) inputs. A SET input   */
/* can be used to initialize the outputs to the all ones state */
/**************************************************************/
/** Inputs **/
PIN 1          = clock;        /* Register Clock              */
PIN [2..9]     = [D7..0];      /* Data Inputs                 */
PIN [10,11,14] = [S2..0];      /* Shift Count Select Inputs   */
PIN 13         = !out_enable;  /* Register Output Enable      */
PIN 23         = SET;          /* Set to Ones Input           */
/** Outputs **/
PIN [15..22] = [Q7..0];        /* Register Outputs            */
/** Declarations and Intermediate Variable Definitions **/
field shift  = [S2..0];        /* Shift Width Field           */
field input  = [D7..0];        /* Inputs Field                */
field output = [Q7..0];        /* Outputs Field               */
/** Logic Equations **/
output.d = [D7, D6, D5, D4, D3, D2, D1, D0] & shift:0
         # [D0, D7, D6, D5, D4, D3, D2, D1] & shift:1
         # [D1, D0, D7, D6, D5, D4, D3, D2] & shift:2
         # [D2, D1, D0, D7, D6, D5, D4, D3] & shift:3
         # [D3, D2, D1, D0, D7, D6, D5, D4] & shift:4
         # [D4, D3, D2, D1, D0, D7, D6, D5] & shift:5
         # [D5, D4, D3, D2, D1, D0, D7, D6] & shift:6
         # [D6, D5, D4, D3, D2, D1, D0, D7] & shift:7;
output.sp = SET;               /* synchronous preset for SET  */
output.oe = out_enable;        /* tri-state control           */
output.ar = 'b'0;              /* asynchronous reset not used */
```

Practical Example 6.11: Digital Waveform Synthesizer/Generator

Sine wave signals are often generated using oscillators that are driven by passive components. Oscillators may make use of crystals or phase-locked loops for stability. Another method might be to use active filters or switched-capacitor filters in order to obtain stable, variable frequency signals. Consider another method: digital waveform synthesis. People knowledgeable about music synthesis are familiar with the application of digital waveform

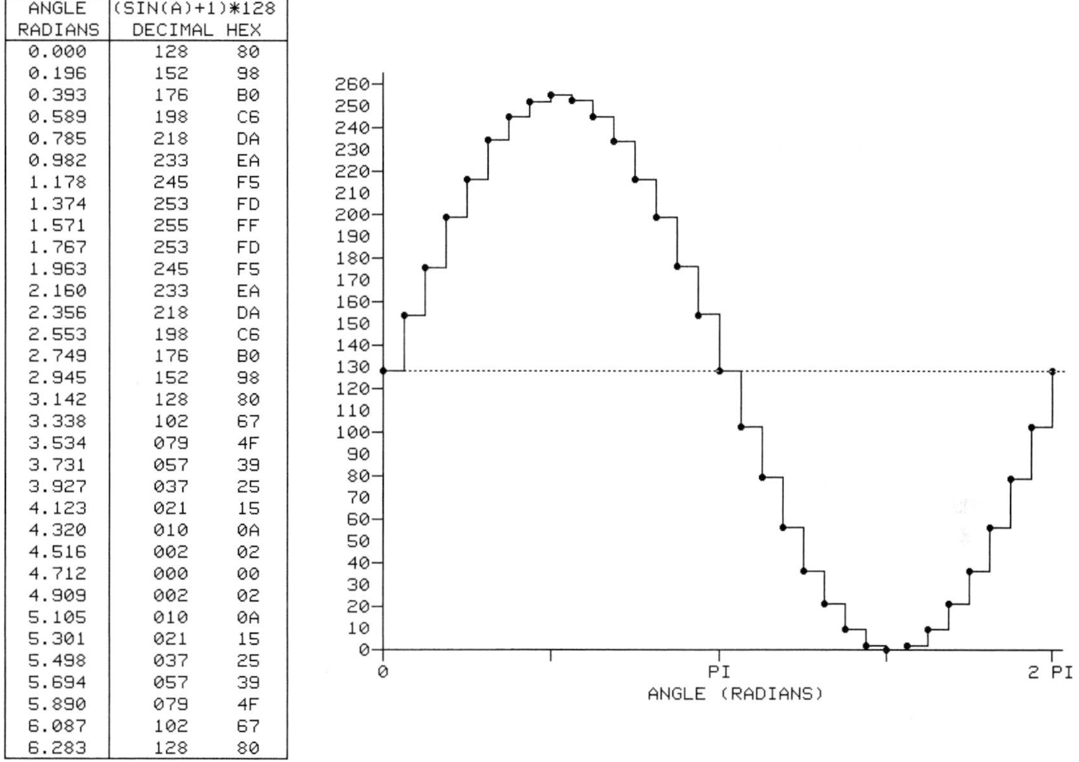

ANGLE RADIANS	(SIN(A)+1)*128 DECIMAL	HEX
0.000	128	80
0.196	152	98
0.393	176	B0
0.589	198	C6
0.785	218	DA
0.982	233	EA
1.178	245	F5
1.374	253	FD
1.571	255	FF
1.767	253	FD
1.963	245	F5
2.160	233	EA
2.356	218	DA
2.553	198	C6
2.749	176	B0
2.945	152	98
3.142	128	80
3.338	102	67
3.534	079	4F
3.731	057	39
3.927	037	25
4.123	021	15
4.320	010	0A
4.516	002	02
4.712	000	00
4.909	002	02
5.105	010	0A
5.301	021	15
5.498	037	25
5.694	057	39
5.890	079	4F
6.087	102	67
6.283	128	80

FIGURE 6.33 Practical Example 6.11, sine wave envelope.

synthesis. The method of signal generation is as follows: A table of numbers defines the envelope of the signal over a single cycle, as shown in Figure 6.33.

The value of the sine from 0 to 2 pi radians was calculated and scaled to fit the range from 0 to 255. In the example illustrated in Figure 6.33, the waveform was divided into 32 equal segments. Consider programming a 32 x 8 PROM with the table of calculated numbers illustrated in the figure, starting with address 0, and storing each successive number in an ascending address. If this is the case, address 0 = 128, address 1 = 152, and the like. If a five-bit binary counter is used to drive the ROM and the output of the ROM is converted to an analog signal using an digital to analog converter (DAC), the waveform of Figure 6.33 results.

It is evident that the waveform has poor resolution because only 32 samples were taken. If the number of samples is increased from 32 to 256, a much smoother output waveform is realized. This would require the application of a 256 x 8 PROM. Such a device, a National DM74S571, is included in Appendix A and is supported by CUPL. Figure 6.34 illustrates the complete schematic diagram of the sine waveform synthesizer.

An eight-bit binary counter is driven by a clock signal of a known frequency. The output of the counter drives the address inputs of the PROM. The PROM is using a table

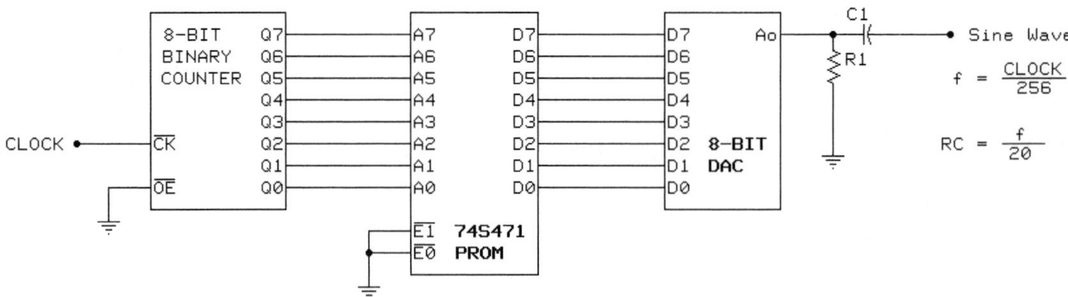

Figure 6.34 Practical Example 6.11, sine wave synthesizer schematic diagram.

look-up method to convert the binary address into a binary number that represents the amplitude of the sine signal using the methodology discussed above. Each time the binary counter completes a cycle, the output of the PROM completes a single sine wave cycle. Therefore, since we are using an eight-bit counter, the sine wave frequency will be 1/256th of the clock frequency. The output of the PROM drives a digital-to-analog converter that will generate the actual analog sine wave signal. Since the sine wave amplitude was calculated at 256 points (approximately every 1.4 degrees), the output signal will have the stair-step appearance of Figure 6.33. Note that an RC filter was added to the output of the DAC to smooth out the stair-step appearance of the signal. The values for RC are very approximate and assume that the circuit is driving a high-impedance load.

Note that any wave shape can be generated by storing the envelope information in the PROM.

```
Name      SINEWAVE;
Date      01/01/96;
Designer  J.W. Carter;
Company   Logical Devices, Inc.;
Device    RA8P8;
/******************************************************************/
/* 8-Bit Sine Wave Synthesis                                      */
/*                                                                */
/* This programs a DM74LS471 PROM such that the output follows    */
/* a sine amplitude of 0 to 255 as the angle varies in 256        */
/* steps from 0 to 2 pi radians. If driven by an 8-bit binary     */
/* clock, the output will generate a sine wave when using a DAC */
/******************************************************************/
/** Inputs **/
PIN [1..5]    = [I0..4];           /* Angle Inputs              */
PIN [17..19] = [I5..7];
/** Outputs **/
PIN [6..9]    = [O0..3];           /* Waveform Outputs          */
PIN [11..14] = [O4..7];
FIELD input  = [I7..I0];
FIELD output = [O7..O0];
TABLE input  =>output {
'H'0 =>'H'80;   'H'2B=>'H'EE;   'H'56=>'H'ED;   'H'81=>'H'7C;   'H'AC=>'H'F ;   'H D7=>'H'14.
'H'1 =>'H'83;   'H'2C=>'H'F0;   'H'57=>'H'EB;   'H'82=>'H'79;   'H'AD=>'H'E ;   'H'D8=>'H'15;
'H'2 =>'H'86;   'H'2D=>'H'F1;   'H'58=>'H'EA;   'H'83=>'H'76;   'H'AE=>'H'C ;   'H'D9=>'H'17;
'H'3 =>'H'89;   'H'2E=>'H'F3;   'H'59=>'H'E8;   'H'84=>'H'73;   'H'AF=>'H'B ;   'H'DA=>'H'19;
```

```
'H'4 =>'H'8C;
'H'5 =>'H'8F;
'H'6 =>'H'92;
'H'7 =>'H'95;
'H'8 =>'H'98;
'H'9 =>'H'9B;
'H'A =>'H'9E;
'H'B =>'H'A2;
'H'C =>'H'A5;
'H'D =>'H'A7;
'H'E =>'H'AA;
'H'F =>'H'AD;
'H'10=>'H'B0;
'H'11=>'H'B3;
'H'12=>'H'B6;
'H'13=>'H'B9;
'H'14=>'H'BC;
'H'15=>'H'BE;
'H'16=>'H'C1;
'H'17=>'H'C4;
'H'18=>'H'C6;
'H'19=>'H'C9;
'H'1A=>'H'CB;
'H'1B=>'H'CE;
'H'1C=>'H'D0;
'H'1D=>'H'D3;
'H'1E=>'H'D5;
'H'1F=>'H'D7;
'H'20=>'H'DA;
'H'21=>'H'DC;
'H'22=>'H'DE;
'H'23=>'H'E0;
'H'24=>'H'E2;
'H'25=>'H'E4;
'H'26=>'H'E6;
'H'27=>'H'E8;
'H'28=>'H'EA;
'H'29=>'H'EB;
'H'2A=>'H'ED;
'H'2F=>'H'F4;
'H'30=>'H'F5;
'H'31=>'H'F6;
'H'32=>'H'F8;
'H'33=>'H'F9;
'H'34=>'H'FA;
'H'35=>'H'FA;
'H'36=>'H'FB;
'H'37=>'H'FC;
'H'38=>'H'FD;
'H'39=>'H'FD;
'H'3A=>'H'FE;
'H'3B=>'H'FE;
'H'3C=>'H'FE;
'H'3D=>'H'FF;
'H'3E=>'H'FF;
'H'3F=>'H'FF;
'H'40=>'H'FF;
'H'41=>'H'FF;
'H'42=>'H'FF;
'H'43=>'H'FF;
'H'44=>'H'FE;
'H'45=>'H'FE;
'H'46=>'H'FE;
'H'47=>'H'FD;
'H'48=>'H'FD;
'H'49=>'H'FC;
'H'4A=>'H'FB;
'H'4B=>'H'FA;
'H'4C=>'H'FA;
'H'4D=>'H'F9;
'H'4E=>'H'F8;
'H'4F=>'H'F6;
'H'50=>'H'F5;
'H'51=>'H'F4;
'H'52=>'H'F3;
'H'53=>'H'F1;
'H'54=>'H'F0;
'H'55=>'H'EE;
'H'5A=>'H'E6;
'H'5B=>'H'E4;
'H'5C=>'H'E2;
'H'5D=>'H'E0;
'H'5E=>'H'DE;
'H'5F=>'H'DC;
'H'60=>'H'DA;
'H'61=>'H'D7;
'H'62=>'H'D5;
'H'63=>'H'D3;
'H'64=>'H'D0;
'H'65=>'H'CE;
'H'66=>'H'CB;
'H'67=>'H'C9;
'H'68=>'H'C6;
'H'69=>'H'C4;
'H'6A=>'H'C1;
'H'6B=>'H'BE;
'H'6C=>'H'BC;
'H'6D=>'H'B9;
'H'6E=>'H'B6;
'H'6F=>'H'B3;
'H'70=>'H'B0;
'H'71=>'H'AD;
'H'72=>'H'AA;
'H'73=>'H'A7;
'H'74=>'H'A5;
'H'75=>'H'A2;
'H'76=>'H'9E;
'H'77=>'H'9B;
'H'78=>'H'98;
'H'79=>'H'95;
'H'7A=>'H'92;
'H'7B=>'H'8F;
'H'7C=>'H'8C;
'H'7D=>'H'89;
'H'7E=>'H'86;
'H'7F=>'H'83;
'H'80=>'H'7F;
'H'85=>'H'70;
'H'86=>'H'6D;
'H'87=>'H'6A;
'H'88=>'H'67;
'H'89=>'H'64;
'H'8A=>'H'61;
'H'8B=>'H'5D;
'H'8C=>'H'5A;
'H'8D=>'H'58;
'H'8E=>'H'55;
'H'8F=>'H'52;
'H'90=>'H'4F;
'H'91=>'H'4C;
'H'92=>'H'49;
'H'93=>'H'46;
'H'94=>'H'43;
'H'95=>'H'41;
'H'96=>'H'3E;
'H'97=>'H'3B;
'H'98=>'H'39;
'H'99=>'H'36;
'H'9A=>'H'34;
'H'9B=>'H'31;
'H'9C=>'H'2F;
'H'9D=>'H'2C;
'H'9E=>'H'2A;
'H'9F=>'H'28;
'H'A0=>'H'25;
'H'A1=>'H'23;
'H'A2=>'H'21;
'H'A3=>'H'1F;
'H'A4=>'H'1D;
'H'A5=>'H'1B;
'H'A6=>'H'19;
'H'A7=>'H'17;
'H'A8=>'H'15;
'H'A9=>'H'14;
'H'AA=>'H'12;
'H'AB=>'H'11;
'H'B0=>'H'A ;
'H'B1=>'H'9 ;
'H'B2=>'H'7 ;
'H'B3=>'H'6 ;
'H'B4=>'H'5 ;
'H'B5=>'H'5 ;
'H'B6=>'H'4 ;
'H'B7=>'H'3 ;
'H'B8=>'H'2 ;
'H'B9=>'H'2 ;
'H'BA=>'H'1 ;
'H'BB=>'H'1 ;
'H'BC=>'H'1 ;
'H'BD=>'H'0 ;
'H'BE=>'H'0 ;
'H'BF=>'H'0 ;
'H'C0=>'H'0 ;
'H'C1=>'H'0 ;
'H'C2=>'H'0 ;
'H'C3=>'H'0 ;
'H'C4=>'H'1 ;
'H'C5=>'H'1 ;
'H'C6=>'H'1 ;
'H'C7=>'H'2 ;
'H'C8=>'H'2 ;
'H'C9=>'H'3 ;
'H'CA=>'H'4 ;
'H'CB=>'H'5 ;
'H'CC=>'H'5 ;
'H'CD=>'H'6 ;
'H'CE=>'H'7 ;
'H'CF=>'H'9 ;
'H'D0=>'H'A ;
'H'D1=>'H'B ;
'H'D2=>'H'C ;
'H'D3=>'H'E ;
'H'D4=>'H'F ;
'H'D5=>'H'11;
'H'D6=>'H'12;
'H'DB=>'H'1B;
'H'DC=>'H'1D;
'H'DD=>'H'1F;
'H'DE=>'H'21;
'H'DF=>'H'23;
'H'E0=>'H'25;
'H'E1=>'H'28;
'H'E2=>'H'2A;
'H'E3=>'H'2C;
'H'E4=>'H'2F;
'H'E5=>'H'31;
'H'E6=>'H'34;
'H'E7=>'H'36;
'H'E8=>'H'39;
'H E9=>'H'3B;
'H'EA=>'H'3E;
'H'EB=>'H'41;
'H'EC=>'H'43;
'H'ED=>'H'46;
'H'EE=>'H'49;
'H'EF=>'H'4C;
'H'F0=>'H'4F;
'H'F1=>'H'52;
'H'F2=>'H'55;
'H'F3=>'H'58;
'H'F4=>'H'5A;
'H'F5=>'H'5D;
'H'F6=>'H'61;
'H'F7=>'H'64;
'H'F8=>'H'67;
'H'F9=>'H'6A;
'H'FA=>'H'6D;
'H'FB=>'H'70;
'H'FC=>'H'73;
'H'FD=>'H'76;
'H'FE=>'H'79;
'H'FF=>'H'7C;
}
```

6.3 SUMMARY

In this chapter we have looked closely at the development of nine PAL applications. The variety of possible PAL applications is limited only by the designer's imagination and knowledge of the resources that commercially available PALs have to offer.

These examples have made use of PAL and GAL devices. It seems reasonable to repeat here that the predominant difference between PAL and GAL devices is in their programming environment, not in their basic architecture. PAL devices are programmed by blowing fuses. Once a PAL device is programmed, it cannot be reprogrammed. In contrast, a GAL device uses electrically alterable diode junctions, so the chip is reprogrammable. The GAL devices used in this text can be bulk erased prior to programming with a new fuse pattern. Consequently, in a development environment, it may be advantageous to do all design and development using the more expensive GAL devices, and once development is complete, switch to the PAL device for manufacturing.

A short review of some of the application notes from PLD vendors reveals a little more of the scope of these applications. Consider the following list of circuits that some of the vendors have developed and documented:

Six-bit shift register

Control store sequencer

Memory-mapped I/O (Detailed in Chapter 8)

Control logic for CPU board

Hexadecimal decoder/lamp driver

Between limits comparitor/register

Quad three-line/one-line data selector multiplexer

Four-bit counter with multiplexing

Four-bit up/down counter with shift

Hex keyboard scanner

ALU accumulator

Floppy drive control logic

The most important applications of PAL devices are those that are unique to the design and development environment. Those applications reviewed in this text are relatively general in nature. However, do not forget the availability of such design methods in the real world.

QUESTIONS AND PROBLEMS

1. Why was a Set-Reset type flip-flop used to drive the outputs of pin 13 and pin 12 of Practical Example 6.1?
2. Modify the design of Practical Example 6.1 to include an additional error output, ERR, which is set to a logic 1 if the serial input pattern is 11111100_2. Include a CUPL program that uses function statements to implement the input-forming logic.
3. Write a CUPL program for Problem 2 that uses the SEQUENCE statement to implement the input-forming logic.
4. Modify the design of Practical Example 6.2 to include two more products with product codes of 110_2 and 111_2, which are to be directed to destinations 2 and 3, respectively.
5. Repeat the design of Practical Example 6.3 using a different numeric sequence, yet retaining its glitch-free property.
6. Using a design method similar to that used in Practical Example 6.3, design a synchronous counter that provides eight unique glitch-free states.
7. Describe the purpose and operation of the stepper motor controller of Practical Example 6.4.

8. Describe in detail the design of the stepper motor controller of Practical Example 6.4.
9. Describe the purpose and operation of the DTACK generator of Practical Example 6.5.
10. Describe in detail the design of the DTACK generator of Practical Example 6.5.
11. Describe the purpose and operation of the priority interrupt encoder of Practical Example 6.6.
12. Describe in detail the design of the priority interrupt encoder of Practical Example 6.6.
13. Describe the purpose and operation of the programmable retriggerable monostable multivibrator of Practical Example 6.7.
14. Describe in detail the design of the programmable retriggerable monostable multivibrator of Practical Example 6.7.
15. Using a design similar to that used in Practical Example 6.8, design a four-bit combinational shifter.
16. Using a design similar to that used in Practical Example 6.8, design a six-bit combinational shifter.
17. Using a design similar to that used in Practical Example 6.8, design a four-bit combinational shifter.
18. What is the purpose of a shift linkage multiplexer, as described in Practical Example 6.8?
19. Using a design similar to Practical Example 6.11, design a digital waveform synthesizer/generator that generates a sawtooth wave.

CHAPTER 7

ROM-Based Controller Architecture

OBJECTIVES

After completing this chapter, you should be able to:

- Describe the advantage of using a ROM instead of a PAL or FPLA in a state machine design.
- Describe the general architecture of a ROM-based state machine.
- Design a ROM-based binary counter.
- Design a ROM-based state machine that responds to input decisions.
- Create a state table describing a defined state machine sequence for a ROM-based state machine.
- Convert a state table into a CUPL program using a ROM as the target device.
- Describe the operation and purpose of the pipeline register.
- Define the following terms:
  - Opcode
  - Instruction register
  - Microprogram
  - Microprogram ROM
  - Microprogram Word

7.1 OVERVIEW

The primary purpose of this text is to facilitate an understanding of how to use digital logic circuits for monitoring and controlling events. Up to this point we have investigated the use of state machines implemented with PAL, FPLA, and ROM devices in order to accomplish these activities. These devices provide the controller portion of circuits too complex to reasonably implement with discrete logic. However, as the complexity of the controller increases, the number of product terms needed to control a larger state machine increases dramatically. The primary feature of the PAL or FPLA is its ability to create logical expressions that have few minterms, though each minterm can have a large number of arguments. Recall that PALs and FPLAs are similar to ROMs, except they generate functions with far fewer minterms. When the complexity of the system becomes too complex for the PAL or FPLA devices, it is possible to move up in device complexity by using a ROM for the controller. If we base our understanding of programmable logic devices on the PAL or FPLA we can state that a ROM is similar to a PAL or FPLA and has the additional feature of generating all of the possible minterms of its input arguments.

The ROM-based controller fills the complexity gap between designs that can utilize a PAL or FPLA and designs that require the use of a microprocessor or microcontroller. The microprocessor is a state machine. Its clock-cycle-by-clock-cycle control is facilitated by a ROM-based controller rather than by the synchronous counter-based controller often used in simpler state machine designs. The purpose of this chapter is to introduce the use of ROM-based control instead of the PAL- or FPLA-based control of simpler state machines. We will observe example state machines and some of the inputs that stimulate them to execute defined operations.

The primary purpose of this chapter is to study the use of a ROM and a pipeline register to create state machines rather than discrete logic PALs or FPLAs.

7.2 ROM-BASED DESIGN

Recall from Chapter 1 that a ROM is actually nothing more than an AND-OR gate array that generates all possible minterms of the input arguments. The PAL or FPLA devices are very similar, except that they generate far fewer minterms. The P16R8 device illustrated in Figure 7.1 generates eight functions, each with a maximum of seven minterms. Each function is then presented to the inputs of a register of D-type flip-flops. If the gate array were to be replaced with a ROM, the only difference to the gate array circuitry would be to increase the number of minterms of each function from seven to the maximum, 2^N where N refers to the number of inputs to that array. The P16R8 generates a maximum of seven minterms at each function output. If a ROM were used for the gate array, each function would support 2^{16}, or 65,536 minterms at each function output. This implies that to emulate this PAL, the ROM architecture would have to be 65,536 bits by 8 bits, or 64 kilobytes, quite realistic by today's fabrication standards. Therefore, the minterm limit of PAL and FPLA devices is far exceeded when a ROM is used for the AND-OR gate array.

If the AND-OR gate array of Figure 7.1 is replaced with a ROM, the device of Figure 7.2 is realized.

Note the following characteristics of this device. The inputs to the address lines A_0 through A_7 come from external stimuli. The inputs to address lines A_8 through A_{15} come from the stable outputs of the D-type register. When the register is loaded with stable data, that data can be fed back to the address inputs to the ROM to look up the next ROM address while the register output identifies current data. This is the concept of a pipeline register. A **pipeline register** is a latch that has been placed within a closed loop in order to allow concurrent operations. In this case there are two operations that can take place at the same time: (1) the device outputs can control some process while (2) the feedback to the address inputs can be looking up the next ROM address. This phenomenon can be used to

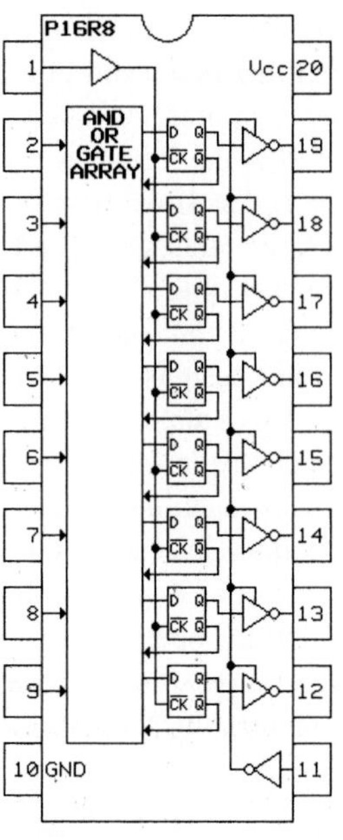

FIGURE 7.1 P16R8 programmable logic array.

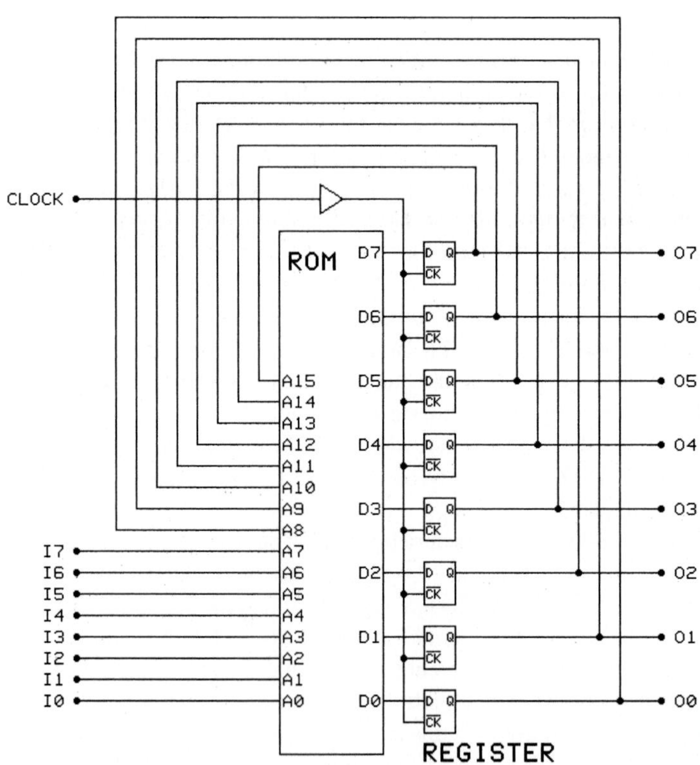

FIGURE 7.2 ROM/Register emulation of a 16R8 PAL device.

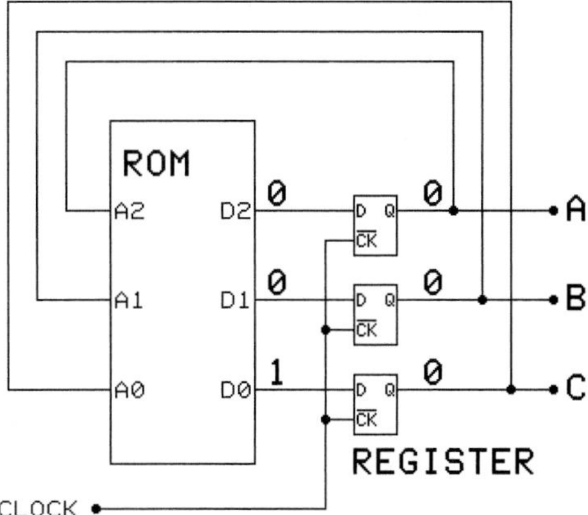

FIGURE 7.3 Three-bit ROM-based binary counter.

produce a very effective state machine design. It might be useful to note at this point that this is the same circuit that is used to control the circuitry within a microprocessor.

7.2.1 ROM-Based Binary Counters

Consider a simple ROM that contains the data shown in Table 7.1 and the ROM shown in Figure 7.3, which is programmed with that data.

Note that when the register is loaded with the value ABC = 000_2, that same value appears on the address inputs of the ROM. After a propagation period (the ROM access time), the data stored in the ROM at address 000_2 appears on its outputs, and that data is 001_2 according to Table 7.1. When the next asserted clock edge arrives, the register loads with that value, ABC = 001_2. Again, this value is fed back to the address inputs, causing the ROM to look up the data stored there: 010_2. As the register continues to receive clock pulses, the value in the register continues through the binary count, per the stored program data sequence. The basic rule for this design process is simple: **Each *present state* in the sequence is represented as a ROM address. The data stored at that address is the numeric value of the *next state* in the sequence.**

Table 7.2 illustrates how the ROM program of Table 7.1 might be reorganized to reflect this design method.

TABLE 7.1 Simple ROM program.

Address	*Data*
000	001
001	010
010	011
011	100
100	101
101	110
110	111
111	000

TABLE 7.2 Simple ROM program represented as a state table.

Present State (Address)	*Next State (Data)*
A B C	*A B C*
0 0 0	0 0 1
0 0 1	0 1 0
0 1 0	0 1 1
0 1 1	1 0 0
1 0 0	1 0 1
1 0 1	1 1 0
1 1 0	1 1 1
1 1 1	0 0 0

Practical Example 7.1: Simple ROM-Based State Machine Design

Consider the simple state machine defined by the state diagram of Figure 7.4. This is the same state machine described in Practical Example 2.2.

If the previously described rule of design is applied to this three-bit counter design, the state table of Table 7.3 is realized.

Recall that the design method requires us to store the next state as data at the address defined by the present state. That is, the counter design solution is simply defined by re-stating the state table as a ROM program. This ROM program is illustrated in Table 7.4.

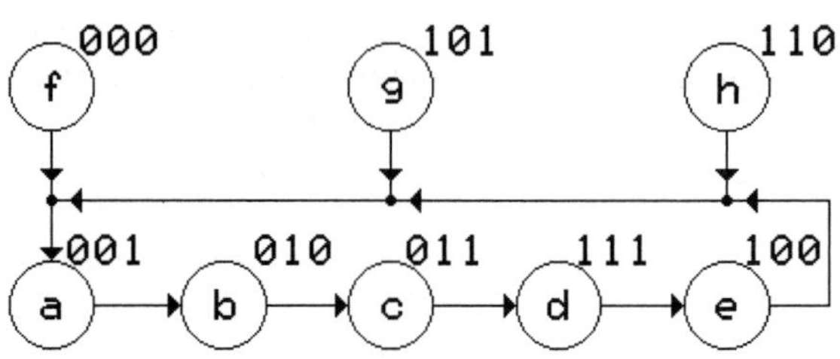

FIGURE 7.4 Practical Example 7.1, state diagram.

TABLE 7.3 Practical Example 7.1, state table.

Present State	*Next State*
A B C	*A B C*
0 0 0	0 0 1
0 0 1	0 1 0
0 1 0	0 1 1
0 1 1	1 1 1
1 0 0	0 0 1
1 0 1	0 0 1
1 1 0	0 0 1
1 1 1	1 0 0

TABLE 7.4 Practical Example 7.1, ROM program.

Address	*Data*
0 0 0	0 0 1
0 0 1	0 1 0
0 1 0	0 1 1
0 1 1	1 1 1
1 0 0	0 0 1
1 0 1	0 0 1
1 1 0	0 0 1
1 1 1	1 0 0

Practical Exercise 7.1: Simple ROM-Based Counter

Consider the state machine defined by Figure 7.5. Draw the state table, ROM program, and a schematic diagram of a ROM-based state machine that exhibits these characteristics.

7.2.2 Decision Making in ROM-Based Counters

When discrete logic PALs or FPLAs were used to design state machines, input stimuli were used to determine which of a multiple of next states would be selected. The same principle applies using ROM-based designs. The method used here will be to assign input stimuli to low-order address bits. For example, consider Figure 7.6. A single input to the state machine is assigned to a single low-order address bit. This truly implies that each state will have two possible next states by virtue of that low-order bit: It decodes to two separate addresses that contain two unique next states. The next state will be identical for both permutations of the low-order address when an unconditional state is encountered. This principle will be demonstrated in Practical Example 7.2.

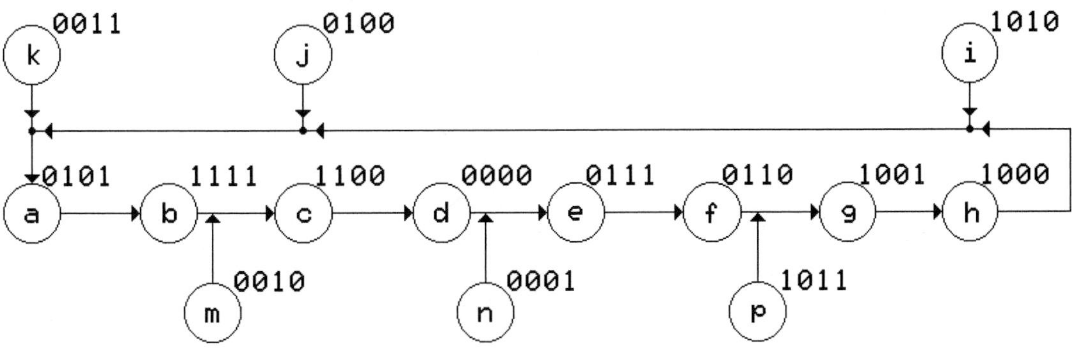

FIGURE 7.5 Practical Exercise 7.1, state diagram.

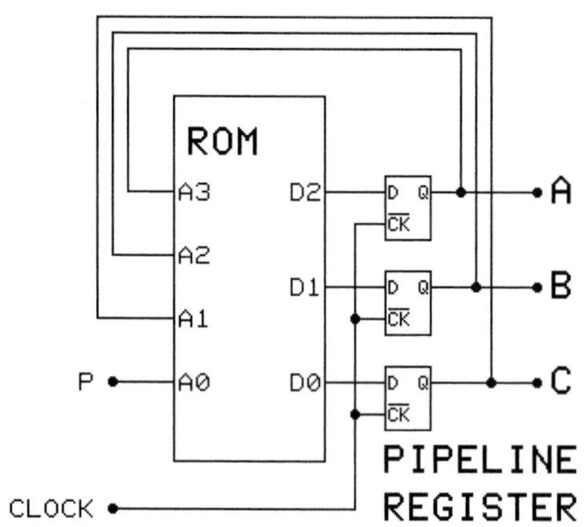

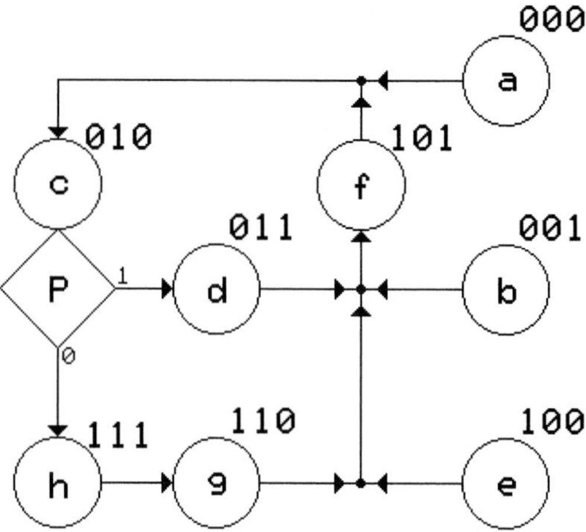

FIGURE 7.6 Schematic and state diagrams.

Practical Example 7.2: Decision Making in ROM-Based Counters

Consider the state machine defined in Figure 7.6. Determine the ROM-based design that will implement this state machine.

As with each of the other design methods, the first step in the solution is the development of a state table. In a ROM-based design we will include the minimum information illustrated in Table 7.5.

Review the state table of Table 7.5, comparing it with the state diagram. The translation of the state table to a ROM program is rather straightforward. Using the input stimuli as a least-significant address bit, simply replace all of its "don't care" states with each of its possible permutations, and keep the same next state for both of those addresses. Consider the solution in Table 7.6.

TABLE 7.5 Practical Example 7.2, state table.

	Present	*Input*	*Next*
	A B C	*P*	*A B C*
a	0 0 0	X	0 1 0
b	0 0 1	X	1 0 1
c	0 1 0	0	1 1 1
		1	0 1 1
d	0 1 1	X	1 0 1
e	1 0 0	X	1 0 1
f	1 0 1	X	0 1 0
g	1 1 0	X	1 0 1
h	1 1 1	X	1 1 0

TABLE 7.6 Practical Example 7.2, ROM program.

	Address	*Data*
	$A_3 A_2 A_1 A_0$	$D_2 D_1 D_0$
a	0 0 0 0	0 1 0
	0 0 0 1	0 1 0
b	0 0 1 0	1 0 1
	0 0 1 1	1 0 1
c	0 1 0 0	1 1 1
	0 1 0 1	0 1 1
d	0 1 1 0	1 0 1
	0 1 1 1	1 0 1
e	1 0 0 0	1 0 1
	1 0 0 1	1 0 1
f	1 0 1 0	0 1 0
	1 0 1 1	0 1 0
g	1 1 0 0	1 0 1
	1 1 0 1	1 0 1
h	1 1 1 0	1 1 0
	1 1 1 1	1 1 0

Another way of looking at the use of the ROM in this configuration is to note that it contains a "look-up" table. That is, we apply the address to the ROM for the purpose of looking up the next state in the controller. The ROM has been relegated to the simple and mundane function of a "look-up" table. Actually, any design that makes use of a ROM can be viewed in this way. Also, if a PAL or FPLA device is simply a subset of the ROM circuit, it can be viewed in the same way. When applied in a state machine, the gate array circuit is being used to look up the state of the system for the next clock cycle while the current clock cycle status is stabilized by a register of some form.

Carefully compare Table 7.5 with Table 7.6. Note how the input stimulus column of the state table was added to the ROM address in the least significant bit position. When the next state was conditional, the data representing the next state reflected that condition.

Practical Example 7.3: Handling Multiple Inputs

Let's consider the solution of a state machine problem that exhibits multiple input stimuli. The design method is the same. Practical Example 3.4 provided a good solution using discrete logic devices in the solution. Consider its solution using a ROM-based state machine. First, consider Figure 7.7.

When converted from a state table to a ROM look-up table, the table of Figure 7.8 results. Note that, with three input stimuli, each state has a possibility of as many as eight destination states. When input stimuli are assigned to low-order address bits, this also results

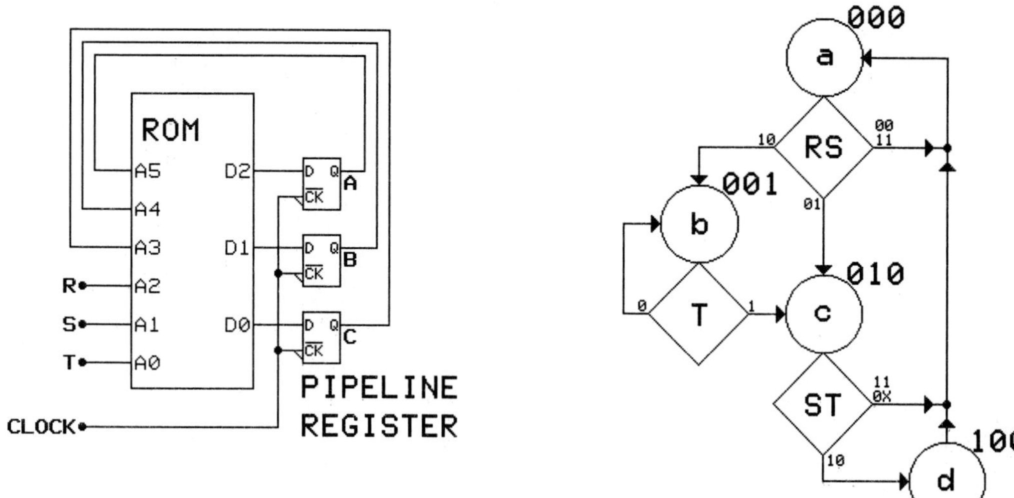

FIGURE 7.7 Practical Example 7.3, schematic and state diagrams.

in a commensurate increase in needed gate array complexity. Each address defined as a state in the state machine will have eight possible next addresses. Though the translation of the state table to a ROM program becomes exponentially more difficult as more inputs are added, it is still a simple, yet tedious process. **When input stimuli are added to the design of a state machine, the gate array complexity needed to generate minterms increases dramatically. The minterm-generating limits of PAL and FPLA devices may prove insufficient, resulting in the need of a ROM/pipeline register solution.**

Consider the solution shown in Table 7.7. Note that only those states that were defined in the state diagram are included in the ROM program. The "don't care" stable states in the state table require no ROM entries, since those states are never encountered in the

FIGURE 7.8 Practical Example 7.3, state table.

STATE	N	PRES ABC	INPUT RST	NEXT ABC
a	0	000	00*	000
			01*	010
			10*	001
			11*	000
b	1	001	**0	001
			**1	010
c	2	010	*00	000
			*01	000
			*10	100
			*11	000
	3	011	***	***
d	4	100	***	000
	5	101	***	***
	6	110	***	***
	7	111	***	***

* = Don't Care

TABLE 7.7 Practical Example 7.3, ROM program.

	Address	*Data*
	$A_5\ A_4\ A_3\ A_2\ A_1\ A_0$	$D_2\ D_1\ D_0$
a	0 0 0 0 0 0	0 0 0
	0 0 0 0 0 1	0 0 0
	0 0 0 0 1 0	0 1 0
	0 0 0 0 1 1	0 1 0
	0 0 0 1 0 0	0 0 1
	0 0 0 1 0 1	0 0 1
	0 0 0 1 1 0	0 0 0
	0 0 0 1 1 1	0 0 0
b	0 0 1 0 0 0	0 0 1
	0 0 1 0 0 1	0 1 0
	0 0 1 0 1 0	0 0 1
	0 0 1 0 1 1	0 1 0
	0 0 1 1 0 0	0 0 1
	0 0 1 1 0 1	0 1 0
	0 0 1 1 1 0	0 0 1
	0 0 1 1 1 1	0 1 0
c	0 1 0 0 0 0	0 0 0
	0 1 0 0 0 1	0 0 0
	0 1 0 0 1 0	1 0 0
	0 1 0 0 1 1	0 0 0
	0 1 0 1 0 0	0 0 0
	0 1 0 1 0 1	0 0 0
	0 1 0 1 1 0	1 0 0
	0 1 0 1 1 1	0 0 0
d	1 0 0 0 0 0	0 0 0
	1 0 0 0 0 1	0 0 0
	1 0 0 0 1 0	0 0 0
	1 0 0 0 1 1	0 0 0
	1 0 0 1 0 0	0 0 0
	1 0 0 1 0 1	0 0 0
	1 0 0 1 1 0	0 0 0
	1 0 0 1 1 1	0 0 0

environment. Again, compare the state table and the ROM program carefully. You will note that essentially the ROM program is the same as the state table when all of the "don't care" logic values on the input stimuli have been expanded for all of their possible permutations. Note also that each output function, D2, D1, and D0, is made up of a selection from 32 minterms (four states with eight minterms each). If the true minterms for each output are summated and expressed as a Boolean function, the same result as described in Practical Example 4.4 will be realized.

This design method bypasses the need for minimizing the output functions by keeping the state table in its original form, translating it to a ROM look-up table. Had we tried minimizing each function, we would often find that the resulting expressions are far too complex to implement with any device other than a ROM. Also, we will find that this step becomes a tedious, error-prone, and probably unnecessary task. Instead of determining the output expressions, we will take advantage of the TABLE statement in the CUPL language to define the look-up table to be programmed into the ROM.

The ROM defined in Table 7.7 has six address lines and three data lines. Therefore, we will need at least a 64 x 3 architecture. The word size and address range of commercial ROM devices are always multiples of the exponent of two. If we consider this, we will look for a 64 x 4 device for implementation. The smallest available ROM in Appendix A is the RA8P4 architecture, available from National Semiconductor as the DM74S387. Using the TABLE statement in conjunction with the FIELD statement in the CUPL language, the following CUPL program results:

```
Name      EXP0703a;
Date      01/01/96;
Designer  J.W. Carter;
Device    RA8P4;
/*****************************************************************/
/* Practical Example 7.3                                         */
/*   State machine with three inputs, uses DM74S387 256x4 PROM   */
/*   All ROM data is manually included in the TABLE construct    */
/*****************************************************************/
/** Inputs **/
PIN [3..1] = [A2..A0];              /* Inputs                    */
PIN [6..4] = [A5..A3];              /* Present State             */
/** Outputs **/
PIN [11..9]  = [D2..D0];            /* Next State                */
FIELD input  = [A5..A0];
FIELD output = [D2..D0];
TABLE input =>output {
   /* State-a */           /* State-b */           /* State-c */           /* State-d */
'B'000000 => 'B'000;  'B'001000 => 'B'001;  'B'010000 => 'B'000;  'B'100000 => 'B'000;
'B'000001 => 'B'000;  'B'001001 => 'B'010;  'B'010001 => 'B'000;  'B'100001 => 'B'000;
'B'000010 => 'B'010;  'B'001010 => 'B'001;  'B'010010 => 'B'100;  'B'100010 => 'B'000;
'B'000011 => 'B'010;  'B'001011 => 'B'010;  'B'010011 => 'B'000;  'B'100011 => 'B'000;
'B'000100 => 'B'001;  'B'001100 => 'B'001;  'B'010100 => 'B'000;  'B'100100 => 'B'000;
'B'000101 => 'B'001;  'B'001101 => 'B'010;  'B'010101 => 'B'000;  'B'100101 => 'B'000;
'B'000110 => 'B'000;  'B'001110 => 'B'001;  'B'010110 => 'B'100;  'B'100110 => 'B'000;
'B'000111 => 'B'000;  'B'001111 => 'B'010;  'B'010111 => 'B'000;  'B'100111 => 'B'000;
}
```

Another method can be used to create the CUPL program. This method takes advantage of the ability to represent the "don't care" state in the CUPL language with the character "X." The ROM program was developed by expanding all of the "don't care" states. If these are left intact, we can write the CUPL program from the state table directly. Consider the following solution:

```
Name      EXP0703b;
Date      01/01/96;
Designer  J.W. Carter;
Device    RA8P4;
```

```
/******************************************************************/
/* Practical Example 7.3                                          */
/*   State machine with three inputs, uses DM74S387 256x4 PROM    */
/*   TABLE construct is built from the state table directly.      */
/******************************************************************/
/** Inputs **/
PIN [3..1] = [A2..A0];        /* Inputs                           */
PIN [6..4] = [A5..A3];        /* Present State                    */
/** Outputs **/
PIN [11..9]  = [D2..D0];      /* Next State                       */
FIELD input  = [A5..A0];
FIELD output = [D2..D0];
TABLE input =>output {
/* State-a */ 'b'00000X => 'b'000;
              'b'00001X => 'b'010;
              'b'00010X => 'b'001;
              'b'00011X => 'b'000;
/* State-b */ 'b'001XX0 => 'b'001;
              'b'001XX1 => 'b'010;
/* State-c */ 'b'010X0X => 'b'000;
              'b'010X10 => 'b'100;
              'b'010X11 => 'b'000;
/* State-d */ 'b'100XXX => 'b'000;
}
```

Compare carefully this CUPL program with the state table of Figure 7.8. Of the two programming methods, the latter is probably the preferable one. It is both simpler to write and its appearance is closer to the state table, making it easier to relate to the problem. Since the state table was written from information in the state diagram, it is possible (after some experience) to write the CUPL program directly from the state diagram.

Figure 7.9 illustrates the final schematic diagram of this binary counter. Note how the unused address lines were tied to ground to ensure the use of the lower ROM addresses. The unused data line is left open. An external register of D-type flip-flops is used

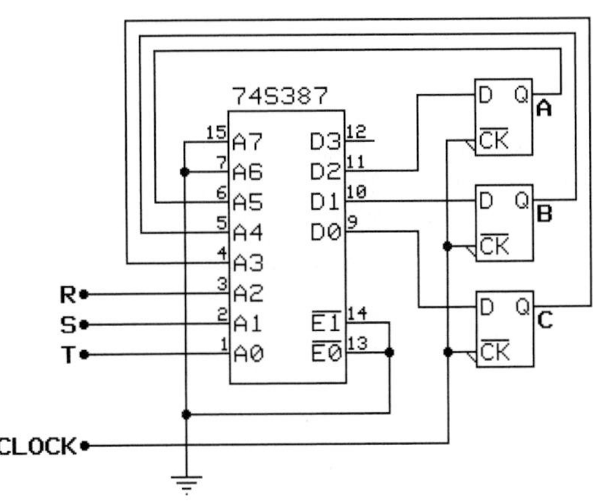

FIGURE 7.9 Practical Example 7.3, final schematic diagram.

to hold the current state stable while the ROM is looking up the data for the next state. Again, a register used in this fashion is referred to as a pipeline register.

Practical Exercise 7.2: Handling Multiple Outputs in ROM-Based Designs

Consider the state machine defined by Figure 7.9. Develop the state diagram, schematic diagram, state table, and ROM program for a ROM-based solution for this state machine. The state diagram is shown in Figure 7.10.

7.2.3 Output-Forming Logic Design

The implementation of output-forming logic in a ROM-based state machine design is really quite simple if we limit the transition definitions to the same limits placed on other programmable logic devices. This limit states that the output-forming logic transitions can take place only at the beginning and ending of any state. If we want to generate the signals at other transition points, such as DSB, DSE, and DDSE, we will need to add the circuitry to do so. The design of that circuitry is discussed in Chapter 4. Output-forming arguments are implemented in a ROM-based state machine by the use of an additional data bit. The logical value of the output-forming logic for any given state is simply assigned to the respective bit in the respective address. That is, if in state n, output X is to be a logic 1, then in the ROM locations for state n, an additional data bit, X, will be set to a logic 1. Conditional outputs will be set to their respective values when the proper conditions on the input stimuli are met.

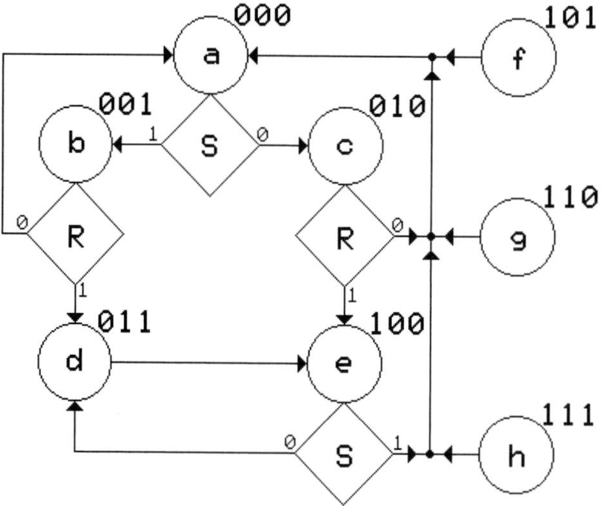

FIGURE 7.10 Practical Exercise 7.2, state diagram.

Practical Example 7.4: A Simple Soda Machine Controller

Consider the state diagram of Figure 7.11. This is the state diagram for the soda machine controller of Practical Example 4.5. A review of that example might be useful at this point. The controller has four inputs, CR_i, Q_i, C_i, and D_i, which respectively refer to "coin return input," "quarter input," "cola input," and "diet input." These are used to drive three output signals, Q^o, C^o, and D^o, which respectively refer to "quarter output," "cola output," and "diet output."

Note that the state diagram defines the output-forming logic in a way that each pulse is TRUE from DSB to SE. In order to accomplish this, we will enable the output-forming logic using the clock, thus delaying the signal by a half clock cycle. (This is alternate-state timing or AST as discussed in previous chapters.) Again, the output signals are generated by adding data bits to the ROM.

Consider Figure 7.12. There are four inputs to this circuit too. These are assigned to the four least significant address bits. Three outputs from the circuit will be used to control the soda machine outputs. These are assigned to the three least significant data bits of the ROM. Finally, the three-bit feedback counter is assigned to the next significant bits of the

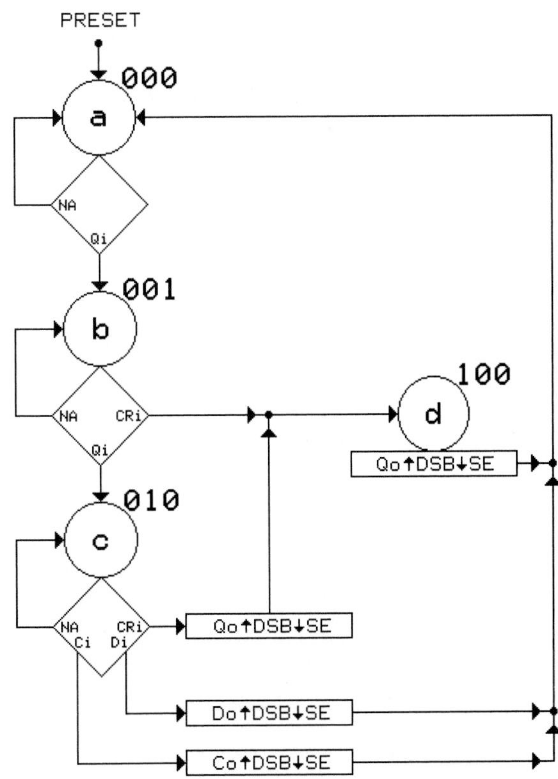

FIGURE 7.11 Practical Example 7.4, state diagram.

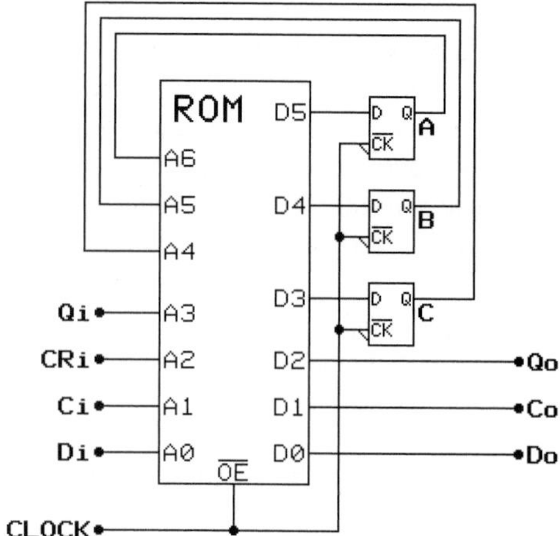

FIGURE 7.12 Practical Example 7.4, block diagram.

address and data lines. This assignment pattern is a convenient convention, as will be evident when we look at the state table. This table is illustrated in Figure 7.13.

Note that in creating the state table we determined the next state for each present state using the same method as with other programmable devices. However, since we will be using a look-up table rather than designing feedback logic, it is not necessary to determine the minimized functions for D_A, D_B, and D_C. (See Practical Example 4.5 to review the minimization.) Eliminating this step from the design process eliminates a lot of opportunity for error.

Recall the method for creating the ROM program from the state table. Though simple, it is tedious. We simply assign the state table columns to the address and data lines as prescribed and list all permutations of bits described in the state table, as shown in Table 7.8.

One of the features of the soda machine controller was that no two inputs could be active at the same time. These input permutations could be ignored, so the data stored at

	PRES ABC	INPUTS Qi	Cri	Ci	Di	NEXT ABC	Qo	Co	Do
a	000	0	X	X	X	000	0	0	0
		1	X	X	X	001			
b	001	0	0	X	X	001	0	0	0
		1	X	X	X	010			
		X	1	X	X	100			
c	010	X	0	0	0	010	0	0	0
		X	1	X	X	100	↑DSB↓SE	0	0
		X	X	1	X	000	0	↑DSB↓SE	0
		X	X	X	1	000	0	0	↑DSB↓SE
	011	X	X	X	X	XXX	X	X	X
d	100	X	X	X	X	000	↑DSB↓SE	0	0
	101	X	X	X	X	XXX	X	X	X
	110	X	X	X	X	XXX	X	X	X
	111	X	X	X	X	XXX	X	X	X

FIGURE 7.13 Practical Example 7.4, state table.

the addresses defined by those inputs are recorded in the table as "don't care" states. Carefully examine how the state table is used to determine each line of the ROM program. Again, each "don't care" in the state table input columns is expanded for all of its permutations. The next states of the three counter bits A, B, and C are assigned to the most significant data bits of the ROM and programmed to match the state table. Finally, the output bits are assigned to the least significant data bits, and assigned a logical 1 whenever the input bit pattern dictates. Again, the DSB to SE transition will be controlled externally.

TABLE 7.8 Practical Example 7.4, ROM program. (continued on next page)

	Address	*Data*
	$A\ B\ C\ Q_i\ Cr_i\ C_i\ D_i$	$D_A\ D_B\ D_C\ Q_o\ C_o\ D_o$
	$A_6\ A_5\ A_4\ A_3\ A_2\ A_1\ A_0$	$D_5\ D_4\ D_3\ D_2\ D_1\ D_0$
a	0 0 0 0 0 0 0	0 0 0 0 0 0
	0 0 0 0 0 0 1	0 0 0 0 0 0
	0 0 0 0 0 1 0	0 0 0 0 0 0
	0 0 0 0 0 1 1	0 0 0 0 0 0
	0 0 0 0 1 0 0	0 0 0 0 0 0
	0 0 0 0 1 0 1	0 0 0 0 0 0
	0 0 0 0 1 1 0	0 0 0 0 0 0
	0 0 0 0 1 1 1	0 0 0 0 0 0
	0 0 0 1 0 0 0	0 0 1 0 0 0
	0 0 0 1 0 0 1	0 0 1 0 0 0
	0 0 0 1 0 1 0	0 0 1 0 0 0
	0 0 0 1 0 1 1	0 0 1 0 0 0
	0 0 0 1 1 0 0	0 0 1 0 0 0
	0 0 0 1 1 0 1	0 0 1 0 0 0
	0 0 0 1 1 1 0	0 0 1 0 0 0
	0 0 0 1 1 1 1	0 0 1 0 0 0
b	0 0 1 0 0 0 0	0 0 1 0 0 0
	0 0 1 0 0 0 1	0 0 1 0 0 0
	0 0 1 0 0 1 0	0 0 1 0 0 0
	0 0 1 0 0 1 1	0 0 1 0 0 0
	0 0 1 0 1 0 0	1 0 0 0 0 0
	0 0 1 0 1 0 1	1 0 0 0 0 0
	0 0 1 0 1 1 0	1 0 0 0 0 0
	0 0 1 0 1 1 1	1 0 0 0 0 0
	0 0 1 1 0 0 0	0 1 0 0 0 0
	0 0 1 1 0 0 1	0 1 0 0 0 0
	0 0 1 1 0 1 0	0 1 0 0 0 0
	0 0 1 1 0 1 1	0 1 0 0 0 0
	0 0 1 1 1 0 0	X X X X X X
	0 0 1 1 1 0 1	X X X X X X
	0 0 1 1 1 1 0	X X X X X X
	0 0 1 1 1 1 1	X X X X X X

TABLE 7.8 (continued)

	Address							Data					
	A	B	C	Q_i	Cr_i	C_i	D_i	D_A	D_B	D_C	Q_o	C_o	D_o
	A_6	A_5	A_4	A_3	A_2	A_1	A_0	D_5	D_4	D_3	D_2	D_1	D_0
c	0	1	0	0	0	0	0	0	1	0	0	0	0
	0	1	0	0	0	0	1	0	0	0	0	0	1
	0	1	0	0	0	1	0	0	0	0	0	1	0
	0	1	0	0	0	1	1	X	X	X	X	X	X
	0	1	0	0	1	0	0	1	0	0	1	0	0
	0	1	0	0	1	0	1	X	X	X	X	X	X
	0	1	0	0	1	1	0	X	X	X	X	X	X
	0	1	0	0	1	1	1	X	X	X	X	X	X
	0	1	0	1	0	0	0	0	1	0	0	0	0
	0	1	0	1	0	0	1	0	0	0	0	0	1
	0	1	0	1	0	1	0	0	0	0	0	1	0
	0	1	0	1	0	1	1	X	X	X	X	X	X
	0	1	0	1	1	0	0	1	0	0	1	0	0
	0	1	0	1	1	0	1	X	X	X	X	X	X
	0	1	0	1	1	1	0	X	X	X	X	X	X
	0	1	0	1	1	1	1	X	X	X	X	X	X
d	1	0	0	0	0	0	0	0	0	0	1	0	0
	1	0	0	0	0	0	1	0	0	0	1	0	0
	1	0	0	0	0	1	0	0	0	0	1	0	0
	1	0	0	0	0	1	1	0	0	0	1	0	0
	1	0	0	0	1	0	0	0	0	0	1	0	0
	1	0	0	0	1	0	1	0	0	0	1	0	0
	1	0	0	0	1	1	0	0	0	0	1	0	0
	1	0	0	0	1	1	1	0	0	0	1	0	0
	1	0	0	1	0	0	0	0	0	0	1	0	0
	1	0	0	1	0	0	1	0	0	0	1	0	0
	1	0	0	1	0	1	0	0	0	0	1	0	0
	1	0	0	1	0	1	1	0	0	0	1	0	0
	1	0	0	1	1	0	0	0	0	0	1	0	0
	1	0	0	1	1	0	1	0	0	0	1	0	0
	1	0	0	1	1	1	0	0	0	0	1	0	0
	1	0	0	1	1	1	1	0	0	0	1	0	0

The CUPL program can be written from either the state table or the ROM program. The previous example illustrated that it is advantageous to write the CUPL program from the state table. When this is done, the following CUPL program is realized:

```
Name       EXP0704;
Date       01/01/96;
Designer   J.W. Carter;
Device     RA8P8;
```

```
/*****************************************************************/
/* Practical Example 7.4                                         */
/*    Soda Machine Controller using a DM74S471 256 x 8 PROM      */
/*****************************************************************/
/** Inputs **/
PIN [5..1]   = [A4..A0];          /* Present State & Inputs      */
PIN [18..17] = [A6..A5];
/** Outputs **/
PIN [9..6]   = [D3..D0];          /* Next State                  */
PIN [12..11] = [D5..D4];          /* Outputs                     */
FIELD input  = [A6..A0];
FIELD output = [D5..D0];
TABLE input =>output {
/* State-a */ 'b'0000XXX => 'b'000000;
              'b'0001XXX => 'b'001000;
/* State-b */ 'b'00100XX => 'b'001000;
              'b'0011XXX => 'b'010000;
              'b'001X1XX => 'b'100000;
/* State-c */ 'b'010X000 => 'b'010000;
              'b'010X1XX => 'b'100100;
              'b'010XX1X => 'b'000010;
              'b'010XXX1 => 'b'000001;
/* State-d */ 'b'100XXXX => 'b'000100;
}
```

Figure 7.14 is an illustration of the final schematic diagram of the soda machine controller that uses a 74S471 PROM and a pipeline register.

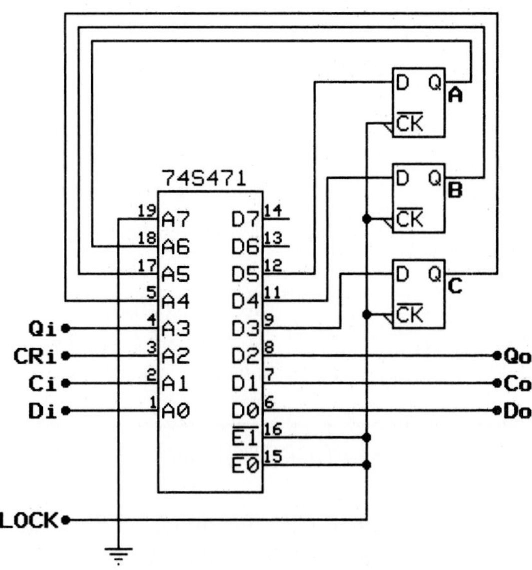

FIGURE 7.14 Practical Example 7.4, final schematic diagram.

Practical Exercise 7.3: A Simple Soda Machine Controller

Consider the state machine defined by Practical Example 7.4. Develop the CUPL program from the ROM program. This is probably an exercise in tedium, but the effort will help to hone the skill of program writing.

Practical Exercise 7.4: A Simple Candy Machine Controller

Using the same approach as that in Practical Example 7.3, design a candy machine controller with the following constraints. Use any convenient ROM architecture.

Inputs:	CR_i, Press coin return	
	Q_i, Input quarter	
	CG_i, Select chewing gum	(25¢ cost)
	PC_i, Select potato chips	(50¢ cost)
	CB_i, Select chocolate bar	(50¢ cost)
Outputs:	Q_o, Dispense quarter	
	CG_o, Dispense chewing gum	
	PC_o, Dispense potato chips	
	CB_o, Dispense chocolate bar	

Practical Exercise 7.5: A Moderate Candy Machine Controller

Using the same approach as that used in Practical Example 7.3, design a candy machine controller that has the constraints of Practical Exercise 7.4. In addition, implement the following inputs and outputs:

Inputs:	D_i, Input dime	
	CC_i, Select cheese crackers	(40¢ cost)
Outputs:	D_o, Dispense dime	
	N_o, Dispense nickel	
	CC_o, Dispense cheese crackers	

7.3 MICROPROGRAMMED DESIGN

Up to this point in our state machine designs, the devices developed executed only a single function. That is, the PAL, FPLA, or ROM was programmed to perform a single task defined by a single state diagram. Consider the following concept: **When the device used to implement the state machine is a ROM, the function executed by that state machine is defined by a stored program. Additional input (address) lines can be used to point to multiple stored programs. The result is a system where multiple functions can be executed with the same hardware.**

We have just made a quantum leap in the power of the state machine that can be designed using the methods previously discussed in this text. Consider Figure 7.15. We find that this circuit is very similar to those previously discussed, with two primary changes. First, we have added three significant address bits, and we have assigned a three-bit latch to stabilize the course for those address bits. This latch, referred to as an **instruction register,** holds the binary pattern that defines which, of a multiple of state machine programs, is being executed. The binary value that defines this is referred to as an **operation code** or **opcode.** Second, the lines that drive the external circuitry under control are buffered by the pipeline register. Therefore, the contents of the pipeline register will always contain the current status of the control lines and the address of the next state to be executed following the clock edge.

By adding a three-bit instruction register to the most significant three bits of the ROM address, we have produced a state machine that can execute any of eight state machine programs. Each pattern in the instruction register defines a memory partition within which the state machine program for one application is stored. Since there are three bits in the instruction register, eight state machine patterns can be stored in the ROM. The example shown in Figure 7.15 uses a 1024 x 8 ROM to contain the binary patterns that define the state machine function. Figure 7.16 illustrates a memory map of the ROM as defined by its address line connections.

The memory map is a graphic linear list of memory addresses with the most significant address at the top. Since this example uses a 10-bit address, the most significant address is $11{,}1111{,}1111_2$, and the least significant address is $00{,}0000{,}0000_2$.

The most significant three bits, A_9, A_8, and A_7, stored in the instruction register, divide the ROM into eight partitions, each containing a state machine program that can be defined by a unique state diagram or state table.

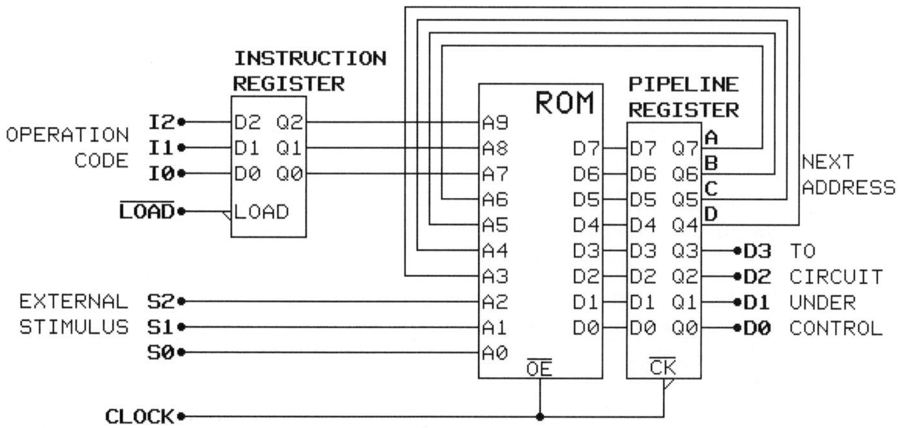

FIGURE 7.15 Microprogrammed control unit.

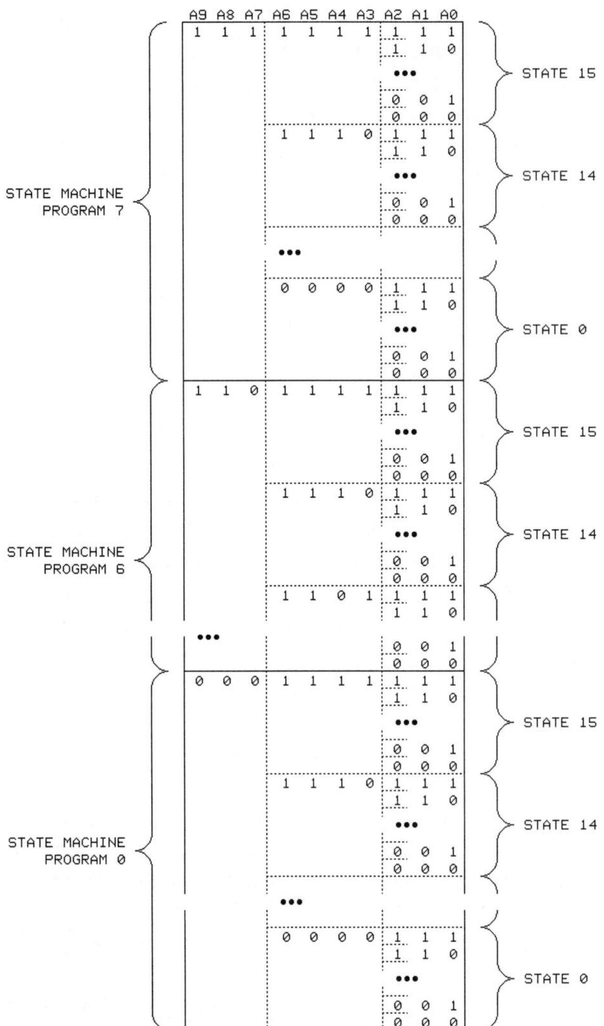

FIGURE 7.16 State machine memory map.

This example also has four bits allotted to the binary counter assigned to the next most significant four bits, A_6, A_5, A_4, and A_3. This provides for the implementation of state machines with up to 16 stable states.

Finally, the least significant three bits, A_5, A_4, and A_3, provide for individual patterns based on external stimulus.

Again, the instruction register contains a binary pattern, or an opcode, that defines which ROM partitition is addressed. Within each partition is the binary program for a defined state machine function. This binary program may be referred to as a microprogram. The **microprogram** is the set of binary words in the ROM that executes an instruction defined in the instruction register. When a ROM is used to store a microprogram, it is referred to as a **microprogram ROM,** and the data word it generates is referred to as the **microprogram word.**

This architecture, with its three-bit instruction register, allows for storage of eight microprograms. A common application of a microprogram ROM is in the computer control unit portion of a monolithic complex instruction set (CISC) processor, such as a Motorola 6800 or 68000, an Intel 80X86 or Pentium, or any of many other processors. In these applications, the control lines from the pipeline register are used to define the sequence of operations performed within the processor for any given instruction, where that instruction is stored in the instruction register. We will discuss this concept in more detail later in the text.

Practical Example 7.5: Part Pattern Cutter Controller

Consider for a moment a piece of industrial equipment that is designed to follow a state machine program in the process of cutting a geometric pattern on blank stock. The design of such a system is similar in form to that which this text has been presenting and can involve the definition of a state machine using either a state diagram or state table. Let's assume that the state machine to run the cutter drives four control lines: GO_LEFT, GO_RIGHT, GO_UP, and GO_DOWN. The state machine also responds to limit alarm inputs, MAX_LEFT, MAX_RIGHT, MAX_UP, and MAX_DOWN.

Figure 7.17 is a physical sketch of such a system. Note first the manual switch at the top of the figure. When the CUTTER HOME switch is closed, the buffers to the controller are disabled and the DC motors move the cutter to the left and down until they hit the limits. A HOME signal is also provided to the controller indicating that the homing process is being executed. The operator is then free to place one of 16 pattern codes on the four switches P_3, P_2, P_1, and P_0. These switches indicate which of 16 state machine programs will be run. These programs will provide signals to the motors in order to move the cutter head around the table. Another output from the state machine toggles the cutter head between up and down positions. Let's assume that the cutter contacts and cuts the stock when it is down and rides above the stock when it is up. The motors are geared DC motors that will run at a constant rate of speed when provided a constant voltage. Therefore, control of the cutter head can be attained by providing positive DC signals to the motor control lines. That is, if a signal is applied to the GO_LEFT line, the horizontal motor will rotate, pulling the cutter to the left.

We will use a ROM to provide the combinational arguments for up to 16 state machine functions. Each one is designed to cut a unique pattern in the stock. Figure 7.18 illustrates a block diagram of the state machine that would be used to drive this hardware. Note the application of the input and output lines of the pattern cutter to the state machine. The four-bit code identifying the pattern to be cut is placed on the four most significant address lines. The input stimuli, coming from the four alarms, are placed on the least significant address lines. The remaining address lines are used to control the counter. The least significant bits of the output data are used to control the cutter motors and up/down functions. The most significant bits of the output data are used to determine the next ROM address to be executed.

Let's consider an example where the pattern code is 0000_2. The microprogram for that pattern is stored in ROM addresses 000_{16} through $0FF_{16}$. Figure 7.19 illustrates this example pattern that cuts a rectangle out of the stock. Note that the motors are displayed in

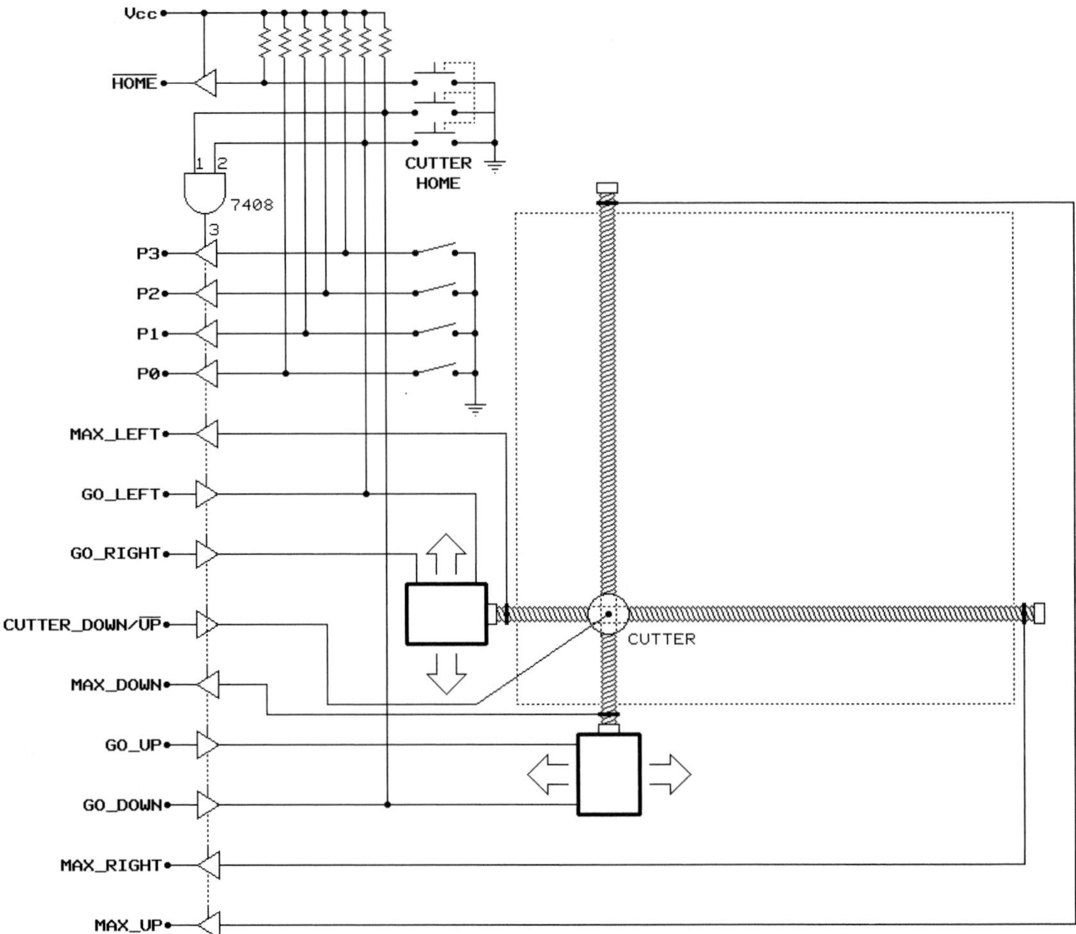

FIGURE 7.17 Part pattern cutter schematic diagram.

the home position. The cutter surface is marked in segments such that the motors will move one segment per state machine cycle. When the respective motor control line is a logic 1, the motor will move in the indicated direction. Consider the following cutter pattern algorithm based on this example:

State 0: CUTTER_UP; next 1

State 1: CUTTER_UP; GO_UP;
if MAX_UP = 1 then next 13 else next 2;

State 2: CUTTER_UP; GO_UP; GO_RIGHT;
if MAX_UP = 1 or MAX_RIGHT = 1 then next 13 else next 3;

State 3: CUTTER_DOWN; GO_UP;
if MAX_UP = 1 then next 13 else next 4;

State 4: CUTTER_DOWN; GO_RIGHT;
if MAX_RIGHT = 1 then next 13 else next 5;

State 5: CUTTER_DOWN; GO_RIGHT;
if MAX_RIGHT = 1 then next 13 else next 6;

State 6: CUTTER_DOWN; GO_RIGHT;
if MAX_RIGHT = 1 then next 13 else next 7;

State 7: CUTTER_DOWN; GO_DOWN;
if MAX_DOWN = 1 then next 13 else next 8;

State 8: CUTTER_DOWN; GO_LEFT;
if MAX_LEFT = 1 then next 13 else next 9;

State 9: CUTTER_DOWN; GO_LEFT;
if MAX_LEFT = 1 then next 13 else next 10;

State 10: CUTTER_DOWN; GO_LEFT;
if MAX_LEFT = 1 then next 13 else next 11;

State 11: CUTTER_UP; GO_DOWN;
if MAX_DOWN = 1 then next 13 else next 12;

State 12: CUTTER_UP; GO_DOWN; GO_LEFT; next 13;

State 13: CUTTER_UP; next 13;

Note that when the motor is moved, the status of the limit switches is tested. If the output of the limit switch is a logic 1, an error has occurred, and the state machine moves to state 12, where the cutter is raised and the motors are not moved.

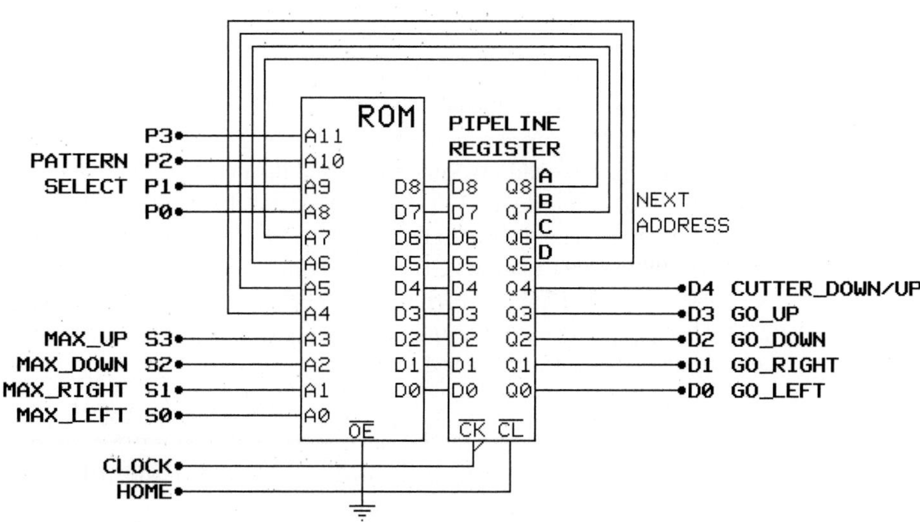

FIGURE 7.18 Practical Example 7.5, state machine.

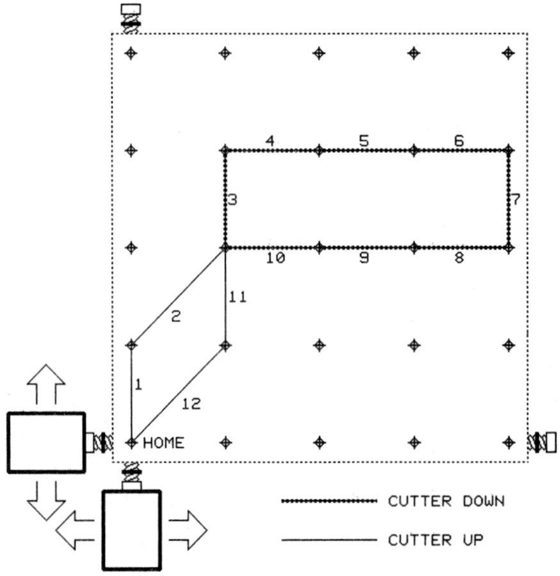

FIGURE 7.19 Practical Example 7.15, pattern 0, cutter path.

Figure 7.20 illustrates the state diagram of this example. Though the state diagram is not really necessary for the completion of this design, its development may improve understanding of the state machine sequence, particularly as it executes decisions.

Since we are using a ROM in this example, we cannot take advantage of the SEQUENCE statement in CUPL when we get to the program-writing point. This is because the SEQUENCE statement requires that its counter ports be registered macrocells. Recall that our register is external to the ROM. From the information developed so far, we are ready to design the state table. Figure 7.21 illustrates the state table for this example.

Note that the development methodology used to create the state table is identical to that used with other programmable devices. The most significant bits, P_3–P_0, are all set to 0 since this is partition 0 in the ROM. The next four significant bits, A, B, C, and D, identify the state number. Finally, the next four significant bits, S_3–S_0, identify the input arguments. Together, these 12 bits make up the ROM address. The ROM data stored at each address is identified by the next address information and the output-forming logic. According to the schematic diagram of Figure 7.18, the next address data make up the four most significant data bits, with the output-forming logic making up the least significant bits.

We are now ready to write our CUPL program for a target device. However, before we do, let's consider another part pattern to be stored in the same ROM. Consider Figure 7.22. This pattern is slightly simpler and will require fewer states. As in pattern 0, we will consider the limit sensors in our design.

Practical Exercise 7.6: Cutter Controller, Pattern 1 Algorithm

Write the cutter control algorithm for cutter pattern 1 as illustrated in Figure 7.22 and demonstrated in the previous discussion.

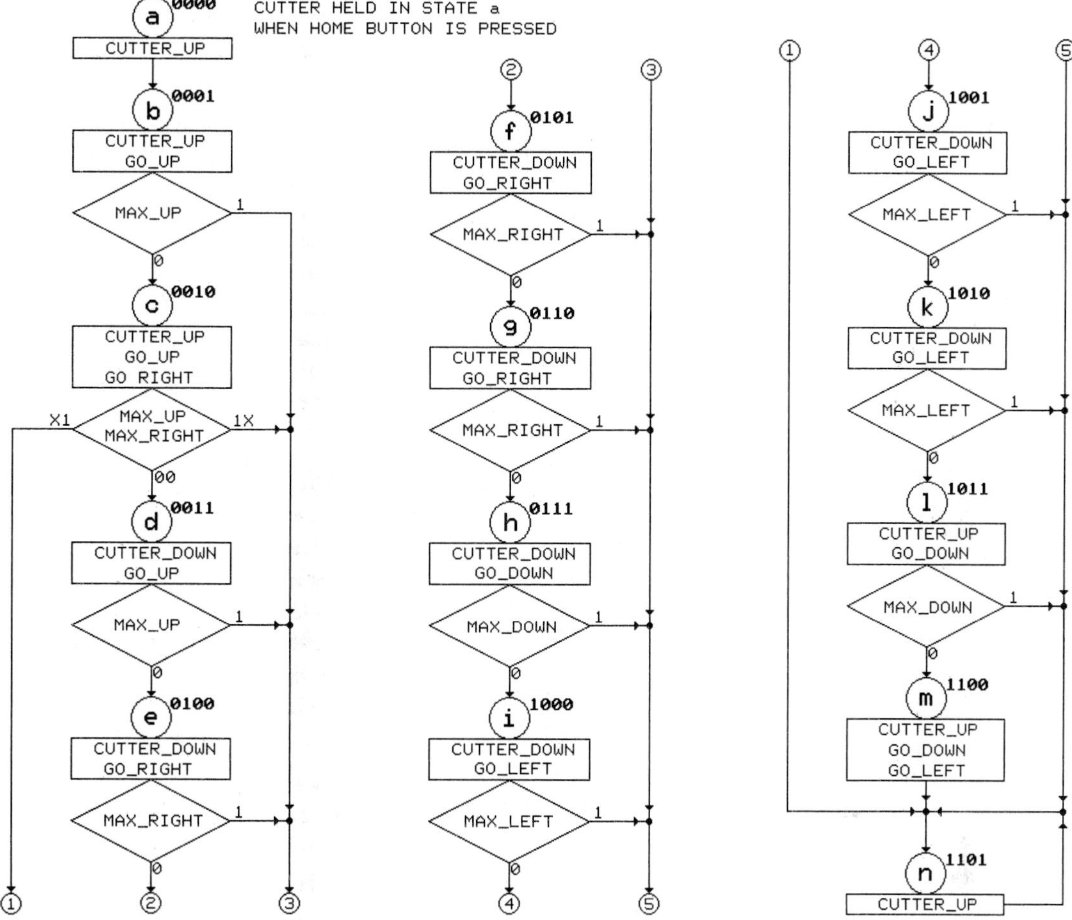

FIGURE 7.20 Practical Example 7.5, pattern 0, state diagram.

Practical Exercise 7.7: Cutter Controller, Pattern 1 State Diagram

Draw the state diagram for cutter pattern 1 as illustrated in Figure 7.22 as part of the previous discussion.

Practical Example 7.5: Continued

With the resources described above, we are ready to draw the state table for cutter pattern 1. Consider the solution illustrated in Figure 7.23. With a little experience, this state table can be developed directly from the problem as stated in Figure 7.22. However, it is probably

	PARTITION P3210	PRESENT ABCD	INPUTS S3210	NEXT ABCD	CUTTER C_DN	MOTORS G_UP	G_DN	G_RT	G_LF
a	0000	0000	XXXX	0001	0	0	0	0	0
b	0000	0001	0XXX 1XXX	0010 1101	0	1	0	0	0
c	0000	0010	0X0X 1XXX XX1X	0011 1101 1101	0	1	0	1	0
d	0000	0011	0XXX 1XXX	0100 1101	1	1	0	0	0
e	0000	0100	XX0X XX1X	0101 1101	1	0	0	1	0
f	0000	0101	XX0X XX1X	0110 1101	1	0	0	1	0
g	0000	0110	XX0X XX1X	0111 1101	1	0	0	1	0
h	0000	0111	X0XX X1XX	1000 1101	1	0	1	0	0
i	0000	1000	XXX0 XXX1	1001 1101	1	0	0	0	1
j	0000	1001	XXX0 XXX1	1010 1101	1	0	0	0	1
k	0000	1010	XXX0 XXX1	1011 1101	1	0	0	0	1
l	0000	1011	X0XX X1XX	1100 1101	0	0	1	0	0
m	0000	1100	XXXX	1101	0	0	1	0	1
n	0000	1101	XXXX	1101	0	0	0	0	0

FIGURE 7.21 Practical Example 7.5, pattern 0, state table.

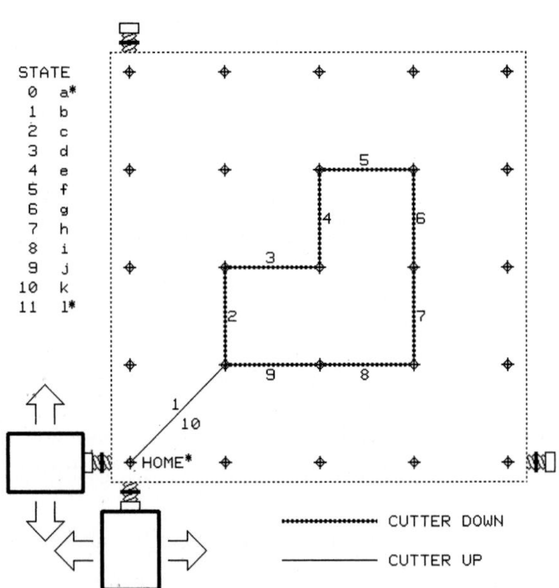

FIGURE 7.22 Practical Example 7.5, pattern 1, cutter path.

	PARTITION P3210	PRESENT ABCD	INPUTS S3210	NEXT ABCD	CUTTER C_DN	MOTORS G_UP	G_DN	G_RT	G_LF
a	0001	0000	XXXX	0001	0	0	0	0	0
b	0001	0001	0X0X 1XXX XX1X	0010 1011 1011	0	1	0	1	0
c	0001	0010	0XXX 1XXX	0011 1011	1	1	0	0	0
d	0001	0011	XX0X XX1X	0100 1011	1	0	0	1	0
e	0001	0100	0XXX 1XXX	0101 1011	1	1	0	0	0
f	0001	0101	XX0X XX1X	0110 1011	1	0	0	1	0
g	0001	0110	X0XX X1XX	0111 1011	1	0	1	0	0
h	0001	0111	X0XX X1XX	1000 1011	1	0	1	0	0
i	0001	1000	XXX0 XXX1	1001 1011	1	0	0	0	1
j	0001	1001	XXX0 XXX1	1010 1011	1	0	0	0	1
k	0001	1010	XXXX	1011	0	0	1	0	1
l	0001	1011	XXXX	1011	0	0	0	0	0

FIGURE 7.23 Practical Example 7.5, pattern 1, state table.

wiser to avoid skipping design steps so that the chances for error are lessened, and a better paper trail is provided.

Let's now consider the CUPL program to implement these two cutter patterns in a single ROM. Again, though our example involves the programming of two patterns, there is a capacity for 16 patterns in the ROM specified in this example. Consider the following CUPL program.

```
Name       EXP0704;
Date       01/01/96;
Designer   J.W. Carter;
Device     RA12P9;
/*****************************************************************/
/* Practical Example 7.5:Pattern Cutter using a 4096 x 9 PROM   */
/*****************************************************************/
/** Inputs **/
PIN [01..04] = [A11..A08];      /* Partition Inputs P3..P0      */
PIN [05..08] = [A07..A04];      /* Counter Bits A, B, C, D      */
PIN [09..10] = [A03..A02];      /* Input Stimulus S3..S0        */
PIN [13..14] = [A01..A00];
/** Outputs **/
PIN [15..19] = [D00..D04];      /* Output Forming Logic         */
PIN [20..23] = [D05..D08];      /* Next Address Bits            */
```

```
FIELD input  = [A11..A00];
FIELD output = [D08..D00];
TABLE input =>output {
*/ PATTERN 0 */
       /*                PPPP    SSSS             DDDDD */
       /*                3210ABCD3210         ABCD43210 */
       /* State-a */ 'b'00000000XXXX => 'b'000100000;
       /* State-b */ 'b'000000010XXX => 'b'001001000;
                     'b'000000011XXX => 'b'110101000;
       /* State-c */ 'b'000000100X0X => 'b'001101010;
                     'b'000000101XXX => 'b'110101010;
                     'b'00000010XX1X => 'b'110101010;
       /* State-d */ 'b'000000110XXX => 'b'010011000;
                     'b'000000111XXX => 'b'110111000;
       /* State-e */ 'b'00000100XX0X => 'b'010110010;
                     'b'00000100XX1X => 'b'110110010;
       /* State-f */ 'b'00000101XX0X => 'b'011010010;
                     'b'00000101XX1X => 'b'110110010;
       /* State-g */ 'b'00000110XX0X => 'b'011110010;
                     'b'00000110XX1X => 'b'110110010;
       /* State-h */ 'b'00000111X0XX => 'b'100010100;
                     'b'00000111X1XX => 'b'110110100;
       /* State-i */ 'b'00001000XXX0 => 'b'100110001;
                     'b'00001000XXX1 => 'b'110110001;
       /* State-j */ 'b'00001001XXX0 => 'b'101010001;
                     'b'00001001XXX1 => 'b'110110001;
       /* State-k */ 'b'00001010XXX0 => 'b'101110001;
                     'b'00001010XXX1 => 'b'110110001;
       /* State-l */ 'b'00001011X0XX => 'b'111000100;
                     'b'00001011X1XX => 'b'110100100;
       /* State-m */ 'b'00001100XXXX => 'b'110100101;
       /* State-n */ 'b'00001101XXXX => 'b'110100000;
*/ PATTERN 1 */
       /* State-a */ 'b'00010000XXXX => 'b'000100000;
       /* State-b */ 'b'000100010X0X => 'b'001001010;
                     'b'000100011XXX => 'b'101101010;
                     'b'00010001XX1X => 'b'101101010;
       /* State-c */ 'b'000100100XXX => 'b'001111000;
                     'b'000100101XXX => 'b'101111000;
       /* State-d */ 'b'00010011XX0X => 'b'010010010;
                     'b'00010011XX1X => 'b'101110010;
       /* State-e */ 'b'000101000XXX => 'b'010111000;
                     'b'000101001XXX => 'b'101111000;
       /* State-f */ 'b'00010101XX0X => 'b'011010010;
                     'b'00010101XX1X => 'b'101110010;
       /* State-g */ 'b'00010110X0XX => 'b'011110100;
                     'b'00010110X1XX => 'b'101110100;
       /* State-h */ 'b'00010111X0XX => 'b'100010100;
                     'b'00010111X1XX => 'b'101110100;
```

```
        /* State-i */ 'b'00011000XXX0 => 'b'100110001;
                      'b'00011000XXX1 => 'b'101110001;
        /* State-j */ 'b'00011001XXX0 => 'b'101010001;
                      'b'00011001XXX1 => 'b'101110001;
        /* State-k */ 'b'00011010XXXX => 'b'101110101;
        /* State-l */ 'b'00011011XXXX => 'b'101100000;
}
```

Note that a 4K x 9-bit ROM is specified as the mnemonic RA12P9 in the CUPL form. (Don't bother searching the library in Appendix B for this device, as there is none there.) Observe carefully the use of the TABLE statement, which is used to relate the output function of the PROM to the input addresses. Also observe how the address and data bits are related to the state tables.

Practical Exercise 7.8: Cutter Controller, Pattern 2

Develop the state diagram, state table, and CUPL program for the cutter pattern illustrated in Figure 7.24.

Practical Exercise 7.9: Cutter Controller, Pattern 3

Develop the state diagram, state table, and CUPL program for the cutter pattern illustrated in Figure 7.25.

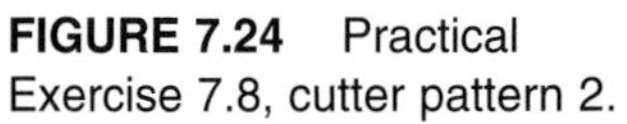

FIGURE 7.24 Practical Exercise 7.8, cutter pattern 2.

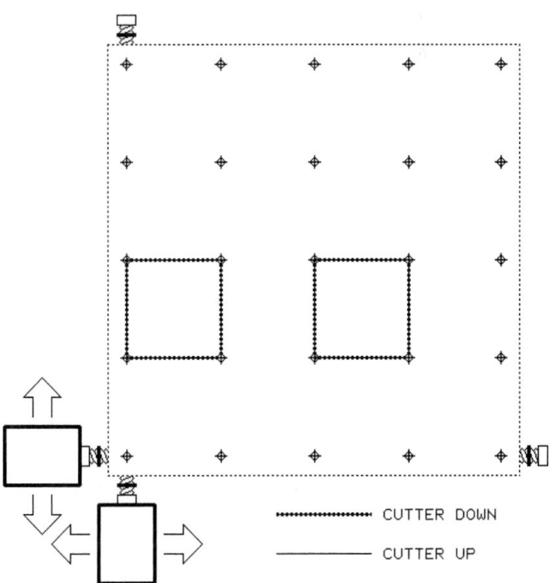

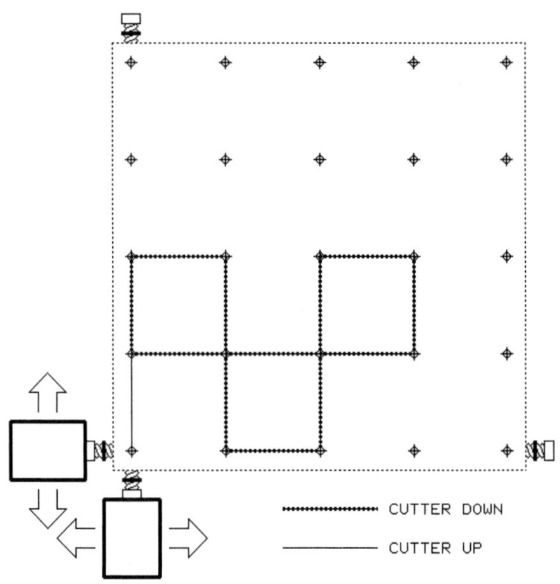

FIGURE 7.25 Practical Exercise 7.9, cutter pattern 3.

QUESTIONS AND PROBLEMS

Define each of the terms listed in 1–6.

1. Pipeline register
2. Instruction register
3. Operation code or opcode
4. Microprogram
5. Microprogram ROM
6. Microprogram word

7. What are the primary differences between a ROM and a PAL?
8. What advantages are realized when a ROM is used to control a state machine instead of a PAL or an FPLA?
9. How many minterms can be contained in an expression generated by a 4K x 8-bit ROM?
10. How many logical expressions can be generated by a 4K x 8-bit ROM?
11. How are decisions made in a ROM-based state machine?
12. How are input stimulus signals connected to the ROM in a ROM-based state machine?
13. Which ROM pins are used to determine the next state in a ROM-based state machine?
14. Which ROM pins are used to generate output-forming logic in a ROM-based state machine?
15. What is the purpose of the pipeline register in the ROM-based state machine?
16. How are multiple state machine programs possible in a ROM-based state machine?
17. What is the purpose of the instruction register in the ROM-based state machine?
18. What is the relationship among a microprogram, the microprogram ROM, and a microprogram word?

19. Design a ROM-based binary counter that counts in the repeating sequence 0 - 2 - 4 - 6 - 7 - 5 - 3 - 1 - 0 - ...
20. Design a ROM-based binary counter that counts in the repeating sequence 0 - 5 - 10 - 15 - 20 - ... - 35 - 0 - ...
21. Design a ROM-based solution for Practical Example 6.1.
22. Design a ROM-based solution for Practical Example 6.2.
23. Design a ROM-based solution for Practical Example 6.3.
24. Design a ROM-based solution for Practical Example 6.4.
25. Design a ROM-based solution for Practical Example 6.5.
26. Design a ROM-based solution for Practical Example 6.6.
27. Design a ROM-based solution for Practical Example 6.7.
28. Design a ROM-based solution for Practical Example 6.8.
29. Design a ROM-based solution for Practical Example 6.9.
30. Design a ROM-based solution for Practical Example 6.10.
31. Design a digital waveform synthesizer/generator similar to that described in Practical Example 6.11, but substitute a triangular wave for the sine wave, maintaining the same frequency and amplitude.
32. Design a digital waveform synthesizer/generator similar to that described in Practical Example 6.11, but substitute a rising ramp wave for the sine wave, maintaining the same frequency and amplitude.
33. Design a digital waveform synthesizer/generator similar to that described in Practical Example 6.11, but substitute a falling ramp wave for the sine wave, maintaining the same frequency and amplitude.
34. Combine the ROM programs of Practical Exercise 6.11 and Problems 31, 32, and 33 above to create a digital waveform synthesizer that will generate a sine, triangle, rising ramp, or falling ramp signal as a function of two input control lines. (Use a two-bit operation code to select the signal to be generated.)

CHAPTER 8

Microprocessor Architecture

OBJECTIVES

After completing this chapter you should be able to:

- Describe the overall architecture of a microprocessor.
- Describe the general functions of the following units within the microcontroller:
  - Computer control unit
  - Arithmetic logic unit
  - Status and shift control unit
  - Program control unit
- Describe the function of a pipeline register within a microprocessor.
- Describe how data is moved between the microprocessor and the external memory.
- Describe the purpose of the following microprocessor registers:
  - Accumulators
  - Program counter
  - Stack pointer
  - Index registers
- Describe the operation and use of the condition code register.
- Describe the difference between a microprogram and a macroprogram.
- Describe data movements within the processor using a register transfer language.
- Describe the fetch-execute cycle.
- Describe what takes place during the fetch portion of the fetch-execute cycle.
- Describe the following addressing modes:
  - Inherent
  - Implied
  - Direct
  - Indexed
  - Relative
- Write microprograms for a variety of macroinstructions in a variety of addressing modes.

8.1 OVERVIEW

By this point in our study we should have a pretty solid idea of what a state machine is. We have observed digital circuits that, when given a clock pulse signal, execute a predefined, finite number of logical states. The repetition sequence of those states can be altered by external stimulus, and the logical bit pattern of counter bits at each state can be used to generate output signals that may be used to control some digital system. The examples we have observed up to this point include, at the simplest level, binary counters, and at a more complex level, ROM-based multiple-microprogram controllers that are designed to generate the control signals for circuitry other than the counter. We observed the use of PAL devices to create simple state machines. We also looked into more complex state machines requiring the use of FPLA devices. Still more complex state machines were designed using a ROM and a pipeline register. These different devices were selected based on the complexity of the circuitry to be controlled with the state machine. At the more complex end of this spectrum, where the design is too intricate to apply a ROM and pipeline register, the microprocessor or microcontroller device was used.

A microprocessor or microcontroller is a state machine. Furthermore, it is a ROM-based state machine that uses a multiple-microprogram configuration similar to that of the last example in Chapter 7. The microprograms in the microprocessor or microcontroller are used to control on-chip resources such as:

- an arithmetic logic unit (ALU), which performs arithmetic and logical operations on binary numbers
- a status and shift control unit (SSCU), which performs multiple shift operations on binary numbers as well as managing a register of status flags in order to enable conditional microprogram operations
- a program control unit (PCU) or memory management unit (MMU), which manages the movement of data between the microprocessor and external memory
- a vectored-interrupt control unit (VICU), which manages interrupts of the microprogram sequence that are stimulated by external logic signals
- a computer control unit (CCU), which generates addresses for the microprogram ROM
- a microprogram ROM

Usually, a **microprocessor** is identified as a single-chip (or monolithic) state machine that contains the ALU, SSCU, PCU, VICU, CCU, and microprogram ROM. Because of the large number of circuits needed to implement such a large state machine, microprocessors often use CMOS technology because of its small transistor size. When higher speeds are needed, bipolar circuits are often used. However, because of their larger transistor size and larger power dissipation requirements, these are often implemented using several chips instead of a single monolithic device.

Microprocessors typically contain a few on-chip registers for the temporary storage of data, but any additional program RAM, ROM, or input/output (I/O) interfacing must be implemented as additional external circuitry. Because of the need for external RAM, ROM, and I/O circuitry, a fully developed microprocessor-based system can require a minimum of several chips. A microprocessor-based system usually contains a fully populated circuit board, or motherboard, which contains the processor and all of its support circuitry. This is particularly true when the processing power of the microprocessor significantly exceeds that of a ROM-based state machine. Large microprocessors often push the integration limits of the technology, and adding the support circuitry that is normally implemented outside the chip is impractical or impossible. However, when the processing power of the device is not as significant, the state machine circuitry may be simple enough to allow the addition of RAM, ROM, or I/O circuitry to the monolithic device. When these resources are added to the microprocessor design, we can refer to the device as a **microcontroller.** Because of the additional chip space needed to implement the on-board RAM, ROM, and I/O circuitry, the processing power of the most common microcontrollers is considerably less than for microprocessors of a similar size and cost that ship without these resources.

Microcontrollers often contain enough on-board RAM, ROM, and I/O functions (often including both parallel and serial binary data transfer and support for analog interfacing) that very little external circuitry is needed to develop a completely operational controller. For example, the Motorola M68HC11 microcontroller, which is based on the simple M6800 architecture, can be implemented with just a few additional passive components that provide reset and clock functions.

This chapter will not cover in detail all of the architectural details of the microprocessor. Such discussion would require a complete text or two, and other books available in the literature go into this level of detail. This chapter will illustrate the development of the microcontroller architecture as a state machine in a simple form. This will provide the reader with a unique perspective on the microprocessor or microcontroller that should enhance the reader's ability to select and apply such devices. We will also overview the Motorola M68HC11 microcontroller, observing its architecture, instruction set, and input/output capabilities.

8.2 THE MICROPROCESSOR AS A STATE MACHINE

Up to this point in this text, we have developed a basic structure for a binary state machine. A microprocessor or microcontroller is a ROM-based multiple-microprogram state machine similar to that described in Chapter 7. The primary difference is the complexity of the controlled circuitry of the microprocessor. Where we have been using the controlled binary counter to control simple devices, the controlled binary counter of the microprocessor controls circuitry far more complex than that we have already observed. However, the complexity of that circuitry should be well within the scope of the reader. If we spend some time looking at the microprocessor as a state machine, our understanding of both state machines and the microprocessor should be greatly enhanced.

Consider Figure 8.1. This is a basic block diagram of the ROM-based state machine as implemented in Chapter 7. Review for a moment each of the components of this circuit. The pipeline register maintains the current stable state of the system. That state includes the logic values of the control signals that drive the controlled circuitry, as well as the address of the next microprogram word that will be loaded into the pipeline register.

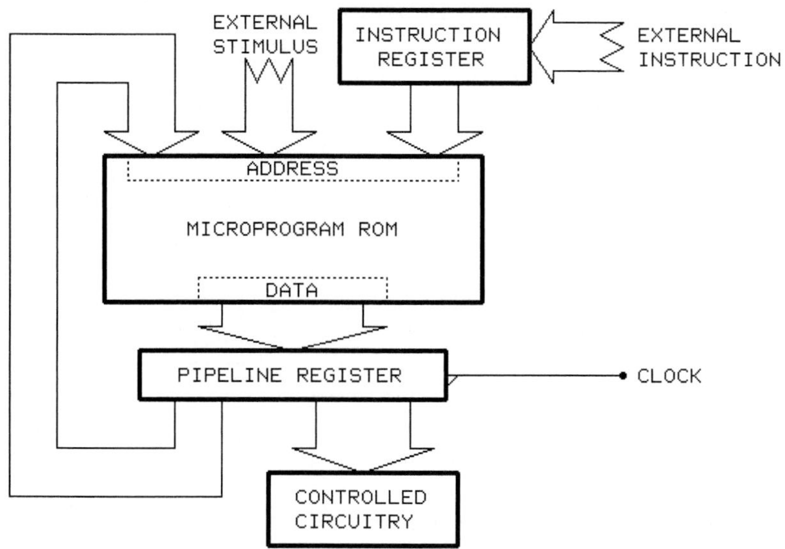

FIGURE 8.1 ROM-based state machine block diagram.

The microprogram ROM was partitioned by virtue of the assignment of most significant address bits from an instruction register. Also, additional least significant bits of the ROM address were generated by external stimulus to enable conditional microprogram sequences. Very few adjustments need to be made to this circuit in order to support the more complex controlled circuitry of the microprocessor.

8.2.1 The Computer Control Unit (CCU)

Consider Figure 8.2. The primary change made from Figure 8.1 to Figure 8.2 is the addition of a multiplexer immediately preceding the microprogram ROM address inputs. This multiplexer, referred to as the CCU multiplexer, is the focal point of the computer control unit. The function of the **computer control unit (CCU)** is to determine the address of the next microprogram data word to be loaded into the pipeline register. The **CCU multiplexer** is used to select one of several microprogram address sources. That is, the data on the S_C input to the multiplexer, coming from the pipeline register, determines the address of the next microinstruction to be loaded into the pipeline register. In this example processor, the next state can be addressed from any of four sources. A commercially available processor would have several more. Consider the following four microprogram address sources:

S = 0: Go Fetch. Selecting this CCU microprogram input causes the address of the microprogram ROM to go to 0. Stored at address 0 in the microprogram ROM is the initial state of the processor followed by the microprogram sequence that will go out to the external memory and obtain an instruction (opcode), placing it in the instruction register. This microprogram is referred to as the instruction **fetch microprogram.** This CCU multiplexer input is selected at the end of the execution of each instruction so that the next instruction can be fetched.

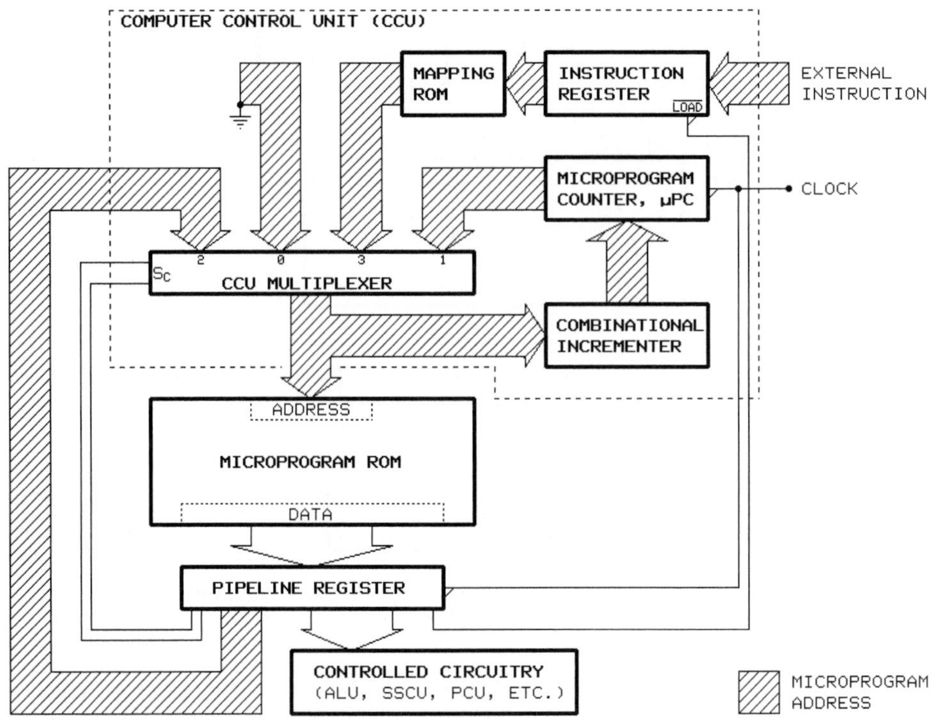

FIGURE 8.2 A simple microprocessor as a state machine.

S = 1: Increment. The current microprogram address is sent through a combinational incrementer so that when the pipeline register loads with the current data, the address of the next microprogram word is stored in the microprogram counter (μPC). Therefore, when this input to the CCU multiplexer is selected, the microprogram address increments.

S = 2: Jump Pipeline. When this input to the CCU multiplexer is selected, the address stored in a reserved field of the pipeline register is selected. This results in a direct microprogram jump to the address stored in the branch address field of the pipeline register.

S = 3: Jump Map. When this input to the CCU multiplexer is selected, the next microprogram address is taken from the mapping ROM. The **mapping ROM** converts the opcode, which is stored in the instruction register, to a specific starting address in the microprogram ROM. That is, stored in the mapping ROM is a list of microprogram addresses. The instruction register contents (the opcode) are used to address the mapping ROM. For example, if the microprogram for opcode 01_{16} is at microprogram address $B01_{16}$, the value $B01_{16}$ will be stored in the mapping ROM at address 01_{16}. This information is used at the end of the microprogram that fetches an instruction from external memory.

The method used to write the microprograms for the microprogram ROM is different from the methods previously used. As daunting as this circuit might currently appear, writing microprograms for it is actually a simple task, taking much less analysis or design theory than previously required. From a known current state, all that is needed to go to the next state is to provide the microprogram address of the microprogram word defining that state to the microprogram ROM. While the current state is being executed, the next state data is being looked up in the microprogram ROM and is stored in the pipeline register at the next clock edge. Therefore, part of the pipeline register is reserved for determining the next microprogram address.

As was implied by the preceding list, there are four possible next addresses. The most common next state is in the next microprogram address. This requires the ability to increment the microprogram address using a counter. This is done by selecting the S = 1 input to the CCU multiplexer. If it is desired to jump to a known microprogram address, that address is stored in the pipeline register and the S = 2 input to the CCU multiplexer is selected. The remaining two CCU multiplexer inputs are used to control the fetch-execute cycle.

When a computer executes a program, it does so by executing a sequence of machine language (macroprogram) instructions that are loaded into the instruction register from external memory. The machine language program is a sequence of opcodes and operands. The following sequence repeats in a continuing cycle:

1. An instruction opcode is read from external memory (the fetch cycle).
2. The microprogram for the selected opcode is executed (the execute cycle).
3. Go to 1.

This fetch-execute cycle has defined the basic structure of the computer for many years. Actually, from the time the fetch-execute processor was designed in the 1940s, there were few changes in processor architecture until the late 1980s. Through those years the electronic technology used to implement the computer circuitry became smaller and faster, but the basic architecture changed little. It wasn't until the late 1980s, when computer memory became inexpensive and plentiful, that significant changes in processor architecture started taking place. We will investigate the fetch cycle in more detail later.

Again, the purpose of the computer control unit is to determine the sequence of memory addresses presented to the microprogram ROM. This simple CCU has the ability to go to the next address, to the fetch microprogram, to the address on the output of the mapping ROM, or to the address contained in the assigned pipeline register data field.

8.2.2 The Pipeline Register (PL)

Figure 8.3 illustrates a block diagram of the pipeline register used in our example microprocessor.

There is little difference between this pipeline register architecture and that described in Chapter 7. Recall that some bits were fed back to the address inputs in order to establish the next microprogram address. Since this is the function of the computer control unit, the bits that drive the computer control unit (illustrated on the left side of the pipeline register of Figure 8.3) serve this purpose. This function is divided into two components. As already described, two bits from our pipeline register will select one of four sources of microprogram addresses using the CCU multiplexer. This field may be referred to as the "load control" field. In a commercial processor, this field may be from four to six bits

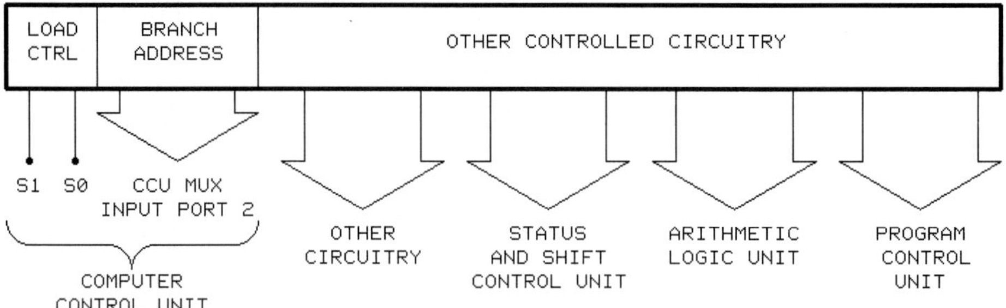

FIGURE 8.3 Microprocessor pipeline register.

wide. The second field used to control the CCU is shown in Figure 8.3 as the branch address field. This field will have the same number of bits as the width of the microprogram address. As one of the operands of the CCU multiplexer, the branch address field provides a hard-coded branch destination for direct jumping within the microprogram ROM. For example, if there is a microprogram at microprogram address $A13_{16}$ that is to be executed next, the value $A13_{16}$ would be placed in the branch address field, and the load control field S1,S0 would be set to 1,0.

The remaining bits would all be considered output-forming logic and are used to control the circuitry within the processor. That circuitry can be organized into several functional units, including the program control unit, the arithmetic logic unit, the status and shift control unit, the vectored interrupt control unit, and others. We will look at the function of a few of these in a little detail.

8.2.3 The Program Control Unit (PCU)

The **program control unit (PCU),** also referred to as a **memory management unit (MMU),** contains those resources needed to move information between the microprocessor and external memory. Consider Figure 8.4.

The primary function of the program control unit is the management of program and data information movement between the processor and the external memory. Data coming into the processor from external memory can be loaded into the instruction register or the data bus buffer (DBB). (Data loaded into the DBB from external memory is also loaded in a latch in the PCU for use when calculating addresses using the index registers.) From the data bus buffer data can be placed on the internal data bus for use throughout the processor. Also, data in the data bus buffer can be written back out to memory since the external data bus is bidirectional. Memory addresses are placed on the internal data bus (IDB) by the **IDB multiplexer.** The internal data bus is the primary data communications link within the processor.

In order for the memory unit to execute a read or write operation, it must be provided with a memory address and some handshaking signals. The first handshaking signal is provided by the **read/write (R/$\overline{\text{W}}$) line.** This bit, generated by a bit in the pipeline register, identifies whether the memory unit is in a read mode or a write mode. The second memory handshake line illustrated is the **valid memory address (VMA) line.** This line, also generated by the pipeline register, is asserted when memory activity is requested by the processor

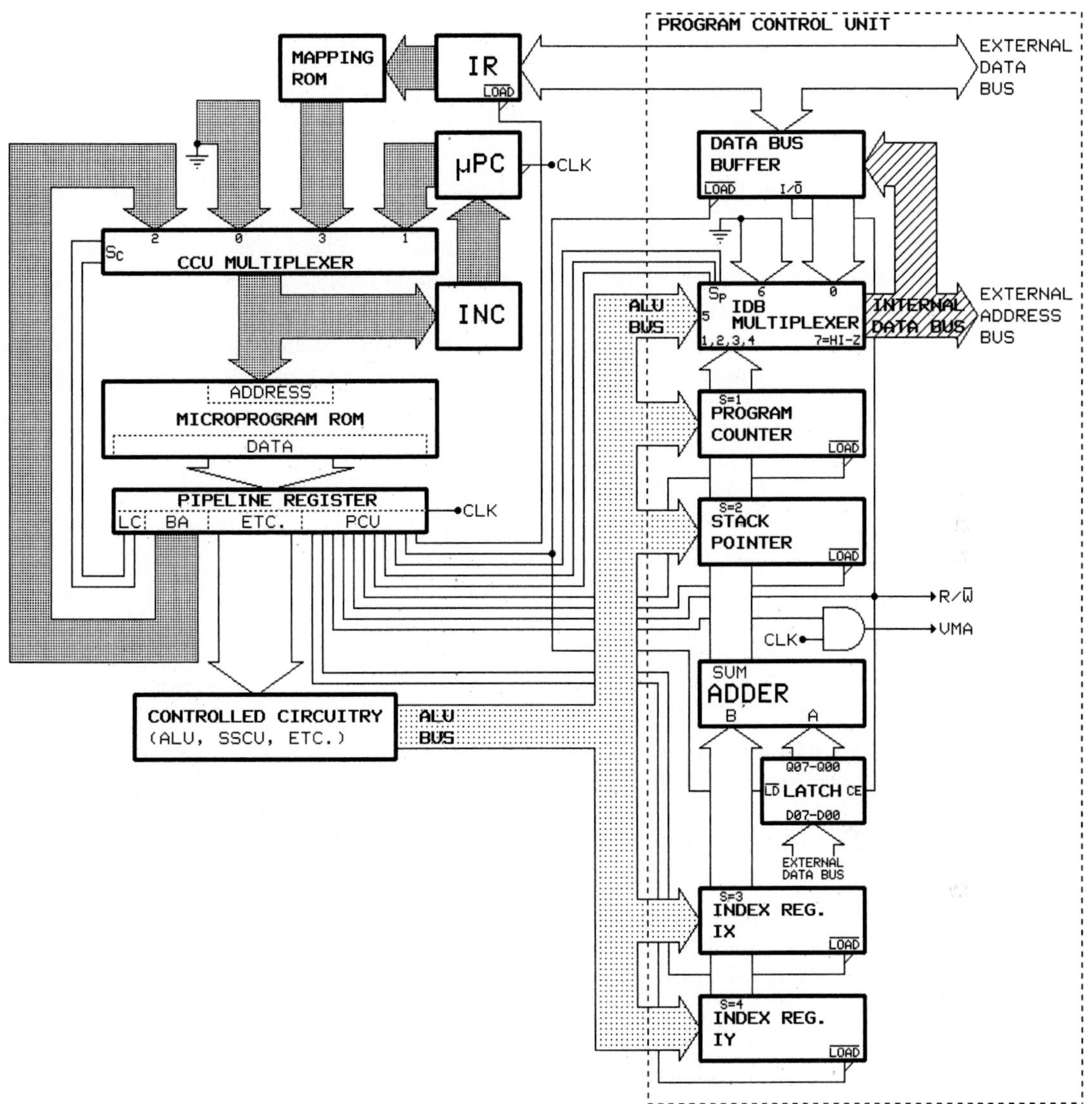

FIGURE 8.4 Addition of the program control unit to the state machine.

and a valid address has been placed on the internal data bus. Note that the assertion of the VMA line is delayed by a half clock cycle to allow memory activity in the same clock cycle that the memory address is generated. Some processors use individual read-enable and write-enable bits instead of the read/$\overline{\text{write}}$/VMA protocol shown here.

Also included in the program control unit is a set of registers that are used to maintain memory addresses. The **program counter (PC)** is a latch that contains the address of the next machine language (macroprogram) word to be fetched from memory. It is always incremented as soon as a word is read from memory. Therefore, the program counter is used to point into the program being executed, always pointing at the next word to be read. When the processor is reset, the PC must point to the opcode of the first instruction to be executed following that reset. In our example processor, that address is 0.

The **stack pointer (SP)** is a latch that contains the address of the top of the stack in memory. The stack is simply a one-dimension array where temporary data is stored. Data is written into the array using a PUSH operation—that is, the data to be stored is placed into the data bus buffer, the stack pointer contents (the stack address) is placed on the external address bus, and a memory write operation is executed. The stack pointer is then decremented, moving the pointer to a lower address. Data is read from the stack using a POP or PULL operation—that is, the stack pointer is incremented so that it points to the last data item stored in the stack; then the stack pointer address is placed on the external address bus and a memory read takes place. These two operations can be described using a register transfer language as follows:

PUSH t_0: External address bus ← (SP)
(DBB) ← Data to be pushed
Memory write

t_1: (SP) ← (SP) – 1

POP t_0: (SP) ← (SP) + 1
Memory read

t_1: (DBB) ← Memory data

This stack protocol may be referred to as a post-decrement PUSH protocol, since the stack pointer is decremented after the memory write operation takes place. Another common protocol is the pre-decrement PUSH, where the stack pointer is decremented prior to writing the data to memory. Both of these protocols require the stack pointer to be initialized at the top of the stack in memory, with the stack growing downward as it fills with data. The other two possible protocols, pre-increment and post-increment, are sometimes encountered in the industry, but are not very common. The stack pointer may be initialized upon processor startup through either hardware or software means. In our example processor, we will assume that the SP must be manually initialized using macroprogram instructions.

Two additional pointer registers are included in this program control unit and are referred to as **index registers IX and IY.** These registers are latches. Their outputs pass through an adder, along with the instruction operand, resulting in the calculation of a memory address equal to the sum of the two. A latch is placed immediately prior to the adder, acting like the data bus buffer. However, this latch only loads when an operand is fetched from memory. We will look at this concept in more detail when we investigate the use of the index registers. For now, we should understand that these index registers are used simply to maintain pointers for use by the programmer.

8.2.4 The Arithmetic Logic Unit (ALU)

The function of the **arithmetic logic unit (ALU)** is to execute the arithmetic and logical operations defined by the instruction set of the processor. In some cases, as in our example, it also provides the increment and decrement functions of the program counter and stack pointer. Figure 8.5 illustrates the addition of the ALU circuitry to our example

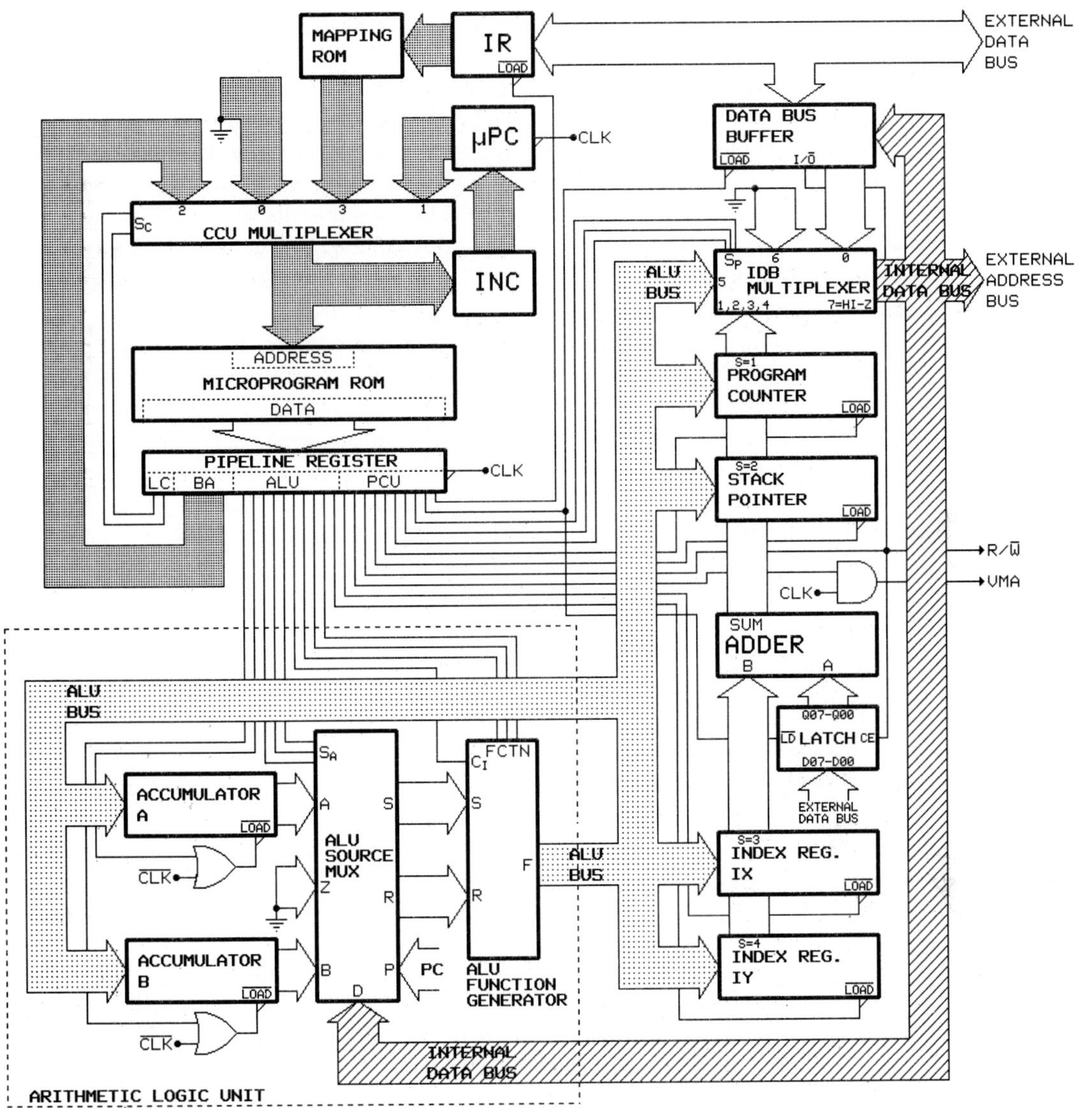

FIGURE 8.5 Addition of the arithmetic logic unit to the state machine.

processor. This processor has two accumulators, A and B, that may be combined with other operands, including the program counter contents, the data on the internal data bus, or the value 0, in order to generate arithmetic and logical functions. The operands to be processed by the ALU function generator are selected by the ALU source multiplexer. Our ALU function generator has three function control lines, implying that there are up to eight different arithmetic and logic functions. Commercial devices typically have many more.

The output of the ALU function generator is placed on the ALU bus. The ALU bus provides the generated arithmetic and logic function to the program counter and stack pointer so that increment and decrement functions are available. The bus also is presented to the IDB multiplexer so that the ALU function generator output can be placed into the data bus buffer so that the data can be written to memory. The ALU bus also drives the inputs to the two latches, accumulators A and B, so that they can be loaded with the generated function. Note also that the load of the accumulators is delayed by a half clock cycle by inhibiting the load signal with the clock. During the first half of the clock cycle, the inverted clock is at a logical 1 value, keeping the load input to the accumulators from being asserted. This is done to enable the loading of the accumulators with data during the same clock cycle as when that data is generated. This allows many of the arithmetic logic functions to be executed in a single clock cycle.

In order for us to make some use of our example processor, we must define the ALU source multiplexer and ALU function generator definitions. Consider Figure 8.6.

Both the ALU source multiplexer and ALU function generator get a three-bit command from the pipeline register. In order to execute an arithmetic operation, the source of the operation must be selected during the same clock cycle as when the function is selected. For example, to add the contents of accumulator A to accumulator B, the following sequence would be executed in two successive clock cycles:

t_0: ALU source = AB, ALU function = ADD, CI = 0

t_1: Load B

In order to effect these operations, the respective bits in the pipeline register must be set accordingly. We will look at this in more detail later.

8.2.5 The Status and Shift Control Unit (SSCU)

The status and shift control unit (SSCU) manages the shifting of ALU function data left and right and manages the use of the condition code register. Figure 8.7 illustrates the addition of the status and shift control unit to our microprocessor.

ALU Shifter. Our ALU shifter is a combinational circuit that shifts the ALU function data left or right depending on the status of the shift-enable (SE) and right/left ($R/\overline{L}$) lines. If SE = 0, the data passes through the shifter unchanged. If SE = 1, then the data is shifted left when $R/\overline{L} = 0$ and the data is shifted right when $R/\overline{L} = 1$.

ALU SOURCE MULTIPLEXER			
I2 1 0	MNEM	R	S
0 0 0	DZ	D	Z
0 0 1	AZ	A	Z
0 1 0	AB	A	B
0 1 1	AD	A	D
1 0 0	DP	D	PC
1 0 1	BZ	B	Z
1 1 0	BA	B	A
1 1 1	BD	B	D

ALU FUNCTION GENERATOR		
I2 1 0	MNEM	FUNCTION
0 0 0	PAR	R + CI
0 0 1	PAS	S + CI
0 1 0	ADD	S + R + 1 − CI
0 1 1	SUB	S − R − 1 + CI
1 0 0	AND	$S \wedge R$
1 0 1	OR	$S \vee R$
1 1 0	XOR	$S \veebar R$
1 1 1	NOT	$\overline{R}$

FIGURE 8.6 ALU sources and functions.

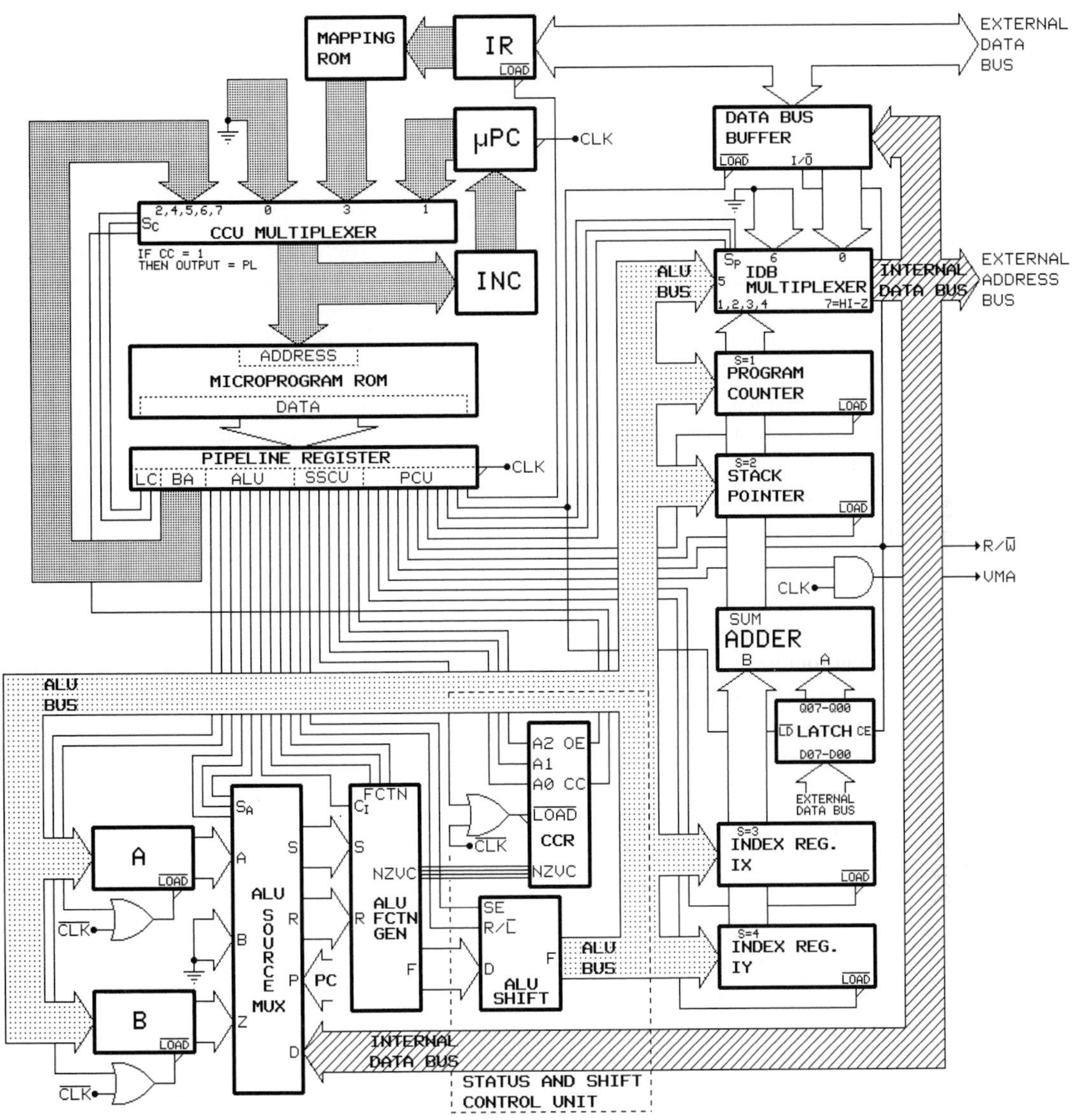

FIGURE 8.7 Addition of the status and shift control unit to the state machine.

Condition Code Register (CCR). The condition code register, referred to as a status register in some processors, is used to store the binary states of status flags generated by the ALU function generator. These flags include:

N bit. Set when the ALU function output is a negative number.

Z bit. Set when the ALU function output is 0.

V bit. Set when an overflow occurred in an ALU addition or subtraction operation.

C bit. Set when a carry occurred in an ALU addition or subtraction operation.

Note also that the load of the CCR is delayed by a half clock cycle by inhibiting the load signal with the inverted clock. During the first half of the clock cycle, the inverted clock is at a logical 1 value, keeping the load input to the CCR from being asserted. This is done to enable the loading of the CCR with the flag status during the same clock cycle as when the flags are generated.

The output of the condition code register is used to enable conditional operations. The condition code register output, CC, is sent to the CCU multiplexer. Anytime CC = 1, the multiplexer selects the PL input, causing a microprogram branch to the address specified in the branch address field of the pipeline register. The output is enabled by setting the OE input to the CCR to a logical 1 and selecting the desired flag and flag polarity within the CCR. Figure 8.8 illustrates the internal circuitry of the CCR and may help facilitate understanding of what is taking place within it.

Note that when the logical value of the load input drops from a logical 1 to a logical 0, the four-bit latch within the CCR loads with the N, Z, V, and C flags that are generated by the ALU function generator. Inputs A_2 and A_1 act as a data selector/multiplexer, selecting one of the four bits, N, Z, V, or C. A_0 acts as a programmable inverter so that each of the two polarities of a selected flag are available on the output of the CCR. For example, if the carry flag (C) is to be presented at the output of the CCR, then the inputs A_2, A_1 would be set to 0, 0, respectively. A_0 would be set to 0, and OE would be set to 1. Figure 8.9 illustrates the different forms of flag selection.

Again, the purpose of the condition code register is to provide the hardware needed to do decision making. Decisions are made based on the status of a selected CCR flag. The actual decision process is executed by enabling the output of the CCR when the desired flag is selected. This causes the CCR multiplexer to select the pipeline register branch address field if the selected flag is TRUE. If the selected flag is FALSE, the CCR multiplexer selects one of its four inputs as if the CCR were not enabled.

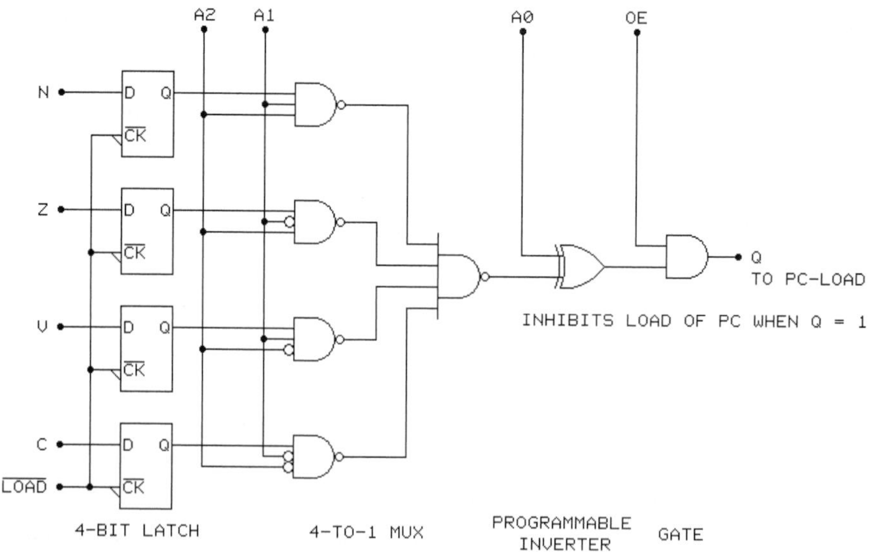

FIGURE 8.8 Condition code register architecture.

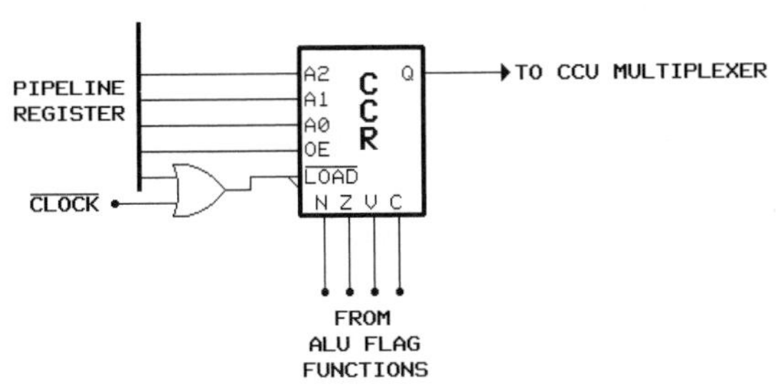

A2	1	0	OE	Q	USE	
X	X	X	0	0	--	OUTPUT = 0
0	0	0	1	C	CS	CARRY SET
0	0	1	1	$\overline{C}$	CC	CARRY CLEAR
0	1	0	1	V	VS	OVERFLOW SET
0	1	1	1	$\overline{V}$	VC	OVERFLOW CLEAR
1	0	0	1	Z	EQ	EQUAL
1	0	1	1	$\overline{Z}$	NE	NOT EQUAL
1	1	0	1	N	MI	MINUS
1	1	1	1	$\overline{N}$	PL	PLUS

FIGURE 8.9 Condition code register use.

8.2.6 The Pipeline Register, Revisited

Let's take another look at the pipeline register. Having covered descriptions of all of the circuits driven by the pipeline register, we can now describe the register in considerable detail. Consider Figure 8.10. Across this figure are indicated each of the bits in the pipeline register as they are oriented in Figure 8.7. They are organized into the control units they drive. Within each control unit a mnemonic description is assigned to each bit. Beneath each description is the binary default value of each bit. The default value is the defined binary state when not otherwise stated. For example, all active-low edge-triggered lines (such as the load lines of the IR, DBB, PC, SP, IX, IY, A, and B registers) are initialized to a logical 1 so they can be dropped to 0 when a load operation is to take place. The multiplexers are defaulted to their most common usage. The arithmetic logic unit and status and shift control unit pass IDB data through unchanged. The computer control unit is defaulted to increment the microprogram ROM address.

For the remainder of this chapter we will refer to the bits in the pipeline register by the subscripted variable P_N. Since each of the bits in the pipeline register is used to control a part of our microprocessor, and since all parts of the microprocessor are controlled by bits in the pipeline register, it would be worthwhile to review the use of each of these bits, looking at each microprogramming feature of the processor. We will look at them in sequence, unit by unit.

Program Control Unit. Bits P_0 through P_{10} are used to control the resources within the program control unit.

> **P_0: IR Load.** When this bit drops from a logical 1 to a logical 0, the instruction register will load with data present on the external memory data bus. It is assumed

UNIT	COMPUTER CONTROL UNIT										ARITHMETIC LOGIC UNIT								
DESCRIPTION	LOAD CTRL		BRANCH ADDRESS								LDB	LDA	SOURCE			CI	FCTN		
	I1	I0	A7	A6	A5	A4	A3	A2	A1	A0			I2	I1	I0		I2	I1	I0
REGISTER BIT	36	35	34	33	32	31	30	29	28	27	26	25	24	23	22	21	20	19	18
DEFAULT VALUE	0	1	X	X	X	X	X	X	X	X	1	1	0	0	0	0	0	0	0

UNIT	STATUS AND SHIFT							PROGRAM CONTROL UNIT										
DESCRIPTION	R/L	SE	LD	FLAG SELECT			OE	IY	IX	VMA	R/W	SP	PC	IDB			DBB	IR
				A2	A1	A0								S2	S1	S0		
REGISTER BIT	17	16	15	14	13	12	11	10	09	08	07	06	05	04	03	02	01	00
DEFAULT VALUE	X	0	1	X	X	X	0	1	1	0	1	1	1	0	0	0	1	1

FIGURE 8.10 Pipeline register default values.

that a memory read operation has just taken place with the program counter pointing at an instruction opcode. This opcode, which is loaded into the instruction register, is used to address the mapping ROM, in which are stored the starting microprogram addresses of the microprograms for each respective opcode. The default value for P_0 is a logical 1.

P_1: DBB Load. When this bit drops from a logical 1 to a logical 0, the data bus buffer loads with data either from the external memory unit if the read/$\overline{\text{write}}$ line is a logical 1 or from the internal data bus if the read/$\overline{\text{write}}$ line is a logical 0. The purpose of the data bus buffer is to hold data following a memory read operation, or prior to a memory write operation. Figure 8.11 illustrates a detailed schematic diagram of the circuitry of the data bus buffer. Note that it provides a bidirectional connection to the external data bus as well as provides the necessary multiplexing needed to load the DBB from the two sources. Again, the data direction is specified by the read/$\overline{\text{write}}$ line, consistent with memory read and write operations. The default value for P_1 is a logical 1.

$P_{4,3,2}$: IDB MUX. The IDB multiplexer is used to specify the data source for the internal data bus. The IDB is the primary data path within the processor and can only have one data source at a time. That source is determined by the logical value of the three bits, $P_{4,3,2}$. Figure 8.12 illustrates the relationship between these bit patterns and the resource placed on the IDB.

P_5: PC Load. When this bit drops from a logical 1 to a logical 0, the program counter will load with data present on the ALU bus. The most common operation

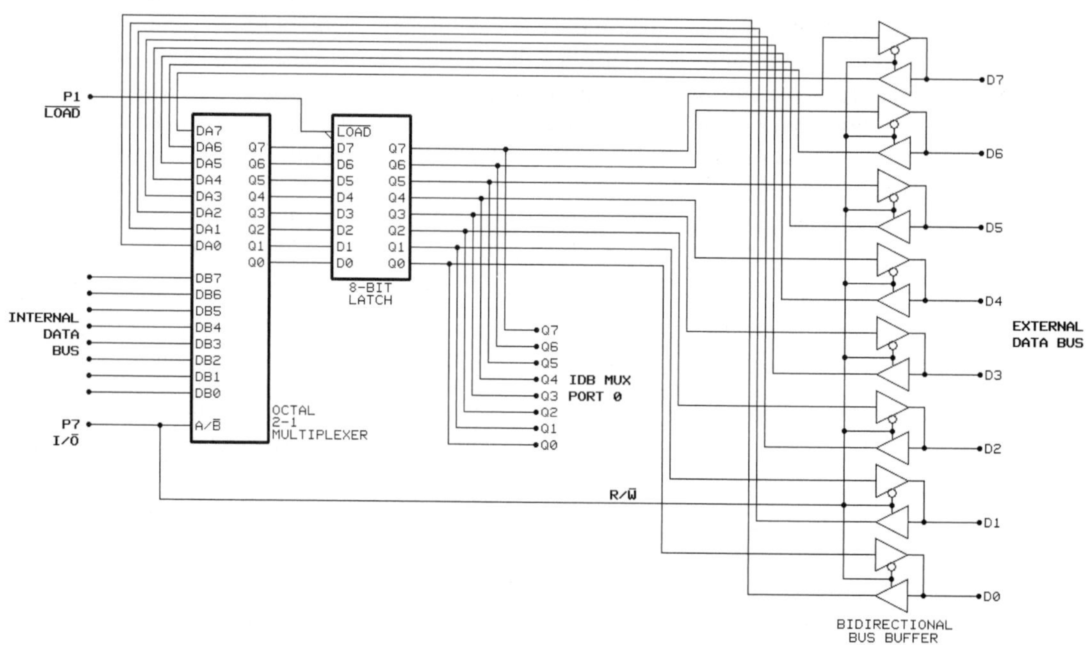

FIGURE 8.11 Data bus buffer.

FIGURE 8.12 IDB multiplexer selection.

S2	1	0	IDB MUX OUTPUT
0	0	0	DBB
0	0	1	PC
0	1	0	SP
0	1	1	IX + OPERAND
1	0	0	IY + OPERAND
1	0	1	ALU BUS
1	1	0	0
1	1	1	HI-Z

for the program counter is an increment. This is done by placing it on the IDB using the IDB multiplexer, passing the data through the ALU, adding 1 to it. In the next clock cycle the PC is loaded by dropping line P_5. The PC is also loaded with addresses found at other sources. To load the PC with an address, it must first be passed through the ALU. The default value for P_5 is a logical 1.

P_6: SP Load. When this bit drops from a logical 1 to a logical 0, the stack pointer will load with data present on the ALU bus. The SP is also loaded with addresses found at other sources. To load the SP with an address, it must first be passed through the ALU. The default value for P_6 is a logical 1.

P_7: R/$\overline{W}$ Mode. This bit determines the direction of intended data movement between the processor and external memory. It drives both the external memory unit and the direction input of the data bus buffer in order to define that direction. When this bit is a logical 1, it places the external memory unit in the read mode and prepares the DBB to be loaded from the external data bus. When this bit is a logical 0, it places the external memory unit in the write mode and prepares the DBB to be loaded from the internal data bus. Though the external memory reference mode is defined, no external memory activity takes place unless the VMA line, P_8, is asserted. The default value for P_7 is a logical 1, the memory read mode.

P_8: VMA. This bit, referred to as the valid memory address bit, is used to provide a minimum of handshaking with the external memory unit. This bit is set to a logical 1 when it is desired to enable external memory for a read or write operation. It is used with P_7, the read/$\overline{\text{write}}$ line, to control memory use. The default value for P_8 is a logical 0, deasserting the external memory unit. Note that the assertion of the VMA line is delayed by a half clock cycle by virtue of its being ANDed with the clock. During the first half of the clock cycle, the clock is a logical 0, disabling the memory. During this time the memory address can be calculated, propagating through the circuits used to generate it. This way, memory access can be made during the same clock cycle as when the needed memory addresses are defined.

P_9: IX Load. When this bit drops from a logical 1 to a logical 0, index register IX will load with data present on the ALU bus. The index register is used to generate memory addresses for the manipulation of data. In our example processor, the contents of the index register are added to the contents of the data bus buffer before its output is presented to the internal data bus. To load the index register IX with an address, it must first be passed through the ALU. Consequently, it can be loaded with data from several sources, including itself. This latter form is useful for incrementing or decrementing the register. The default value for P_9 is a logical 1.

P_{10}: IY Load. When this bit drops from a logical 1 to a logical 0, index register IY will load with data present on the ALU bus. The index register is used to generate memory addresses for the manipulation of data. In our example processor, the contents of the index register are added to the contents of the data bus buffer before its output is presented to the internal data bus. To load the index register IY with an address, it must first be passed through the ALU. Consequently, it can be loaded with data from several sources, including itself. This latter form is useful for incrementing or decrementing the register. The default value for P_{10} is a logical 1.

Status and Shift Control Unit. Bits P_{11} through P_{17} are used to control the resources within the status and shift control unit.

P_{11}: CCR OE. When this bit is a logical 1, the output of the condition code register (CCR) is enabled. When enabled, the CCU multiplexer is forced to present the pipeline branch address field (bits P_{34} through P_{27}) to the microprogram ROM. This forces a microprogram jump to the address specified in that field. The state of the output of the condition code register is dependent on the condition of the selected CCR flag N, Z, V, or C. This results in the capability of executing conditional instructions such as conditional branches and jumps. The default value for P_{11} is a logical 0. This disables the CCR output and causes the CCU multiplexer to respond to the load control field of the pipeline register, bits $P_{36,35}$. The default value for P_{11} is a logical 0.

$P_{14,13,12}$: FLAG Select. These three bits are used to select the flag and flag polarity to be output from the CCR when its output is asserted by setting $P_{11} = 1$. The flag to be selected for each of the eight possible bit patterns is indicated in Figure 8.9. The status of these lines is not required to be defined when the output of the CCR is deasserted.

P_{15}: CCR Load. When this bit drops from a logical 1 to a logical 0, the condition code register will load with the condition flags present on the ALU function generator. Also, the load of the CCR is delayed by a half clock cycle by inhibiting the load signal with the clock. (See Figure 8.9.) During the first half of the clock cycle, the clock is at a logical 1 value, keeping the load input to the CCR from falling to a logical 0. This is done to enable the loading of the CCR with the flag status during the same clock cycle as when the flags are generated in the ALU. The default value for P_{15} is a logical 1.

P_{16}: SE, Shift Enable. When this bit is asserted with a logical 1, the ALU shifter is instructed to shift its input data to either the right or the left, depending on the state of P_{17}, prior to outputting that data onto the ALU bus. A logical shift is executed. That is, a logical 0 is shifted into the register. When the shifter is deasserted by setting $P_{16} = 0$, the output of the shifter follows its input. The default value for P_{16} is a logical 0.

P_{17}: Right/$\overline{\text{Left}}$. This bit is used to define the direction of the logical shift executed in the ALU shifter when P_{16} is asserted. When the ALU shifter is asserted, the data passing through it is shifted to the right one bit if $P_{17} = 1$ or shifted to the left one bit if $P_{17} = 0$. When P_{16} is deasserted, the value of P_{17} does not have to be defined.

Arithmetic Logic Unit. Bits P_{26} through P_{18} are used to control the resources within the arithmetic logic unit.

$P_{20,19,18}$: FCTN, ALU Function Select. These three bits are used to select the arithmetic or logical function that is generated by the ALU function generator. The ALU functions generated by this three-bit pattern are indicated in Figure 8.6. The function is generated using inputs R and S to the ALU function generator, and the function output is presented at the F output. (See Figure 8.7.) The ALU function generator also generates the status flags N, Z, V, and C, which describe the output function. The default value of $P_{20,19,18} = 0,0,0$ passes the data on the R input of the ALU function generator to its output without change.

P_{21}: CI, Carry Input. A short perusal of the ALU function generator operations reveals that all of the arithmetic functions operate on two operands and a carry input, CI. The ability to specify the carry input will allow us to conveniently increment and decrement the value passed through the ALU. It also provides the facility for other arithmetic algorithms that will not be considered in this text. The default value for P_{21} is a logical 0. In this state, the arithmetic functions will pass, add, or subtract without the carry. Take a moment to look at the arithmetic functions of Figure 8.6 and observe the effect of the two logic states of the CI input.

$P_{24,23,22}$: SOURCE, ALU Source Select. These three bits are used to select the source operands that will be processed by the ALU function generator. The different patterns of these three bits are described in Figure 8.6. The default value of $P_{24,23,22} = 0,0,0$ presents the contents of the IDB on the R input of the ALU function generator and a logical 0 on the S input of the ALU function generator. These sources allow the contents of the IDB to be passed through the ALU function generator for many operations; incrementing, decrementing, complementing, or passing the data through unchanged are the most common of these operations.

P_{25}: LDA, Accumulator A Load. When this bit drops from a logical 1 to a logical 0, accumulator A will load with data present on the ALU bus. The accumulator is used to store the results of memory, arithmetic, and logical operations. Also, the inverted load of accumulator A is delayed by a half clock cycle by inhibiting the load signal with the clock. (See Figure 8.9.) During the first half of the clock cycle, the inverted clock is at a logical 1 value, keeping the LOAD input to the accumulator from falling to a logical 0. This is done to enable the loading of the accumulator with the ALU function generator output during the same clock cycle as when its sources are defined. This allows many arithmetic operations to be completed in a single clock cycle. The default value for P_{25} is a logical 1.

P_{26}: LDB, Accumulator B Load. When this bit drops from a logical 1 to a logical 0, accumulator B will load with data present on the ALU bus. The accumulator is used to store the results of memory, arithmetic, and logical operations. Also, the load of accumulator B is delayed by a half clock cycle by inhibiting the load signal with the clock. (See Figure 8.9.) During the first half of the clock cycle, the inverted clock is at a logical 1 value, keeping the load input to the accumulator from falling to a logical 0. This is done to enable the loading of the accumulator with the ALU function generator output during the same clock cycle as when its sources are defined. This allows many arithmetic operations to be completed in a single clock cycle. The default value for P_{26} is a logical 1.

Computer Control Unit. Bits P_{36} through P_{27} are used to control the CCU multiplexer within the computer control unit in order to determine the next microinstruction (microprogram ROM memory data word) to be loaded into the pipeline register.

$P_{34\text{–}27}$: PL, Pipeline Register Branch Address Field. These eight bits are used to define the microprogram ROM address to be used when the PL input to the CCU multiplexer is selected. When the load control bits $P_{36,35}$ are set to 1,0, the CCU multiplexer selects its input from these eight bits in the pipeline register. The result will be a jump to the microprogram ROM address identified by these eight bits. This can be used to allow microprogram jumps for loop control and other purposes. When the load control bits $P_{36,35}$ are any pattern other than 1,0, the data in the pipeline register branch address field, $P_{34\text{–}27}$, does not need to be defined.

$P_{36,35}$: LOAD CTRL, Load Control. These two bits determine the source operand for the CCU multiplexer. Consequently, they determine where the next microprogram word (or microinstruction) will be read from. There are four permutations of these two bits, allowing any of four sources to be used as the next microprogram address. These sources and the bit permutations were discussed in Section 8.2.1. Refer to this section of the text for a detailed explanation of the use of these three bits. The default value for these bits is $P_{36,35}$ = 0,1. This causes the microprogram memory address to increment to the next address.

8.2.7 Architecture Summary

We have observed the internal design of our microprocessor in quite a bit of detail. With this information, we should be able to analyze every bit that flips within it as it executes an operation. We will review the development of the microprogram ROM contents for some typical microprocessor assembly language (or machine language or macro) instructions. This should give us a considerable understanding of how the processor executes instructions. The thesis of this chapter is that one can develop applications and software for a microcontroller or microprocessor much more effectively if the internal operations of the device are understood.

It would be useful at this point to review the architecture of each unit of the microprocessor and how the bits in the pipeline register are used to control those units.

8.3 PROGRAMMING THE PROCESSOR

We are concerned with the programming of a state machine, and in this case that state machine is a simple microprocessor. The previous text discussions imply that this processor is microprogrammable. That is, we have the capability of changing the programming in the microprogram ROM in order to create microprograms that suit our own needs.

Most commercial microprocessors are not microprogrammable. Their microprograms have been hard-coded (mask-programmed) into their microprogram ROMs, and they are programmable only in macrocode (machine code), usually through an assembly language compiler (an assembler) or a high-level language compiler.

In order to program this microprogrammable microprocessor, we must become familiar with programming at several levels. It is assumed that the reader is already somewhat familiar with assembly language programming, and so this subject will not be covered in detail in this text. However, we will use assembly language as the highest level of programming for our needs and work down through several lower levels. The programming levels we are concerned with can be summarized as follows:

The top two levels are concerned with macroprogram instruction generation. The first is the mnemonic form of the second.

First level: **Assembly language programming.** Use of mnemonics to represent machine language (macroprogram) instructions.

Second level: **Machine code,** or **macrocode, programming.** Use of the actual binary opcodes and operands in programming.

The next two levels are concerned with microprogram instruction generation. The first is the mnemonic form of the second.

Third level: **Register transfer language (RTL) programming.** Use of mnemonics to represent transitions in the microprogram word.

Fourth level: **Binary microprogramming.** The translation of RTL into actual microprogram words.

Note that the lowest level of programming for a nonmicroprogrammable processor, such as the Motorola M68HC11 or the Motorola M68000 or the Intel 80486 or Pentium®, is machine language programming. The lowest level of programming for a microprogrammable processor, such as the American Micro Devices, Inc. Am2900 series, is microprogramming. There are some instances where devices driven by the pipeline register are themselves programmable. Some authors refer to programming at this level as "nanoprogramming."

Fifth level: **Binary nanoprogramming.** The act of programming devices that themselves are driven by the microprogram word in the pipeline register.

In order to begin programming our processor, we will do a quick review of assembly language programming as it applies to this architecture. Our main objective will be to write microprograms that execute machine language programs. In the next chapter we will be concerned with the programming of the MC68HC11 microprocessor, so we will write instructions in an assembly language that is compatible with that device. We will write assembly language programs, placing their machine language equivalents in external memory so that we can exercise and test our microprograms for proper, and expected, operation.

8.3.1 Basic Macroinstruction Execution

Recall from previous study of microprocessors that program execution at its simplest level takes place in a fetch-execute cycle.

The Fetch Microprogram

- A binary macroinstruction opcode is read from the external memory unit using the address stored in the program counter.
- The PC is incremented to point to the next macroprogram word.
- The opcode is loaded into the instruction register where the mapping ROM translates it to a microprogram beginning address.
- The computer control unit jumps to the address on the output of the mapping ROM, causing the execution of the microprogram corresponding to the opcode.

The Execute Microprogram

- A sequence of microprogram words is presented to the pipeline register, and these words define the macroprogram instruction being executed. More information may be fetched from memory when the macroinstruction includes one or more operands.
- The computer control unit jumps to the fetch microprogram, obtaining the next microprogram opcode.

8.3.2 Macroinstruction Formats

Our example processor uses eight-bit address and data buses throughout its design. Consequently, macroinstruction words will be eight bits in length. In order to refer to macroinstructions at this point in our discussion, we will use an assembly language similar to that used in the Motorola M68HC11 and similar devices.

A macroprogram (machine language) instruction may be from one to several words in length, depending on the activity defined during the execute cycle. Figure 8.13 illustrates two such instruction formats. All instructions are at least one word in length, and the first word is always the opcode to be loaded into the instruction register during the fetch microprogram.

Zero-Operand Instruction. This instruction format can be used whenever no operand is required during the execution of the instruction. Such is the case for any operations that take place completely within the processor, such as register-to-register operations, clearing of processor registers, etc. The litmus test for this type of instruction is that there be no requirement to get any more addresses or data information from the program memory during the execute cycle.

Examples:	RTS	Return from subroutine
	CLRA	Clear accumulator A

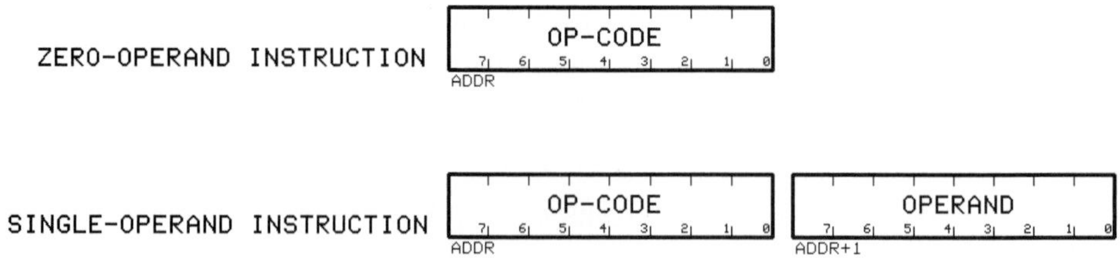

ZERO-OPERAND INSTRUCTION — OP-CODE (bits 7 6 5 4 3 2 1 0) ADDR

SINGLE-OPERAND INSTRUCTION — OP-CODE (bits 7 6 5 4 3 2 1 0) ADDR; OPERAND (bits 7 6 5 4 3 2 1 0) ADDR+1

FIGURE 8.13 Example instruction formats.

No memory reference or additional data is required to execute these instructions. Therefore, all that is necessary is the opcode.

Single-Operand Instruction. This instruction format can be used whenever an operand is required during the macroprogram instruction execution. During the fetch cycle, the first word is fetched into the IR and the execute cycle commences. During the execute cycle, the processor goes back out to the external machine language program to obtain the operand. This operand may be data to be processed, or it may be an address of data to be processed. Moreover, that address may be manipulated in any of several addressing protocols or addressing modes. These addressing modes will be discussed shortly. Obviously, a single-operand instruction requires two memory words.

Example: LDAA #$F0 Load accumulator A with the value $F0

In this example, there is an eight-bit operand. (The "$" symbol identifies the number as hexadecimal, the "#" symbol indicates that the following numbers are the data to be processed.) During the execute cycle this instruction will use the program counter to go back out to memory and fetch the operand, $F0, place that data on the internal data bus, and then load it into the accumulator.

Example: LDAA $80 Load accumulator A with data stored at address $80

There is an operand in this example also. In this case, the format of the assembly language identifies that the operand is a memory address. During the execute cycle, the processor will use the program counter to fetch the operand. It will then use the operand to go back out to memory and fetch the data pointed to by the operand.

8.3.3 Addressing Modes

The method used to obtain the address in memory where data processing is to take place is referred to as an **addressing mode.** The **effective address** is the memory address where data processing is to take place. Often the effective address is implied in the instruction. Other times it is contained in the operand. Sometimes it must be calculated on the basis of values of operands or processor registers. There are many methods of determining effective addresses. For our first processor, the example, we will be concerned with seven different addressing modes, as shown in Table 8.1.

TABLE 8.1 Example addressing modes.

Addressing Mode	*Description*	*Assembly Language*	*Effective Address*
Inherent (or implied)	The opcode is self-explanatory. No operand is required.	No operand	None
Immediate	The operand is an eight-bit data constant	#$NN	PC
Direct	The operand is an eight-bit address.	$NN	Operand
Indexed	The address is calculated by adding the index register contents to the operand.	$NN,X $NN,Y	Operand + X
Relative	The address is calculated by adding the program counter contents to the operand.	*+$NN	Operand + PC

Inherent Addressing. This addressing mode is often referred to as **implied addressing.** The instruction opcode is all that is necessary for the instruction execution to be completed. That is, there is no operand of any kind. Instructions that use this addressing mode typically make reference only to resources within the processor.

Example: CLRA Clear accumulator A

Note that this instruction clears the contents of the accumulator. This instruction can be executed without making any references outside of the arithmetic logic unit. The ALU can be set to output a value of 0 by executing an AND DZ function and source. This value of 0 can then be loaded into the accumulator. Figure 8.14 illustrates the effect on the processor registers and memory of this zero-operand instruction.

Prior to instruction execution, the contents of the instruction register and the accumulator are arbitrary, since they will both be modified by this process. The program counter is pointing to the instruction to be executed. Note that the instruction is shown to be in memory address \$24. The opcode for the instruction, 01001111_2, is assigned by the processor designer. Recall that the opcode is converted to a microprogram starting address by the mapping ROM in the computer control unit. The examples in this chapter will use the opcode numbers assigned to the Motorola M68HC11 microcontroller chip since this is the commercial device discussed in the next chapter.

During the fetch cycle, the instruction is read from memory and placed in the instruction register. The program counter is incremented, causing it to point to the next memory location where the next instruction is stored.

During the execute cycle, the contents of the accumulator are cleared.

Basically, anytime an instruction can be executed without making any reference to the macroprogram, the addressing mode is **inherent.**

Immediate Addressing. This addressing mode requires a single-operand format, and during the execute cycle the processor has to go back out to the external memory and fetch the operand. Immediate addressing signifies that the operand that follows the opcode in the macroprogram is the data to be processed. The operand is assembled into a single byte, forming the second word of the instruction.

Example: LDAA #$63 Initialize accumulator A to the value $63

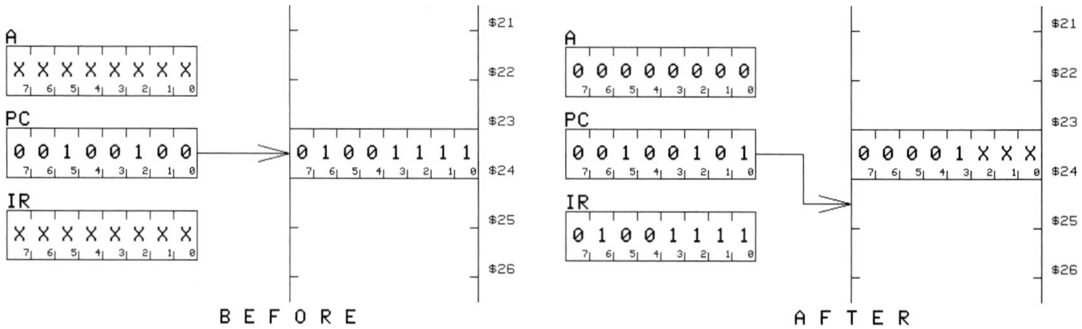

FIGURE 8.14 Implied addressing instruction: CLRA.

Note that this instruction initializes accumulator A to the hexadecimal value $63. (Refer to Figure 8.7.) Once the opcode has been fetched, this instruction must go back out and fetch the operand using the program counter, placing the value into the data bus buffer. The DBB contents can be placed on the IDB by the IDB multiplexer. Then the data can be passed unchanged through the ALU by executing an ADD DZ function and source. The data can then be loaded into the accumulator. Figure 8.15 illustrates the effect on the processor registers and memory of this single-operand instruction.

Prior to instruction execution, the contents of the instruction register and the accumulator are arbitrary since both are modified by this process. The program counter is pointing to the instruction to be executed. Note that the instruction is shown to be in memory address $26.

During the fetch cycle, the instruction is read from memory and placed in the instruction register. The program counter is incremented, causing it to point to the operand of the instruction $63, stored at address $27.

During the execute cycle, the program counter is again used to access memory. The operand is read and loaded into the DBB and the program counter is again incremented, now pointing to the next instruction. The contents of the DBB can be passed through the ALU and loaded into the accumulator.

Direct Addressing. This addressing mode utilizes an operand following the opcode, in a manner similar to immediate addressing. However, whereas in immediate addressing the operand is the data to be processed, in direct addressing the operand is a pointer to the data to be processed. That is, when direct addressing is used, the operand is the effective address. For example, consider the following example:

Example: LDAA $C3 Initialize the accumulator with the data stored at memory location $C3

Note that this instruction initializes the accumulator with the data at address $C3. Once the opcode has been fetched, this instruction must go back out to memory during the execute cycle and fetch the operand using the address stored in the program counter, placing the value in the data bus buffer. With the data bus buffer now containing the address of the data to be obtained, it is placed on the IDB, and a memory read cycle is again executed, obtaining the data. Then the data can be placed in the DBB, sent through the

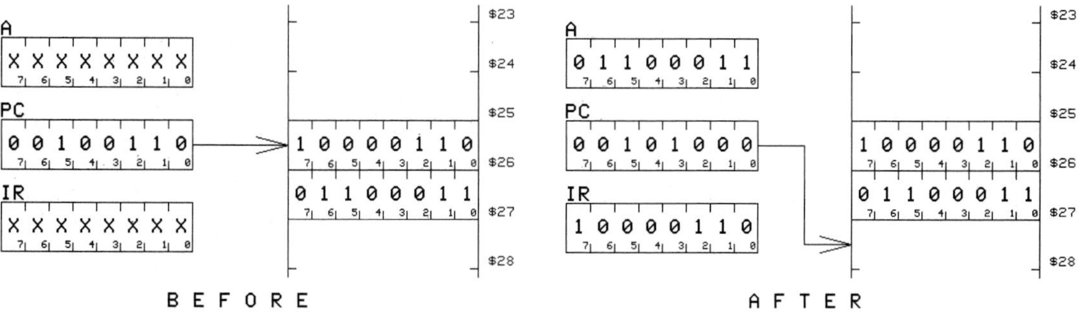

FIGURE 8.15 Immediate addressing instruction: MOVE #$63,A.

arithmetic logic unit, and loaded into the accumulator. Figure 8.16 illustrates the effect on the processor registers and memory of this single-operand instruction.

Prior to execution of the instruction, the program counter points to the instruction to be executed. Note that in this example the instruction is at memory location $28.

During the fetch cycle, the instruction is read from memory and placed in the instruction register. The program counter is incremented, causing it to point to the operand of the instruction, a value of $C3, stored at memory location $29.

During the execute cycle, the program counter is again used to access memory. The operand is read and loaded into the DBB, and the program counter is incremented to point to the next macroprogram word to be fetched. The operand is then used to access memory in order to fetch the data to be placed in the accumulator. To do this, the operand stored in the DBB is placed on the IDB, and a memory read cycle is executed. Following the read cycle, the data is loaded into the DBB, placed on the IDB, run through the ALU without change, and then loaded into the accumulator.

Indexed Addressing. This addressing mode is very similar to direct addressing, but before the instruction operand is used to access memory, the contents of the index register are added to it. That is, the sum resulting from the addition of the instruction operand and the index register is used to address memory. Note that in our example processor, the operand is stored in a latch that holds the inputs to the adder stable. This latch is loaded whenever information is loaded into the data bus buffer from the external data bus. Indexed addressing is convenient for accessing tables of data, where the operand points to the base of the table and the index register is used to move the pointer up and down through it (or vice versa). Consider the following example:

Example: LDAA $C0,X Initialize the accumulator with the data stored at the memory location obtained by adding the operand, $C0, to the contents of the index register, which in this example is $18. The effective address is $D8.

Figure 8.17 illustrates the effect this instruction has on the processor registers and memory.

Prior to execution of the instruction, the program counter points to the instruction to be executed. Note that in this example the instruction is at memory location $2A.

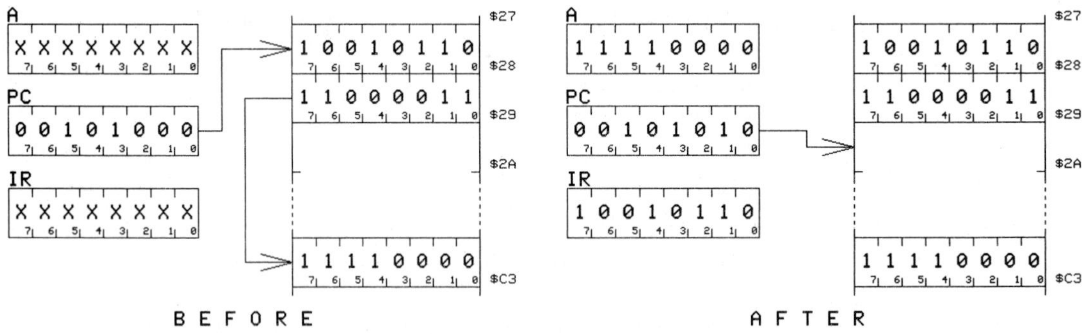

FIGURE 8.16 Direct addressing instruction: LDAA $C3.

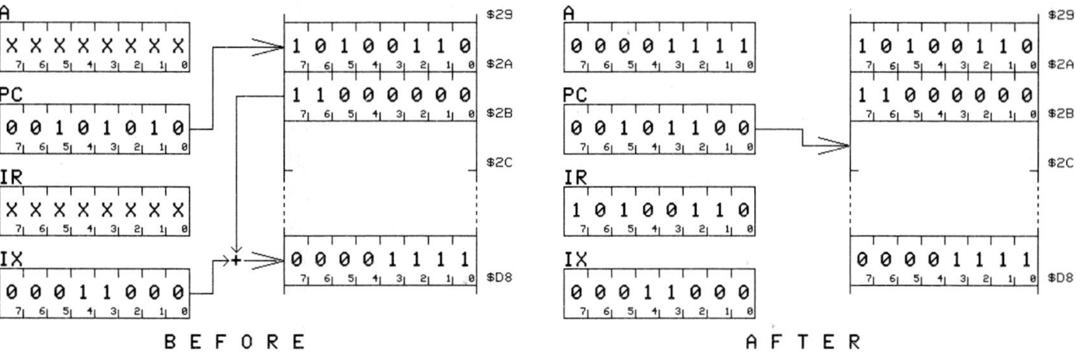

FIGURE 8.17 Direct indexed addressing instruction: LDAA $C0,X.

During the fetch cycle, the opcode, $A6, is read from memory and placed in the instruction register. The program counter is incremented, causing it to point to the operand of the instruction, a value of $C0, stored at memory location $2B.

During the execute cycle:

- The program counter is again used to access memory. The operand is read and loaded into the latch that drives the adder assigned to the index registers, and the program counter is incremented to point to the next instruction to be executed (address $2C).
- The sum of these operands, $C0 + $18, is then placed on the IDB and a memory read cycle is executed. At the end of the read cycle, the data found, $0F, is loaded into the DBB, placed on the IDB, passed through the ALU, and loaded into the accumulator.

Relative Addressing. This addressing mode is very similar to indexed addressing, but the contents of the program counter are used in the calculation of the effective address instead of an index register. The contents of the program counter are added to the operand in order to determine the effective address. The difference between the program counter value and the effective address is often referred to as a **displacement** and is defined by the operand of the instruction. When this addressing mode is used, program and data references (addresses) are always relative to the position of the program counter. Therefore, the operands used are not dependent on the absolute address at which the program itself is loaded. This allows the program to be **relocatable,** that is, the program can be loaded at any starting address in memory without affecting its operation.

Example: BRA *+$40 Branch always. Add the operand, $40, to the program counter, causing a jump to a memory address relative to the position of the program counter. In this example PC = $2E + $02 + $40 = $70.

Figure 8.18 illustrates the effect this instruction has on the processor registers and memory.

Prior to execution of the instruction, the program counter points to the instruction to be executed. Note that in this example the instruction is at memory location $2E. During

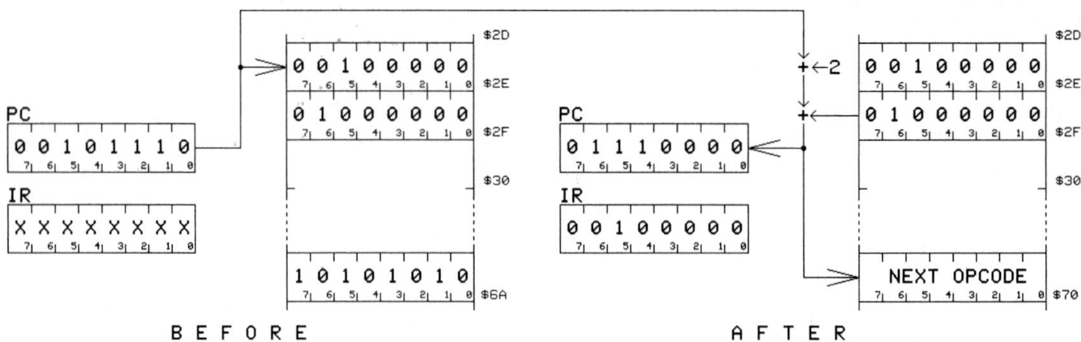

FIGURE 8.18 Relative addressing instruction: BRA *+$40.

the fetch cycle, the instruction is read from memory and placed in the instruction register. The program counter is incremented, causing it to point to the operand of the instruction, a value of $40, stored at memory location $2F.

During the execute cycle:

- The program counter is used to access the operand, which is read and loaded into the DBB. The program counter is incremented to point to the next instruction to be executed (address $30). The operand is added to the PC to determine the effective address.
- The DBB, containing the operand, is placed on the IDB, and the ALU function and source are set to ADD DP. The sum ($30 + $40 = $70) is then loaded into the program counter. The next instruction executed will be the instruction stored in memory address $70.

Addressing Modes Summary. The following is a summary of the addressing modes currently included in our discussion.

Inherent addressing. In inherent addressing the opcode contains all of the information needed to execute the instruction. There is no operand and no necessity to access the machine language program during the execute cycle. Inherent addressing is sometimes referred to as **implied addressing.**

Immediate addressing. This mode of addressing is used to initialize or apply constants to internal processor registers. The operand follows the opcode in the program and must be fetched into the DBB during the execute cycle. The contents of the DBB are placed on the IDB in order to facilitate the completion of the instruction. The effective address is literally the program counter, because it points to the operand.

Direct addressing. In this mode, the operand of the instruction contains the effective address. During the execute cycle, the operand must be loaded into the DBB and then must be placed on the IDB, where it is used as an address to access the memory location to be processed. Direct addressing is also referred to as **absolute addressing.**

Indexed addressing. In indexed addressing, the operand of the instruction is added to the contents of the index register in order to determine the effective address. In typical application, the operand is used as a base address of a table of data, and the index register is used as a displacement that can be moved up and down the table. In our example processor, the operand and the index register are both eight bits in length, and 2's complement arithmetic is used, allowing the displacement to range from +127 to –128 bytes. During the execute cycle, the operand must be added to the index register. The sum then must be placed in the address bus buffer for the subsequent read or write operation. This form of indexed addressing is also referred to as **direct indexed addressing** or **absolute indexed addressing.**

Relative addressing. In relative addressing the operand of the instruction is added to the contents of the program counter in order to determine the effective address. During the execute cycle, the program counter is used in a read cycle to fetch the operand into the DBB. The DBB contents are then passed through the ALU with the contents of the program counter added to them. The sum is the effective address. When executing relative jump instructions, this value is loaded back into the program counter.

8.4 MICROPROGRAMMING PRIMER

There is probably no better way to gain an understanding of the internal operation of a microprocessor than to write microprograms for it. Microprogramming has become somewhat of a rarity in the industry, so our purpose here is not to teach the discipline, but to use it as a tool to enhance the understanding of microprocessor operation. This is attempted with the thesis that one who specifies and makes use of microprocessors and microcontrollers can do so more effectively if their internal operation is fundamentally understood. We will approach this task by observing the binary microprogram for the fetch cycle as well as a few macroinstructions, including at least one example from each addressing mode.

Recall that the creation of machine language programs is accomplished by the use of mnemonic languages such as assembly language or high-level languages such as C or Pascal. We will use the same mnemonic approach in the writing of microprograms. Our mnemonic language is referred to as register transfer language (RTL). We will use a two-step approach to writing microprograms. First, RTL will be used to describe the processes that take place at each clock cycle of the microprogram. Then, a microprogram worksheet will be completed that describes the state of each bit in the pipeline register for each respective clock cycle. If we were actually programming a processor, the binary patterns described in the worksheet would be loaded into the microprogram ROM at the addresses identified on that worksheet. Figure 8.19 is an example of the microprogramming worksheet. Note that this worksheet is nothing more than a state table which describes the state of each processor resource during each discrete state, or clock cycle, in the microprogram.

8.4.1 Register Transfer Language

Register transfer language (RTL) is to microprogramming what assembly language is to macroprogramming. RTL is a mnemonic form of microprogramming whereby the state of pipeline register bits is illustrated by a meaningful text statement.

NAME: DATE:	µP ROM ADDRESS	COMPUTER CONTROL UNIT										ARITHMETIC LOGIC UNIT									STATUS AND SHIFT							PROGRAM CONTROL UNIT										
MNEMONIC: OPCODE:		LOAD CTRL		BRANCH ADDRESS								LDB	LDA	SOURCE			CI	FCTN			R/L	SE	LD	FLAG SELECT			OE	IY	IX	VMA	R/W	SP	PC	IDB			DBB	IR
		I1	I0	A7	A6	A5	A4	A3	A2	A1	A0			I2	I1	I0		I2	I1	I0				A2	A1	A0								S2	S1	S0		
		36	35	34	33	32	31	30	29	28	27	26	25	24	23	22	21	20	19	18	17	16	15	14	13	12	11	10	09	08	07	06	05	04	03	02	01	00
T0																																						
T1																																						
T2																																						
T3																																						
T4																																						
T5																																						

FIGURE 8.19 Microprogramming worksheet.

Example: The instruction register is loaded from the external data bus when pipeline register bit P0 goes from a high to a low logic level from one clock cycle to the next.

This can be stated as:

$$P0 = 0$$

or

$$(IR) \leftarrow (EDB)$$

Certainly, the second form is easier to relate to the function involved. Translating the RTL to pipeline register bit states is only a matter of assigning the binary state of the pipeline register bits for each defined statement.

There is an RTL statement that coincides with each bit or group of bits in the pipeline register. Before we get too serious about writing microprograms, we will investigate this language and its application on the pipeline. Table 8.2 illustrates many of the available RTL statements and their pipeline register bit-state equivalents. They are arranged by pipeline register bit locations from the least significant to the most significant.

In order to provide some RTL consistency, we will consider a few types of statements: assignment statements, data path statements, function statements, and CCU statements.

TABLE 8.2 RTL statement summary.

Pipeline	*Example RTL*	*Description*
$P_0 = 0$	(IR) ← (EDB)	Load the IR with the macroinstruction opcode on the external data bus.
$P_1 = 0$, $P_7 = 1$	(DBB) ← (EDB)	Load the DBB with the information in the EDB following a read cycle.
$P_1 = 0$, $P_7 = 0$	(DBB) ← (IDB)	Load the DBB with the information on the IDB prior to a write cycle.

TABLE 8.2 RTL statement summary (continued).

Pipeline	*Example RTL*	*Description*
$P_{4,3,2} = 0,0,0$	DBB → IDB path	Place the contents of the data bus buffer on the IDB.
$P_{4,3,2} = 0,0,1$	PC → IDB path	Place the contents of the program counter on the IDB.
$P_{4,3,2} = 0,1,0$	SP → IDB path	Place the contents of the stack pointer on the IDB.
$P_{4,3,2} = 0,1,1$	IX → IDB path	Place the sum of the IX register and the instruction operand on the IDB.
$P_{4,3,2} = 1,0,0$	IY → IDB path	Place the sum of the IY register and the instruction operand on the IDB.
$P_{4,3,2} = 1,0,1$	ALU bus → IDB path	Place the contents of the ALU bus on the IDB.
$P_{4,3,2} = 1,1,0$	Gnd → IDB path	Set the internal data bus to a logical 0.
$P_{4,3,2} = 1,1,1$	Hi-Z → IDB path	Set the internal data bus to a high impedance state.
$P_5 = 0$	(PC) ← (ALU bus)	Load the PC with the data on the ALU bus.
$P_6 = 0$	(SP) ← (ALU bus)	Load the SP with the data on the ALU bus.
$P_{8,7} = 1,1$	Memory read	Enable the external memory system to execute a read cycle.
$P_{8,7} = 1,0$	Memory write	Enable the external memory system to execute a write cycle.
$P_9 = 0$	(IX) ← (ALU bus)	Load the index register IX with the data on the ALU bus.
$P_{10} = 0$	(IY) ← (ALU bus)	Load the index register IY with the data on the ALU bus.
$P_{11} = 1$	Enable flags	Enable the output of the CCR.
$P_{14,13,12}$	cc = CC	Select the desired CCR flag from Figure 8.9.
$P_{15} = 0$	Load flags	Load the condition codes register with the flags from the ALU.
$P_{17,16} = 0,1$	Shift left	Shift the data passing through the ALU shifter one bit to the left.
$P_{17,16} = 1,1$	Shift right	Shift the data passing through the ALU shifter one bit to the right.
P_{24}–P_{18}	ALU = ADD DZ+1 ALU = SUB AX ALU = OR AZ ALU = EXNOR	These bits define the ALU function, source, and CI input values. Functions are selected from bits $P_{20,19,18,}$ and include PAR, PAS, ADD, SUB, AND, OR, XOR and NOT. Sources are DZselected from bits $P_{24,23,22}$, and include DZ, AZ, AB, AD, ZD,BZ, BA, and BD. The status of the CI input to the ALU is defined by bit P_{21}.
$P_{25} = 0$	(A) ← (ALU)	Load accumulator A with the data on the ALU bus.
$P_{26} = 0$	(B) ← (ALU)	Load accumulator B with the data on the ALU bus.
P_{34-27}		Branch address indicated as a hexadecimal $NN in the JPL instruction.
$P_{36,35} = 0,0$	[Z]	Set the next microinstruction address to the beginning of the fetch cycle.
$P_{36,35} = 0,1$	[μPC]	Set the next microinstruction address to follow the current one.
$P_{36,35} = 1,0$	[JPL $NN]	Set the next microinstruction address to the branch address field, P_{34-27}.
$P_{36,35} = 1,1$	[MAP]	Set the next microinstruction address to the mapping ROM output value.

Assignment Statements. Anytime we desire to illustrate the movement of information from one device in the processor to another, that movement may be referred to as an *assignment.* In the previous example, we stated that the IR was loaded with the contents of the external data bus: this is an assignment. Assignments between two devices are made with a data path between those devices. When an assignment statement is written, the assignment of the data is always from right to left. Some typical assignment statements are:

```
(IR) ← (EDB)       ;Load instruction register with external data bus
                    contents.
(DBB) ← (EDB)      ;Load data bus buffer with external data bus
                    contents.
(PC) ← (PC) + 1    ;Increment the program counter.
(SP) ← (SP) - 1    ;Decrement the stack pointer.
```

The parentheses in these statements refer to the contents of the indicated devices, not to the devices themselves.

Data Path Statements. Anytime we desire to alter the current data path arrangement in the processor we do so by altering multiplexer control lines. For clarity, data path assignments are written with the flow from left to right, and the statement is terminated with the word "path." Because we are concerned with the paths between devices, and not the data traveling on the paths, no parentheses are used. Some typical data path statements are:

```
PC → IDB Path      ;Place the program counter output on the
                    internal data bus.
DBB → IDB Path     ;Place the data bus buffer output on the
                    internal data bus.
```

The only programmable data path in our example processor is the selection of the device placed on the internal data bus. A commercial processor would have many such programmable data paths.

Function Statements. These statements are used to identify the functions assigned to devices that are programmed by pipeline register bits. The programmable devices in the example processor include the condition code register, the arithmetic logic unit, and the external memory unit. Some typical function statements are:

Condition Codes Register

```
cc = NE      ;CCR enabled, selecting the not equal to zero flag.
cc = CS      ;CCR enabled, selecting the carry-set flag.
Load Flags   ;Load the CCR with flags from the ALU.
```

Arithmetic Logic Unit

```
ALU = ADD DZ + 1   ;Set ALU function = ADD, source = DZ, CI = 1.
ALU = SUB AB       ;Set ALU function = SUB, source = AB, CI = 0.
ALU = XOR AB       ;Set ALU function = XOR, source = AB.
```

External Memory Unit

```
Memory Read    ;VMA = TRUE, read/write = 1, executing read cycle.
Memory Write   ;VMA = TRUE, read/write = 0, executing write cycle.
```

CCU Statements. These statements are used to identify where the next microprogram word to be executed is located. Recall from Chapter 2 that the most significant field of bits in the pipeline register is fed back to the computer control unit so that, while a process is taking place as defined by the other bits, these bits are setting up the data path in the CCU to generate the next microprogram word. The CISC01 processor has only three possible next address CCU instructions. At the end of the fetch cycle, a jump to the address present at the output of the mapping ROM can be done. At the end of the execute cycle, a jump back to the fetch cycle (at address 0) can be done. In all other instances, a jump to the next microprogram address is executed, incrementing the program through sequential addresses. Some typical CCU statements are:

```
[Z]         ;Jump to microprogram ROM address zero.
[μPC]       ;Increment to next address using the microprogram
             counter.
[JPL $NN]   ;Jump to pipeline register branch address $NN
             (hexadecimal).
[MAP]       ;Jump to the address on the output of the mapping ROM.
```

8.4.2 The Fetch Microprogram

Recall that this type of processor operates in a fetch-execute cycle wherein an instruction is fetched from memory and then executed, another instruction is fetched and executed, and so on, until an instruction is encountered that suspends or terminates the cycle. This cycle initiates with the fetch half of the cycle.

In order for the fetch half of this cycle to commence, the program counter must contain the memory address of the instruction to be fetched. This is the primary function of the PC, and the PC should normally be used for no other purpose.

System RESET. Note that when the example processor is reset, the program counter is cleared. This implies that macroprogram execution will commence with an instruction found in external memory at location $00. Also, when the processor is reset, the output of the CCU multiplexer is disabled, causing an automatic jump to microprogram ROM address zero. For this reason, the microprogram for the fetch half of the fetch-execute cycle will be stored at microprogram ROM address $00.

In general, the fetch microprogram must obtain the macroinstruction opcode from external memory using the PC contents as the address. The macroinstruction word that has been read must be loaded into the instruction register. After the PC is used, it must be incremented. Also, in order to facilitate a smooth transition to the execute cycle, the memory unit executes another read with the incremented PC value while the IR is being loaded, so that if the DBB has to be loaded with an operand, it will be waiting on the external data bus at the end of the fetch microprogram. This concept of reading the next macroinstruction word while the current one is still being processed is referred to as a *look-ahead fetch* and is used to speed up processor throughput.

The fetch of the macroinstruction from memory requires execution of the following sequence (refer to Figure 8.7):

t_0: PC → IDB path, Memory read

Because the program counter contains the address of the instruction to be read from memory, it must be placed on the IDB. With it there, the external memory must be commanded to execute a read cycle. Recall that the VMA line goes TRUE halfway through the clock cycle, allowing the data in the PC to propagate through the IDB multiplexer and onto the external address bus before the memory cycle starts.

ALU = PAR D + 1

Every time the program counter is used to access a macroprogram word, it must be incremented to point to the next one. Note that the current PC contents are on the IDB and attached to the D input of the ALU source multiplexer. If the ALU is set to PAR DZ + 1, the output of the ALU function generator will be equal to PC + 1 after a period of propagation. Consequently, the PC "sees" its incremented value on its inputs, ready to be loaded in the next clock cycle. Note: ADD DZ + 1 will also work.

[μPC]

We are not yet done with the fetch microprogram, so we must advance to the next microprogram ROM address because nothing else can be done until the next clock edge.

t_1: (IR) ← (EDB), (PC) ← (PC) + 1

Following the read cycle of clock cycle 0, the macroinstruction word is present on the external data bus, ready to be used by the processor. Consequently, in this clock cycle we will load it into the instruction register where it belongs. Also, because we generated the incremented PC value in the previous clock cycle, we can load that incremented value into the PC now. Note that changing the output of the PC at this point has no effect on the load of the IR, since any changes caused by the command to load the PC must propagate through the PC. By the time the output of the PC changes, the edge that stimulated the change is long gone, and the IR has been loaded on that edge. This is the feature of edge-triggered circuitry.

PC → IDB path, Memory read

Although we have accomplished the task of reading the macroinstruction opcode into the IR and incrementing the PC, because both of these actions take place on the leading edge of this clock cycle, we still have the remainder of the clock cycle to do something useful. In this case we are going to execute a "look-ahead fetch" of the operand of the instruction. If there is an operand associated with this instruction, it is being read from memory during this cycle, because the PC has been incremented. It can then be loaded into the DBB on the next clock cycle, which is the first clock cycle of the execute microprogram. If there is no operand for the current instruction, we will simply fail to load the DBB.

[MAP]

The last RTL statement identifies the location of the next microinstruction word to be executed. Because the purpose of the fetch microprogram was to obtain the opcode and place it in the instruction register, it is appropriate that we take a look at the output of the mapping ROM at this point. Stored in the mapping ROM at the address identified by the opcode is the beginning address of the corresponding microprogram. Consequently, at the end of the fetch microprogram, we will switch the CCU multiplexer to the mapping ROM output to cause the CCU to look up the microprogram word at that address. By the time the next clock edge comes along, the address will have propagated through the CCU multiplexer and the microprogram ROM, and the data for the first microinstruction of the selected macroinstruction will be ready to be loaded into the pipeline register .

Normally it would not be suitable to consider such a lengthy discussion in the development of a microprogram, but certainly all of the discussed factors must be understood. The development process, however, starts with the generation of this RTL program. The following is a summary of that RTL:

t_0: PC → IDB path, Memory read, ALU = PAR D + 1, [μPC]

t_1: (IR) ← (EDB), (PC) ← (PC) + 1, PC → IDB path, Memory read, [MAP]

With the RTL program in hand, the next step is to translate it into the pipeline register bit patterns that will actually execute the functions. From Figure 8.7 and other resources, including the circuit definitions discussed to this point, we can determine the following:

t_0: Microprogram ROM Address $00

PC → IDB path	P4,3,2 = 0,0,1
Memory read	P8,7 = 1,1
ALU = PAR D + 1	P24,23,22 = 0,0,0; P21 = 1; P20,19,18 = 0,0,0
[μPC]	P36,35 = 0,1

t_1: Microprogram ROM address $01

(IR) ← (EDB)	P0 = 0
(PC) ← (PC) + 1	P5 = 0
PC → IDB path	P4,3,2 = 0,0,1
Memory read	P8,7 = 11
[MAP]	P36,35 = 1,1

We have completed the writing of the microprogram for the fetch cycle. However, it may be better understood and better documented if we state it as a state table or as a binary microprogram that defines all of the microprogram ROM bits. This is done using the worksheet of Figure 8.19. Figure 8.20 illustrates the completion of the worksheet for the fetch microprogram.

As you observe Figure 8.20, note that those bits that are maintained in their default state are left blank on the worksheet so that we can better see what is going on. The microprogram is in ROM addresses $00 and $01 since we will be executing this microprogram on RESET as well as at the end of each instruction execution following a jump zero [Z] CCU load control instruction. At the end of the fetch microprogram, the CCU load control instruction is a jump map [MAP] so that the next microprogram word to be presented to the pipeline register will be the first word in the microprogram specified by the fetched opcode.

NAME: DATE:

MNEMONIC: FETCH CYCLE OPCODE:

Step	Description	μP ROM ADDRESS		COMPUTER CONTROL UNIT: LOAD CTRL I1 (36)	LOAD CTRL I0 (35)	BRANCH ADDRESS A7 (34)	A6 (33)	A5 (32)	A4 (31)	A3 (30)	A2 (29)	A1 (28)	A0 (27)
T0	PC→IDB Path, Memory READ,	0	0	0	1								
	ALU = PAR D+1, [μPC]			μPC									
T1	(IR)←(EDB), (PC)←(PC)+1,	0	1	1	1								
	PC→IDB Path, Memory READ, [MAP]			MAP									

Step	ARITHMETIC LOGIC UNIT: LDB (26)	LDA (25)	SOURCE I2 (24)	SOURCE I1 (23)	SOURCE I0 (22)	CI (21)	FCTN I2 (20)	FCTN I1 (19)	FCTN I0 (18)	STATUS AND SHIFT: R/L (17)	SE (16)	LD (15)	FLAG SELECT A2 (14)	A1 (13)	A0 (12)	OE (11)
T0			0	0	0	1	0	0	0							
				D		C		PAR								
T1																

Step	PROGRAM CONTROL UNIT: IY (10)	IX (09)	VMA (08)	R/W (07)	SP (06)	PC (05)	IDB S2 (04)	IDB S1 (03)	IDB S0 (02)	DBB (01)	IR (00)
T0			1	1			0	0	1		
			RD					PC			
T1			1	1		0	0	0	1		0
			RD			P		PC			I

FIGURE 8.20 Fetch microprogram.

During the first clock cycle of the fetch microprogram, the PC was used to read the opcode from memory. In the second clock cycle, the IR was loaded with the opcode, the program counter was incremented, and then the PC was used again to read from the external memory. If the loaded instruction has an operand, it will be ready to load into the data bus buffer in the next clock cycle, the first clock cycle of the execute microprogram. If the instruction has no operand, we will ignore this look-ahead fetch.

8.4.3 Execute Microprograms

The fetch microprogram was loaded into the microprogram ROM at addresses $00 and $01. The remainder of the microprogram ROM will contain microprograms for instructions fetched during the fetch cycle. At the end of the fetch cycle, the computer control unit presents to the microprogram ROM the address of the first microprogram word in the instruction fetched. The execute microprogram may be one or more clock cycles in length. At the termination of the execute microprogram, the CCU will be instructed to return to the fetch cycle so that the next macroinstruction can be fetched and executed.

We will look at several execute microprograms, using the addressing modes previously discussed: inherent, immediate, direct, indexed, and relative.

Practical Example 8.1: CLAA ;Clear Accumulator, Inherent

Recall from our previous discussion that inherent addressing implies that the instruction has no operand. This example instruction, CLAA, is to clear accumulator A, resetting all of its bits. We will select an opcode of $4F, the same used in the Motorola M68HC11 microcontroller.

By observing Figure 8.6 and Figure 8.7, we can determine the sequence of pipeline register bits needed to affect the reset of accumulator A. Consider the following events:

- Obtain a logical 0 on the output of the ALU function generator. This can be done by setting the ALU function and source to AND DZ.
- This value will propagate through the ALU function generator and onto the ALU bus. From there it is presented to the inputs of the accumulators. From this point all we need to do is drop the load control line that drives accumulator A. Note that the assertion of this line is delayed by a half clock cycle, allowing time for the data to propagate through the ALU source multiplexer and ALU function generator and

NAME: DATE:

MNEMONIC: CLAA OPCODE: 4F

	μP ROM ADDRESS		COMPUTER CONTROL UNIT: LOAD CTRL		BRANCH ADDRESS							
			I1	I0	A7	A6	A5	A4	A3	A2	A1	A0
			36	35	34	33	32	31	30	29	28	27
T0 ALU = AND DZ, (A)←(ALU),	0	2	0	0								
Set Flags, [Z]			Z									

	ARITHMETIC LOGIC UNIT: LDB	LDA	SOURCE			CI	FCTN		
			I2	I1	I0		I2	I1	I0
	26	25	24	23	22	21	20	19	18
T0 ALU = AND DZ, (A)←(ALU),		0	0	0	0		1	0	0
Set Flags, [Z]	LDA		DZ				AND		

	STATUS AND SHIFT: R/L	SE	LD	FLAG SELECT			OE
				A2	A1	A0	
	17	16	15	14	13	12	11
T0 ALU = AND DZ, (A)←(ALU),			0				
Set Flags, [Z]			L				

	PROGRAM CONTROL UNIT: IY	IX	VMA	R/W	SP	PC	IDB			DBB	IR
							S2	S1	S0		
	10	09	08	07	06	05	04	03	02	01	00
T0 ALU = AND DZ, (A)←(ALU),											
Set Flags, [Z]											

FIGURE 8.21 CLAA instruction microprogram.

onto the ALU bus. Therefore, the accumulator can be loaded in the same clock cycle as when the ALU function generator produces the data.

- Since this is an arithmetic/logical instruction, the CCR flags must be loaded with the flag functions at the same time accumulator A is loaded with the arithmetic function. This will allow the program to respond to the instruction result conditions at a later time.

The RTL for this instruction would be:

t_0:	ALU = AND DZ	Set the output of the ALU function generator to zero.
	(A) ← (ALU)	Load the value zero into accumulator A.
	Set flags	Set the flags (N = 0, Z = 1, V = 0, C = 0).
	[Z]	Jump to the fetch microprogram.

Figure 8.21 illustrates the microprogram for the CLAA instruction.

Note that stored in address $4F of the mapping ROM would be the value $02, the starting address of the microprogram for the CLAA instruction.

Practical Exercise 8.1: TAB ;Transfer Accumulators, Inherent

Using the method just described, write a microprogram that will copy the contents of accumulator A into accumulator B. Use an opcode of $16. Place the microprogram at microprogram ROM address $04. Document both the RTL and the binary microprogram. This microprogram should take a single clock cycle.

Hint: (B) ← (A). Pass accumulator A through the ALU unchanged and load into accumulator B.

How would the mapping ROM be programmed to provide for this instruction?

Practical Exercise 8.2: SBA ;Subtract Accumulators, Inherent

Using the method just described, write a microprogram that will subtract the contents of accumulator B from the contents of accumulator A. Use an opcode of $10. Place the microprogram at microprogram ROM address $06. Document both the RTL and the binary microprogram. This microprogram should take a single clock cycle.

Hint: (A) ← (A) – (B). Use the ALU source multiplexer and ALU function generator to calculate (A) – (B) and load the result into accumulator A.

How would the mapping ROM be programmed to provide for this instruction?

Practical Example 8.2: LDAA #$NN ;Load Accumulator, Immediate

Recall that when a macroinstruction uses immediate addressing, it fetches a second program word from memory during the execute cycle and uses it as data. For example, consider the following instruction:

LDAA #$3F ;Load accumulator A with the hex value $3F.

Recall also that the operand of the instruction was fetched during the second clock cycle of the fetch microprogram and is waiting on the external data bus to be used, if needed. We need the operand when using immediate addressing.

To complete this instruction we must load the operand into the data bus buffer, send the operand through the ALU and on into accumulator A, and increment the program counter so that it points to the opcode of the next macroinstruction. Both the load of the accumulator and the increment of the program counter require the use of the ALU, so this instruction will take a minimum of two clock cycles. Consider the following RTL sequence:

t_0:	(DBB) ← (EDB)	Load operand into the data bus buffer.
	PC → IDB path	Prepare the path for program counter increment.
	ALU = PAR D + 1	ALU bus = (PC) + 1.
	[μPC]	Increment to the next microprogram word.
t_1:	(PC) ← (PC) + 1	Load the PC with ALU bus to point to next opcode.
	(DBB) → IDB path	Prepare for loading accumulator A with the operand.
	ALU = PAR D	Pass the operand through the ALU unchanged.
	(ACC) ← (ALU)	Load the operand into the accumulator.
	[Z]	Jump to the fetch microprogram.

Figure 8.22 illustrates the microprogram for this instruction, using $86 as the opcode. We will place the microprogram into the microprogram ROM at address $08.

Practical Exercise 8.3: ADDB #$NN ;Add to Accumulator, Immediate

Write a microprogram that will add the macroinstruction operand to the current contents of accumulator B. Assign an opcode of $CB and place the microprogram in the microprogram ROM at address $0A. Document both the RTL and the binary microprogram.

NAME:	DATE:	µP ROM ADDRESS		COMPUTER CONTROL UNIT										ARITHMETIC LOGIC UNIT									STATUS AND SHIFT							PROGRAM CONTROL UNIT										
MNEMONIC: LDAA #$NN	OPCODE: 86			LOAD CTRL		BRANCH ADDRESS								LDB	LDA	SOURCE			CI	FCTN			R/L	SE	LD	FLAG SELECT			OE	IY	IX	VMA	R/W	SP	PC	IDB			DBB	IR
				I1	I0	A7	A6	A5	A4	A3	A2	A1	A0			I2	I1	I0		I2	I1	I0				A2	A1	A0								S2	S1	S0		
				36	35	34	33	32	31	30	29	28	27	26	25	24	23	22	21	20	19	18	17	16	15	14	13	12	11	10	09	08	07	06	05	04	03	02	01	00
T0	(DBB)←(EDB), PC→IDB Path,	0	8	0	1											0	0	0	1	0	0	0														0	0	1	0	
	ALU = PAR D+1, [µPC]			µPC												D			C	PAR																PC			D	
T1	(PC)←(PC)+1, DBB→IDB Path,	0	9	0	0										0	0	0	0	0	0	0	0													0	0	0	0		
	ALU = PAR D, (A)←(ALU), [Z]			Z										LDA		D				PAR															P	DBB				

FIGURE 8.22 LDAA #$NN microprogram worksheet.

Hint: (B) ← (B) + Operand. Get the operand from memory and place into the DBB. Increment the PC. Place the DBB on the IDB and use the ALU source multiplexer and ALU function generator to calculate the required sum, placing it into accumulator B. Since this is an arithmetic instruction, do not forget to set the flags.

How would the mapping ROM be programmed to provide for this instruction?

Practical Exercise 8.4: LDS #$NN ;Load Stack Pointer, Immediate

Write a microprogram that will load the stack pointer with the data contained in the macroinstruction operand. Assign an opcode of $8E and place the microprogram in the microprogram ROM at address $0C. Document both the RTL and the binary microprogram.

Hint: (SP) ← Operand. Get the operand from memory and place into the DBB. Increment the PC. Load the SP with the operand that is stored in the DBB. This instruction should take no more than three clock cycles.

How would the mapping ROM be programmed to provide for this instruction?

Practical Example 8.3: LDAA $NN ;Load Accumulator, Direct

Recall that direct addressing identifies that the operand is the address of the data to be processed. Another way of saying this is that the operand is the effective address. In this instruction, we are to obtain the data stored in memory location $NN and load it into accumulator A. We will use an opcode of $96 and place the microprogram in the microprogram ROM at address $10.

In order to accomplish direct addressing we will need to obtain the macroinstruction operand, place it on the IDB, and use it to go back out to memory in order to obtain the data that is to be loaded into accumulator A. The program must be incremented to point to the opcode of the next instruction. Consider the following RTL definitions:

t_0:	(DBB) ← (EDB)	Load operand into the data bus buffer.
	PC → IDB path	Prepare the path for program counter increment.
	ALU = PAR D + 1	ALU bus = (PC) + 1
	[μPC]	Increment to the next microprogram word.
t_1:	(PC) ← (PC) + 1	Load the PC with ALU bus to point to next opcode.
	(DBB) → IDB path	Prepare use of operand as a memory address.
	Memory read	Read data at memory pointed to by operand.
	[μPC]	Increment to the next microprogram word.
t_2:	(DBB) ← (EDB)	Load memory data into the data bus buffer.
	(DBB) → IDB path	Prepare for loading accumulator A with the data.
	ALU = PAR D	Pass the memory data through the ALU unchanged.

NAME: DATE: MNEMONIC: LDAA $NN OPCODE: 96		µP ROM ADDRESS		COMPUTER CONTROL UNIT										ARITHMETIC LOGIC UNIT									STATUS AND SHIFT							PROGRAM CONTROL UNIT										
				LOAD CTRL I1	I0	BRANCH ADDRESS A7	A6	A5	A4	A3	A2	A1	A0	LDB	LDA	SOURCE I2	I1	I0	CI	FCTN I2	I1	I0	R/L	SE	LD	FLAG SELECT A2	A1	A0	OE	IY	IX	VMA	R/W	SP	PC	IDB S2	S1	S0	DBB	IR
				36	35	34	33	32	31	30	29	28	27	26	25	24	23	22	21	20	19	18	17	16	15	14	13	12	11	10	09	08	07	06	05	04	03	02	01	00
T0	(DBB)←(EDB), PC→IDB Path,	1	0	0	1											0	0	0	1	0	0	0														0	0	1	0	
	ALU = PAR D+1, [µPC]			µPC													D		C		PAR																PC		D	
T1	(PC)←(PC)+1, DBB→IDB Path,	1	1	0	1																											1	1		0	0	0	0	1	
	Memory Read, [µPC]			µPC																												RD			P		DBB			
T2	(DBB)←(EDB), DBB→IDB Path,	1	2	0	0											0	0	0	0	0	0	0	0													0	0	0	0	
	ALU = PAR D, (A)←(ALU), [Z]			Z										LDA			D				PAR																DBB		D	

FIGURE 8.23 LDAA $NN microprogram worksheet.

(ACC) ← (ALU) Load the data into the accumulator.
[Z] Jump to the fetch microprogram.

Figure 8.23 illustrates the microprogram worksheet for the LDAA $NN instruction.

Practical Exercise 8.5: ADDB $NN ;Add to Accumulator, Direct

Write a microprogram that will add the data pointed to by the macroinstruction operand to the current contents of accumulator B. Assign an opcode of $DB and place the microprogram in the microprogram ROM at address $14. Document both the RTL and the binary microprogram.

Hint: (B) ← (B) + (Operand). Get the operand from memory and place it into the DBB. Increment the PC. Place the DBB on the IDB and execute a memory read cycle. Go back to memory to get the data pointed to by the operand and use the ALU source multiplexer and ALU function generator to calculate the required sum, placing it into accumulator B. Since this is an arithmetic instruction, do not forget to set the flags.

How would the mapping ROM be programmed to provide for this instruction?

Practical Exercise 8.6: LDS $NN ;Load Stack Pointer, Direct

Write a microprogram that will load the stack pointer with the memory data pointed to by the macroinstruction operand. Assign an opcode of $9E and place the microprogram in the microprogram ROM at address $18. Document both the RTL and the binary microprogram.

Hint: (SP) ← (Operand). Get the operand from memory and place it into the DBB. Increment the PC. Place the DBB on the IDB and execute a memory read cycle. Load the SP with the memory data that is stored in the DBB.

How would the mapping ROM be programmed to provide for this instruction?

Practical Example 8.4: LDAA $NN,X ;Load Accumulator, Indexed

Indexed addressing is very similar to direct addressing. The macroinstruction operand is used as a pointer to memory, but, prior to use, the contents of an index register are added to it. That is, the effective address of the instruction is the sum of the operand and the index register. This example is to obtain the data stored in memory at the indicated effective address and place that data into the accumulator. We will assign the instruction an opcode of \$A6 and place the microprogram in the microprogram ROM at address \$1C.

In order to accomplish indexed addressing we will need to obtain the macroinstruction operand, place it on the IDB, and use the sum of it and index register X to go back out to memory in order to obtain the data that is to be loaded into accumulator A. The program must be incremented to point to the opcode of the next instruction. Consider the following RTL definitions:

t_0:	(DBB) ← (EDB)	Load operand into the data bus buffer.
	PC → IDB path	Prepare the path for program counter increment.
	ALU = PAR D + 1	ALU bus = (PC) + 1
	[μPC]	Increment to the next microprogram word.
t_1:	(PC) ← (PC) + 1	Load the PC with ALU bus to point to next opcode.
	(IX) + \$NN → IDB path	Use sum of operand and IX as a memory address.
	Memory read	Read data at memory pointed to by operand.
	[μPC]	Increment to the next microprogram word.
t_2:	(DBB) ← (EDB),	Load memory data into the data bus buffer.
	(DBB) → IDB path	Prepare for loading accumulator A with the data.
	ALU = PAR D	Pass the memory data through the ALU unchanged.
	(ACC) ← (ALU)	Load the data into the accumulator.
	[Z]	Jump to the fetch microprogram.

Figure 8.24 illustrates the microprogram worksheet for the LDAA \$NN,X instruction.

Practical Exercise 8.7: ADDB \$NN,Y ;Add to Accumulator, Indexed

Write a microprogram that will add the data pointed to by the sum of index register Y and the macroinstruction operand to the current contents of accumulator B. Assign an opcode

NAME: DATE:

MNEMONIC: LDAA \$NN,X OPCODE: A6

		μP ROM ADDRESS		COMPUTER CONTROL UNIT: LOAD CTRL I1 (36)	LOAD CTRL I0 (35)	BRANCH ADDRESS A7 (34)	A6 (33)	A5 (32)	A4 (31)	A3 (30)	A2 (29)	A1 (28)	A0 (27)	ARITHMETIC LOGIC UNIT: LDB (26)	LDA (25)	SOURCE I2 (24)	SOURCE I1 (23)	SOURCE I0 (22)	CI (21)	FCTN I2 (20)	FCTN I1 (19)	FCTN I0 (18)
T0	(DBB)←(EDB), PC→IDB Path,	1	C	0	1											0	0	0	1	0	0	0
	ALU = PAR D+1, [μPC]			μPC												D			C	PAR		
T1	(PC)←(PC)+1, X+OPERAND→IDB Path,	1	D	0	1																	
	Memory Read, [μPC]			μPC																		
T2	(DBB)←(EDB), DBB→IDB Path,	1	E	0	0										0	0	0	0	0	0	0	0
	ALU = PAR D, (A)←(ALU), [Z]			Z										LDA		D				PAR		

	STATUS AND SHIFT: R/L (17)	SE (16)	LD (15)	FLAG SELECT A2 (14)	A1 (13)	A0 (12)	OE (11)	PROGRAM CONTROL UNIT: IY (10)	IX (09)	VMA (08)	R/W (07)	SP (06)	PC (05)	IDB S2 (04)	IDB S1 (03)	IDB S0 (02)	DBB (01)	IR (00)
T0														0	0	1	0	
															PC		D	
T1										1	1		0	0	1	1	1	
										RD			P		X+O			
T2														0	0	0	0	
															DBB		D	

FIGURE 8.24 LDAA \$NN,X microprogram worksheet.

of \$18 and place the microprogram in the microprogram ROM at address \$20. Document both the RTL and the binary microprogram.

Hint: (B) ← (B) + (Y + Operand). Get the operand from memory and place it into the DBB. Increment the PC. Place the sum of the operand and the Y register on the IDB and execute a memory read cycle. Go back to memory to get the data pointed to by the operand and use the ALU source multiplexer and ALU function generator to calculate the required sum, placing it into accumulator B. Since this is an arithmetic instruction, do not forget to set the flags.

How would the mapping ROM be programmed to provide for this instruction?

Practical Exercise 8.8: LDS \$NN,Y ;Load Stack Pointer, Indexed

Write a microprogram that will load the stack pointer with the memory data pointed to by the sum of index register Y and the macroinstruction operand. Assign an opcode of \$AE and place the microprogram in the microprogram ROM at address \$24. Document both the RTL and the binary microprogram.

Hint: (SP) ← (Operand). Get the operand from memory and place it into the DBB. Increment the PC. Place the sum of the operand and index register Y on the IDB and execute a memory read cycle. Load the SP with the memory data which is stored in the DBB.

How would the mapping ROM be programmed to provide for this instruction?

Practical Example 8.5: JMP \$NN ;Unconditional Direct Jump

Frequently, at the end of a program segment it is desired to jump to an instruction other than the one following the current instruction. Such a jump can be accomplished by loading the destination address into the program counter and jumping to the fetch cycle. There are two types of jumps: relative and direct. A *relative jump* is accomplished by adding the operand of the instruction to the program counter. A *direct jump* is accomplished by placing the operand into the program counter. We will first examine the unconditional direct jump. The jump instruction is unconditional because it will always jump to the destination, regardless of the status of the flags in the condition code register. We will look at conditional instructions shortly.

The microprogram for the direct jump is rather straightforward: The operand must be placed into the program counter. Since we are initializing the PC, there is no need to increment it, even though we have fetched an operand. Consider the following RTL program:

t_0:	(DBB) ← (EDB)	Load operand into the data bus buffer.
	DBB → IDB path	Prepare the path for program counter load.
	ALU = PAR D	Pass the operand through the ALU unchanged.
	[μPC]	Increment to the next microprogram word.
t_1:	(PC) ← (ALU)	Load the PC with the operand.
	[Z]	Jump to the fetch microprogram.

NAME: DATE:

MNEMONIC: JMP $NN OPCODE: 7E

		μP ROM ADDRESS	COMPUTER CONTROL UNIT: LOAD CTRL I1 (36)	I0 (35)	BRANCH ADDRESS A7 (34)	A6 (33)	A5 (32)	A4 (31)	A3 (30)	A2 (29)	A1 (28)	A0 (27)	ARITHMETIC LOGIC UNIT: LDB (26)	LDA (25)	SOURCE I2 (24)	I1 (23)	I0 (22)	CI (21)	FCTN I2 (20)	I1 (19)	I0 (18)
T0	(DBB)←(EDB), DBB→IDB Path,	30	0	1											0	0	0	0	0	0	0
	ALU = PAR D, [μPC]		μPC												D				PAR		
T1	(PC)←(ALU), [Z]	31	0	0																	
			Z																		

	STATUS AND SHIFT: R/L (17)	SE (16)	LD (15)	FLAG SELECT A2 (14)	A1 (13)	A0 (12)	OE (11)	PROGRAM CONTROL UNIT: IY (10)	IX (09)	VMA (08)	R/W (07)	SP (06)	PC (05)	IDB S2 (04)	S1 (03)	S0 (02)	DBB (01)	IR (00)
T0														0	0	0	0	
														DBB			D	
T1													0					
													P					

FIGURE 8.25 JMP $NN, microprogram worksheet.

We will assign this instruction an opcode of $7E, and place it in the microprogram ROM at address $30. Figure 8.25 illustrates the microprogram for the JMP $NN instruction.

Practical Example 8.6: BRA *+NN ;Unconditional Relative Branch

It is often quite advantageous to be able to jump to a program destination that is located at a memory address that is relative to the program counter. For example, if all of the jumps within a program were relative to the position of the program counter, that program could be loaded into the memory at any address and still run properly. A program that can be located anywhere in the memory is referred to as *relocatable.* A program that has even a single direct jump must be loaded at the predetermined starting address in order to operate properly. Such a program is referred to as *absolute.* Consequently, using relative jumps provides the program with the property of relocatability. There are other advantages to using relative addressing also.

Typically, a relative jump is referred to as a *branch* and is accomplished by adding the macroinstruction operand to the contents of the program counter. Since the arithmetic logic unit executes 2's complement arithmetic, the operand is a signed number and allows branches to either higher or lower addresses. The branch microprogram is similar to the jump microprogram except that the current value of the program counter is added to the operand as it passes through the ALU. (In the direct jump, the operand passed through the ALU unchanged.) Consider the following RTL program.

t_0:	(DBB) ← (EDB)	Load operand into the data bus buffer.
	DBB → IDB path	Prepare the path for program counter load.
	ALU = ADD DP	Calculate the sum of the PC and the operand.
	[μPC]	Increment to the next microprogram word.
t_1:	(PC) ← (ALU)	Load the PC with the sum of the PC and the operand.
	[Z]	Jump to the fetch microprogram.

We will assign this instruction an opcode of $20 and place it in the microprogram ROM at address $32. Figure 8.26 illustrates the microprogram for the BRA *+$NN instruction.

Note that when two objects are added using the ALU, the function is ADD R + S + 1 – CI. Therefore, to avoid incrementing the sum one more than desired, it is necessary to set CI = 1.

NAME:	DATE:	μP ROM		COMPUTER CONTROL UNIT										ARITHMETIC LOGIC UNIT									STATUS AND SHIFT							PROGRAM CONTROL UNIT										
MNEMONIC: BRA *+$NN	OPCODE: 20	ADDRESS		LOAD CTRL I1	LOAD CTRL I0	BRANCH ADDRESS A7	A6	A5	A4	A3	A2	A1	A0	LDB	LDA	SOURCE I2	I1	I0	CI	FCTN I2	I1	I0	R/L	SE	LD	FLAG SELECT A2	A1	A0	OE	IY	IX	VMA	R/W	SP	PC	IDB S2	S1	S0	DBB	IR
				36	35	34	33	32	31	30	29	28	27	26	25	24	23	22	21	20	19	18	17	16	15	14	13	12	11	10	09	08	07	06	05	04	03	02	01	00
T0	(DBB)←(EDB), DBB→IDB Path,	3	2	0	1											1	0	0	1	0	1	0														0	0	0	0	
	ALU = ADD DP, [μPC]			μPC												DP			C	ADD																DBB			D	
T1	(PC)←(ALU), [Z]	3	3	0	0																														0					
				Z																															P					

FIGURE 8.26 BRA *+NN, microprogram worksheet.

The value $NN used in the relative branch is often referred to as a displacement. This displacement is simply the difference between the destination address and the current address of the program counter (after the operand has been fetched.) The assembler or compiler program that generates the macroprogram calculates these displacements, so the programmer is usually unconcerned with them. The only time they would need to be calculated is if the program were being assembled by hand.

Practical Example 8.7: BEQ *+NN ;Branch if Equal to Zero, Relative

A fundamental process of the computer is its ability to make decisions. Processors make decisions based on the recorded results of arithmetic and logical operations. This is done by setting the condition code register flags (or status flags) as arithmetic and logical functions are generated. The four status flags in our processor, N, Z, V, and C, are set by the ALU function generator when its output function is negative, zero, there was an overflow, or a carry was generated, respectively. When these flags are loaded into the condition code register, they can be observed later so that a decision can be made.

The decision is made by enabling the output of the CCR, selecting a desired status flag. If the output of the CCR is TRUE, the computer control unit automatically jumps to the microprogram address indicated in the branch address of the pipeline register. If the output of the CCR is FALSE, the condition is ignored. There are several conditions that can be tested based on the four status flags. Eight of these are indicated in Figure 8.9.

In this example we want to branch to the destination address if the result of a previous operation was equal to 0. If we enable the output and select the Z bit at its output, the next microprogram address will be taken from the pipeline register branch address if Z = 1. Otherwise, the next microprogram address will be selected based on the CCU load control command. Consider the following RTL:

t_0:	(DBB) ← (EDB)	Load operand into the data bus buffer.
	PC → IDB path	Prepare the path for program counter increment.
	ALU = PAR D + 1	Calculate the incremented PC value.
	cc = EQ, PL = T2	Enable CCR output, jump to T2 if Z = 1.
	[μPC]	Go to the next microprogram word if Z = 0.
t_1:	(PC) ← (ALU)	Increment PC to point to next instruction.
	[Z]	Jump to the fetch microprogram.

NAME: DATE:

MNEMONIC: BEQ *+$NN OPCODE: 27

		µP ROM ADDRESS	COMPUTER CONTROL UNIT: LOAD CTRL I1 (36)	LOAD CTRL I0 (35)	BRANCH ADDRESS A7 (34)	A6 (33)	A5 (32)	A4 (31)	A3 (30)	A2 (29)	A1 (28)	A0 (27)
T0	(DBB)←(EDB), PC→IDB Path,	34	0	1	0	0	1	1	0	1	1	0
	ALU = PAR D+1, cc=EQ, PL=T2, [µPC]		µPC		JPL $36 if Z=1							
T1	(PC)←(ALU), [Z]	35	0	0								
			Z									
T2	DBB→IDB Path, ALU = ADD DP, [µPC]	36	0	1								
			µPC									
T3	(PC)←(ALU), [Z]	37	0	0								
			Z									

	ALU: LDB (26)	LDA (25)	SOURCE I2 (24)	SOURCE I1 (23)	SOURCE I0 (22)	CI (21)	FCTN I2 (20)	FCTN I1 (19)	FCTN I0 (18)	STATUS AND SHIFT: R/L (17)	SE (16)	LD (15)	FLAG SELECT A2 (14)	FLAG SELECT A1 (13)	FLAG SELECT A0 (12)	OE (11)	PROGRAM CONTROL UNIT: IY (10)	IX (09)	VMA (08)	R/W (07)	SP (06)	PC (05)	IDB S2 (04)	IDB S1 (03)	IDB S0 (02)	DBB (01)	IR (00)
T0			0	0	0	1	0	0	0				1	0	0	1							0	0	1	0	
				D		C		PAR						cc=EQ										PC		D	
T1																						0					
																						P					
T2			1	0	0	1	0	1	0														0	0	0		
				DP		C		ADD																DBB			
T3																						0					
																						P					

FIGURE 8.27 BEQ *+NN, microprogram worksheet.

t_2:	DBB → IDB path	Prepare to add the operand to the PC.
	ALU = ADD DP	Calculate the branch address.
	[µPC]	Go to the next microprogram word.
t_3:	(PC) ← (ALU)	Load the PC with the sum of the PC and the operand.
	[Z]	Jump to the fetch microprogram.

If Z = 0, the microprogram executes cycles t_0 and t_1, incrementing the PC and jumping to the fetch cycle. If Z = 1, the microprogram executes cycles t_0, t_2, and t_3, adding the operand to the PC (without incrementing the PC).

We will assign this instruction an opcode of $27 and place it in the microprogram ROM at address $34. Figure 8.27 illustrates the microprogram for the BEQ *+$NN instruction.

Practical Exercise 8.9: BCC *+$NN ;Branch if Carry Clear, Relative

Using the methodology of the previous example, write the microprogram for a BCC *+$NN instruction that adds $NN to the PC if the carry flag in the condition code register is clear. If the carry flag is set, increment the PC to point to the next macroinstruction.

Assign an opcode of $24 and place the microprogram in the microprogram ROM at address $37.

8.5 MICROCONTROLLERS

A microcontroller is simply a microprocessor that includes additional circuitry that will handle a limited amount of memory and input/output functions. The microprocessor illustrated in this chapter contains no such additional circuitry. A typical microcontroller might contain from 128 bytes to several kilobytes of RAM and ROM, several digital I/O ports, one or more analog I/O ports, and possibly a serial communications port. Often the digital I/O ports are organized into parallel data ports that can manipulate single and multibyte data. Chapter 9 describes such a microcontroller.

8.6 SUMMARY

This chapter has been by no means an exhaustive treatment of microprocessor architecture, but it should begin to shed some light on what is happening inside the typical (and simple) complex instruction set computer (CISC). The architecture used in this chapter lacks a few fundamental resources that are necessary for a commercial-grade product. We have made little reference to the capabilities and application of the status and shift control unit and have completely eliminated discussion of several other processor circuits that are quite common in the field. Our purpose here was to introduce some of the more basic architectural forms and microprogramming methodology through the application of a simple, less than commercial grade, processor.

Though commercial devices are more powerful and complex, the principles under which they operate are the same. CISC computers operate under the fetch-execute cycle and have been since the late 1940s. The predominant change in computer technology over the years has been the electronics used to fabricate the processors. Moving from vacuum tubes to transistors to integrated circuits has allowed the design of faster and more powerful processors, but there has been very little substantial change in architectures. However, only in recent years has the fetch-execute architecture been significantly and successfully challenged. Now with memory systems becoming significantly larger and less expensive than in the past, new concepts in architecture have been developed. One of the most significant of these changes eliminates the microprogram ROM and the computer control unit by using external memory to store the microprograms. This concept eliminates the fetch portion of the fetch-execute cycle, increasing the throughput of the processor by an order of two or three times when running the same clock speed. Also, by using memory caching schemes (reading data faster than the processor can use it), throughput is increased from five to ten times, again running the same clock speed. These devices, referred to as reduced instruction set computers (RISC), threaten the long dynasty held by the CISC processor.

Understanding the internal architecture of a processor at the microprogramming level is not necessary for its programming. However, those who do become involved with programming computers should not be ignorant of processor architecture since, when the processor resources are well understood, decisions can be made in software design that result in more efficient application.

For a more detailed discussion of microprocessor architecture at the microprogramming level, refer to: Carter, J.W. (1996) *Microprocessor Architecture and Microprogramming: A State Machine Approach,* Englewood Cliffs, NJ: Prentice Hall, Inc. 400pp.

QUESTIONS AND PROBLEMS

Describe the function of each of the segments of a typical microprocessor listed in Problems 1–4.

1. Computer control unit
2. Program control unit

3. Arithmetic logic unit
4. Status and shift control unit

5. What is the purpose of the pipeline register?
6. How are the read/$\overline{\text{write}}$ and VMA lines used together to control memory data transfer?
7. What is the purpose of the program counter in a simple computer processor?
8. What is the purpose of the stack pointer in a simple computer processor?
9. What is the purpose of an index register in a simple computer processor?
10. How are arithmetic functions performed in a simple computer processor?
11. How does the shifter in an ALU perform its operation?
12. What is the purpose of the condition code register?
13. What arithmetic properties are described by the four status bits in the condition code register described in this chapter?
14. Describe the function of each of the pipeline register bits that drive the program control unit of the example processor of this chapter.
15. Describe the function of each of the pipeline register bits that drive the status and shift control unit of the example processor of this chapter.
16. Describe the function of each of the pipeline register bits that drive the arithmetic logic unit of the example processor of this chapter.
17. Describe the function of each of the pipeline register bits that drive the computer control unit of the example processor of this chapter.
18. What is the purpose of the data bus buffer?
19. How are memory addresses provided to the external memory by our example processor?
20. What makes a computer microprogrammable?
21. Are typical monolithic microprocessors found in the market today microprogrammable?
22. What is the difference between assembly language and machine code (or macroprogramming) language?
23. What is register transfer language?
24. What is the relationship between register transfer language and the binary microprogram that is stored in the microprogram ROM?
25. What is the difference between a macroprogram word and a microprogram word?
26. Describe in general terms what takes place during the fetch portion of the fetch-execute cycle.
27. What is the purpose of the mapping ROM?
28. How does the mapping ROM convert binary opcodes found in the macroprogram word to a microprogram ROM address?
29. Describe zero- and single-operand macroinstruction formats.
30. What is an addressing mode?
31. Describe how inherent (or implied) addressing is accomplished.
32. Describe how immediate addressing is accomplished.
33. Describe how direct addressing is accomplished.
34. Describe how indexed addressing is accomplished.
35. Describe how relative addressing is accomplished.
36. What is meant by an effective address?

37. Why is a register transfer language used in the development of microprograms?
38. Describe the typical statement types used in a register transfer language.
39. Describe, in detail, what takes place during the fetch portion of the fetch-execute cycle.
40. How is a direct macroprogram jump accomplished?
41. How is a relative macroprogram jump accomplished?
42. How is a conditional macroprogram jump accomplished?
43. Describe, in table form, the contents of the mapping ROM if it were programmed to execute the microprograms illustrated in all of the practical examples and practical exercises of this chapter.

For Problems 44–74, describe the addressing mode, write the RTL and the microprogram for the indicated instructions.

	Mnemonic	*Operand*	*Description*	*Opcode*	*ROM Address*
44.	ADDA	#$NN	Add Memory to A	$8B	$40
45.	ADDA	$NN	Add Memory to A	$9B	$44
46.	ADDA	$NN,X	Add Memory to A	$E9	$48
47.	ANDB	#$NN	AND B with Memory	$C4	$4C
48.	ANDB	$NN	AND B with Memory	$D4	$50
49.	ANDB	$NN,X	AND B with Memory	$E4	$54
50.	BCS	*+$NN	Branch if Carry Set	$25	$58
51.	BNE	*+$NN	Branch if Not Equal	$26	$5C
52.	BVC	*+$NN	Branch if Overflow Clear	$28	$60
53.	BVS	*+$NN	Branch if Overflow Set	$29	$64
54.	COMA		1's Complement A	$43	$68
55.	COMB		1's Complement B	$53	$6A
56.	DECA		Decrement A	$4A	$6C
57.	DECB		Decrement B	$34	$6E
58.	DES		Decrement Stack Pointer	$34	$70
59.	EORA	#$NN	Exclusive OR A with Memory	$88	$72
60.	EORB	$NN	Exclusive OR B with Memory	$D8	$74
61.	JMP	$NN,X	Jump	$6E	$76
62.	LDS	#$NN	Load Stack Pointer	$8E	$7A
63.	LDS	$NN	Load Stack Pointer	$9E	$7C
64.	LDX	$NN,X	Load Index Register	$EE	$80
65.	LSLA		Logical Shift Left A	$48	$84
66.	LSRB		Logical Shift Right B	$54	$86
67.	NEGA		2's Complement A	$40	$88
68.	ORAA	$NN	OR A with Memory	$9A	$8A
69.	ORAB	$NN,X	OR B with Memory	$EA	$8D
70.	SUBA	#$NN	Subtract Memory from A	$80	$92
71.	SUBB	$NN	Subtract Memory from B	$D0	$96
72.	TBA		Load B into A	$17	$9A
73.	TSX		Load SP into X	$30	$9C
74.	TXS		Load X into SP	$35	$9E

CHAPTER 9

The MC68HC11 Microcontroller

OBJECTIVES

After completing this chapter you should be able to:

- Describe the overall architecture of the MC68HC11 microcontroller.
- Describe the functional units within the MC68HC11 microcontroller.
- Describe the functional capabilities of the five input/output ports of the MC68HC11 microcontroller.
- Describe the operational modes (single-chip, extended) of the microcontroller.
- Describe the instruction formats of the MC68HC11 family devices.
- Describe the addressing modes of the MC68HC11 family devices.
- Describe the basic instructions in the MC68HC11 instruction set.
- Describe the basic concepts of pulse accumulation, input data capture, and output data capture.
- Describe the basic concepts of asynchronous and synchronous serial data communication.
- Describe the basic concepts of the analog-to-digital converter subsystem.
- Write MC68HC11 assembly language program segments that use the fundamental capabilities of the microcontroller's I/O ports, including:
    - Use of Port A for pulse accumulation, input capture, output capture, and real-time interrupts
    - Use of Port B for digital output and pulsed handshaking
    - Use of Port C for digital I/O
    - Use of Port D for serial communications
    - Use of Port E for analog input

9.1 OVERVIEW

We stated in earlier chapters that progressively more complex circuits are required to monitor and control progressively more complex systems. The simplest circuits in the spectrum

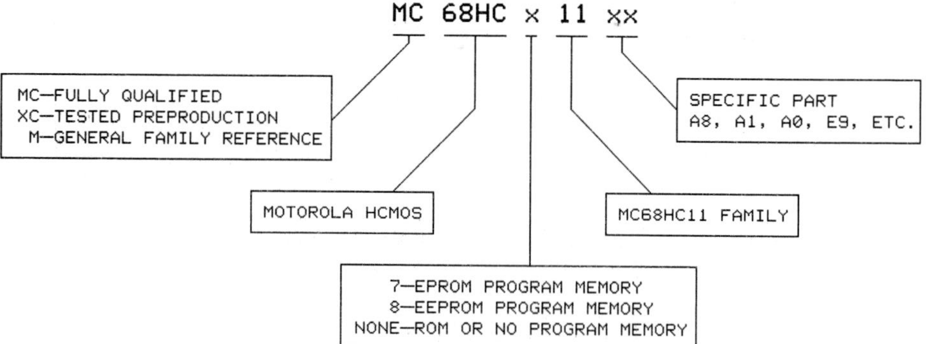

FIGURE 9.1 MC68HC11 family naming convention.

were discrete logic devices, followed by PALs, then FPLAs and ROMs. When the system gets too complex for these devices, a microcontroller can often be applied to the situation. A microcontroller is a state machine, and it is certainly a programmable logic device. From Chapter 8 we find that a microprocessor is a relatively complex state machine, where included with the controlled binary counter are circuits that perform the computing functions. These required circuits, which interface with external memory (PCU), provide some arithmetic and logic functions (ALU and SSCU). A microcontroller is a microprocessor that includes additional circuitry that provides analog and digital input/output functions.

There are several fine microcontrollers in the marketplace at the time of this writing, and few have found the penetration of the MC68HC11 microcontroller produced by Motorola, Inc. This microcontroller, based on the earlier M6800 design, contains the typical microprocessor components discussed in the previous chapter, plus input/output circuitry that provides the capability of single- and multibit digital and analog I/O, synchronous and asynchronous serial communications, pulse accumulation, and various timer functions. It also contains some ROM, RAM, EEPROM, and clock oscillator circuitry. Because of all these resources, an MC68HC11 microcontroller can often be placed into an application with very little support circuitry.

Several versions of the MC68HC11 microcontroller are available, and their architectures are identified in their product name. Figure 9.1 illustrates the naming convention this family of microcontrollers.

Table 9.1 compares some of the features of this family of microcontrollers. The examples for this chapter will use the MC68HC11A8 device.

This chapter will present an overview of the architecture of the MC68HC11A8 microcontroller and introduce its assembly language, programming, and port utilization. Detailed coverage of the device is beyond the scope of this chapter. For a more detailed study, refer to Motorola device documentation or texts that are dedicated to this device.*

*The technical information that describes the MC68HC11 in this chapter is current with "MC68HC11 Reference Manual," Rev. 3, ©1991, Motorola Inc.

TABLE 9.1 MC68HC11 family devices.

Part Number	*EPROM*	*ROM*	*EEPROM*	*RAM*	*CONFIG*[1]	*Comments*
MC68HC11A8		8K	512	256	$0F	Family built around this device
MC68HC11A1			512	256	$0D	MC68HC11A8 with ROM disabled
MC68HC11A0				256	$0C	MC68HC11A8 with ROM and EEPROM disabled
MC68HC811A8			8K + 512	256	$0F	EEPROM emulator for MC68HC11A8
MC68HC11E9		12K	512	512	$0F	Four-input capture
MC68HC11E1			512	512	$0D	MC68HC11E9 with ROM disabled
MC68HC11E0				512	$0C	MC68HC11E9 with ROM and EEPROM disabled
MC68HC811E2			2K[2]	256	$FF	No ROM part for expanded systems
MC68HC711E9	12K		512	512	$0F	One-time programmable version of MC68HC11E9
MC68HC11D3		4K		192	N/A	Low-cost 40-pin version
MC68HC711D9	4K			192	N/A	One-time programmable version of MC68HC11D3
MC68HC11F1			512[2]	1K	$FF	High-performance non-multiplexed
MC68HC11K4		24K	640	768	$FF	>1 MB memory space, PWM, CS
MC68HC711K4	24K		640	768	$FF	One-time programmable version of MC68HC11K4
MC68HC11L6		16K	512	512	$0F	MC68HC11E9 with more ROM and more I/O
MC68HC711L6	16K		512	512	$0F	One-time programmable version of MC68HC11L4

[1]CONFIG register values in this table reflect the value programmed prior to shipment from Motorola.
[2]The EEPROM is relocatable to the top of any 4K memory page. Relocation is done with the upper four bits of the control register.

9.2 GENERAL DESCRIPTION

The MC68HC11 microcontroller contains all of the features of a main processor and several additional functional units that provide digital serial, digital parallel, digital binary, and analog input/output functions. The I/O functions are provided through five data ports labeled Port A, Port B, Port C, Port D, and Port E. The chip has two primary modes of operation. When in the single-chip mode, all of the designed port facilities are available. In this mode there is no externally available address or data bus, so the environment cannot be expanded to include more RAM, ROM, I/O devices, etc. The second primary mode of operation, the expanded mode, sacrifices the facilities of Port B and Port C in order to pin out the address and data buses. When the chip is being operated in the expanded mode, not all of the I/O functions of Port B and Port C are available. However, Motorola provides a supplemental I/O chip, the MC68HC24, which contains these resources. When the two chips are used together, all of the features of the MC68HC11 are available, as well as the address and data buses, so that the system can be expanded. Figure 9.2 illustrates a block diagram of this microcontroller.

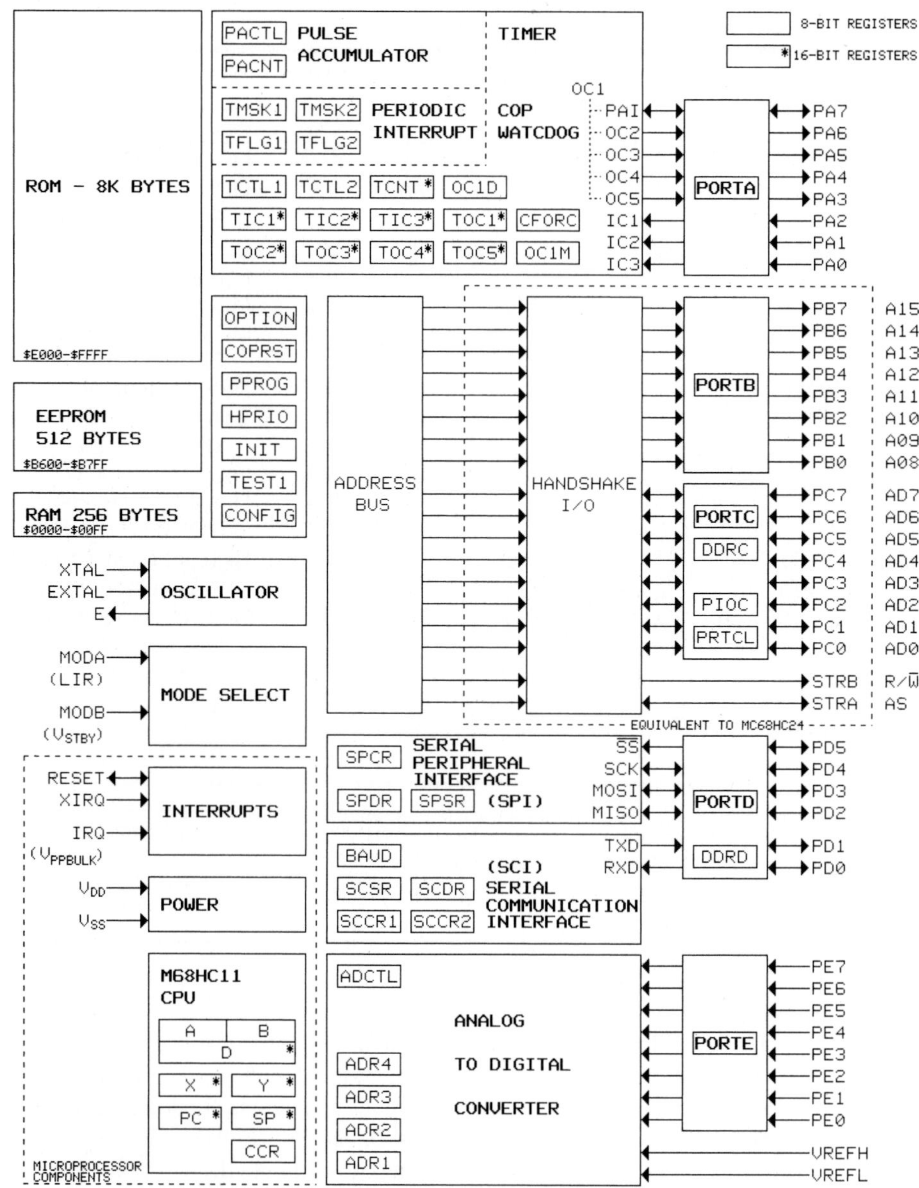

FIGURE 9.2 MC68HC11 block diagram.

The block diagram of Figure 9.2 may appear a little imposing at first, but a review of each of its functional units should serve to help tame the beast. These functional units include:

- Main processor
- RAM
- ROM

- EEPROM
- Port A: Digital I/O and timer functions
- Port B: Digital output/address bus bits A15–A08
- Port C: Digital I/O/address bus bits A07–A00, data bus bits D7–D0
- Port D: Digital I/O/synchronous and asynchronous serial communication
- Port E: Digital I/O/multibit analog input

Note that the five I/O ports are multifunctional. Essentially, each port serves one of two programmable functions. We will observe each of these functional units individually.

9.2.1 Main Processor

As already noted, the main processor portion of the MC68HC11 microcontroller is very similar to the M6800 microprocessor. The microcontroller is a member of that microprocessor family and represents a superset of the microprocessor. Consider first the programmer's model of the MC68HC11 microcontroller illustrated in Figure 9.3.

This programmer's model defines each of the registers in the main processor that may be referenced using the processor's assembly language. This model is similar to the M6800 microprocessor. The D register and Y register have been added to form the main processor portion of the MC68HC11 microcontroller. The MC68HC11 microcontroller drives an eight-bit data bus and a 16-bit address bus. All internal I/O resources are mapped into the 16-bit address space between addresses $1000 and $103F, inclusive. These resources appear as eight- and 16-bit registers and are accessible at these addresses using

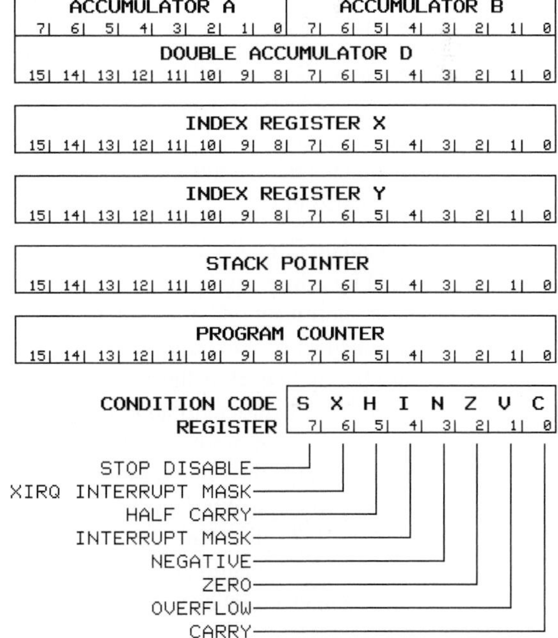

FIGURE 9.3 MC68HC11 programmer's model.

standard memory reference instructions. We will look at this in more detail shortly. Consider the following description of the main processor registers:

Accumulator A. This is an eight-bit general-purpose register that is primarily used to hold one of the two operands used in any arithmetic or logical operation.

Accumulator B. This is an eight-bit general-purpose register that is primarily used to hold the other of the two operands used in any arithmetic or logical operation.

Accumulator D. This is a 16-bit general-purpose register that is formed by concatenating (placing end-to-end) the two accumulators A and B. Its purpose is to provide a resource for the execution of some 16-bit arithmetic operations, including multiplication and division.

Index Register X. This 16-bit register is used to address memory using the indexed addressing mode. When applied, the effective address is calculated by adding the contents of index register X with the macroinstruction operand.

Index Register Y. This 16-bit register is used to address memory using the indexed addressing mode. When applied, the effective address is calculated by adding the contents of index register Y with the macroinstruction operand.

Stack Pointer. This 16-bit register is used to address the top of the system stack in external memory. This processor uses a post-decrement PUSH protocol. That is, when data is pushed onto the stack, the stack pointer value is first used to address the memory unit, the data is then written, and the stack pointer is then decremented to point to the new top of the stack. When data is read from the stack, the stack pointer is first incremented. It is then used to address memory for the subsequent read operation. The stack pointer is typically initialized in the power-up software sequence by placing the address of the top of the RAM segment used for the stack. A convenient value for the MC68HC11A8 microcontroller would be $00FF, the top of the on-chip RAM.

Program Counter. This 16-bit register is used to point into the macroprogram being executed. Since this device uses a 16-bit address bus, there can be up to 65,536 memory locations assigned in the environment. Much of this address space is already used by resources within the microcontroller. The remaining space is available when the chip is configured in expanded mode.

Condition Code Register. This eight-bit register contains five status flags that are generated by arithmetic operations and three flag bits used to define interrupt status. The least four significant bits of the CCR contain the arithmetic/logic function flags N, Z, V, and C. Their function and purpose are identical to that described in the previous chapter. Another arithmetic flag, H, is the half-carry flag, which is set when there is a carry from bit 3 to bit 4 during a byte-length addition or subtraction operation. This flag is used to assist in the management of nibble-length data such as binary coded decimal numbers. When asserted to a logical 1, the I flag disables interrupts from the external IRQ line and from internal interrupt sources. When asserted to a logical 1, the X flag disables interrupts from the external XIRQ line. The most significant bit in the CCR is the S bit. When asserted, it disables the execution of the STOP instruction. This instruction stops the processor clocks, and its inadvertent execution may cause problems in the environment.

Mode-Select Pins MODB/V_{STBY} and MODA/LIR. The Mode B/V standby RAM supply pin functions as both a mode-select input pin and a standby power-supply pin. The Mode A load instruction register pin is used to select the microcontroller operating mode while the microcontroller is in RESET, and it operates as a diagnostic output signal while the microcontroller is executing instructions.

The hardware mode-select mechanism starts with the logic levels on the MODA and MODB pins. The levels on these lines determine the operating mode the microcontroller assumes at RESET. After the RESET occurs, the mode-select inputs no longer influence microcontroller operation. The MODA line is asserted low by the microcontroller during the first E cycle of each executed instruction. This signal can be used for troubleshooting purposes. The MODB line provides the capability of operating in a power standby mode. Refer to Motorola literature for a detailed explanation of standby operation.

Table 9.2 illustrates the configuration options of these two pins. While in the normal single-chip mode, the I/O facilities of Port B and Port C are available to the system designer and the address and data buses are not available. When in the normal expanded mode, the I/O facilities of Port B and Port C are sacrificed to make room for the pin-out of the address and data buses. The special bootstrap mode allows programs to be downloaded through the on-chip serial communications interface (SCI) into the internal RAM. The special test mode is rarely used by the user.

Crystal Oscillator and Clock Pins EXTAL, XTAL, and E. The oscillator pins can be used with an external crystal network or CMOS-compatible clock source. The frequency applied to these pins is four times higher than the rate of the internal clock, the E clock. The microcontroller is completely static, so clock rates from 0 through 8 MHz can be used. Programming of the EEPROM requires an E clock rate of at least two megahertz. Figure 9.4 illustrates a possible crystal oscillator configuration. If clock rates less than 1 MHz are desired, a resistor may have to be inserted between the network and the XTAL input to the microcontroller. Typical component values for a crystal frequency of 8 MHz are 22pF for C1 and C2 and 10MΩ for R1. Such a system would have an E clock rate of 2 MHz, slow enough to support inexpensive memory devices without wait states and fast enough to respond to a wide variety of process control and monitoring needs.

RESET Pin. This active-low, bidirectional control signal is used as an input to initialize the microcontroller to a known startup state. It may also be used as an open drain circuit to indicate an internal failure. Care should be taken to assure that the RESET line is driven low

TABLE 9.2 Hardware mode-select summary.

Inputs			*Control Bits in HPRIO**			
MODB	*MODA*	*Mode Description*	*RBOOT*	*SMOD*	*MDA*	*IRV*
1	0	Normal Single Chip	0	0	0	0
1	1	Normal Expanded	0	0	1	0
0	0	Special Bootstrap	1	1	0	1
0	1	Special Test	0	1	1	1

*Latched at reset.

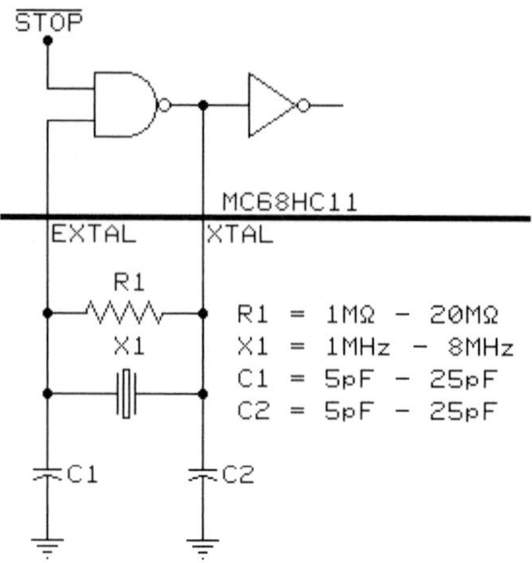

FIGURE 9.4 Crystal oscillator circuit.

when the power supply voltage drops below minimum operating limits. Motorola suggests using an MC34064 or MC34164 low-voltage inhibitor device to generate the RESET signal. Consider the reset circuit illustrated in Figure 9.5.

If manual RESET is desired, the MC34164 may be used as shown in Figure 9.5. Also, manual RESET may be accomplished by eliminating this device and placing the pull-down RC circuit at the RESET input of the microcontroller. However, the circuit shown in Figure 9.5 is more reliable. The MC34064 performs a dual function. First, it acts as a low-voltage inhibitor, placing the microcontroller in the RESET state if the value of V_{DD} drops below acceptable limits. It also acts as a power-up RESET circuit, since the power levels are below acceptable limits when they are not applied. If no manual RESET is desired, the use of the MC34064 may be used to provide the entire RESET circuitry as shown in the figure. Both devices are shipped in small, three-pin TO-92 packages.

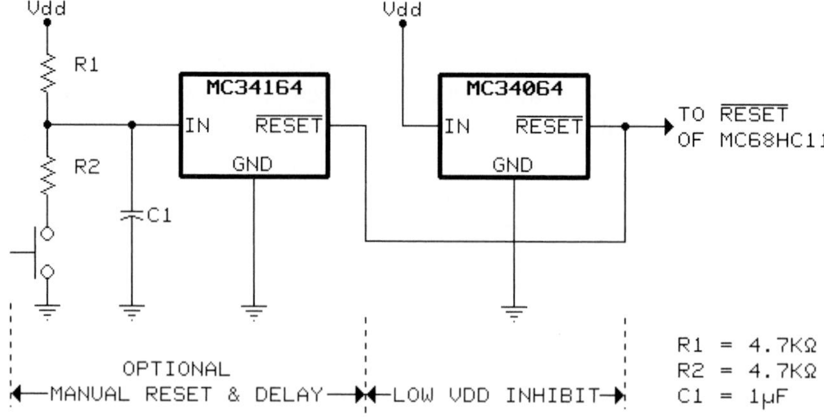

FIGURE 9.5 MC68HC11 RESET circuitry.

Minimum System. We now have enough information to put together a minimum working system. Figure 9.6 illustrates an MC68HC11 configured for single-chip mode, taking advantage of all of its available internal resources. A few additional components will be provided later that enhance the operation of the input/output ports.

Note that Figure 9.6 also illustrates the RC circuit needed to drive the voltage reference input lines V_{RL} and V_{RH}. The use of the ports is also indicated. When configured in this manner, the device will operate in single-chip mode without additional circuitry.

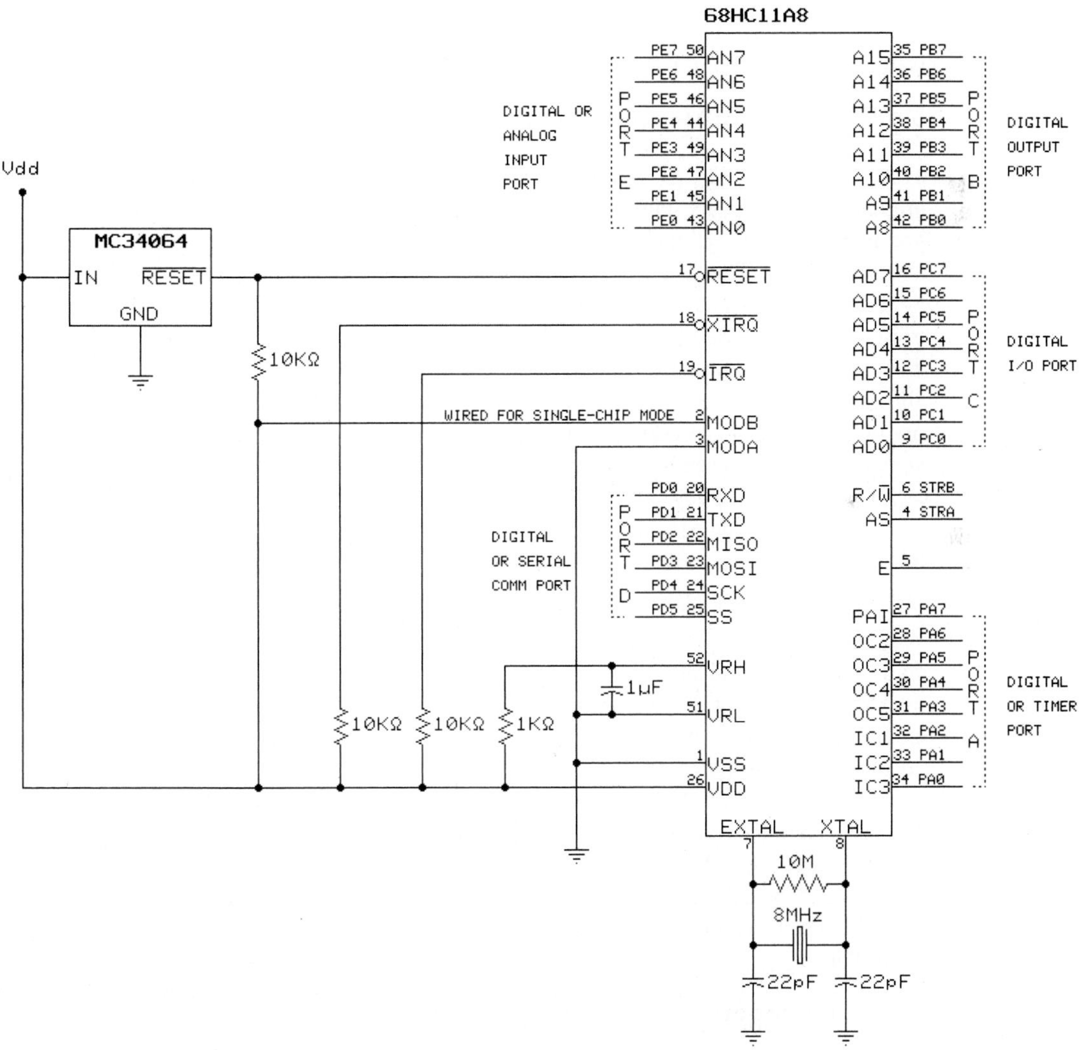

FIGURE 9.6 MC68HC11A8 microcontroller minimum single-chip configuration.

9.2.2 MC68HC11 RAM Segment

The MC68HC11 microcontroller includes a segment of RAM on the chip. Most of the family devices include 256 bytes of RAM allocated at memory addresses \$0000–\$00FF. Some family devices include more—as much as 1024 bytes in the F1 device. The RAM block can be reallocated to the beginning of any 4K memory segment by programming the most significant four bits of the INIT register to the value of the most significant four bits of the desired 16-bit segment address. For example, to reallocate the RAM to addresses \$3000–\$3FFF, the most significant four bits of the INIT register would be set to 0011_2. After RESET, these bits are initialized to 0000_2, replacing the RAM segment at address \$0000. The INIT register is write-protected shortly after RESET, so any change to the RAM segment address must be made quickly following the RESET function.

9.2.3 MC68HC11 ROM Segment

Most of the MC68HC11 family of microcontrollers include a segment of mask-programmed ROM on the chip. Allocated at memory addresses \$E000–\$FFFF, it is programmed by Motorola during manufacturing after being provided with the ROM program by the customer. The primary use of this ROM is the storage of the user's application program instructions. The internal ROM segment is enabled, by default, when the microcontroller is in the single-chip mode. When in the expanded mode, the internal ROM can be disabled by clearing bit 1 in the CONFIG register, located at memory address \$103F. This allows external memory devices such as ROMs and EPROMs to be used within the memory mapped range of \$E000–\$FFFF. This also allows for external definition of the RESET vectors that are allocated at the very top of the memory map. A few of the family devices lack the ROM (see Table 9.1).

One family device, the MC68HC811A8, contains EEPROM in place of the 8K ROM. This device is extremely useful in the prototyping of HC11-based systems.

Interrupt Vector Segment. Addresses \$FFC0–\$FFFF are reserved for specific interrupt vectors. These are memory addresses of interrupt service routines that service interrupts coming from the indicated sources. In order to operate properly, the user must write these interrupt service routines, and they must be loaded in a memory segment (either internal or external). The addresses of those routines must be recorded in the ROM at these interrupt vector memory assignments.

For example, if the interrupt service routine that services the RESET function (the power-up initialization sequence, etc.) is located at address \$E000 in memory, the value \$E000 will be loaded into the ROM at addresses \$FFFE and \$FFFF. Note that the high-order half of the address is loaded in the lower memory byte of the two-byte vector space.

Figure 9.7 is a summary of the interrupt vector addresses in the MC68HC11 microcontroller.

9.2.4 MC68HC11 EEPROM Segment

Most of the MC68HC11 family devices contain 512 bytes of CMOS EEPROM at memory addresses \$B600–\$B7FF. The MC68HC811A8 also contains another 8K bytes of EEPROM

VECTOR ADDRESS	INTERRUPT SOURCE	CCR MASK	LOCAL MASK
\$FFFE:FFFF	$\overline{\text{RESET}}$	NONE	NONE
\$FFFC:FFFD	COP CLOCK MONITOR FAIL (RESET)	NONE	CME
\$FFFA:FFFB	COP FAILURE (RESET)	NONE	NOCOP
\$FFF8:FFF9	ILLEGAL OPCODE TRAP	NONE	NONE
\$FFF6:FFF7	SWI SOFTWARE INTERRUPT	NONE	NONE
\$FFF4:FFF5	$\overline{\text{XIRQ}}$ PIN (PSEUDO NON-MASKABLE INTERRUPT)	X BIT	NONE
\$FFF2:FFF3	$\overline{\text{IRQ}}$ (EXTERNAL PIN OR PARALLEL I/O)	I BIT	
\$FFF0:FFF1	REAL TIME INTERRUPT	I BIT	RTII
\$FFEE:FFEF	TIMER INPUT CAPTURE 1	I BIT	IC1I
\$FFEC:FFED	TIMER INPUT CAPTURE 2	I BIT	IC2I
\$FFEA:FFEB	TIMER INPUT CAPTURE 3	I BIT	IC3I
\$FFE8:FFE9	TIMER OUTPUT CAPTURE 1	I BIT	OC1I
\$FFE6:FFE7	TIMER OUTPUT CAPTURE 2	I BIT	OC2I
\$FFE4:FFE5	TIMER OUTPUT CAPTURE 3	I BIT	OC3I
\$FFE2:FFE3	TIMER OUTPUT CAPTURE 4	I BIT	OC4I
\$FFE0:FFE1	TIMER OUTPUT CAPTURE 5	I BIT	OC5I
\$FFDE:FFDF	TIMER OVERFLOW	I BIT	TOI
\$FFDC:FFDD	PULSE ACCUMULATOR OVERFLOW	I BIT	PAOVI
\$FFDA:FFDB	PULSE ACCUMULATOR INPUT EDGE	I BIT	PAII
\$FFD8:FFD9	SPI SERIAL TRANSFER COMPLETE	I BIT	SPIE
\$FFD6:FFD7	SCI SERIAL SYSTEM	I BIT	
\$FFD4:FFD5 • • \$FFC0:FFC1	RESERVED	—	—

SCI SERIAL SYSTEM:

INTERRUPT CAUSE	LOCAL MASK
RECEIVE DATA REGISTER FULL	RIE
RECEIVER OVERRUN	RIE
IDLE LINE DETECT	ILIE
TRANSMIT DATA REGISTER EMPTY	TIE
TRANSMIT COMPLETE	TCIE

IRQ (EXTERNAL PIN OR PARALLEL I/O):

INTERRUPT CAUSE	LOCAL MASK
EXTERNAL PIN	NONE
PARALLEL I/O HANDSHAKE	STAI

FIGURE 9.7 Interrupt vector ROM segment.

at addresses \$E000–\$FFFF. The EEPROM maintains its programming when the power is removed and can be programmed under software control. Memory bytes can be erased in a byte-by-byte fashion, or the entire segment can be bulk-erased. The device uses an on-chip charge pump to generate the signals needed to program the EEPROM, and in order for the charge pump to operate, the E clock of the microcontroller must be 2 MHz. However, the CSEL bit in the option register can be used to select an on-chip oscillator for this purpose. This internal oscillator runs at about 2.5 MHz. The programming procedure will be reviewed shortly.

9.2.5 MC68HC11 Port Control and Data Register Segment

Located from addresses \$1000 through \$103F are 51 eight- and 16-bit registers that are used to configure and use the I/O ports on the microcontroller as well as to establish operating modes and monitor various aspects of the system status. Figure 9.8 illustrates a detailed memory map of this segment. We will not attempt to describe all of the facilities within these registers, as the use of many of the microcontroller functions is beyond the scope of this text. We will look at some of the simpler port applications and the use of the port control and data registers that are used on those applications.

Note that the CONFIG register at address \$103F is nonvolatile and is reprogrammed through software control. The software program sequence needed to reprogram this register will be illustrated shortly.

The least significant four bits of the INIT register can be used to reallocate the 64-byte register block to the beginning of any 4K-byte memory segment. After RESET, these bits are 0001_2, placing the block at address \$1000. The INIT register is write-protected shortly after RESET, so any change to the register block address must be made quickly following the RESET function.

BIT07	—	—	—	—	—	—	BIT00	$101F	TOC5	OUTPUT COMPARE 5 REGISTER
BIT15	—	—	—	—	—	—	BIT08	$101E		
BIT07	—	—	—	—	—	—	BIT00	$101D	TOC4	OUTPUT COMPARE 4 REGISTER
BIT15	—	—	—	—	—	—	BIT08	$101C		
BIT07	—	—	—	—	—	—	BIT00	$101B	TOC3	OUTPUT COMPARE 3 REGISTER
BIT15	—	—	—	—	—	—	BIT08	$101A		
BIT07	—	—	—	—	—	—	BIT00	$1019	TOC2	OUTPUT COMPARE 2 REGISTER
BIT15	—	—	—	—	—	—	BIT08	$1018		
BIT07	—	—	—	—	—	—	BIT00	$1017	TOC1	OUTPUT COMPARE 1 REGISTER
BIT15	—	—	—	—	—	—	BIT08	$1016		
BIT07	—	—	—	—	—	—	BIT00	$1015	TIC3	INPUT CAPTURE 3 REGISTER
BIT15	—	—	—	—	—	—	BIT08	$1014		
BIT07	—	—	—	—	—	—	BIT00	$1013	TIC2	INPUT CAPTURE 2 REGISTER
BIT15	—	—	—	—	—	—	BIT08	$1012		
BIT07	—	—	—	—	—	—	BIT00	$1011	TIC1	INPUT CAPTURE 1 REGISTER
BIT15	—	—	—	—	—	—	BIT08	$1010		
BIT07	—	—	—	—	—	—	BIT00	$100F	TCNT	TIMER COUNT REGISTER
BIT15	—	—	—	—	—	—	BIT08	$100E		
OC1D7	OC1D6	OC1D5	IC1D4	OC1D3				$100D	OC1D	OC1 ACTION DATA REGISTER
OC1M7	OC1M6	OC1M5	OC1M4	OC1M3				$100C	OC1M	OC1 ACTION MASK REGISTER
FOC1	FOC2	FOC3	FOC4	FOC5				$100B	CFORC	COMPARE FORCE REGISTER
BIT07	—	—	—	—	—	—	BIT00	$100A	PORTE	INPUT PORT E
		BIT05	—	—	—	—	BIT00	$1009	DDRD	DATA DIRECTION FOR PORT D
		BIT05	—	—	—	—	BIT00	$1008	PORTD	I/O PORT D
BIT07	—	—	—	—	—	—	BIT00	$1007	DDRC	DATA DIRECTION FOR PORT C
								$1006		RESERVED
BIT07	—	—	—	—	—	—	BIT00	$1005	PORTCL	ALTERNATE LATCHED PORT C
BIT07	—	—	—	—	—	—	BIT00	$1004	PORTB	OUTPUT PORT B
BIT07	—	—	—	—	—	—	BIT00	$1003	PORTC	I/O PORT C
STAF	STAI	CWOM	HNDS	OIN	PLS	EGA	INVB	$1002	PIOC	PARALLEL I/O CONTROL REGISTER
—	—	—	—	—	—	—	—	$1001		RESERVED
BIT07	—	—	—	—	—	—	BIT00	$1000	PORTA	I/O PORT A

FIGURE 9.8 MC68HC11 Register and control segment, (a): $1000–$101F.

9.2.6 MC68HC11 Memory Map

Figure 9.9 illustrates the memory map of the MC68HC11A8 microcontroller. Note that by using the INIT and CONFIG registers the RAM segment, the register segment, and the EEPROM segment can be reallocated to other addresses. Those unused memory spaces indicated in the memory map are available to the user when the chip is operating in expanded mode. In this mode, the Port B and Port C input/output functions are lost. In their place, the address and data buses are multiplexed. The port functions can be reestab-

				NOSEC	NOCOP	ROMON	EEON	$103F	CONFIG	COP, ROM, AND EEPROM ENABLES
TILOP		OOCR	CBYP	DISR	FCM	FCOP	TCON	$103E	TEST1	FACTORY TEST CONTROL REGISTER
RAM3	RAM2	RAM1	RAM0	RAM3	RAM2	RAM1	RAM0	$103D	INIT	RAM AND I/O MAPPING REGISTER
RBOOT	SMOD	MDA	IRV	PSEL3	PSEL2	PSEL1	PSEL0	$103C	HPRIO	HIGHEST PRIORITY I-BIT INT AND MISC
ODD	EVEN		BYTE	ROW	ERASE	EELAT	EEPGM	$103B	PPROG	EEPROM PROGRAMMING CONTROL REGISTER
BIT07	—	—	—	—	—	—	BIT00	$103A	COPRST	ARM/RESET COP TIMER CIRCUITRY
ADPU	CSEL	IRQE	DLY	CME		CR1	CR0	$1039	OPTION	SYSTEM CONFIGURATION OPTIONS
								$1038		RESERVED
								$1037		RESERVED
								$1036		RESERVED
								$1035		RESERVED
BIT07	—	—	—	—	—	—	BIT00	$1034	ADR4	A/D RESULT REGISTER 4
BIT07	—	—	—	—	—	—	BIT00	$1033	ADR3	A/D RESULT REGISTER 3
BIT07	—	—	—	—	—	—	BIT00	$1032	ADR2	A/D RESULT REGISTER 2
BIT07	—	—	—	—	—	—	BIT00	$1031	ADR1	A/D RESULT REGISTER 1
CCF		SCAN	MULT	CD	CC	CB	CA	$1030	ADCTL	A/D CONTROL REGISTER
BIT07	—	—	—	—	—	—	BIT00	$102F	SCDR	SCI DATA (READ RDR, WRITE TDR)
TDRE	TC	RDRF	IDLE	OR	NF	FE		$102E	SCSR	SCI STATUS REGISTER
TIE	TCIE	RIE	ILIE	TE	RE	RWU	SBK	$102D	SCCR2	SCI CONTROL REGISTER 2
R8	T8		M	WAKE				$102C	SCCR1	SCI CONTROL REGISTER 1
TCLR		SCP1	SCP0	RCKB	SCR2	SCR1	SCR0	$102B	BAUD	SCI BAUD RATE CONTROL
BIT07	—	—	—	—	—	—	BIT00	$102A	SPDR	SPI DATA REGISTER
SPIF	WCOL		MODF					$1029	SPSR	SPI STATUS REGISTER
SPIE	SPE	DWOM	MSTR	CPOL	CPHA	SPR1	SPR0	$1028	SPCR	SPI CONTROL REGISTER
BIT07	—	—	—	—	—	—	BIT00	$1027	PACNT	PULSE ACCUMULATOR COUNT REGISTER
DDRA7	PAEN	PAMOD	PEDGE			RTR1	RTR0	$1026	PACTL	PULSE ACCUMULATOR CONTROL REGISTER
TOF	RTIF	PAOVF	PAIF					$1025	TFLG2	TIMER INTERRUPT FLAG REGISTER 2
TOI	RTII	PAOVI	PAII			PR1	PR0	$1024	TMSK2	TIMER INTERRUPT MASK REGISTER 2
OC1F	OC2F	OC3F	OC4F	OC5F	IC1F	IC2F	IC3F	$1023	TFLG1	TIMER INTERRUPT FLAG REGISTER 1
OC1I	OC2I	OC3I	OC4I	OC5I	IC1I	IC2I	IC3I	$1022	TMSK1	TIMER INTERRUPT MASK REGISTER 1
		EDG1B	EDG1A	EDG2B	EDG2A	EDG3B	EDG3A	$1021	TCTL2	TIMER CONTROL REGISTER 2
OM2	OL2	OM3	OL3	OM4	OL4	OM5	OL5	$1020	TCTL1	TIMER CONTROL REGISTER 1

FIGURE 9.8 Continued, (b) $1020–$103F.

lished if an MC68HC24 chip is allocated onto the external address and data buses. This device is a port emulator.

9.2.7 Expanded Mode of Operation

Figure 9.10 illustrates the application of the MC68HC11 in the expanded mode of operation. Note how the Port B and Port C functions have been displaced by the address and data bus. Note also, the use of the 74HC373 to demultiplex the buses. The address strobe, AS,

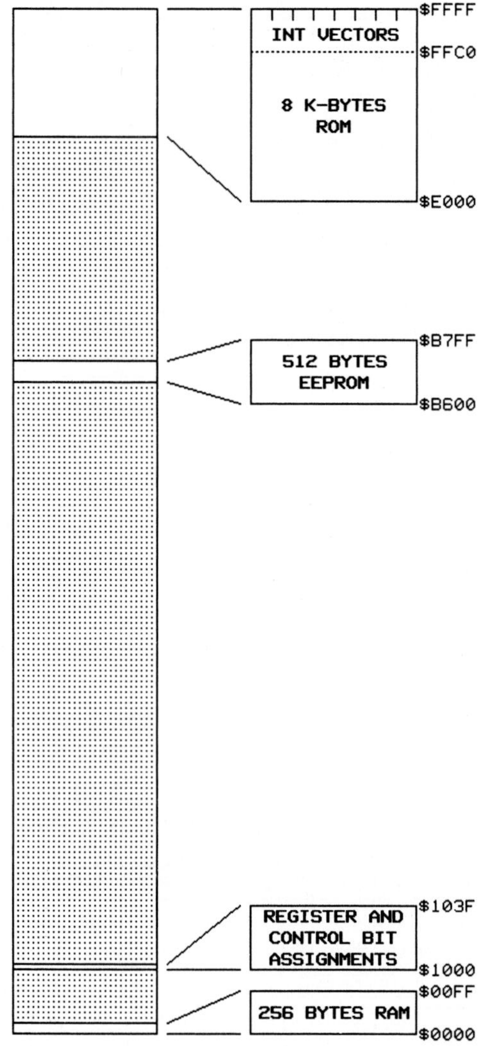

FIGURE 9.9 MC68HC11A8 memory map.

provides the same function as the valid memory address (VMA) signal discussed previously in the text. When active, it denotes that a valid address is on the address bus. In this case, when FALSE, it denotes that Port C is providing the data bus interface.

9.3 MC68HC11 INSTRUCTION FORMATS

Since this microcontroller manages eight- and 16-bit data and 16-bit addresses, operands can be either eight bits or 16 bits in length. The opcode is eight bits in length except when index register Y is used. This register was added when the design of the MC68HC11 was

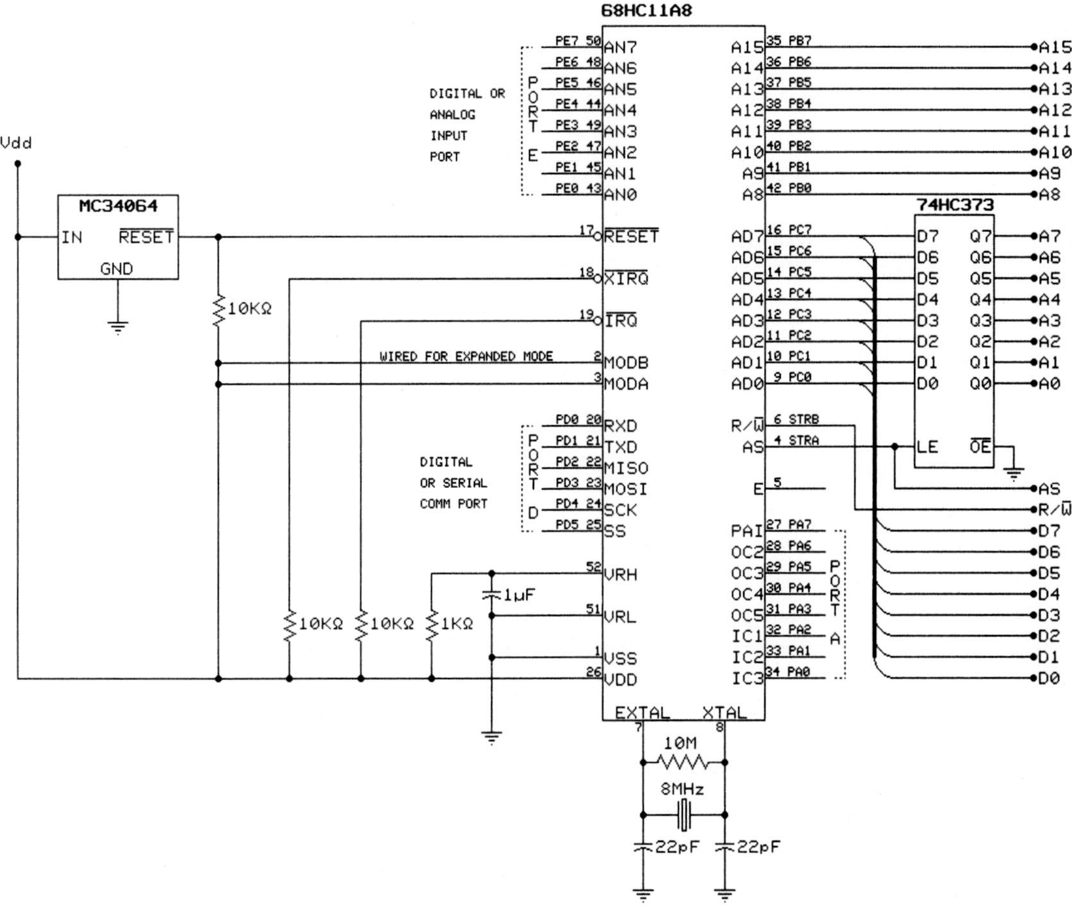

FIGURE 9.10 MC68HC11A8 microcontroller, minimum expanded mode configuration.

upgraded from the M6800. The opcodes for managing the Y register are two bytes in length. The first byte is referred to as a "pre-byte." Consequently, instructions using the X register are one byte shorter than those using the Y register. Hence, they require one less memory fetch operation and operate faster. This might be taken into consideration when choosing the use of the X register over the Y register. Figure 9.11 illustrates the different instruction formats for this microcontroller.

9.4 MC68HC11 ADDRESSING MODES

The addressing modes include those discussed in Chapter 8. However, since this processor manages a 16-bit address bus, the direct addressing mode has two forms, referred to as direct and extended. Table 9.3 illustrates a summary of those addressing modes.

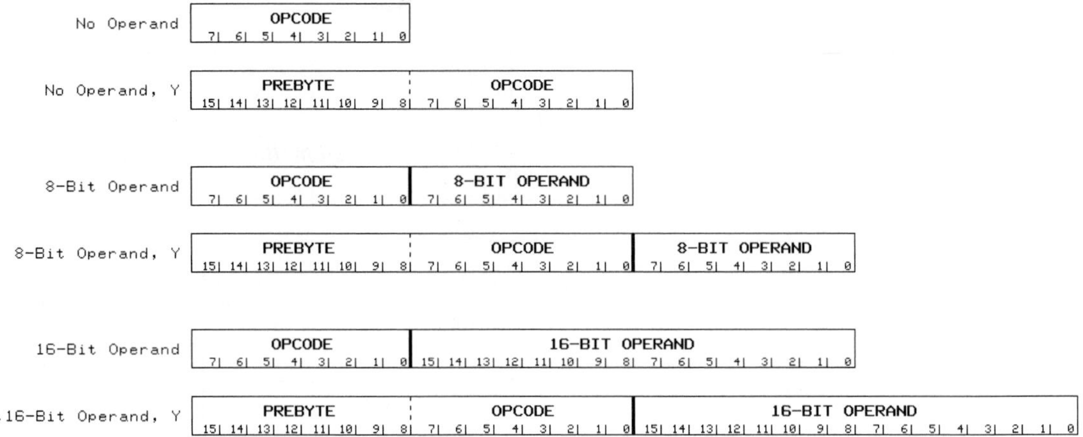

FIGURE 9.11 MC68HC11 instruction formats.

TABLE 9.3 MC68HC11 family addressing modes.

Addressing Mode	*Description*	*Assembly Language*	*Effective Address*
Inherent (or implied)	The opcode is self-explanatory. No operand is required.	No operand	None
Immediate	The operand is an eight- or 16-bit data constant.	#$NN #$NNNN	PC
Direct	The operand is an eight-bit address.	$NN	Operand
Extended	The operand is a 16-bit address.	$NNNN	Operand
Indexed	The address is calculated by adding the index register contents to the eight-bit operand.	$NN,X $NN,Y	Operand + X Operand + Y
Relative	The address is calculated by adding the program counter contents to the eight-bit operand.	*+$NN	Operand + PC

9.5 MC68HC11 INSTRUCTION SET

Table 9.4 illustrates the instruction set for the MC68HC11 microcontroller. Note that each row of the table refers to a single instruction. Included is the assembly language format for the instruction, its description, a list of the valid addressing modes, and the effect of the instruction on the condition code register. The following is a key for the column abbreviations:

Addressing Modes

T	Inherent	E	Extended	R	Relative
M	Immediate	X	Indexed, X		
D	Direct	Y	Indexed, Y		

(Asterisk indicates a valid addressing mode for the instruction. Blank indicates not effected by instruction.)

Condition Codes

Blank	Not effected by the instruction
*	Set or reset based upon the result of the instruction operation.
1	Set to a logical 1 by the instruction.
0	Reset to a logical 0 by the instruction.

TABLE 9.4 MC68HC11 Instruction set.

Instruction	*Operation*	*Addressing Modes* T	M	D	E	X	Y	R	*Condition Codes* H	I	N	Z	V	C
ABA	Add Accumulators	*							*		*	*	*	*
ABX	Add B to X	*												
ABY	Add B to Y	*												
ADCA <ea>	Add with Carry to A		*	*	*	*	*		*		*	*	*	*
ADCB <ea>	Add with Carry to B		*	*	*	*	*		*		*	*	*	*
ADDA <ea>	Add Memory to A		*	*	*	*	*		*		*	*	*	*
ADDB <ea>	Add Memory to B		*	*	*	*	*		*		*	*	*	*
ADDD <ea>	Add 16-Bit Memory to D		*	*	*	*	*				*	*	*	*
ANDA <ea>	AND Memory to A		*	*	*	*	*				*	*	0	
ANDB	AND Memory to B		*	*	*	*	*				*	*	0	
ASL <ea>	Arithmetic Shift Left Memory				*	*	*				*	*	*	*
ASLA	Arithmetic Shift Left A	*									*	*	*	*
ASLB	Arithmetic Shift Left B	*									*	*	*	*
ASLD	Arithmetic Shift Left D	*									*	*	*	*
ASR	Arithmetic Shift Right Memory				*	*	*				*	*	*	*
ASRA	Arithmetic Shift Right A	*									*	*	*	*
ASRB	Arithmetic Shift Right B	*									*	*	*	*
BCC <disp>	Branch if Carry Clear							*						
BCLR <ea>,m	Clear mask bits in memory			*		*	*				*	*	0	
BCS <disp>	Branch if Carry Set							*						
BEQ <disp>	Branch if Equal to Zero							*						
BGE <disp>	Branch if > or = to Zero							*						
BGT <disp>	Branch if > Zero							*						
BHI <disp>	Branch if Higher							*						
BHS <disp>	Branch if Higher or Same							*						
BITA <ea>	Bit Test A AND Memory		*	*	*	*	*				*	*	0	
BITB <ea>	Bit Test B AND Memory		*	*	*	*	*				*	*	0	
BLE <disp>	Branch if < or = to Zero							*						
BLO <disp>	Branch if Lower (Unsigned <)							*						
BLS <disp>	Branch if Lower or Same							*						

TABLE 9.4 MC68HC11 Instruction set (continued).

		Addressing Modes							*Condition Codes*					
Instruction	*Operation*	*T*	*M*	*D*	*E*	*X*	*Y*	*R*	*H*	*I*	*N*	*Z*	*V*	*C*
BLT <disp>	Branch if Less Than Zero							*						
BMI <disp>	Branch if Minus							*						
BNE <disp>	Branch if Not Equal to Zero							*						
BPL <disp>	Branch if Plus							*						
BRA <disp>	Branch Always							*						
BRCLR <ea>,m	Branch if m Bit(s) Clear			*		*	*							
BRN <disp>	Branch Never							*						
BRSET <ea>,m	Branch if m Bit(s) Set							*						
BSET <ea>,m	Set Bits <ea> AND m			*		*	*				*	*	0	
BSR <disp>	Branch to Subroutine							*						
BVC	Branch if Overflow Clear							*						
BVS	Branch if Overflow Set							*						
CBA	Compare A to B (A-B)	*									*	*	*	*
CLC	Clear Carry Bit	*												0
CLI	Clear Interrupt Mask	*								0				
CLR <ea>	Clear Memory Byte				*	*	*				0	1	0	0
CLRA	Clear Accumulator A	*									0	1	0	0
CLRB	Clear Accumulator B	*									0	1	0	0
CLV	Clear Overflow Flag	*											0	
CMPA <ea>	Compare A to Memory (A-M)		*	*	*	*	*				*	*	*	*
CMPB <ea>	Compare B to Memory (B-M)		*	*	*	*	*				*	*	*	*
COM <ea>	1's Complement Memory				*	*	*				*	*	0	*
COMA	1's Complement A	*									*	*	0	*
COMB	1's Complement B	*									*	*	0	*
CPD <ea>	Compare D to Memory (D-M)		*	*	*	*	*				*	*	*	*
CPX <ea>	Compare X to Memory (X-M)		*	*	*	*	*				*	*	*	*
CPY <ea>	Compare Y to Memory (Y-M)		*	*	*	*	*				*	*	*	*
DAA	Decimal Adjust A	*									*	*	*	*
DEC <ea>	Decrement Memory Byte				*	*	*				*	*	*	
DECA	Decrement Accumulator A	*									*	*	*	
DECB	Decrement Accumulator B	*									*	*	*	
DES	Decrement Stack Pointer	*												
DEX	Decrement Index Register X	*										*		
DEY	Decrement Index Register Y	*										*		
EORA <ea>	Exclusive OR Memory to A		*	*	*	*	*				*	*	0	
EORB <ea>	Exclusive OR Memory to B		*	*	*	*	*				*	*	0	
FDIV	Divide X=D/X, D=rem (D<X)	*										*	*	*
IDIV	Divide X=D/X, D=remainder	*										*	0	*
INC <ea>	Increment Memory Byte				*	*	*				*	*	*	
INCA	Increment Accumulator A	*									*	*	*	
INCB	Increment Accumulator B	*									*	*	*	
INS	Increment Stack Pointer	*												
INX	Increment Index Register X	*										*		
INY	Increment Index Register Y	*										*		

TABLE 9.4 MC68HC11 Instruction set (continued).

		Addressing Modes							Condition Codes					
Instruction	*Operation*	*T*	*M*	*D*	*E*	*X*	*Y*	*R*	*H*	*I*	*N*	*Z*	*V*	*C*
JMP <ea>	Jump				*	*	*							
JSR <ea>	Jump to Subroutine			*	*	*	*							
LDAA <ea>	Load Accumulator A		*	*	*	*	*				*	*	0	
LDAB <ea>	Load Accumulator B		*	*	*	*	*				*	*	0	
LDD <ea>	Load Double Accumulator D		*	*	*	*	*				*	*	0	
LDS <ea>	Load Stack Pointer		*	*	*	*	*				*	*	0	
LDX <ea>	Load Index Register X		*	*	*	*	*				*	*	0	
LDY <ea>	Load Index Register Y		*	*	*	*	*				*	*	0	
LSL <ea>	Logical Shift Left Memory				*	*	*				*	*	*	*
LSLA	Logical Shift Left A	*									*	*	*	*
LSLB	Logical Shift Left B	*									*	*	*	*
LSLD	Logical Shift Left D (A:B)	*									*	*	*	*
LSR <ea>	Logical Shift Right Memory				*	*	*				0	*	*	*
LSRA	Logical Shift Right A	*									0	*	*	*
LSRB	Logical Shift Right B	*									0	*	*	*
LSRD	Logical Shift Right D (A:B)	*									0	*	*	*
MUL	Multiply D = A x B	*												*
NEG <ea>	2's Complement Memory				*	*	*				*	*	*	*
NEGA	2's Complement A	*									*	*	*	*
NEGB	2's Complement B	*									*	*	*	*
NOP	No Operation (2 clock cycles)	*												
ORAA <ea>	OR Memory to Acc A		*	*	*	*	*				*	*	0	
ORAB <ea>	OR Memory to Acc B		*	*	*	*	*				*	*	0	
PSHA	Push A onto the Stack	*												
PSHB	Push B onto the Stack	*												
PSHX	Push X onto Stack, Lo First	*												
PSHY	Push Y onto Stack, Lo First	*												
PULA	Pull A from the Stack	*												
PULB	Pull B from the Stack	*												
PULX	Pull X from Stack, Hi First	*												
PULY	Pull Y from Stack, Hi First	*												
ROL <ea>	Rotate Left (with Carry)				*	*	*				*	*	*	*
ROLA	Rotate Left Accumulator A	*									*	*	*	*
ROLB	Rotate Left Accumulator B	*									*	*	*	*
ROR <ea>	Rotate Right (with Carry)				*	*	*				*	*	*	*
RORA	Rotate Right Accumulator A	*									*	*	*	*
RORB	Rotate Right Accumulator B	*									*	*	*	*
RTI	Return from Interrupt	*									*	*	*	*
RTS	Return from Subroutine	*							*	*	*	*	*	*
SBA	Subtract B from A	*												
SBCA <ea>	A = A - <ea> - C-bit		*	*	*	*	*				*	*	*	*
SBCB <ea>	B = B - <ea> - C-bit		*	*	*	*	*				*	*	*	*
SEC	Set Carry Flag	*												1
SEI	Set Interrupt Mask	*								1				

TABLE 9.4 MC68HC11 Instruction set (continued).

		Addressing Modes							*Condition Codes*					
Instruction	*Operation*	*T*	*M*	*D*	*E*	*X*	*Y*	*R*	*H*	*I*	*N*	*Z*	*V*	*C*
SEV	Set Overflow Flag	*											1	
STAA <ea>	Store Accumulator A to Mem			*	*	*	*				*	*	0	
STAB <ea>	Store Accumulator B to Mem			*	*	*	*				*	*	0	
STD <ea>	Store Accumulator D to Mem			*	*	*	*				*	*	0	
STOP	Stop Internal Clocks	*												
STS <ea>	Store Stack Pointer to Mem			*	*	*	*				*	*	0	
STX <ea>	Store Index Register X			*	*	*	*				*	*	0	
STY <ea>	Store Index Register Y			*	*	*	*				*	*	0	
SUBA <ea>	Subtract Memory from A		*	*	*	*	*				*	*	*	*
SUBB <ea>	Subtract Memory from B		*	*	*	*	*				*	*	*	*
SUBD <ea>	Subtract Memory from D		*	*	*	*	*				*	*	*	*
SWI	Software Interrupt (Call ISR)	*								1				
TAB	Transfer A to B	*									*	*	0	
TAP	Transfer A to CCR	*							*	*	*	*	*	*
TBA	Transfer B to A	*									*	*	0	
TPA	Transfer CCR to A	*												
TST <ea>	Test for Zero or Minus				*	*	*				*	*	0	0
TSTA	Test A for Zero or Minus	*									*	*	0	0
TSTB	Test B for Zero or Minus	*									*	*	0	0
TSX	Transfer Stack Pointer to X	*												
TSY	Transfer Stack Pointer to Y	*												
TXS	Transfer X to Stack Pointer	*												
TYS	Transfer Y to Stack Pointer	*												
WAI	Wait for an Interrupt	*												
XGDX	Exchange D with X	*												
XGDY	Exchange D with Y	*												

9.6 PORT A: DIGITAL I/O, PULSE ACCUMULATOR, AND TIMER FUNCTIONS

Shared functions of Port A include general-purpose digital I/O, the main timer system, and the pulse accumulator system.

9.6.1 Port A as Digital I/O

Port A has three fixed-direction input pins, four fixed-direction output pins, and one bidirectional pin. Digital data can be input on pins PA0–PA2 by reading from the PORTA register at address $1000. Digital data can be output on pins PA3–PA6 by writing to address $1000. The direction of the PA7 pin is determined by bit 7 (DDRA) in the pulse accumulator control (PACTL) register, located at address $1026.

Practical Example 9.1: Port A as Digital I/O

Consider the following program segments, which exercise the I/O capabilities of Port A.

```
0000              ; PORT A (DIGITAL I/O PORT)
0000              ;
0000                      .EQU  PORTA,$00     ;Port A Data Register
0026                      .EQU  PACTL,$26     ;Pulse Acc. Control Register
1000                      .EQU  REGLIST,$1000
E100                      .ORG  $E100         ;Locate following code
E100              PREX01  LDX   #REGLIST      ;X points to internal registers
E103              ;
E103              ; 1. LOAD B[7,2,1,0] FROM PORT A
E103              ;
E103 1D2680               BCLR  PACTL,X,$80   ;Clear DDRA for input
E106 E600                 LDAB  PORTA,X       ;Input data on PA7,2,1,0.
E108              ;
E108              ; 2. WRITE B[6,5,4,3] OUT PORT A (PA7 = INPUT)
E108              ;
E108 1D2680               BCLR  PACTL,X,$80   ;Clear DDRA for input
E10B A600                 LDAA  PORTA,X       ;Get current Port A data
E10D 9487                 ANDA  b'10000111    ;Keep only Bits A7,2,1,0
E10F D478                 ANDB  b'01111000    ;Keep only Bits B6,5,4,3
E111 DA00                 ABA                 ;Merge A and B data
E113 A700                 STAA  PORTA,X       ;Output data on PA6,5,4,3.
E115              ;
E115              ; 3. WRITE B[7,6,5,4,3] OUT PORT A (PA7 = OUTPUT)
E115              ;
E115 1C2680               BSET  PACTL,X,$80   ;Set DDRA for output
E118 E700                 STAB  PORTA,X       ;Output data on PA7,6,5,4,3.
E11A                      .END
```

Note in these examples how bit 7 of the A port is used in a bidirectional fashion. The value of bit 7 in PACTL is cleared to initialize A7 as input or set to initialize A7 as output. Once this bit is programmed, it does not need to be changed as long as the data direction on A7 does not change. In the first example, PACTL bit 7 is cleared without affecting the other seven bits. Once this is done, the data can be read from the port. Since all eight bits are read, it will be necessary to isolate which input bits are desired.

In the second example, four bits in accumulator B are written out Port A by first initializing PACTL bit 7. Then, in order to protect the current state of PA7, it is read into accumulator A, merged with the output data, and written out to the port.

In the third example, the data in accumulator B can simply be written to the port since PA7 is an output.

9.6.2 Port A as a Pulse Accumulator

The pulse accumulator is an eight-bit counter/timer system that can be configured to operate in either of two basic modes: event counting and gated time accumulation. In the event

PAMOD=0: EVENT COUNTING MODE

PAMOD=1: GATED TIME ACCUMULATOR MODE

FIGURE 9.12 Pulse accumulator operating modes.

counting mode, the eight-bit counter is clocked to increasing values at each active edge of the PAI pin (PA7). In the gated time accumulation mode, the eight-bit counter is clocked by E ÷ 64 when the PAI pin (PA7) is active. Figure 9.12 illustrates these two possible modes.

There are four registers in the register segment that control the use of the pulse accumulator. These are TMSK2, TFLG2, PACTL, and PACNT, illustrated in Figure 9.8(b) and specifically shown in Figure 9.13.

DDRA7: 0 = Port A bit 7 is configured for input only (output buffer is disabled).
1 = Port A bit 7 is configured for output.

PAEN: 0 = Pulse accumulator disabled.
1 = Pulse accumulator enabled (counter stops counting, interrupts inhibited).

PAMOD: 0 = External event counting mode (pin acts as a clock).
1 = Gated time accumulation mode (pin acts as clock enable for E/64 clock).

PEDGE: 0 = Pulse accumulator responds to falling edges.
1 = Pulse accumulator responds to rising edges.

PAOVI: 0 = Inhibit pulse accumulator overflow interrupt.
1 = Enable pulse accumulator overflow interrupt. Vector = $FFDC:FFDD

PAOVF: Status bit, set when pulse accumulator count rolls over from $FF to $00. This allows the use of polled rather than vectored interrupts. The bit is reset to a 0 when software resets this bit by writing to TFLG2. A BCLR instruction with a mask of 0010000_2 would work.

PAII: 0 = Inhibit pulse accumulator input interrupt.
1 = Enable pulse accumulator input interrupt. Vector = $FFDA:FFDB

PAIF: Status bit, set when pulse accumulator input edge is encountered. This allows the use of polled rather than vectored interrupts. The bit is reset to a 0 when software resets this bit by writing to TFLG2. A BCLR instruction with a mask of 0001000_2 would work.

BIT07	—	—	—	—	—	—	BIT00	$1027	PACNT
DDRA7	PAEN	PAMOD	PEDGE			RTR1	RTR0	$1026	PACTL
TOF	RTIF	PAOVF	PAIF					$1025	TFLG2
TOI	RTII	PAOVI	PAII			PR1	PR0	$1024	TMSK2

FIGURE 9.13 Pulse accumulator control and status registers

Practical Example 9.2: Port A as a Pulse Accumulator

Consider the following segment of code, which exercises the pulse accumulator event counter capabilities of Port A.

```
0000                 ; PORT A (PULSE ACCUMULATOR)
0000                 ;
0000                        .EQU   PORTA,$00         ;Port A Data Register
0024                        .EQU   TMSK2,$24         ;Timer Int Mask Register 2
0025                        .EQU   TFLG2,$25         ;Timer Int Flag Register 2
0026                        .EQU   PACTL,$26         ;Port Acc. Control Register
0027                        .EQU   PACNT,$27         ;Pulse Acc. Count Register
1000                        .EQU   REGLIST,$1000     ;List of internal registers.
E120                        .ORG   $E120             ;Locate following code
E120                 ;
E120                 ; CAUSE AN INTERRUPT AFTER 100 EVENTS.
E120                 ;
E120 CE1000          PREX02 LDX    #REGLIST          ;Point to list of registers.
E123 1D2680                 BCLR   PACTL,X,$80       ;Clear DDRA for input
E126 8664                   LDAA   #100              ;100 events on PA7
E128 40                     NEGA                     ;2's comp..for count down.
E129 A727                   STAA   PACNT,X
E12B 1D2630                 BCLR   PACTL,X,$30       ;PEDGE=0; Active-low edge
E12E                                                 ;PAMOD=0; Ext Event Counting
E12E 1C2420                 BSET   TMSK2,X,$20       ;PAOVI=1; Enable overflow int
E131 1C2640                 BSET   PACTL,X,$40       ;PAEN=1; Enable pulse acc.
E134 39                     RTS
E135                 ;
E135                 ; WAIT FOR 100 EVENTS
E135                 ;
E135 1D2680          WT100  BCLR   PACTL,X,$80       ;Clear DDRA for input
E138 8664                   LDAA   #100              ;100 events on PA7
E13A 40                     NEGA                     ;2's comp. for count down.
E13B A727                   STAA   PACNT,X
E13D 1D2630                 BCLR   PACTL,X,$30       ;PEDGE=0; Active-low edge
E140                                                 ;PAMOD=0; Ext Event Counting
E140 1C2400                 BCLR   TMSK2,X,$20       ;PAOVI=0; Disable overflw int
E143 1C2640                 BSET   PACTL,X,$40       ;PAEN=1; Enable pulse acc.
E146 1F2520FC WT100A BRCLR  TFLG2,X,$20,WT100A ;Repeat if PAOVF=0
E14A 1D2520                 BCLR   TFLG2,X,$20       ;Clear PAOVF
E14D 39                     RTS
E14E                 ;
E14E                 ; COUNT EVENTS FOR 1 SECOND
E14E                 ;
E14E 1D2680          CT1SET BCLR   PACTL,X,$80       ;Clear DDRA for input
E151 6F27                   CLR    PACNT,X           ;Clear event counter
E153 1D2630                 BCLR   PACTL,X,$30       ;PEDGE=0; Active-low edge
E156                                                 ;PAMOD=0; Ext event counting
E156 1C2400                 BCLR   TMSK2,X,$20       ;PAOVI=0; Disable overflw int
E159 1C2640                 BSET   PACTL,X,$40       ;PAEN =1; Enable pulse acc.
E15C BD____                 JSR    SECOND            ;Wait for one second
E15F 1D2640                 BCLR   PACTL,X,$40       ;PAEN =0; Disable pulse acc.
E162 39                     RTS
```

Increasing the Count Limit. Note that the value for PACNT is limited to eight bits, or 256 counts. The pulse accumulator overflow interrupt or pulse accumulator overflow flag can be used to stimulate a routine that increments another counter to scale the figure to a larger number. For example, if an eight-bit register is incremented every time an overflow occurs, 2^{16} ($65{,}536_{10}$) events can be counted. If a 16-bit register such as the D register is incremented, 2^{24} ($16{,}777{,}216_{10}$) events can be counted.

These examples are based on a number of input events. If the events occur at a stable frequency and the period between the events is known, these same routines can be based on lengths of time rather than on the number of input events. The same principles also apply to the gated time accumulation mode.

9.6.3 Port A as Main Timer and Real-Time Interrupt

This timer system is based on a free-running 16-bit counter with a four-stage programmable prescaler. Timer overflow functions are available (as with the pulse accumulator) to extend the length or number of events. Three independent input capture functions can be used to automatically record the time when a selected transition is detected at a respective timer input pin. Five output-compare functions are included for generating output signals or for timing software delays.

A programmable periodic interrupt circuit, or real-time interrupt (RTI), is tapped off of the main 16-bit timer counter. Software can select one of four rates for the RTI, which is often used to pace the execution of software routines. (For example, execute a subroutine that monitors temperature once each ten seconds.)

The timer subsystem involves more registers and control bits than any other subsystem in the microcontroller. It is not within the scope of this text to provide an exhaustive study of each of these functions, but some of the common timer functions will be illustrated. Figure 9.14 illustrates a block diagram of the main timer system.

Input Capture. Input capture functions, used to record the time at which some external event occurred, are accomplished by latching the contents of the free-running counter when a selected edge is detected at the related timer input pin. The time at which the event occurred is saved in the input capture register (16-bit latch TIC1, TIC2, or TIC3).

Output Capture. For each of the five output compare functions, there is a separate 16-bit compare register (TOC1, TOC2, TOC3, TOC4, TOC5) and a dedicated 16-bit comparitor. The value in the compare register is compared to the value of the free-running counter on every bus cycle. When the compare register matches the counter value, an output is generated that sets an output-compare status flag and initiates the automatic actions for that output compare function. Equality can be recognized by polling the status bits or by vectored interrupts.

Free-Running Counter. The central element of the timer subsystem is the 16-bit free-running counter. It is reset to 0 when the microcontroller is reset and counts up continually. When it rolls over from $FFFF to $0000, an overflow flag is set, and the counter continues to count up. There is no facility for modifying the contents of this register. The count rate depends on bits PR1 and PR0 in TMSK2 at address $1024. Table 9.5 illustrates the count rate as a function of the E clock and these two bits.

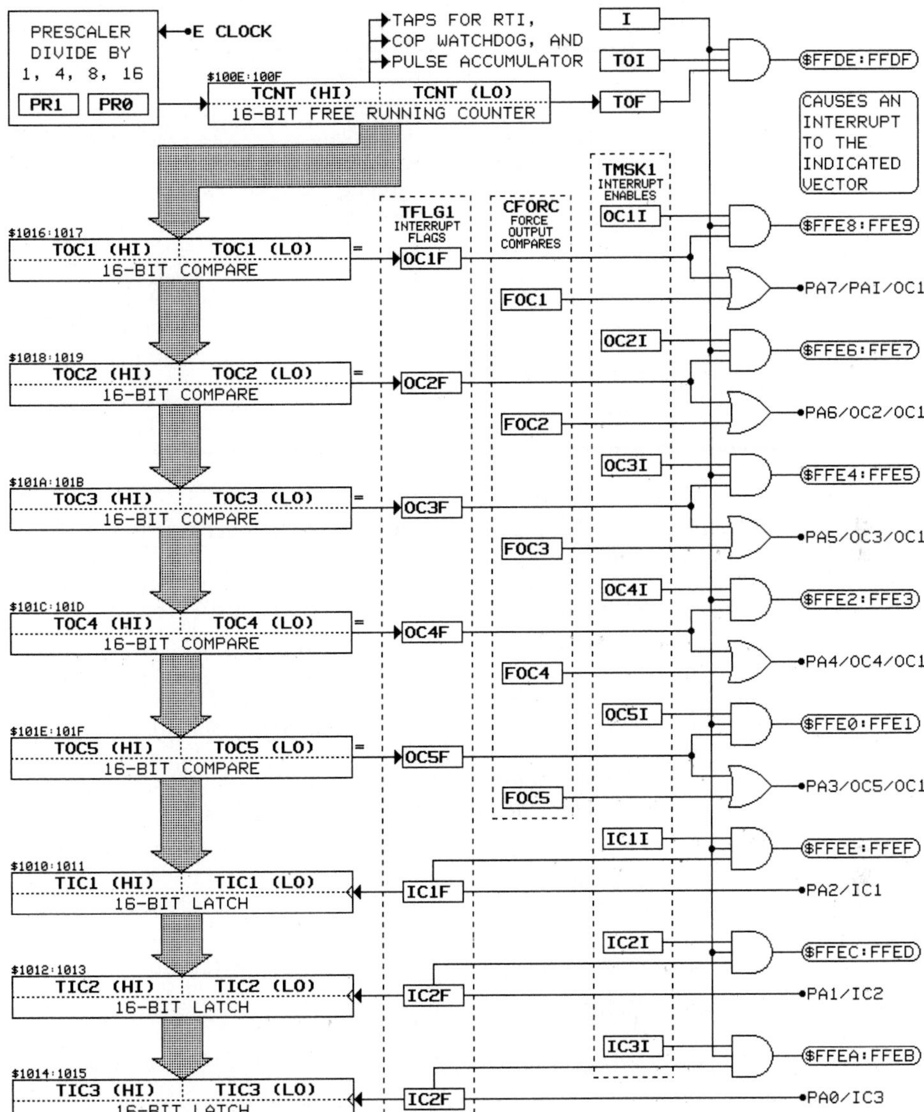

FIGURE 9.14 Main timer system block diagram.

Obviously, by adjusting the external clock frequency, the count rates can be proportionally adjusted. For example, if an external clock rate of 8.4 MHz is used, an E clock rate of 2.1 MHz is generated, resulting in counter rates and overflow rates that conveniently are even numbers.

Figure 9.15 illustrates the control and status registers associated with the main timer subsystem. The counter registers and latches are shown in Figure 9.14, and their memory address allocations are identified on that figure.

TABLE 9.5 Timer prescaler functions.

PR1	*PR0*	*Prescale Division*	*If E = 2 MHz, Count Rate = Overflow Rate =*	*If E = 2.1 MHz, Count Rate = Overflow Rate =*
0	0	÷ 1	500 ns 32.77 ms	477 ns 31.25 ms
0	1	÷ 4	2 μs 131.1 ms	1.91 μs 125 ms
1	0	÷ 8	4 μs 262.1 ms	3.81 μs 250 ms
1	1	÷ 16	8 μs 524.3 ms	8 μs 500 ms

Data Direction on PA7. If PA7 is being used for a timer function, its data direction should be defined as follows:

DDRA7: Data direction bit for PA7 is set to a logical 1 when being used as a timer output.

Timer Overflow. Several timer functions make use of the timer counter overflow functions. The TOI and TOF bits are used to manage this capability:

TOI: 0 = Inhibit timer overflow interrupts.
1 = Enable timer overflow interrupts.

TOF: Set to 1 each time the free-running counter, TCNT, overflows from $FFFF to $0000. Reset by writing a 0 into the bit.

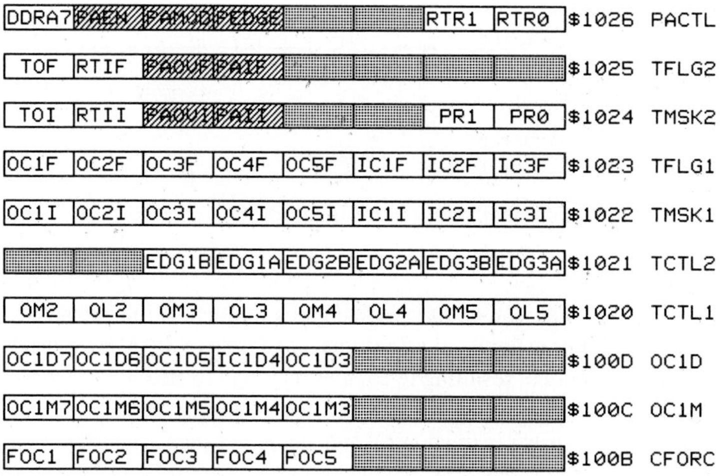

DDRA7	PAEN	PAMOD	PEDGE			RTR1	RTR0	$1026	PACTL
TOF	RTIF	PAOVF	PAIF					$1025	TFLG2
TOI	RTII	PAOVI	PAII			PR1	PR0	$1024	TMSK2
OC1F	OC2F	OC3F	OC4F	OC5F	IC1F	IC2F	IC3F	$1023	TFLG1
OC1I	OC2I	OC3I	OC4I	OC5I	IC1I	IC2I	IC3I	$1022	TMSK1
		EDG1B	EDG1A	EDG2B	EDG2A	EDG3B	EDG3A	$1021	TCTL2
OM2	OL2	OM3	OL3	OM4	OL4	OM5	OL5	$1020	TCTL1
OC1D7	OC1D6	OC1D5	IC1D4	OC1D3				$100D	OC1D
OC1M7	OC1M6	OC1M5	OC1M4	OC1M3				$100C	OC1M
FOC1	FOC2	FOC3	FOC4	FOC5				$100B	CFORC

FIGURE 9.15 Main timer control and status registers.

TABLE 9.6 Real-time interrupt rates as a function of RTR1:RTR0.

RTR1	*RTR0*	*E ÷ 2^{13} Divided by*	*If E = 2 MHz, Interrupt Rate =*	*If E = 2.1 MHz, Interrupt Rate =*
0	0	÷ 1	4.1 ms	3.91 ms
0	1	÷ 2	8.19 ms	7.81 ms
1	0	÷ 4	16.38 ms	15.62 ms
1	1	÷ 8	32.77 ms	31.25 ms

Real-Time Interrupts. Bits RTII and RTIF can be used with RTR1 and RTR0 to create interrupts at regular intervals. RTR1 and RTR0 control a frequency divider circuit that divides the E clock by the rates shown in Table 9.6.

The presence of a real-time interrupt can be determined by polling RTIF or responding to the interrupt by a service routine pointed to by the RTI vector at $FFF0:FFF1.

Programmable Input Signal Options. The user can program each input capture function to detect a particular edge polarity on its corresponding timer input pin. The control bits EDGxB and EDGxA in TCTL2 are used to select the edges based on the states of these bits. Table 9.7 illustrates these definitions.

Practical Example 9.3: Measuring a Pulse Width with Input Capture

The trick to this application is to compare two counters, one of which is stopped on one edge of the pulse and the other of which is stopped on the next edge of the pulse. Let's assume we want to measure the width of an active-high pulse. This means we must stop the first counter on the rising edge of the pulse, and then stop the second counter on the falling edge of the pulse. The width of the pulse is indicated by the difference between the two counters. The process will start by waiting until the free-running counter overflows so that the full period between overflows is available. Using this methodology, a pulse width equal to or less than 524.3 milliseconds (see Table 9.5) can be generated if the E clock is running at 2 MHz and the prescaler is set to divide by sixteen. The smallest possible pulse

TABLE 9.7 Input capture edge polarity as a function of EDGxB:EDGxA.

EDGxB	*EDGxA*	*Configuration*
0	0	Capture disabled
0	1	Capture on rising edges only
1	0	Capture on falling edges only
1	1	Capture on any edge (rising or falling)

width would be determined by the execution times of the program instructions, since, with a prescale factor of 1, the counter increments on each E clock cycle. We will connect the signal to be measured on Port A inputs PA0 and PA1, which correspond to inputs IC3 and IC2, respectively. We will stop the counter TIC3 on the rising edge of the input pulse. We will then stop the counter TIC2 on the falling edge of the input pulse. The pulse width is then determined by calculating the difference between the two counter values. Let's assume that the pulse width is between 50 and 100 milliseconds. If this is the case, we can use a prescale factor of 4, and each count represents a time period of 2μs (again, see Table 9.5). Consider the following program code:

```
0000                   ; PORT A (INPUT CAPTURE): Measure pulse width on PA0 & PA1
0000                   ;
0013                            .EQU   TIC2,$13             ;Input Capture Register 2
0015                            .EQU   TIC3,$15             ;Input Capture Register 3
0021                            .EQU   TCTL2,$21            ;Timer Control Register 2
0022                            .EQU   TMSK1,$22            ;Timer Interrupt Flag Reg 1
0023                            .EQU   TFLG1,$23            ;Timer Interrupt Flag Reg 1
0024                            .EQU   TMSK2,$24            ;Timer Interrupt Mask Reg 2
0025                            .EQU   TFLG2,$25            ;Timer Interrupt Flag Reg 2
1000                            .EQU   REGLIST,$1000        ;Internal 68HC11 Registers
E200                            .ORG   $E200                ;Locate following code
E200                   ;
E200 CE1000   PREX03   LDX    #REGLIST             ;Point to list of registers
E203 86FF              LDAA   #$FF                 ;Disable timer interrupts
E205 A722              STAA   TMSK1,X
E207 8601              LDAA   #$01                 ;Set prescale to divide by 4
E209 A724              STAA   TMSK2,X              ; and disable TOI, RTI
E20B 8609              LDAA   #$09                 ;Set PA0/IC3 rising edge
E20D A721              STAA   TCTL2,X              ; PA1/IC2 falling edge
E20F 1E2302FC PWCAPT1 BRSET  TFLG1,X,$02,PWCAPT1  ;Wait for input to fall
E213 1D2580            BCLR   TFLG2,X,$80          ;Wait for overflow flag
E216 1F2580FC PWCAPT2 BRCLR  TFLG2,X,$80,PWCAPT2
E21A 4F                CLRA                        ;Clear capture registers
E21B 5F                CLRB                        ;
E21C ED15              STD    TIC3,X
E21E ED13              STD    TIC2,X
E220                   ;
E220                   ; Initialization done, input capture counters are counting
E220                   ; up. TIC3 will stop first, then TIC2, so we will wait for
E220                   ; the interrupt flag on TIC2.
E220                   ;
E220 1F2302EB PWCAPT3 BRCLR  TFLG1,X,$02,PWCAPT1  ;Wait for input to fall
E224                   ;
E224 EC13              LDD    TIC2,X               ;Calculate TIC2 - TIC3
E226 A315              SUBD   TIC3,X               ;Accumulator D contains time
E228                                               ; in 2 µs increments.
E228 39                RTS
E229
```

After running the subroutine, the double accumulator D contains the number of 2μs increments between the rising and falling edges of the pulse. Therefore, the pulse width is equal to the number in double accumulator D times 2μs.

Practical Example 9.4: Temperature Alarm with Real-Time Interrupt

Let's assume that our microcontroller is being used to manage the processing within a device that, among other things, must monitor a temperature reading at some point in the environment. Let's also assume that there is an electronic circuit outside the microcontroller that provides it with a logical 1 level on PA7/IC1 if the temperature is too high and with a logical level 0 if the temperature is within limits. The task of our routine is to read that logic level at periodic rates and, if that level is a logical 1, to call an alarm subroutine.

First, it is our intent to use a real-time interrupt to initiate a temperature reading at regular intervals. Recall that the interrupt vector for the real-time interrupt is at $FFF0:FFF1. Stored at this address must be a pointer to the routine that will service this interrupt. Let's place the interrupt service routine at memory address $F000. Therefore, we will store the number $F000 at memory location $FFF0 to set up the pointer to the interrupt service routine.

In order to cause the periodic interrupt, we will have to enable the real-time interrupt (RTI) by clearing the real-time interrupt mask. We will cause interrupts at approximately 33 ms intervals by selecting a prescale factor of 8 (see Table 9.6). Consider the following code:

```
0000              ; PORT A (REAL TIME INTERRUPT):     Monitor and control a temp
0000              ;
0000                      .EQU   PORTA,$00            ;Port A Data Register
0026                      .EQU   PACTL,$26            ;Pulse Acc. Control Register
0024                      .EQU   TMSK2,$24            ;Timer Interrupt Mask Reg 2
1000                      .EQU   REGLIST,$1000        ;Address of register list
FFF0                      .ORG   $FFF0                ;Initialize RTI Vector
FFF0 F000                 .DW    TMPMON
E200                      .ORG   $E200                ;Prog to exercise the system
E200 CE1000      PREX04   LDX    #REGLIST             ;Point to list of registers
E203 1C2440               BSET   TMSK2,X,$40          ;Disable RTI during setup
E206 8683                 LDAA   #$83                 ;Set PA7 direction for input
E208 A726                 STAA   PACTL,X              ; and RTI rate to 8
E20A 1D2440               BCLR   TMSK2,X,$40          ;Enable RTI
E20D 3E          MAIN1    WAI                         ;Wait for RTI
E20E 20FD                 BRA    MAIN1                ;Branch back for next int.
E210              ;
F000                      .ORG   $F000                ;INTERRUPT SERVICE ROUTINE
F000 1F008003    TMPMON   BRCLR  PORTA,X,$80,TMPMON1  ;Exit if temp OK
F004 BD____               JSR    ALARM                ;Call alarm if temp high
F007 3B          TMPMON1  RTI
```

The interrupt service routine TMPMON will execute every 32.77 ms if the microcontroller is running an E clock rate of 2 MHz. The routine looks at the port and exits if the temperature is within limits. If not, a subroutine to respond to the alarm is called, then the routine exits.

We have only introduced a few of the capabilities of the main timer in the MC68HC11. To provide an exhaustive study of this one subsystem would require an entire chapter, and that is not the intent of this study. We have not observed the output capture functions, though, based on the preceding technical description, some of those capabilities can be inferred. For more detailed information on the timer subsystem, please refer to Motorola literature. An excellent source is the reference manual cited in the overview at the beginning of this chapter.

9.7 PORT B: DIGITAL OUTPUT PORT

Like the other ports, Port B has more than one use. If the microcontroller is configured for the expanded mode of operation, the pins assigned to Port B are used for the high-order byte of the address bus, and the digital port capabilities are sacrificed. The Port B digital output port is available only when the processor is in the single-chip mode. However, Motorola provides a device, an MC68HC24 port replacement unit, which can be allocated to the address and data buses of an expanded mode system. When this is done, the port capabilities are restored through the use of that device. Figure 9.16 illustrates the status and control register resources assigned to Port B.

The Port B data register is available to the programmer by accessing it within the control and status register block at address $1004. Writes to the Port B register cause data to be latched and driven out of the port B pins. Reads of Port B return the last data that was written to Port B. When the handshake I/O subsystem is operating in simple strobed mode, writes to Port B automatically cause a pulse on the STRB output in. This mode is selected by clearing the handshake control bit (HNDS) in the parallel I/O control register (PIOC). The following is a summary of the use of these control bits:

HNDS: 0 = Simple strobe mode. The STRB pin is pulsed for two E clock cycles after each write to Port B. The status of PLS is ignored.
1 = Full handshake mode. (For interlocked Port C use.)

PLS: 0 = Interlocked STRB mode. (For interlocked Port C use.)
1 = Pulsed STRB mode.

INVB: 0 = STRB active low.
1 = STRB active high.

Any use of Port B as an output port independent of Port C requires that HNDS = 0. In this mode, the STRB output will pulse for two E clock cycles following a write to the Port B data register.

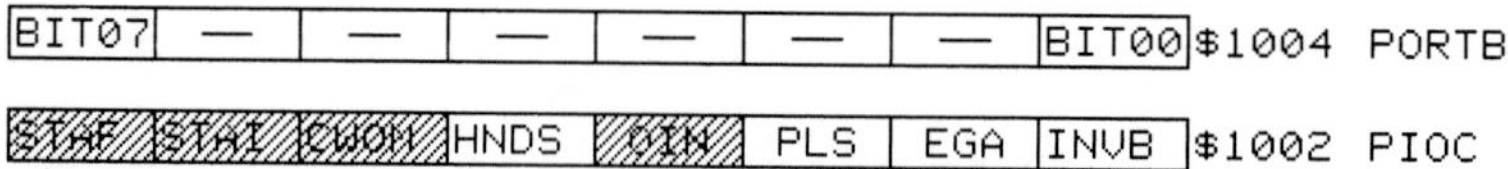

FIGURE 9.16 Port B data and control registers.

Practical Example 9.5: PORT B as Digital Output

Consider the following program example:

```
0000                ; PORT B (DIGITAL OUTPUT)
0000                ;
0004                        .EQU   PORTB,$04          ;Port B Data Register
1000                        .EQU   REGLIST,$1000      ;List of Internal Registers
E210                        .ORG   $E210
E210                ;
E210                ; The following segment writes all bits of A to Port B.
E210                ;
E210 A704           PREX05  STAA   PORTB,X            ;Port B = Accumulator A
E212                ;
E212                ; The following segment sets bit 2 of Port B to a logical 1
E212                ;
E212 1C0404                 BSET   PORTB,X,$04        ;Port B[2] = 1
E215                ;
E215                ; The following segment copies the least significant nibble
E215                ; of Accumulator A to Port B without damaging the other four
E215                ; bits of Port B
                    ;
E215 E604                   LDAB   PORTB,X            ;Get Port B data
E217 C4F0                   ANDB   #$F0               ;Clear ls nibble of B data
E219 840F                   ANDA   #$0F               ;Clear ms nibble of A data
E21B 1B                     ABA                       ;Merge the two together
E21C A704                   STAA   PORTB,X            ;Write back to the port.
E21E                        .END
```

Note that, when writing to a port, the status of any unwritten bits may have to be considered. Also, when reading from a port, it may be necessary to mask out unwanted bits if the targeted data is less than eight bits in length.

9.8 PORT C: DIGITAL I/O

Port C is a bidirectional general-purpose I/O port. When the processor is configured for the expanded mode of operation, the resources of this port are sacrificed so that the respective microcontroller pins can be used to provide the low byte of the memory address when the address strobe (AS) is asserted high, and as a memory data bus when the address strobe is deasserted. The resources of Port C can be realized in expanded mode by properly allocating an MC68HC24 port replacement unit on the address and data buses. The remainder of the discussion of Port C will assume the processor to be in the single-chip mode.

Figure 9.17 illustrate the status, control, and data registers used by the Port C subsystem.

BIT07	—	—	—	—	—	—	BIT00	$1007	DDRC
BIT07	—	—	—	—	—	—	BIT00	$1005	PORTCL
BIT07	—	—	—	—	—	—	BIT00	$1003	PORTC
STAF	STAI	CWOM	HNDS	OIN	PLS	EGA	INVB	$1002	PIOC

FIGURE 9.17 Port C data and control registers.

The primary direction of data flow at each port C pin is independently controlled by a corresponding bit in the data direction register for Port C, DDRC. In addition to normal I/O functions at Port C, there is an independent eight-bit parallel latch (PORTCL) that captures Port C data whenever a selected active edge is detected on the STRA input pin. Reads of data from PORTCL or PORTC return data that is stored in those respective latches. Writes to either PORTCL or PORTC cause the written data to be driven out the Port C pins. However, PORTCL writes trigger handshake sequences, whereas writes to PORTC do not. Writes to Port C pins not configured as outputs do not cause data to be driven out of those pins, but the data is remembered in internal latches. Then, if the pins later become configured as outputs, the last data written to PORTCL or PORTC will be present on the Port C outputs.

Port C can be configured for wired-OR (open drain) operation by setting the Port C wired-OR mode bit (CWOM) in the PIOC register. When configured in an open-drain mode, it is necessary to provide an external pull-up resistor.

Whenever the handshake subsystem is configured for a full handshake mode, Port C is used for parallel data input or output. STRA is a strobe input pin that causes Port C pins to be driven outputs while STRA is in its selected state. STRB is a strobe output pin that can be used in pulsed or interlocked configuration. Again, port activity that utilizes the handshake subsystem moves data through the PORTCL register. Data movements through the PORTC register do not stimulate handshake pulses. For the purposes of this text discussion, we will not be considering the handshake subsystem and will be doing all of our port I/O through PORTC without handshake. For more detailed information on the handshake subsystem, please refer to Motorola literature. An excellent source is the reference manual cited in the overview at the beginning of this chapter.

Data Direction Register C (DDRC). Data direction register C is used to configure the output buffers of Port C for either input data flow or output data flow. Each bit in DDRC corresponds to a single data bit on Port C. If a DDRC bit is a logical 0, the respective Port C data bit is configured as an input. If a DDRC bit is a logical 1, the respective Port C data bit is configured as an output. DDRC must be initialized prior to the use of the port as a general purpose I/O port.

Practical Example 9.6: Port C as Digital I/O

Consider the following program segments, which exercise Port C.

```
0000        ; PORT C (DIGITAL INPUT/OUTPUT)
0000        ;
```

```
0003                  .EQU   PORTC,$03        ;Port C Data Register
0007                  .EQU   DDRC,$07         ;Port C Data Direction Register
1000                  .EQU   REGLIST,$1000    ;List of Internal Registers
E500                  .ORG   $E500
E500            ;
E500            ; The following segment writes all bits of A to Port C.
E500            ;
E500 C6FF       PREX06 LDAB  #$FF            ;Configure Port C for output
E502 E707              STAB  DDRC,X
E504 A703              STAA  PORTC,X         ;Port C = Accumulator A
E506            ;
E506            ; The following segment sets bit 2 of Port C to a logical 1.
E506            ; It assumes that the data direction register has been
E506            ; initialized.
                ;
E506 1C0304            BSET  PORTC,X,$04     ;Port C[2] = 1
E509            ;
E509            ; The following segment copies the least significant nibble
E509            ; of Accumulator A to Port C without damaging the other four
E509            ; output bits
                ;
E509 E603              LDAB  PORTC,X         ;Get Port C data
E50B C4F0              ANDB  #$F0            ;Clear least sig nibble of B data
E50D 840F              ANDA  #$0F            ;Clear most sig nibble of A data
E50F 1B                ABA                   ;Merge the two nibbles together
E510 A703              STAA  PORTC,X         ;Write back to the port.
E512                   .END
```

9.9 PORT D: DIGITAL I/O AND SERIAL COMMUNICATIONS

Port D is a general-purpose six-bit bidirectional data port. Two port pins (PD0, PD1) are alternately used by the asynchronous serial communications interface (SCI) subsystem. The remaining four Port D pins are alternately used by the synchronous serial peripheral interface (SPI) subsystem. The primary direction of data flow at each of the Port D pins is selected by a corresponding bit in the data direction register for Port D (DDRD). Port D can be configured for wired-OR operation by setting the Port D wired-OR mode control bit (DWOM) in the SPI control register (SPCR). Figure 9.18 illustrates the data, status, and control registers that are used by the Port D subsystem.

9.9.1 Port D as Digital I/O

Any or all of the Port D bits can be used for digital I/O. The direction of data transfer is defined by the contents of data direction register D (DDRD). The logical value of each bit in DDRD defines the data direction of its corresponding Port D bit. As with all data direction registers, 0 signifies input and 1 signifies output. It should also be noted that in order to use the port bits for digital I/O, their subsequent serial functions must be turned off (SPE = 0, WAKE = 0). If the serial communication subsystems are enabled, their functions supersede the state of the data direction registers.

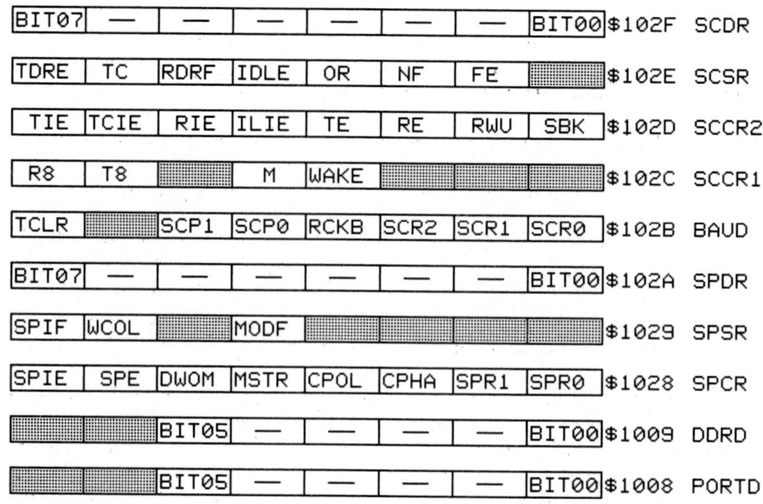

FIGURE 9.18 Port D data, control, and status registers.

Practical Example 9.7: Port D as Digital I/O

Consider the following program segments, which exercise Port D.

```
0000               ; PORT D (DIGITAL INPUT/OUTPUT)
0000               ;
0008                       .EQU   PORTD,$08        ;Port D Data Register
0009                       .EQU   DDRD,$09         ;Port D Data Direction Register
002C                       .EQU   SCCR1,$2C        ;SCI Control Register 1
0028                       .EQU   SPCR,$28         ;SPI Control Register
1000                       .EQU   REGLIST,$1000    ;List of Internal Registers
E520                       .ORG   $E520
E520               ;
E520               ; The following segment writes B[3,2,1] to Port D without
E520               ; damaging the other Port D data bits. It then reads
E520               ; PORTD[5,4,0], writing them into B without destroying the
E520               ; remainder of the accumulator B bits.
                   ;
E520 CE1000 PREX08         LDX    #REGLIST         ;Point to register list.
E523 7F002C                CLR    SCCR1            ;Turn off SCI
E526 7F0028                CLR    SPCR             ;Turn off SPI
E529 8607                  LDAA   #$07             ;Set Port D[3,2,1] as output
E52B 9709                  STAA   DDRD
E52D               ;
E52D               ; Ready to write B[3,2,1] to PORTD[3,2,1]
E52D               ;
E52D 9608                  LDAA   PORTD            ;Get current Port D data
E52F 84F1                  ANDA   #$F1             ;Keep the unchanged bits
E531 C407                  ANDB   #$07             ;Keep the B bits to write
E533 1B                    ABA                     ;Merge them
```

```
E534 9708          STAA  PORTD       ;Write merged data to Port D
E536          ;
E536          ; Read PORTD[5,4,0] to B[5.4.0]
E536          ;
E536 9608          LDAA  PORTD       ;Get Port D data
E538 84F1          ANDA  #$F1        ;Isolate input Bits
E53A C407          ANDB  #$07        ;Clear corresponding bits in B
E53C 1B            ABA               ;Merge them
E53D 16            TAB               ;Write result to B
E53E               .END
```

9.9.2 Port D as an Asynchronous Serial Communications Port

The serial communications interface (SCI) subsystem of the MC68HC11 is a full-duplex UART-type circuit designed to provide asynchronous serial communications with the variety of external devices that support a similar protocol. Asynchronous serial communications involve sending a group of data bits, often a byte, on a single data line, one following another. If the receiving device can be alerted to the beginning of the transmission, and if the transmit bit rate is the same as the receive bit rate (this is known as speed of bit transfer), it is a simple task for the receiver to discern the transmitted bit pattern. The line is normally idle at a known state (usually a logical 0). When the first bit, a logical 1, arrives to the receiver, it is alerted of the incoming character. Figure 9.19 illustrates an example asynchronous serial character protocol.

When the communication line is at an idle state, it is at a high voltage level (often +12 V) signifying that the line is active, but no data is on the line. When the transmitter sends a character, it first drops the line low (often –12V), alerting the receiver that a character is coming. This is referred to as the **start bit.** The line is held low for one predetermined bit period, T, which is a function of the transmission baud rate.

The number of bit periods per second is referred to as the **bit rate,** or the **baud rate.** For example, if a communication line sends data at 2400 bits per second, the line is referred to as a 2400 baud line. (One might note that "baud rate" did not used to be calculated this way, so there might be some variation of definition when compared with old texts.)

After one-half of a bit period (see Figure 9.19) the receiver should see the correct start bit state. If not, the receiver will assume the transition that stimulated it was simply a noise spike. Once the start bit has been established, the receiver continues to take samples at the prescribed bit rate, receiving the character sent by the transmitter, low-order bit first. These characters are shifted into a shift register. The last bit in the received character is the **stop bit.** This bit is the same logic level as the idle line and allows the line to go idle for at least one bit period prior to the transmission of the next character. Sometimes a character

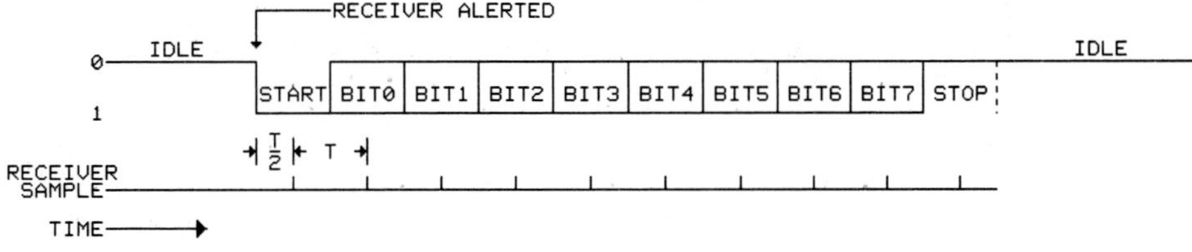

FIGURE 9.19 Asynchronous serial character format.

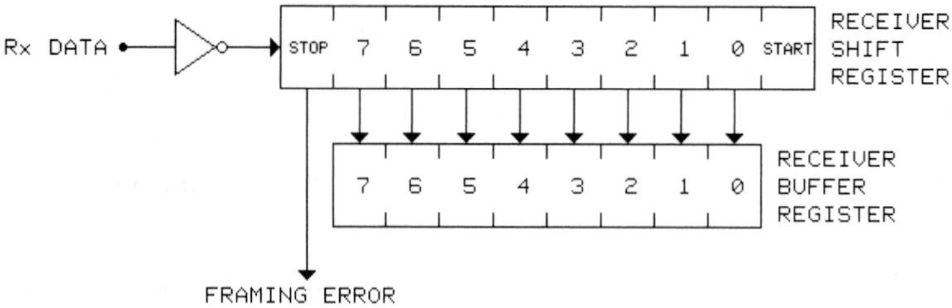

FIGURE 9.20 Asynchronous receiver buffer.

protocol prescribes two stop bits, holding the line idle for a minimum of two bit periods. This is done to allow the receiver time to respond to the receipt of the character. This protocol originated with the use of mechanical teletypes, which required time for the mechanical parts to be prepared to print another character. Obviously, adding a second stop bit adds a bit to the character length, reducing the numbers of characters that can be transmitted per second.

The receiver will place the incoming characters into a shift register buffer and, upon correct receipt of the stop bit, load it into a receive data register. Figure 9.20 illustrates such a receiver circuit.

Figure 9.20 also implies that the character is inverted prior to transmission, since it is inverted upon receipt. Figure 9.19 shows this as a logical 0 in the idle state of the character format. An inverting buffer is used to change the signal from TTL logic states to the voltage signals used by the transmission medium, usually some form of RS-232 protocol. Note also the generation of a framing error flag using the state of the stop bit. The logical state of the stop bit should always be the idle state of the line. (After inversion it is a logic 0.) If the stop bit is in the wrong state, there was some sort of transmission or receipt error that may be responded to. The MC68HC11 uses this form of asynchronous serial communication. However, it provides some flexibility in the character format and bit rate.

Baud Rate Selection. The baud register, allocated at address $1028, is used to configure the baud rate, or bit rate, of character transfer. Table 9.8 illustrates some of the available transmission speeds as a function of the bits in this register. For a more complete list, refer to Motorola documentation.

TABLE 9.8 M68HC11 baud rates by crystal frequency.

Baud Register Contents					*External Crystal Frequency*				
Bit-5 SCP1	*Bit-4 SCP0*	*Bit-2 SCR2*	*Bit-1 SCR1*	*Bit-0 SCR0*	*2^{23} Hz*	*8 MHz*	*4.9152 MHz*	*4 MHz*	*3.6864 MHz*
					Generated Baud Rate in Bits per Second				
0	0	0	0	0	131072	125000	76800	62500	57600
0	0	0	0	1	65536	62500	38400	31250	28800
0	0	0	1	0	32768	31250	19200	15625	14400
0	0	0	1	1	16384	1562.5	9600	7812.5	7200

TABLE 9.8 M68HC11 baud rates by crystal frequency (continued).

Baud Register Contents					*External Crystal Frequency*				
Bit-5 SCP1	*Bit-4 SCP0*	*Bit-2 SCR2*	*Bit-1 SCR1*	*Bit-0 SCR0*	*2^{23} Hz*	*8 MHz*	*4.9152 MHz*	*4 MHz*	*3.6864 MHz*
					Generated Baud Rate in Bits per Second				
0	0	1	0	0	8192	7812.5	4800	3906	3600
0	0	1	0	1	4096	3906	2400	195	1800
0	0	1	1	0	2048	1953	1200	977	900
0	1	0	0	0	43691	41666	25600	20833	19200
0	1	0	0	1	21845	20833	12800	10417	9600
0	1	0	1	0	10923	10417	6400	5208	4800
0	1	0	1	1	5461	5208	3200	2604	2400
0	1	1	0	0	2731	2604	1600	1302	1200
0	1	1	1	0	683	651	400	326	300
1	0	0	0	0	32768	31250	19200	15625	14400
1	0	0	0	1	16384	15625	9600	7812.5	7200
1	0	0	1	0	8192	7812.5	4800	3906	3600
1	0	0	1	1	4096	3906	2400	1953	1800
1	0	1	0	0	2048	1953	1200	977	900
1	0	1	1	0	512	488	300	244	225.5
1	0	1	1	1	256	244	150	122	112.5

Asynchronous Serial Port Control and Status Registers. Figure 9.21 illustrates those registers containing bits used in the asynchronous SCI subsystem.

RxD: **Receive data bit,** bit 0 of the PORTD register. It is through this data port line that asynchronous data characters are received.

TxD: **Transmit data bit,** bit 1 of the PORTD register. It is through this data port line that asynchronous data characters are transmitted.

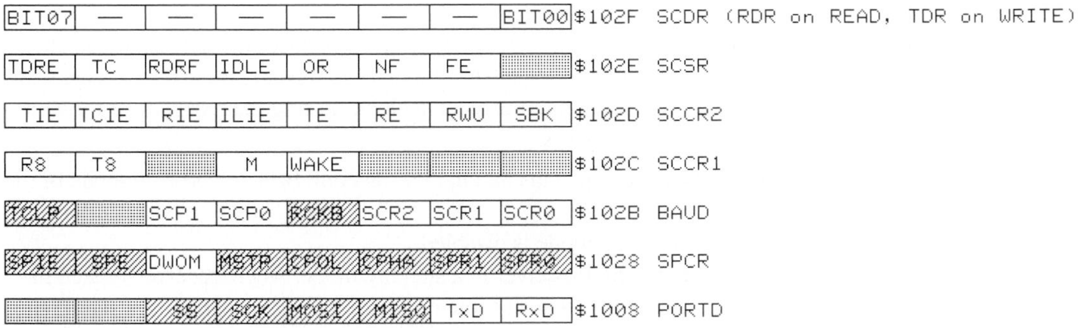

BIT07	—	—	—	—	—	—	BIT00	$102F SCDR (RDR on READ, TDR on WRITE)
TDRE	TC	RDRF	IDLE	OR	NF	FE		$102E SCSR
TIE	TCIE	RIE	ILIE	TE	RE	RWU	SBK	$102D SCCR2
R8	T8		M	WAKE				$102C SCCR1
TCLR		SCP1	SCP0	RCKB	SCR2	SCR1	SCR0	$102B BAUD
SPIE	SPE	DWOM	MSTR	CPOL	CPHA	SPR1	SPR0	$1028 SPCR
		SS	SCK	MOSI	MISO	TxD	RxD	$1008 PORTD

FIGURE 9.21 SCI control and data registers.

SCR2–SCR0: **Baud rate generator configuration,** bits 2,1,0 of the baud register. These three bits are used to configure the baud rate of both transmit and receive data transfers. When used in conjunction with SCP1 and SCP0, a wide range of speeds can be defined. See Table 9.8.

SCP1–SCP0: **Baud rate generator prescale,** bits 5,4 of the SPCR register. These two bits are used to prescale the baud rate generator that drives both transmit and receive data transfers. When used in conjunction with SCR2, SCR1, and SCR0, a wide range of speeds can be defined. See Table 9.8.

SCDR: **Receive data register, RDR** (read only). After receipt of a serial character, that character is placed in this register. If the character is nine data bits in length, the ninth character is placed in R8, in the SCCR1 register, bit 7.

Transmit Data Register, TDR (write only). A serial character is sent by writing the data to be transmitted to this register. If the character is nine data bits in length, the ninth character is placed in T8, in the SCCR1 register, bit 6.

SCI Control Register 1 (SCCR1)

WAKE: **SCI wake-up method select,** bit 3 of SCCR1.
0 = Idle line; detection of at least a full character time of idle line causes the receiver to wake up.
1 = Address mark; a logic 1 in the MSB position (eighth or ninth data bit depending on the character size selected by the M bit) causes the receiver to wake up.

M: **SCI character length,** bit 4 of SCCR1. This bit identifies the character length. A logical 0 configures the port for eight data bits; a logical 1 configures the port for nine data bits.

T8: **Transmit data bit 8,** bit 6 of SCCR1. When the SCI system is configured for nine data bits, this bit acts as the extra (ninth) bit of the transmit data register, TDR.

R8: **Receive data bit 8,** bit 7 of SCCR1. When the SCI system is configured for nine data bits, this bit acts as the extra (ninth) bit of the receive data register, RDR.

SCI Control Register 2 (SCCR2)

SBK: **Send a break.** When set to a logical 1, the transmitter sends a break character, defined by the MC68HC11 as 10 (11 if M=1) logic 0 bits. This word can be used to calibrate the receive clock, to ring the receiver, or for a variety of other purposes.

RWU: **Receiver wake-up.** This bit enables the configurable wake-up mode of the receiver.
0 = The wake-up mode is disabled.
1 = Places the SCI receiver in a standby mode, where receiver-related interrupts are inhibited until some hardware condition is met to wake up the sleeping receiver. This wake up condition depends on the state of the WAKE bit.

RE: **Receive enable.**
0 = SCI receiver disabled. The RDRF, IDLE, OR, NF, and FE status flags cannot become set. They keep their current state if a transition to 0 on RE is encountered.
1 = receiver enabled. The state of DDRD, bit 0 is ignored.

TE: **Transmit enable.**
0 = SCI transmitter disabled.
1 = SCI transmitter enabled. When first enabled, the transmitter holds the transmit line idle for one character time.

ILIE: **Idle line interrupt enable.** Interrupt vector: $FFD6:FFD7
0 = IDLE interrupts disabled (software polling mode).
1 = An SCI interrupt is requested when IDLE is set to 1. The idle-line function is inhibited while the receiver wake-up function is enabled.

RIE: **Receive interrupt enable.** Interrupt vector: $FFD6:FFD7
0 = RDRF and OR interrupts disabled (software polling mode).
1 = An SCI interrupt is requested when either RDRF or OR is set to 1.

TCIE: **Transmit complete interrupt enable.** Interrupt vector: $FFD6:FFD7
0 = TC interrupts disabled (software polling mode).
1 = An SCI interrupt is requested when TC is set to 1.

TIE: **Transmit interrupt enable.** Interrupt vector: $FFD6:FFD7
0 = TDRE interrupts disabled (software polling mode).
1 = An SCI interrupt is requested when DTRE is set to 1.

SCI Status Register (SCSR)

FE: **Framing error.**
0 = No framing error detected.
1 = The stop bit was not a logic high, as expected. The flag is cleared by reading FE = 1 in SCSR, followed by a read of SCDR.

NF: **Noise flag.**
0 = No noise was detected during the receipt of the character.
1 = Data recovery logic detected noise during reception of the character. Contrary to what is implied in Figure 9.20, the receiver takes several samples during the receipt of a bit. If transitions are detected during the bit period, the noise flag is set.

OR: **Overrun error.**
0 = No overrun error was detected.
1 = Another character was serially received and was ready to be transferred to the SCDR, but the previously received character was not yet read. The flag is cleared by reading OR = 1 in SCSR, followed by a read of SCDR.

IDLE: **Idle-line detect.**
0 = The RxD line is either active now or has never been active since IDLE was last cleared.
1 = The RxD line has become idle. That is, there has been at least one character time inactivity on the receive data line.

RDRF: **Receive data register full.**
0 = Not full; nothing has been received since the last read of SCDR.
1 = A character has been received and is in the SCDR, ready to be read. It is best to use some method to get the character quickly in order to avoid an overrun (OR) error.

TC: **Transmit complete.**
0 = The transmitter is busy sending a character.
1 = The transmitter is idle, ready to send another character.

TDRE: **Transmit data register empty.**
0 = Not empty; a character was previously written to SCDR and has not yet been serially sent.
1 = A new character may now be written to the SCDR.

Practical Example 9.8: Asynchronous Serial Transmission

Task: Send a string of characters out the asynchronous serial port at 2400 baud. The string is a list of ASCII characters stored at memory location STRING, and the string is terminated with a $04 character. Assume an external clock speed of 8 MHz. Do not use interrupts.

Consider the following segment of code:

```
0000                         ; PORT D (ASYNCHRONOUS SERIAL TRANSMITTER)
0000                         ;
002B                         .EQU  BAUD, $2B         ;SCI Baud Rate Generator
002C                         .EQU  SCCR1,$2C         ;SCI Control Register 1
002D                         .EQU  SCCR2,$2D         ;SCI Control Register 2
002E                         .EQU  SCSR,$2E          ;SCI Status Register
002F                         .EQU  RDR,$2F           ;SCI Receive Data Register
002F                         .EQU  TDR,$2F           ;SCI Transmit Data Register
1000                         .EQU  REGLIST,$1000     ;List of Internal Registers
E540                         .ORG  $E540
E540                 ;
E540                 ; Configure the Port
E540 CE1000   PREX08   LDX   #REGLIST          ;Point to list of registers
E543 18CEE556          LDY   #STRING           ;Point to string to send
E547 8623              LDAA  #$23              ;Set to 2400 Baud
E549 972B              STAA  BAUD,X
E54B 7F002C            CLR   SCCR1,X           ;8-bit data
E54E 8608              LDAA  #$08              ;Enable the Transmitter
E550 972D              STAA  SCCR2,X
E552 BDE567            JSR   SND_STR
E555 CF                STOP
E556 54686973 STRING   .DB   "This is a string"
E55A 20697320
E55E 61207374
E562 72696E67
E566 04                .DB   $04
E567                 ;
```

```
E567          ; ***********************************************************
E567          ; * SUBROUTINE S N D _ S T R (Y)                            *
E567          ; *                                                         *
E567          ; * This subroutine sends a string of ASCII characters out *
E567          ; * the asynchronous serial port. Y points to the string,  *
E567          ; * the string is terminated with a $04 character.         *
E567          ; ***********************************************************
E567 18A600   SND_STR  LDAA  $00,Y        ;Get a string character
E56A 1808              INY                ;Increment string pointer
E56C 8104              CMPA  #$04         ;End of the string?
E56E 2709              BEQ   SNDSTR2      ; If so, quit.
E570 7D002E   SNDSTR1  TST   SCSR,X       ;Is the transmitter ready?
E573 2AFB              BPL   SNDSTR1      ; If not, wait.
E575 972F              STAA  TDR,X        ;Send character!
E577 20EE              BRA   SND_STR      ;Go send another character
E579 39       SNDSTR2  RTS
E57A
```

Note that the program first initializes the port, establishes the parameter Y so that it points to the string, then it calls the subroutine. The subroutine simply waits to see if the port is available and, if so, sends the character.

Practical Example 9.9: Asynchronous Serial Reception

Task: Use the receive interrupt enable to receive a string of characters, placing that string at an address pointed to by the Y register. The string is terminated with a $04 character.

Consider the following program code:

```
0000          ; PORT D (ASYNCHRONOUS SERIAL RECEIVER)
0000          ;
002B                   .EQU  BAUD, $2B      ;SCI Baud Rate Generator
002C                   .EQU  SCCR1,$2C      ;SCI Control Register 1
002D                   .EQU  SCCR2,$2D      ;SCI Control Register 2
002E                   .EQU  SCSR,$2E       ;SCI Status Register
002F                   .EQU  RDR,$2F        ;SCI Receive Data Register
002F                   .EQU  TDR,$2F        ;SCI Transmit Data Register
1000                   .EQU  REGLIST,$1000  ;List of Internal Registers
FFD6                   .ORG  $FFD6          ;SCI Interrupt Vector
FFD6 E5DC              .DW   REC_CHR        ; points to REC_CHR
E580                   .ORG  $E580
E580          ;
E580          ; Configure the Port
E580          ;
E580 CE1000   PREX09   LDX   #REGLIST       ;Point to list of registers
E583 18CEE59C          LDY   #STRING        ;Point to string buffer
E587 8632              LDAA  #$32           ;Set to 2400 Baud
E589 972B              STAA  BAUD,X
E58B 7F002C            CLR   SCCR1,X        ;8-bit data
E58E 8624              LDAA  #$24           ;Enable the Receiver and
E590 972D              STAA  SCCR2          ; receive interrupts.
```

```
E592 0E                        CLI                       ;Enable interrupts.
E593                           ;
E593                           ; Receive the String
E593                           ;
E593 3E           PREX09A      WAI                       ;Wait for interrupts
E594 8104                      CMPA   #$04               ;End of string?
E596 2702                      BEQ    PREX09B            ;If so, quit
E598 20F9                      BRA    PREX09A
E59A 0F           PREX09B      SEI                       ;Disable interrupts.
E59B CF                        STOP
E59C              STRING       .DS    $40                ;Reserve 64-byte buffer
E59C              ;
E59C              ; ******************************************************
E59C              ; * INTERRUPT SERVICE ROUTINE R E C _ C H R (Y,A)       *
E59C              ; *                                                    *
E59C              ; * This interrupt service routine will receive an 8-bit *
E59C              ; * character from the asynchronous serial port, and then *
E59C              ; * write the character to memory pointed to by Y.      *
E59C              ; ******************************************************
E59C 36           REC_CHR      PSHA                      ;Save current accumulator A
E59D 962F                      LDAA   RDR                ;Get the received character
E59F 18A700                    STAA   $00,Y              ;Store into the buffer
E5A2 32                        PULA                      ;Restore previous A
E5A3 1808                      INY                       ;Increment the string pointer
E5A5 3B                        RTI                       ;Return from Interrupt
E5A6                           .END
```

Note that this example assumes that interrupts were not enabled prior to execution. The interrupt service routine, REC_CHR, compiled at address $E59C to that value was placed at the interrupt vector, $FFD6:FFD7. Once the port is configured, the program sits in a wait loop, waiting for interrupts to generate calls to the interrupt service routine. When an interrupt comes, the routine reads the data from the receive data register (RDR) and writes it to memory pointed to by index register Y. It then increments the index register, preparing it for the next character transfer. This cycle continues until the $04 character is received. The interrupts are then disabled by setting the interrupt mask in the condition code register using the SEI instruction.

This example assumes that the interrupt is generated by the SCI receiver. A better solution would be to verify this first by checking the RDRF flag. Also, there is no attempt to determine if there were any receive framing, overrun, or noise errors. This code could also be added to this example.

9.9.3 Special Bootstrap Mode

Since the MC68HC11 special bootstrap mode makes use of the SCI subsystem, this is a good point in our discussion to observe its purpose and operation. The special bootstrap mode of the processor is accomplished by placing the MODA;MODB bits in the 0,0 state (see Table 9.2). When the chip is RESET in this mode, a special internal **bootstrap ROM,** which is allocated at $BF40–BFFF, is enabled. The RESET vector is fetched from this bootstrap ROM, and the microcontroller proceeds to execute the firmware in this ROM. The program in this ROM initializes the on-chip SCI subsystem, checks for a security option, accepts a 256-byte program through the SCI, and then jumps to the loaded program

TABLE 9.9 Bootstrap mode pseudo-vectors.

Address	*Vector Name*
$00C4:00C6	SCI
$00C7:00C9	SPI
$00CA:00CC	Pulse Accumulator Input Edge
$00CD:00CF	Pulse Accumulator Overflow
$00D0:00D2	Timer Overflow
$00D3:00D5	Timer Output Capture 5
$00D6:00D8	Timer Output Capture 4
$00D9:00DB	Timer Output Capture 3
$00DC:00DE	Timer Output Capture 2
$00DF:00E1	Timer Output Capture 1
$00E2:00E4	Timer Input Capture 3
$00E5:00E7	Timer Input Capture 2
$00E8:00EA	Timer Input Capture 1
$00EB:00ED	Real Time Interrupt
$00EE:00F0	IRQ
$00F1:00F3	XIRQ
$00F4:00F6	SWI
$00F7:00F9	Illegal Opcode
$00FA:00FC	COP Fail
$00FD:00FF	Clock Monitor Fail

at address $0000 on the on-chip RAM. There are almost no limitations on the programs that can be loaded and executed through the bootstrap process.

When operating in the bootstrap mode, the address of the interrupt vectors changes to suit the flexibility of the mode. In order to operate correctly, JMP instructions must be programmed into "pseudo-vector" addresses. Table 9.9 illustrates the interrupt pseudo-vectors available in this mode.

Practical Example 9.10: Microcontroller Bootloader Program

One of the useful features is the ability to use the bootstrap mode to load a program into the EEPROM of the MC68HC11 microcontroller. The following program is in the public domain and provides such a function. It is provided for information purposes. For more information on the operation of the bootstrap mode and programming of the EEPROM, refer to Motorola literature. Note that this example also uses a slightly different assembler syntax than the previous examples.

```
;FILENAME: BOOT8.ASM   (RUN CONVERT WITH SIZE = 1K; OFFSET = 65535)
;MICRO-CONTROLLER BOOT PROGRAM (PROGRAMS A 300 BAUD HEX FILE INTO EEPROM)
; FOR 68HC811E2 WITH 2K EEPROM RUNNING AT 8 MHZ
;4 JULY 1993
; SYSTEM ADDRESSES
```

```
RBASE           .EQU    $1000           ;POWER UP REGISTER BASE ADDRESS
STACK           .EQU    $00FF           ;PUT STACK AT TOP OF RAM
DELAY           .EQU    $0D00           ;10 MS FOR 8 MHZ CLOCK RATE
EEPROM          .EQU    $F800
END_EEPROM      .EQU    $FFFF
ROM_SIZE        .EQU    2048
;
; REGISTERS
;
PORTB           .EQU    RBASE+$04       ;OUTPUT PORT
PORTD           .EQU    RBASE+$08       ;PORTD, SERIAL COMMUNICATIONS
DDRD            .EQU    RBASE+$09       ;DATA DIRECTION REGISTER, PORTD
TCNT            .EQU    RBASE+$0E       ;TIMER COUNT
TOC1            .EQU    RBASE+$16       ;OUTPUT COMPARE, REGISTER 1
TOC2            .EQU    RBASE+$18       ;OUTPUT COMPARE, REGISTER 2
TFLG1           .EQU    RBASE+$23       ;MAIN TIMER INTERRUPT FLAGS
TMSK2           .EQU    RBASE+$24       ;TIMER INTERRUPT MASK REGISTER, PRESCALER
BAUD            .EQU    RBASE+$2B       ;SCI BAUD RATE CONTROL
SCCR2           .EQU    RBASE+$2D       ;SCI CONTROL REGISTER 2
SCSR            .EQU    RBASE+$2E       ;SCI STATUS REGISTER
SCDR            .EQU    RBASE+$2F       ;SCI DATA, READ (RDR) AND WRITE (TDR)
ADCTL           .EQU    RBASE+$30       ;A/D CONTROL REGISTER, NOT USED
ADR1            .EQU    RBASE+$31       ;A/D CHANNEL 1 REGISTER, NOT USED
BPROT           .EQU    RBASE+$35       ;EEPROM BLOCK PROTECT REGISTER (68HC811E2)
OPTION          .EQU    RBASE+$39       ;SYSTEM OPTIONS
PPROG           .EQU    RBASE+$3B       ;EEPROM PROGRAM CONTROL REGISTER
CONFIG          .EQU    RBASE+$3F       ;SYSTEM CONFIGURATION REGISTER
OPTS            .EQU    %00011011       ;A/D OFF, STOP DELAY, CLOCK MONITOR, 1 S COP
TMASK           .EQU    %00000011       ;TIMER PRESCALE MASK: E/16
ONMSK           .EQU    %10000000       ;MASK FOR TIMER FLAG, ON TIME OVER
OFFMSK          .EQU    %11000000       ;MASK FOR TIMER FLAG, OFF TIME OVER
CLRMSK          .EQU    %11000000       ;MASK FOR CLEARING TIMER FLAGS
BDMSK           .EQU    %00110101       ;SCR BITS FOR 300 BAUD @ 8 MHZ
PTDMSK          .EQU    %00000010       ;PORTD DATA DIRECTION: INPUT, EXCEPT TXD
SC2MSK2         .EQU    %00001100       ;MASK TO ENABLE BOTH TRANSMIT AND RECEIVE
TDREMSK         .EQU    %10000000       ;TRANSMITTER READY FLAG
RDRFMSK         .EQU    %00100000       ;RECEIVER READY FLAG
        .ORG    #$FFFF
        .FILL   $FF,1           ;$FF REQUIRED TO START BOOTSTRAP COMM.
INIT:   LDAA    #%00010000      ;ENABLE PROGRAMMING OF EEPROM BUT NOT CONFIG
        STAA    BPROT
        CLR     PORTB
        LDAA    #OPTS           ;LOAD IN SYSTEM OPTIONS
        STAA    OPTION          ;OPTION:=OPTS
        LDS     #STACK          ;INITIALIZE STACK
;                               ;DELAY 5 SECONDS TO KEEP GARBAGE FROM
        LDX     #0100           ; BEING WRITTEN TO EEPROM
IN1:    JSR     DELAY_10_MS
        LDAA    SCDR
        DEX
        BNE     IN1
        LDAA    #PTDMSK         ;PORTD = INPUT EXCEPT TRANSMIT = OUTPUT
        STAA    DDRD
        LDAA    #TMASK          ;TMSK2:=TMASK
        STAA    TMSK2
```

```
        LDAA    #BDMSK          ;BAUD:=BDMSK
        STAA    BAUD
        LDAA    #SC2MSK2
        STAA    SCCR2           ;DISABLE INTERRUPTS, ENABLE RCVR. & XMTR.
        LDAA    SCDR            ;CLEAR OUT RECEIVE BUFFER
        LDAA    SCSR
MAIN:   LDX     #ROM_SIZE-1
        STX     BYTES_LEFT
        LDX     #EEPROM
        STX     NEXT_BYTE
        JSR     ERASEEEPROM
M0:     JSR     GETBYTE
        CMPA    #$FF
        BEQ     M0
M1:     JSR     WRITE
        JSR     GETBYTE
        LDX     BYTES_LEFT
        DEX
        STX     BYTES_LEFT
        BGE     M1
M2:     BRA     M2
;
; PROCEDURE DEFINITIONS:
;
ERASEEEPROM:
; PERFORMS A BULK ERASE ON EEPROM.
        LDAB    #$6
        STAB    PPROG   ;SELECT BULK ERASE MODE
        STAB    EEPROM  ;SELECT ANY EEPROM LOCATION
        LDAB    #$7
        STAB    PPROG   ;TURN ON PROGRAMMING VOLTAGE
        LDAB    #10
EE1:    JSR     DELAY_10_MS    ;100 MS DELAY (USES Y REGISTER)
        DECB
        BNE     EE1
        CLR     PPROG   ;TURN OFF PROGRAMMING VOLTAGE
        RTS
GETBYTE:
; FETCHES ONE BYTE FROM THE SCI,
; AND RETURNS IT IN REGISTER A.
GB1:    LDAA    SCSR            ;GET SCI FLAGS
        ANDA    #RDRFMSK        ;MASK OFF IRRELEVANT BITS
        BEQ     GB1             ;LOOP UNTIL CHARACTER FOUND
        LDAA    SCDR            ;GET CHARACTER
        RTS
; WRITES THE VALUE CONTAINED IN REGISTER A TO THE NEXT EEPROM LOCATION
WRITE:  LDX     NEXT_BYTE
        LDAB    #$02
        STAB    PPROG
        STAA    0,X
        LDAB    #$03
        STAB    PPROG
        JSR     DELAY_10_MS
        CLR     PPROG
        INX
```

```
        STX     NEXT_BYTE
        RTS
DELAY_10_MS:
        LDY     #DELAY
D1:     DEY
        BNE     D1
        RTS
BYTES_LEFT:     RMB     2
NEXT_BYTE:      RMB     2
TEMP:           RMB     1
```

9.9.4 Port D as a Synchronous Serial Communications Port

Synchronous and asynchronous communications protocols are similar except that the latter method uses handshake lines rather than start bits, stop bits, and idle time to synchronize the receiver with the transmitter. The result is an environment where more characters can be transmitted per unit of time at the same clock speed. The disadvantage of the synchronous method is the need for additional handshake lines that contain the synchronization clock. It is quite useful for communications between processors. It is also designed to interface directly with many commercially available product peripherals. The microcontroller can be programmed to act as a master unit (providing the synchronization) or as a slave unit. The maximum bit rate in the master configuration is 1 MHz. The maximum bit rate in the slave configuration is 2 MHz. External serial communication signals are provided on the Port D pins PD5, PD4, PD3, and PD2. Figure 9.22 illustrates the interconnection of four MC68HC11 microcontrollers using the synchronous serial ports.

The detailed discussion of the operation of this port is beyond the scope of this text. For such detailed explanation, refer to Motorola documentation.

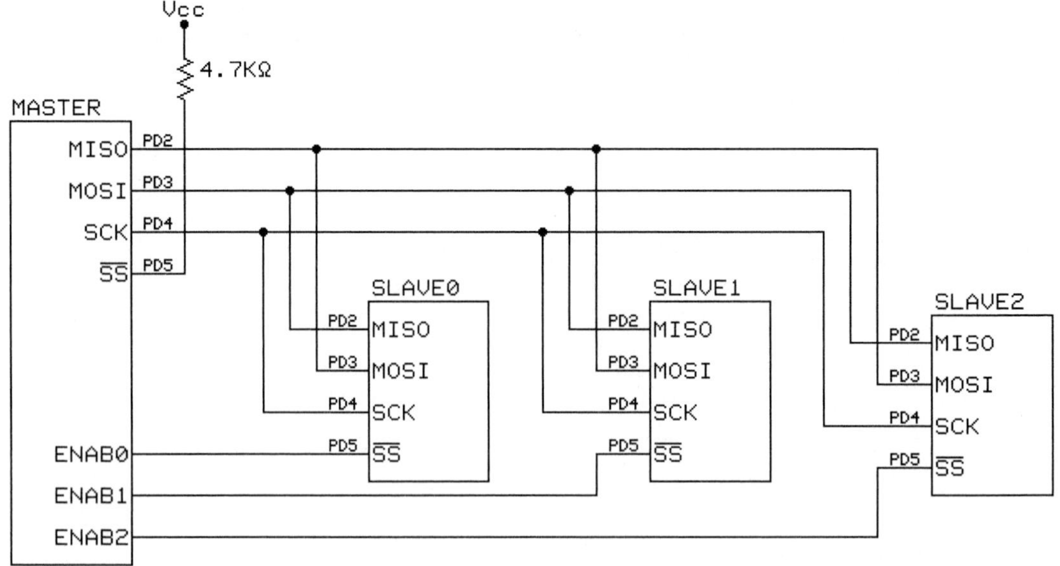

FIGURE 9.22 Synchronous interconnection of four MC68HC11 microcontrollers.

9.10 PORT E: DIGITAL INPUT AND ANALOG INPUT

One of the features of the MC68HC11 microcontroller that makes it particularly useful for such a wide variety of applications is its eight-channel analog-to-digital (A/D) converter subsystem. Port E, like the other ports, has two uses, and these uses can coexist without any configuration needs. A memory read from Port E will always result in reading the TTL byte presented to the eight pins of Port E. However, each bit of the port is also directed to the A/D subsystem, where up to eight analog inputs can be connected. These inputs must be driven with signals that range from 0.00 V to 5.12 V. Input signals can be read individually or in two groups of four. The A/D subsystem uses a successive-approximation/charge-redistribution type converter and requires 32 E clock cycles in order to complete a single conversion. Figure 9.23 illustrates the processor control and status registers that are used to configure and apply Port E.

The ADPU control bit in the OPTION register is used to power-up the A/D subsystem. At RESET, the A/D converter subsystem is not enabled, due to a need for a delay period during which the charge pump and comparitor circuits can stabilize. Therefore, this bit must be set to 1 by the programmer prior to using the A/D converter subsystem. Also, a time delay of at least 100μs must be provided prior to the use of the A/D converter subsystem.

The CSEL control bit in the OPTION register selects an alternate internal RC-driven clock source for the on-chip A/D subsystem and EEPROM charge pump when the E clock is too slow to drive these dynamic circuits. The EEPROM charge pump is separate from the A/D charge pump, but both pumps are selected with the CSEL control bit. In the case of the A/D charge pump, CSEL needs to be 1 when the E clock is too slow to assure that the successive approximation sequence will finish before any significant charge loss. When the E clock is at or above 2 MHz, CSEL should always be 0. When the E clock is below 750 KHz, CSEL should almost always be a 1. At frequencies between these limits, CSEL should be a 0 for A/D operations and a 1 for EEPROM programming. However, the A/D error resulting in having the CSEL bit at 1 within these limits is minimal (1/2 lsb) and may not be of concern.

ADPU	CSEL	IRQE	DLY	CME		CR1	CR0	$1039	OPTION
BIT07	—	—	—	—	—	—	BIT00	$1034	ADR4
BIT07	—	—	—	—	—	—	BIT00	$1033	ADR3
BIT07	—	—	—	—	—	—	BIT00	$1032	ADR2
BIT07	—	—	—	—	—	—	BIT00	$1031	ADR1
CCF		SCAN	MULT	CD	CC	CB	CA	$1030	ADCTL
BIT07	—	—	—	—	—	—	BIT00	$100A	PORTE

FIGURE 9.23 A/D converter subsystem status and control registers.

A/D Result Registers. There are four registers that are used to store the digital results of the analog conversions. Since there are eight analog ports, each register must accommodate two port lines, one at a time. Bits CD, CC, CB, and CA in the ADCTL register are used to select the channel assignments. These are summarized in Table 9.10.

The MULT bit in the ADCTL register determines whether a single conversion is to be made, or if a group of four conversions is to be made. When MULT = 0, a single conversion will be made on the individual Port E input line selected with bits CC, CB, and CA, as shown in Table 9.10. The results of the conversion will be placed in the indicated result register, and the CCF bit will be set when done. When MULT = 1, four conversions are done, either on PE3,2,1,0 or on PE7,6,5,4, depending on the state of the CC bit. The CCF bit will be set after all four conversions have been completed. The results of the four conversions will be placed in the indicated registers.

The SCAN bit in the ADCTL register determines if the A/D subsystem is to take a single reading or to take continuous readings from the specified port or ports. If SCAN = 0, the converter performs a single reading and stops if MULT = 0, or it performs four readings and stops if MULT = 1. If SCAN = 1, the A/D converter takes continuous readings, updating the result registers and setting the CCF bit after each completed conversion.

The CCF bit in the ADCTL register identifies that a requested conversion has been completed. Unlike the other bits in the ADCTL register, it is set by the converter subsystem as a status flag. However, it is cleared by writing to ADCTL.

Conversions are requested by writing to the ADCTL register with CCF = 0, and the desired conversion sequence defined by SCAN, MULT, CD, CC, CB, and CA. The CCF bit must then be polled to determine whether the conversion is complete, or else the program must wait for at least 32 E clock periods per conversion prior to getting the data from the result registers. The A/D subsystem does not support the generation of interrupts. However, when used in conjunction with the timer subsystem, the same need can be accomplished.

Resolution. The input voltage on the analog ports must be in the range of 0.00 V to 5.12 V. This range of voltages is digitized into 256 binary patterns from 00000000_2 to 11111111_2. This results in a resolution of 20 mV per bit.

TABLE 9.10 Result register assignments for A/D converter subsystem.

			MULT = 0				*MULT = 1*	
CC	*CB*	*CA*	*Channel*	*Result Register*	*CD*	*CC*	*Channel*	*Result Register*
0	0	0	PE0	ADR1	0	0	PE0	ADR1
0	0	1	PE1	ADR2			PE1	ADR2
0	1	0	PE2	ADR3			PE2	ADR3
0	1	1	PE3	ADR4			PE3	ADR4
1	0	0	PE4	ADR1	0	1	PE4	ADR1
1	0	1	PE5	ADR2			PE5	ADR2
1	1	0	PE6	ADR3			PE6	ADR3
1	1	1	PE7	ADR4			PE7	ADR4

Practical Example 9.11: Port E Application

Consider the following code segments, which make various uses of the Port E digital and analog-to-digital subsystem. Assume that the E clock is running at 2 MHz.

```
0000                    ; PORT E (DIGITAL AND ANALOG INPUT)
0000                    ;
0039                            .EQU   OPTION,$39      ;Configuration Option Reg.
0034                            .EQU   ADR4,$34        ;A/D Result Register 4
0033                            .EQU   ADR3,$33        ;A/D Result Register 3
0032                            .EQU   ADR2,$32        ;A/D Result Register 2
0031                            .EQU   ADR1,$31        ;A/D Result Register 1
0030                            .EQU   ADCTL,$30       ;A/D Control Register
000A                            .EQU   PORTE,$0A       ;Port E Data Register
1000                            .EQU   REGLIST,$1000   ;List of Internal Registers
E5B0                            .ORG   $E5B0
E5B0 CE1000     PREX11   LDX    #REGLIST        ;Point to list of registers
E5B3                    ;
E5B3                    ; Power up the A/D converter subsystem
E5B3                    ;
E5B3 1C3980              BSET   OPTION,X,$80    ;ADPU = 1, power up A/D
E5B6 1D3940              BCLR   OPTION,X,$40    ;CSEL = 0, select E clock
E5B9                                            ; if E < 750 KHz set CSEL = 1
E5B9 8628                LDAA   #40             ;Wait 100 microseconds.
E5BB 4A         PREX11A  DECA                   ; (Not usually necessary
E5BC 26FD                BNE    PREX11A         ; because of time taken
E5BE                                            ; for following code.)
E5BE                    ;
E5BE                    ; Read digital byte on Port E into accumulator A.
E5BE                    ;
E5BE A60A                LDAA   PORTE,X         ;Get digital byte from E
E5C0                    ;
E5C0                    ; Get single reading from PE0 and place in accumulator A.
E5C0                    ;
E5C0 6F30                CLR    ADCTL,X         ;SCAN=0, MULT=0, CD-CA = 0000
E5C2 1F3080FC PREX11B    BRCLR  ADCTL,X,$80,PREX11B ;Wait for conversion
E5C6 A631                LDAA   ADR1,X
E5C8                    ;
E5C8                    ; Start the A/D converter taking continuous readings on
E5C8                    ; PE7,6,5,4 so that they can be obtained by the system at any
E5C8                    ; later time.
E5C8 9634                LDAA   $34             ;SCAN=1, MULT=1, CC=1
E5CA A730                STAA   ADCTL,X
E5CC 1F3080FC PREX11C    BRCLR  ADCTL,X,$80,PREX11C ;Wait for first conv.
E5D0                    ;        ...
E5D0                            .END
```

Note that digital data can be read from the port at any time. If analog signals are present on the port, the data retrieved will depend on its voltage value when tested against TTL voltage-level definitions. That is, voltages higher than a maximum LOW logic level and lower than a minimum HIGH logic level may return undefined results. The use of the Port E bits can be mixed between analog and digital usage, since both subsystems are separate yet always available to the programmer.

QUESTIONS AND PROBLEMS

1. Describe the processor registers available to the assembly language programmer.
2. What is the purpose of the mode-select pins on the MC68HC11 microcontroller?
3. What is the typical E clock frequency of an MC68HC11-based system?
4. Describe the pins on an MC68HC11A8 microcontroller.
5. What are some of the differences among MC68HC11 family devices?
6. Describe the memory map of an MC68HC11A8 microcontroller.
7. Where is the internal RAM located in an MC68HC11A8 microcontroller?
8. Where is the internal EEPROM located in an MC68HC11A8 microcontroller?
9. Where is the internal ROM located in an MC68HC11A8 microcontroller?
10. Where are the internal port control, status, and data registers located in an MC68HC11A8 microcontroller?
11. What is the "expanded mode" of MC68HC11A8 microcontroller operation?
12. What is the "single-chip mode" of MC68HC11A8 microcontroller operation?
13. What are the differences between the "single-chip mode" and the "expanded mode" of the MC68HC11A8 microcontroller?
14. Describe the instruction formats of the MC68HC11A8 microcontroller.
15. Describe the addressing modes of the MC68HC11A8 microcontroller.
16. Describe each of the MC68HC11A8 microcontroller instructions that are used to copy data from one point to another (data transfer instructions).
17. Describe each of the MC68HC11A8 microcontroller instructions that perform arithmetic operations (arithmetic instructions).
18. Describe each of the MC68HC11A8 microcontroller instructions that perform jumping and branching operations (program control instructions).
19. Describe each of the MC68HC11A8 microcontroller instructions that perform bit manipulation (bit manipulation instructions).
20. Describe each of the MC68HC11A8 microcontroller instructions that perform shift and rotate operations (shift and rotate instructions).
21. What two functions are performed by Port A of the MC68HC11A8 microcontroller?
22. Describe how Port A of the MC68HC11A8 microcontroller is used for digital I/O.
23. Write a segment of MC68HC11A8 microcontroller assembly language that will read three bits of data on PA2, PA1, and PA0, complement those bits, and write them back to PA5, PA4, and PA3 without damaging any other bits in the Port A data register.
24. How can bits in the Port A data register be changed without affecting other bits in that register?
25. Describe the pulse accumulator function of Port A of the MC68HC11A8 microcontroller.
26. Describe the use of each of the pulse accumulator control and status registers illustrated in Figure 9.13.
27. What is the difference between event counting and gated time accumulation as it relates to the use of a pulse accumulator?
28. Write a segment of MC68HC11A8 microcontroller assembly language that will count events on the PA7 input line for a period of 20 seconds, placing the number of events in accumulator B.

29. Write a segment of MC68HC11A8 microcontroller assembly language that will cause an interrupt after 300 events on the pulse accumulator input.
30. Write a segment of MC68HC11A8 microcontroller assembly language that will wait for 300 events on the pulse accumulator input.
31. How can the count limit of 256 events in the pulse accumulator be increased?
32. What is meant by "input capture" as it relates to the Port A main timer system?
33. What is meant by "output capture" as it relates to the Port A main timer system?
34. Describe the basic structure of the Port A main timer system.
35. Describe the function of the free-running counter in the Port A main timer system.
36. How is the count rate of the free-running counter in the Port A main timer system determined?
37. What is a "real time interrupt"?
38. How are bits in the PACTL register used to determine real-time interrupt rates?
39. How are bits in the TCTL2 register used to determine the input capture edge polarity?
40. Write a segment of MC68HC11A8 microcontroller assembly language that will measure the width of an active-low pulse on PA1 and PA0 using the Port A input capture function.
41. Write a segment of MC68HC11A8 microcontroller assembly language that will generate a real-time interrupt every second.
42. What are the functions of Port B of the MC68HC11 microcontroller?
43. How can the digital output function of Port B be restored if the processor is operating in expanded mode?
44. Write a segment of MC68HC11A8 microcontroller assembly language that will write four bits of data to PB5, PB4, PB3, and PB2 without changing the data on the other port lines.
45. What are the functions of Port C of the MC68HC11 microcontroller?
46. Describe in detail the use of the status and control bits in the Port C status and control registers illustrated in Figure 9.17
47. What is the purpose of a data direction register, and how is it programmed?
48. Write a segment of MC68HC11A8 microcontroller assembly language that will write four bits of data to PC5, PC4, PC3, and PC2 without changing the data on the other port lines.
49. Write a segment of MC68HC11A8 microcontroller assembly language that will read four bits of data from PC7, PC6, PC1, and PC0 and record that data into accumulator A without changing the data in the other accumulator bits.
50. Write a segment of MC68HC11A8 microcontroller assembly language that will read a byte of data from Port C, complement it, and write it back to Port C.
51. What are the available functions on Port D of the MC68HC11 microcontroller?
52. How many bits are available for digital I/O on Port D of the MC68HC11 microcontroller?
53. What is the function of register DDRD?
54. Write a segment of MC68HC11A8 microcontroller assembly language that will read data on PD0, PD3, and PD5, placing the data into accumulator A without changing the other bits in accumulator A.
55. Write a segment of MC68HC11A8 microcontroller assembly language that will write data to PD1, PD2, and PD4 without changing the other data bits on Port D.
56. What is meant by "asynchronous serial" communication?

57. Describe the format of an asynchronous serial data character.
58. What is the purpose of the start bit in an asynchronous serial data character?
59. What is the purpose of the stop bit in an asynchronous serial data character?
60. Define "baud rate."
61. How are serial data baud rates determined by the MC68HC11?
62. Where are the asynchronous serial transmit and receive lines tied to the MC68HC11?
63. How is a data byte sent out the asynchronous serial port?
64. How is a data byte received on the asynchronous serial port?
65. Describe the use of each of the bits in the SCI control register 1 (SCCR1).
66. Describe the use of each of the bits in the SCI control register 2 (SCCR2).
67. Describe the use of each of the bits in the SCI status register (SCSR).
68. Write a segment of MC68HC11A8 microcontroller assembly language that will receive characters on the asynchronous serial port of the MC68HC11, send the byte out Port C, and transmit the character out the asynchronous serial port. Do not use interrupts.
69. Describe the special bootstrap mode of the MC68HC11 microcontroller.
70. Describe, in general, the synchronous serial communication port of the MC68HC11.
71. How do asynchronous and synchronous serial communication differ?
72. Describe the functions of Port E of the 68HC11 microcontroller.
73. Write a segment of MC68HC11A8 microcontroller assembly language that will read four bits on PE7, PE6, PE5, and PE4 and place them in accumulator A without affecting the other bits in accumulator A.
74. Describe the function of the ADPU and CSEL bits in the OPTION register.
75. Describe the function of the A/D result registers (ADR1, ADR2, ADR3, and ADR4) in the A/D subsystem of the MC68HC11
76. Describe the function of each of the bits in the ADCTL register in the A/D subsystem of the MC68HC11.
77. What is the difference in the A/D function when MULT = 1 compared to MULT = 0 in the ADCTL register?
78. Write a segment of MC68HC11A8 microcontroller assembly language that will read a single analog value on PE7 and place it in accumulator B.
79. Write a segment of MC68HC11A8 microcontroller assembly language that will obtain 128 analog readings from PE6 and place their average in accumulator B.
80. Write a segment of MC68HC11A8 microcontroller assembly language that will take a single reading on the four analog ports PE7, PE6, PE5, and PE4 and place the average of those readings in accumulator B.

CHAPTER 10

Address Bus Decoding and Logic Design

OBJECTIVES

After completing this chapter you should be able to:

- Describe the purpose of address bus decoding logic.
- Define the following terms:
    - Memory map
    - Memory-mapped I/O
    - Contiguous memory block
    - General memory map
    - Binary memory map
    - Chip-select field
    - Free-allocation field
    - Free-allocation table
    - Free-allocation logic
    - Chip-select table
- Design the minimum memory decoding logic needed to allocate devices to the memory map of a microprocessor- or microcontroller-based design.
- Write a CUPL program to implement address bus decoding circuitry.

10.1 OVERVIEW

Up to this point in the text, programmable logic devices have been used to implement a variety of digital applications. There are many other common applications of PLDs, and there are many varieties of programmable devices. In this chapter we will review one of the more common and necessary tasks encountered when designing embedded-processor systems: address bus decoding. We have left this application alone up to this point so that an entire chapter could be devoted to it. The intent of the chapter is to present a method for address bus decoding logic design that produces an efficient, flexible, and somewhat minimized result that can be easily implemented using programmable logic devices. When applied, this method can often reduce the address bus decoding circuitry of an embedded-processor system from several chips, including logic gates and MSI (medium-scale integration) decoders, to one or two programmable logic devices. It can also reduce propagation delays and address bus loading, enabling faster system clock speeds.

10.2 ADDRESS BUS DECODING

Often, microcontrollers or microprocessors are embedded in digital electronic process, control, and instrumentation systems. One of the necessary tasks that must be completed when developing such a system is the design of the address bus decoding logic that connects any RAM, ROM, or I/O devices to the microprocessor or microcontroller. Let's review a few basic concepts before continuing.

First, the electronic communication in and out of a microcontroller or microprocessor takes place through wires that can be organized into three groups—the address bus, the data bus, and the control bus. The address bus is used to select an external device or memory location for communication. The data is passed back and forth between the microprocessor and the external devices over the data bus. The signals that control this communication sequence, as well as many other control features, are part of the control bus. Figure 10.1 is a very simplified illustration of this concept.

The purpose of the address bus is to provide unique addresses for every device, or portion of a device, that is connected to the data bus. Since there is only a single data path for data to move into and out of the processor, only one external device may have access to that path at any given time. Consequently, each device is given a unique address. Many devices contain several internal addressable locations that must also be given unique addresses.

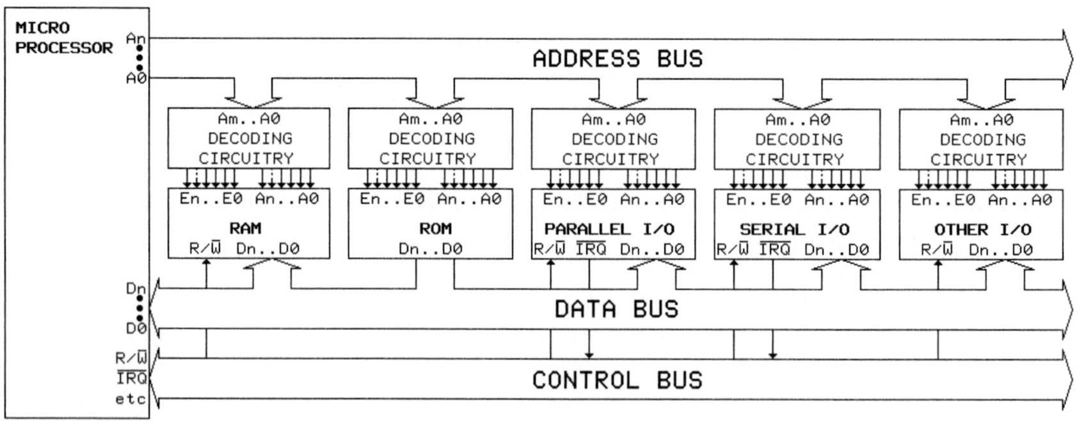

FIGURE 10.1 Microprocessor-based computer system.

When a microprocessor-based system is designed, each external device is assigned a unique address. The comprehensive list of device addresses is referred to as a **memory map.** Once the memory map for the system is defined, it is necessary to design decoding circuitry that enables each respective device when its unique address is placed on the address bus by the processor. The decoding circuitry is combinational in form and represents an excellent application for programmable logic.

Note that Figure 10.1 illustrates the connection of several devices to the processor. The external RAM and ROM provide program and data memory space. The input/output (I/O) devices provide electronic interfaces to parallel, serial, and other types of external devices typically associated with process monitoring and control, as well as typical computer functions. When I/O devices are connected to the address and data buses using this approach, it is referred to as **memory-mapped I/O**. Some advantages of using memory-mapped I/O are its implementation simplicity and the ability to use all of the memory reference instructions of the processor for I/O operations. A disadvantage is that the devices must be mapped to hardwired addresses, placing permanent restrictions on the use of the address space.

10.3 CIRCUIT DESIGN METHODOLOGY

A method for the design of memory-mapped systems is presented here that includes the following steps:

1. Define the general memory map.
2. Create a set of binary memory maps.
3. Create a free-allocation table.
4. Create a set of chip-select tables.
5. Design the logic circuitry defined by the chip-select tables.
6. Implement the logic circuitry with programmable logic.

In order for a device to be memory mapped it must be possible to connect it using the scheme of Figure 10.1. To do this it must have one or more chip-select inputs so that the device can be addressed. Most digital devices can be configured to emulate this characteristic.

In order to demonstrate the memory-allocation method, let's assume the external devices described in Table 10.1 are available for use. Figure 10.2 illustrates the chip outline of each of these devices and includes identifiers for those connections that are made to the processor when using the memory-mapping scheme. The outlines for the I/O devices do not illustrate those connections made with the external environment. Again, only those connections made with the processor are illustrated. The control bus connections include the read/write line, the interrupt request lines, and the clock enable line, E. The read/write lines are tied to the processor read/write line. The interrupt request lines are open-collector outputs that are tied to the processor interrupt request input using a pull-up resistor. The clock enable line is tied to the processor clock in order to provide alternate-state timing to eliminate errors created by glitches generated at the beginning of each state. The data lines are connected to the data bus.

TABLE 10.1 Example devices.

Device	*Architecture*	*Chip-Select Labels*
2K RAM	2048 x 8 bits	CS_0, CS_1, CS_2
4K ROM	4096 x 8 bits	OE_1, $!OE_2$
M6820 PIA (Parallel I/O)	4 x 8 bits	CS_0, $!CS_1$, $!CS_2$
M6850 ACIA (Serial I/O)	2 x 8 bits	CS_0, $!CS_1$, $!CS_2$
Octal Buffer	1 x 8 bits	G_0,$!G_1$

The remaining lines are the address lines and chip-enable lines. These are used to connect the devices to the processor's address bus through decoding circuitry that will establish unique memory addresses for each device. It is these lines, and these lines only, that concern us when we are designing that decoding circuitry.

Now, with these chips in hand, let's consider the design of a microprocessor-based control system that has the following characteristics:

- 8K of RAM allocated at memory address 'h'0000.
- 8K of ROM allocated at memory address 'h'E000.
- Two PIAs allocated at memory address 'h'8000.
- Two ACIAs allocated at memory address 'h'9000.
- One octal buffer located at memory address 'h'C000.

The microprocessor communicates over a 16-bit address bus and an eight-bit data bus.

Before we approach the solution, observe that we are using a microprocessor that has a 16-bit address bus and an eight-bit data bus. Five different memory segments are defined, each containing one or more identical devices. We will refer to a memory segment with contiguous similar devices as a **contiguous memory block.** That is, a contiguous block is a memory segment wherein one or more similar devices are allocated in successive addresses with no unallocated locations between them.

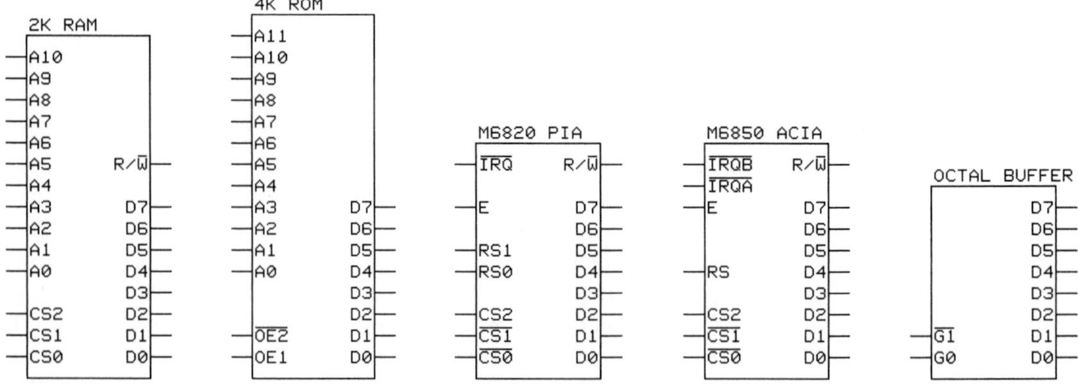

FIGURE 10.2 Example device outlines and processor connections.

10.3.1 General Memory Map

The first step in this method is the development of a **general memory map.** This is a graphical representation of the address space of the processor and includes the device-allocation information that was defined in the original problem.

Consider Figure 10.3, and compare it with the problem. Each of the five contiguous blocks are identified in the memory map, with their address limits defined. The low-order addresses were taken from the original problem. Their high-order addresses were calculated by adding the size of the block to the low-order address and subtracting 1. The 1 must be subtracted since the first location in the block is location 0, not location 1.

This general map provides us with an overall view of the memory environment and is very useful documentation when writing programs for the system and when making hardware changes later on.

10.3.2 Binary Memory Maps

The **binary memory map** is a detailed binary representation of the address line assignments for a single contiguous block of devices. It contains three important pieces of information, as appropriate: (1) those address lines that are decoded by the device, (2) those address lines that will be used to select the individual devices in the block, and (3) those address lines that will be used to address the contiguous block. Its construction is rather straightforward. For each contiguous block:

1. Draw the memory address limits of the block in binary, one above the other.
2. Draw a line partitioning off those address lines decoded by the device.
3. Draw a line partitioning off those address lines used to select each device.

FIGURE 10.3 Example general memory map.

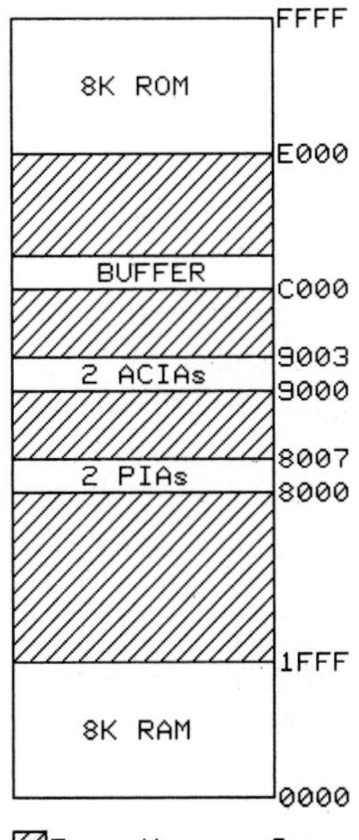

Consider Figure 10.4. The first vertical line (starting from the least significant address bit on the right) is drawn to the left of address line A_{10}. This denotes that the RAM chips will decode address lines A_0 through A_{10}, and will be connected to those respective lines on the address bus. How do we know that the RAM decodes the least significant 11 bits of the address bus? There are two simple ways to determine the number of lines decoded by a device. The simplest method is to observe the device architecture. Figure 10.2 reveals that our RAM chip has 11 address lines, labeled A_0 through A_{10}. Obviously, the field of bits decoded by the RAM chip is already given to us. A second method can be used when the outline drawing is not readily available. Consider the calculation performed in Expression Set 10.1:

$$\text{Number of device address lines} = \log_2(\text{device address range}) \quad \textbf{10.1}$$
$$\text{Number of device address lines} = \log_2(2048)$$
$$\text{Number of device address lines} = 11 \qquad \text{Note: } 2^{11} = 2048$$

The second vertical line drawn on Figure 10.4 marks the field of address bits used to select the individual chips in the contiguous block. The size of the RAM block is 8K (or 8192) words. Since the memory chip we are using contains only 2K (or 2048) words, four chips will be needed. This simple calculation can be expressed as follows:

$$\text{Allocated chips per contiguous block} = \frac{\text{block address range}}{\text{chip address range}} \quad \textbf{10.2}$$

$$\text{Allocated chips per contiguous block} = \frac{8\text{K}}{2\text{K}} = \frac{8192}{2048}$$

$$\text{Allocated chips per contiguous block} = 4$$

With the number of allocated chips known, the number of address lines used to select the chips is easily determined. It takes two address lines to address four devices. This simple equation can be expressed as follows:

$$\text{Number of C/S address lines} = \log_2(\text{allocated chips}) \quad \textbf{10.3}$$
$$\text{Number of C/S address lines} = \log_2(4)$$
$$\text{Number of C/S address lines} = 2 \qquad \text{Note: } 2^2 = 4$$

These two address lines comprise the **chip-select field** (C/S) in the binary memory map of Figure 10.4. Address lines A_{11} and A_{12} were selected to the immediate left of the address lines decoded by the device. This is done so that the individual devices will be allocated in successive memory locations, creating a contiguous block. It is no coincidence that the line marking the limit of the chip-select field of the binary memory map also marks the limit of the address range of the contiguous block. If this is not the case, an error has been made.

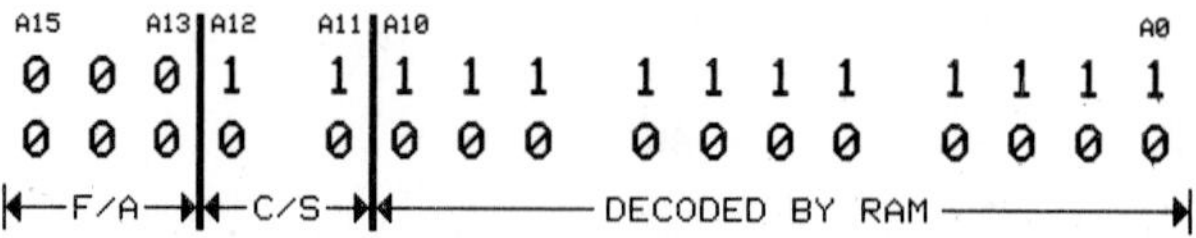

FIGURE 10.4 Binary memory map of the RAM block.

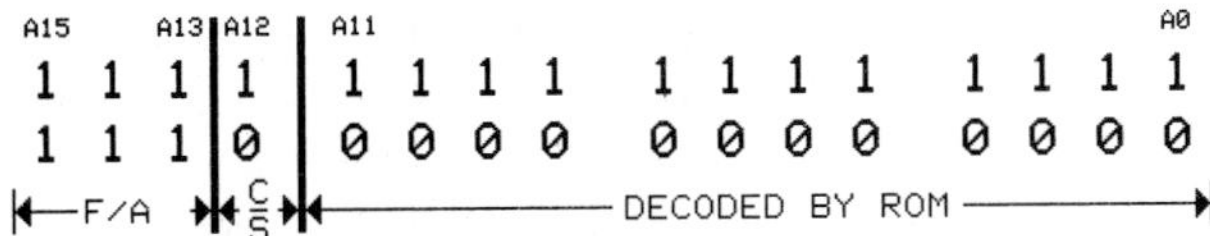

FIGURE 10.5 Binary memory map of the ROM block.

The remaining field in the binary map is referred to as the **free-allocation field (F/A).** This field identifies the address bits that will be used to select the contiguous block. The high-order address and the low-order address will always contain the same bit pattern in this field.

Those address lines defined by the chip-select field will have to be decoded to select the individual chips. A simple solution to this would be the use of a 2-to-4 decoder, addressed by the bits in the chip-select field. The decoder can then be enabled by the address bits in the free-allocation field. This solution would require the use of at least one binary decoder per contiguous block. The method described here will produce a simpler, more cost-effective solution.

As a review, consider the remaining binary memory maps in this problem.

Figure 10.5 illustrates the binary memory map of the ROM block. Each ROM chip decodes the least significant 12 address lines of the address bus, A_0 through A_{11}. Since the contiguous block has a 4K address range and each ROM chip has a 2K address range, two chips will be allocated. One additional address line is needed to select the two chips, so A_{12} is assigned. This leaves address lines A_{13} through A_{15} to select the contiguous block.

Figure 10.6 illustrates the binary memory map of the PIA block. The PIA has two register-select lines that address four logical registers within the device. (There are actually six physical registers, allocated at four contiguous addresses.) This looks to the processor like a RAM chip with four internal memory locations. The two register-select lines RS_0 and RS_1 are tied to address lines A_0 and A_1. This defines the first field in the binary memory map. Since there are two devices allocated, the next address line, A_2, is used to select them. The remaining bits, A_3 through A_{15}, represent the free-allocation field.

Figure 10.7 illustrates the binary memory map of the ACIA block. The ACIA has one register-select line that addresses two logical registers within the device. (There are actually four physical registers. The read/write line is used with the register-select line to access the four registers.) This looks to the processor like a RAM chip with two internal memory locations. The register select line RS is tied to address line A_0. This defines the first field in the binary memory map. Since there are two devices allocated, the next address line, A_1, is used to select them. The remaining bits, A_2 through A_{15}, represent the free-allocation field.

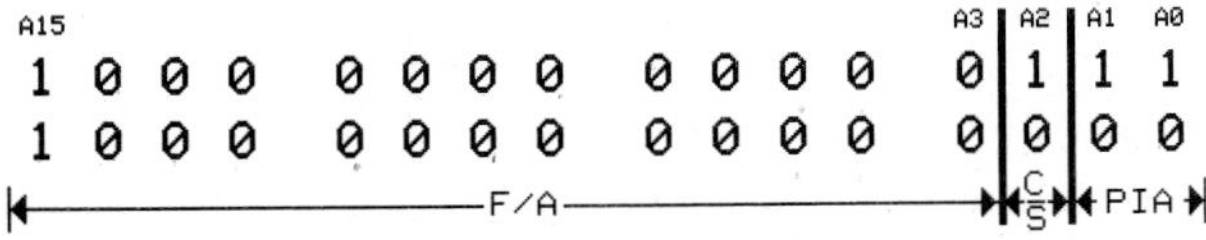

FIGURE 10.6 Binary memory map of the PIA block.

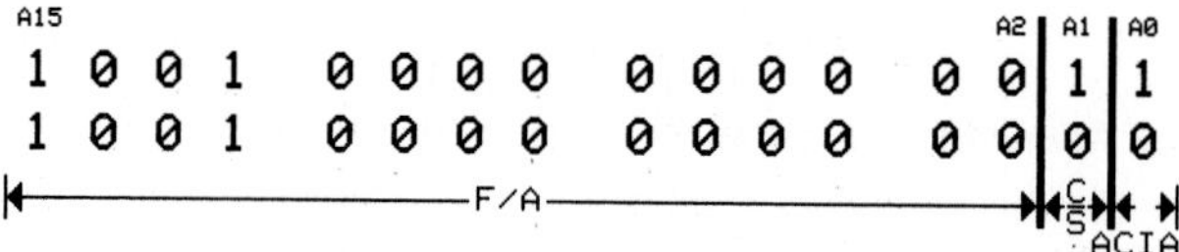

FIGURE 10.7 Binary memory map of the ACIA block.

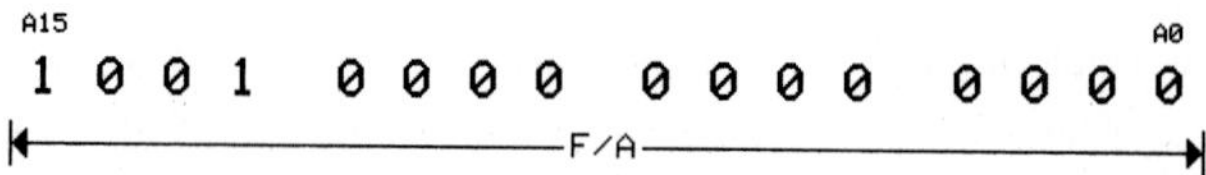

FIGURE 10.8 Binary memory map of the buffer block.

Figure 10.8 illustrates the binary memory map of the buffer block. It differs from the previous binary memory map examples in two ways, because of two respective differences in this block. First, the buffer does not have multiple internal registers or memory locations to select from. Therefore, there is no field of bits in the binary memory map that is used for this purpose. Second, there is only one device in the block. Since this is the case, no address lines need to be used to specify which of the devices in the block are being accessed. Therefore, when there is only one device in the contiguous block, there is no chip-select field in the binary memory map. What we are left with is a single address, assigned to the device, and defined as the free-allocation field. The device will be enabled anytime the required 16-bit address is on the bus.

Figure 10.9 is a summary of the five binary memory maps in this example. Spend a few moments reviewing these maps. The method for their construction should be evident, even if the purpose for constructing them is not yet clear.

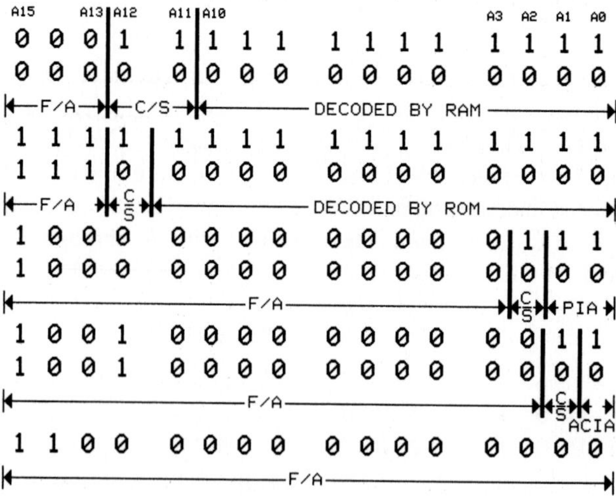

FIGURE 10.9 Summary of binary memory maps.

10.3.3 The Free-Allocation Table

As already stated, the binary field of address lines assigned to a device is decoded by that device. The chip-select field is used to enable the individual devices within the contiguous block, and the free-allocation field is used to enable the respective contiguous block. Observation of Figure 10.9 reveals, for example, that we want the RAM block enabled when the most significant three bits of the address bus are all logical zero. If we extrapolate this concept over all five maps, we come up with the following logic equations:

$$\begin{aligned}
\text{RAM enable} &= \overline{A_{15}} \bullet \overline{A_{14}} \bullet \overline{A_{13}} \\
\text{ROM enable} &= A_{15} \bullet A_{14} \bullet A_{13} \\
\text{PIA enable} &= A_{15} \bullet \overline{A_{14}} \bullet \overline{A_{13}} \bullet \overline{A_{12}} \bullet \overline{A_{11}} \bullet \overline{A_{10}} \bullet \overline{A_9} \bullet \overline{A_8} \bullet \overline{A_7} \bullet \overline{A_6} \bullet \overline{A_5} \bullet \overline{A_4} \bullet \overline{A_3} \\
\text{ACIA enable} &= A_{15} \bullet \overline{A_{14}} \bullet \overline{A_{13}} \bullet A_{12} \bullet \overline{A_{11}} \bullet \overline{A_{10}} \bullet \overline{A_9} \bullet \overline{A_8} \bullet \overline{A_7} \bullet \overline{A_6} \bullet \overline{A_5} \bullet \overline{A_4} \bullet \overline{A_3} \bullet \overline{A_2} \\
\text{Buffer enable} &= A_{15} \bullet A_{14} \bullet \overline{A_{13}} \bullet \overline{A_{12}} \bullet \overline{A_{11}} \bullet \overline{A_{10}} \bullet \overline{A_9} \bullet \overline{A_8} \bullet \overline{A_7} \bullet \overline{A_6} \bullet \overline{A_5} \bullet \overline{A_4} \bullet \overline{A_3} \bullet \overline{A_2} \bullet \overline{A_1} \bullet \overline{A_0}
\end{aligned} \qquad \textbf{10.4}$$

It doesn't take a great deal of inspection to note that the expressions in Expression Set 10.4 imply the use of some AND logic gates with a pretty wide input architecture, up to 16 bits for the buffer enable. Using such gates is impractical when the memory map is "cast in concrete" and no future changes are intended. When this is the case, it is not necessary to decode all of the free-allocation logic bits. Instead, we need only to determine the minimum binary pattern that is unique to each expression and use it to turn on the block.

Consider Figure 10.10. This **free-allocation table** is a summary of the free-allocation fields of the binary maps, much like that of Expression Set 10.4. First, the binary values of the free-allocation fields of each block are recorded on the table. The remaining bits are shown in Figure 10.10 as $^0/_1$, which implies that the bit will be taking on values of both 0 and 1 when the block is accessed. When creating free-allocation tables, a "don't care" symbol may be used instead of $^0/_1$. However, it should be noted that the use of the "don't care" symbol does not change the design method.

The purpose of the table is to determine, by inspection, the minimum logic needed to enable each respective block. Since each block is enabled by its free-allocation field bits, we are looking for a subset of those bits in order to generate some level of minimization. The purpose is to select the smallest subset of the free-allocation bits unique to the block.

DEVICE	A15	A14	A13	A12	A11	A10	A9	A8	A7	A6	A5	A4	A3	A2	A1	A0	MINIMUM LOGIC
RAM	⓪	0	0	$^0/_1$	$^0/_1$	$^0/_1$	$^0/_1$	$^0/_1$	$^0/_1$	$^0/_1$	$^0/_1$	$^0/_1$	$^0/_1$	$^0/_1$	$^0/_1$	$^0/_1$	$\overline{A15}$
ROM	1	1	①	$^0/_1$	$^0/_1$	$^0/_1$	$^0/_1$	$^0/_1$	$^0/_1$	$^0/_1$	$^0/_1$	$^0/_1$	$^0/_1$	$^0/_1$	$^0/_1$	$^0/_1$	A13
PIA	①	⓪	0	⓪	0	0	0	0	0	0	0	0	0	$^0/_1$	$^0/_1$	$^0/_1$	A15 $\overline{A14}$ $\overline{A12}$
ACIA	①	⓪	0	①	0	0	0	0	0	0	0	0	0	0	$^0/_1$	$^0/_1$	A15 $\overline{A14}$ A12
Buffer	1	①	⓪	0	0	0	0	0	0	0	0	0	0	0	0	0	A14 $\overline{A13}$

FIGURE 10.10 Free-allocation table.

For example, consider the RAM line in Figure 10.10. By observation we can see that of the three free-allocation bits, A_{15} is uniquely low. That is, A_{15} is high in all other blocks. Therefore, we can use $\overline{A_{15}}$ to enable the RAM block, instead of $\overline{A_{15}} \bullet \overline{A_{14}} \bullet \overline{A_{13}}$.

Consider the ROM line in Figure 10.10. There is also a single bit in the free-allocation logic field that uniquely identifies this line. Upon inspection we find that bit A_{13} is uniquely TRUE only on the ROM line. Therefore, A_{13} can be used to enable the ROM block.

Consider the PIA block of Figure 10.10 By inspection we find that there are no single-bit or two-bit patterns that are unique. As bits in the free-allocation field are inspected, remember that the $^0/_1$ symbol identifies that the bit takes on both values, 0 and 1. This must be taken into consideration when determining unique patterns. We do find, however, that the two-bit pattern of $A_{15} \bullet \overline{A_{14}}$ is unique to the PIA and ACIA blocks. When the two lines are compared, we find we can add $\overline{A_{12}}$ to the pattern to come up with a minimum three-bit argument of $A_{15} \bullet \overline{A_{14}} \bullet \overline{A_{12}}$.

Consider the ACIA block of Figure 10.10. We have already noted that the pattern of $A_{15} \bullet \overline{A_{14}}$ is unique to the PIA and ACIA blocks. In this case we can add A_{12} to the pattern to come up with a minimum three-bit argument of $A_{15} \bullet \overline{A_{14}} \bullet A_{12}$.

Consider the buffer block of Figure 10.10. We find no single bit pattern that is unique to this line, but by inspection we do find that it is the only line with the two-bit pattern $A_{14} \bullet \overline{A_{13}}$. Therefore, we can use this pattern to enable the block.

These five logical arguments, summarized in the right column of the free-allocation table of Figure 10.10, will be used to enable the contiguous blocks of devices and will be referred to as **free-allocation logic.**

10.3.4 Chip-Select Tables

We are now ready to develop the logical arguments needed to drive the chip-select inputs of the allocated devices in a manner that will enable each device when (and only when) the proper address is on the address bus. The instrument for developing these arguments is the chip-select table. The **chip-select table** is a summary of the connections between the address bus and the chip-select inputs to the devices. Note that, although we use the term "chip-select," this pin or function is often called by other names by different product vendors. We are referring to the input or set of inputs to a logical device that enable its use. Often these will be labeled CS for chip select, RS for register select, E for enable, OE for output enable, CE for chip enable, etc. For consistency in our designs, we will refer to all of these forms as "chip-select" inputs.

A chip-select table will be created for each contiguous block of devices. For example, Figure 10.11 illustrates the chip-select table for the RAM block. The table is designed using the following steps:

1. Make one row for each chip in the contiguous block. The RAM block has four chips, so there are four rows. The first column simply denotes the RAM number from 0 to 3, decimal.
2. The next set of columns is a comprehensive listing of all of the permutations of bit patterns in the chip-select field of the respective binary memory map. Since two address lines, A_{12} and A_{11}, are used to select the four chips, there are four permutations of combinations of logical values of these two bits. Since the number of address lines is determined by the number of devices selected, there should never be a disagreement

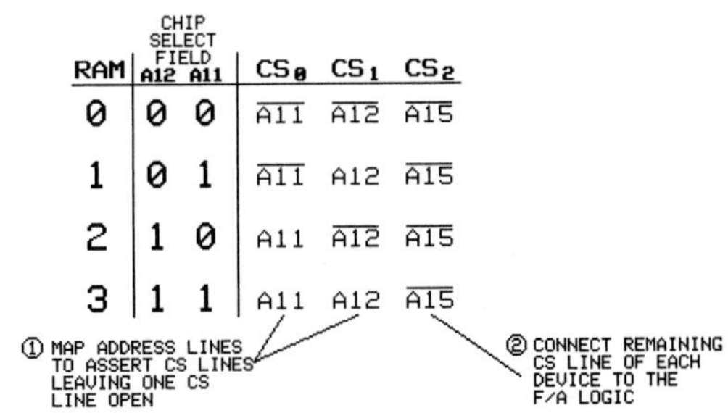

RAM	CHIP SELECT FIELD A12	A11	CS_0	CS_1	CS_2
0	0	0	$\overline{A11}$	$\overline{A12}$	$\overline{A15}$
1	0	1	$\overline{A11}$	A12	$\overline{A15}$
2	1	0	A11	$\overline{A12}$	$\overline{A15}$
3	1	1	A11	A12	$\overline{A15}$

FIGURE 10.11 Chip-select table for the RAM block.

between the number of permutations and the number of devices. If there is, either the number of devices is less than the number of permutations and is not a power of 2 or an error has been made.

3. The last set of columns is where the design method comes in. These columns define where each chip-select input of each chip is assigned. There are several ways of assigning the address lines to chip-select lines.
   (a) When there is more than one chip-select input to the device, connect a single address line to a similar chip-select line of each chip, asserting active, leaving one chip-select line unassigned. Note that this was done in this example by mapping A_{11} to CS_0 and A_{12} to CS_1.
   (b) If all address lines have been mapped over to chip-select inputs, tie the free-allocation logic to the remaining chip-select line. Otherwise, use the free-allocation logic to enable a combinational decoder that decodes the remaining address lines, driving the final chip-select line with the decoder output. This method will work for most combinations of chip-select table architectures.
   (c) Connect as many chip-select lines as possible to single address lines in order to minimize the number of combinational arguments needed.

All four RAMs are connected directly to the address bus with only inverters used. Note that no additional decoding circuitry was necessary with this methodology.

Let's review the method by looking at the chip-select table of the ROM block illustrated in Figure 10.12.

From the free-allocation table, we find that A_{13} can be used to turn the block on. From the binary memory map, we find that A_{12} is used to select one of two chips. Therefore, A_{12} is mapped to OE_1 in its active asserted state. OE_1 was chosen for the A_{12} address line because OE_1 is asserted active high, as is the free-allocation logic. This saves the use of a single inverter. Therefore, A_{13} is tied in its asserted state to each OE_0.

We have now connected the RAM and ROM blocks and have yet to use any additional decoding circuitry. Consider the chip-select table for the PIAs in Figure 10.13.

FIGURE 10.12 Chip-select table of the ROM block.

ROM	A12	OE_0	$\overline{OE_1}$
0	0	A13	A12
1	1	A13	$\overline{A12}$

FIGURE 10.13 Chip-select table for the PIA block.

PIA	A2	CS_0	$\overline{CS_1}$	$\overline{CS_2}$
0	0	$\overline{A2}$	$\overline{A15 \bullet \overline{A14}}$	A12
1	1	A2	$\overline{A15 \bullet \overline{A14}}$	A12

Address line A_2 is taken from the chip-select field in the binary memory map, and each of its two permutations is assigned in its asserted state to CS_0. There are two chip-select lines left with which to implement the free-allocation logic. This time, the free-allocation logic has three variables, $A_{15} \bullet \overline{A_{14}} \bullet \overline{A_{12}}$. Since the combination $A_{15} \bullet \overline{A_{14}}$ is shared by both the PIA and ACIA free-allocation logic arguments, this portion of the logic is inverted and then assigned to $\overline{CS_1}$. The remaining variable, $\overline{A_{12}}$, is inverted to A_{12} and is assigned to $\overline{CS_2}$. Both of these latter terms were inverted because of the active-low assertion levels of the chip-select lines. Recall that it is advantageous to connect as many chip-select lines to a single address line as possible. In this case only a single gate is needed, and this two-input AND logic gate can be shared by both the PIA and the ACIA.

Consider a similar situation with the ACIA chip-select table illustrated in Figure 10.14. Address line A_1 is taken from the chip-select field in the binary memory map, and each of its two permutations is assigned in its asserted state to CS_0. There are two chip-select lines left with which to implement the free-allocation logic. This time, the free-allocation logic has three variables, $A_{15} \bullet \overline{A_{14}} \bullet A_{12}$. Since the combination $A_{15} \bullet \overline{A_{14}}$ is shared by both the PIA and ACIA free-allocation logic arguments, this portion of the logic is inverted and then assigned to $\overline{CS_1}$. The remaining variable, A_{12}, is inverted to $\overline{A_{12}}$ and is assigned to $\overline{CS_2}$. Both of these latter terms were inverted because of the active-low assertion levels of the chip-select lines.

Finally, consider the chip-select table of the buffer block illustrated in Figure 10.15. Since there is only a single device in the buffer block, that device can be enabled directly by the minimum free-allocation logic. It is a convenient coincidence that the assertion levels and the number of chip-select inputs match the free-allocation logic. Consequently, each free-allocation variable is assigned to its similarly asserted chip-select input to the buffer.

10.3.5 Schematic Diagram

The information we have obtained from the chip-select tables enables us to develop the detailed schematic diagram of the solution. Consider Figure 10.16.

FIGURE 10.14 Chip-select table of the ACIA block.

ACIA	A1	CS_0	$\overline{CS_1}$	$\overline{CS_2}$
0	0	$\overline{A1}$	$\overline{A15 \bullet \overline{A14}}$	$\overline{A12}$
1	1	A1	$\overline{A15 \bullet \overline{A14}}$	$\overline{A12}$

FIGURE 10.15 Chip-select table for the buffer block.

BUFFER	G_0	$\overline{G_1}$
0	A14	A13

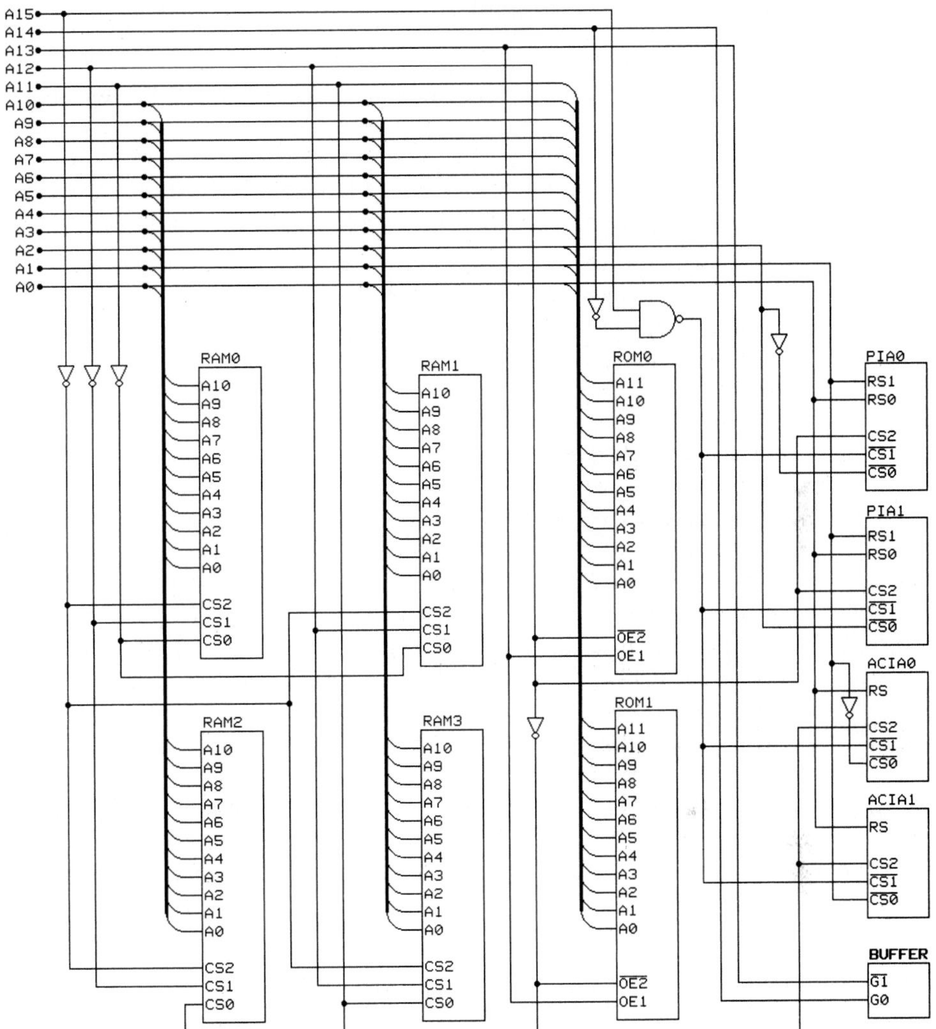

FIGURE 10.16 Address bus connections to memory and I/O devices.

Figure 10.16 is a detailed schematic diagram of the connections between the processor address bus and the devices allocated on it. By observation we see that combinational circuitry was required to generate six logical arguments:

$$\overline{A_{15}} \quad \overline{A_{15} \bullet \overline{A_{14}}} \quad \overline{A_{12}} \quad \overline{A_{11}} \quad \overline{A_{10}} \quad \overline{A_2} \quad \overline{A_1}$$

Implementation of these functions would require seven inverters and a two-input NAND logic gate. A preferred solution might be to use a PAL chip to generate these functions. The advantage to using a PAL for implementation is that some flexibility in future designs is possible.

10.3.6 CUPL Program

Consider the following CUPL program which uses a 16L8 PAL device to implement all of the combinational logic necessary to connect all of the memory and I/O devices to the address bus in this example.

```
/***************************************************************/
/* CUPL program to implement address bus decoding logic for */
/*                                                           */
/*     8K ROM    @ 'h'E000      ROM:    2Kx8,  CS0,CS1,CS2      */
/*      2 PIAs   @ 'h'8000      PIA:     4x8,  CS0,!CS1,!CS2    */
/*      2 ACIAs  @ 'h'9000      ACIA:    2x8,  CS0,!CS1,!CS2    */
/*      1 Buffer @ 'h'C000      Buffer: 1x8,   G0,!G1           */
/*     8K RAM    @ 'h'0000      RAM:    1Kx8,  OE1,!OE2         */
/***************************************************************/
Name         busdcod1;
Revision     01;
Date         04/07/96;
Designer     J. Carter;
Company      University of North Carolina, Charlotte;
Device P16L8;
/***************************************************************/
/* ADDRESS BUS DECODE LOGIC                                    */
/***************************************************************/
/* Inputs from the Address Bus */
Pin  1 =  A1;
Pin  2 =  A2;
Pin  3 = A10;
Pin  4 = A11;
Pin  5 = A12;
Pin  6 = A14;
Pin  7 = A15;
/* Outputs to respective chip-select pins of memory & I/O */
Pin 12 =  !A1;            /* ACIA0:CS0           */
Pin 13 =  !A2;            /* PIA0:CS0            */
Pin 14 = !A10;            /* RAM0:CS1 RAM2:CS1 */
Pin 15 = !A11;            /* RAM0:CS0 RAM1:CS0 */
Pin 16 = !A12;            /* ROM1:OE1            */
Pin 17 = !(A15 & !A14);  /* PIA0:CS1 PIA1:CS1 ACIA0:CS1 ACIA1:CS1 */
Pin 18 = !A15;            /* RAM0:CS0 RAM1:CS0 RAM2:CS0  RAM3:CS0 */
```

Practical Example 10.1: Address Bus Decoding Logic Design

Consider the following memory map description for a serial communications controller:

- 256 bytes of high-speed (15 ns) static RAM at 'h'0000
- 4K of low-speed (60 ns) static RAM at memory address 'h'1000

TABLE 10.2 Example devices.

Device	*Architecture*	*Chip-Select Labels*
2K RAM	2048 x 8 bits	CS_0, CS_1, CS_2
4K ROM	4096 x 8 bits	OE_1, $!OE_2$
M6820 PIA (Parallel I/O)	4 x 8 bits	CS_0, $!CS_1$, $!CS_2$
M6850 ACIA (Serial I/O)	2 x 8 bits	CS_0, $!CS_1$, $!CS_2$
High-Speed RAM	256 x 8 bits	$!CS_0$, $!CS_1$

- 16 ACIAs at memory address 'h'8000
- One PIA at memory address 'h'A000
- 16K ROM at memory address 'h'C000

Complete the design of the address bus decoding logic, including the general memory map, binary memory maps, free-allocation table, chip-select tables, schematic diagram, and CUPL program for implementation. Table 10.2 illustrates the architecture of the devices to be used.

General Memory Map. Figure 10.17 illustrates the general memory map of Practical Example 10.1.

Binary Memory Maps. Figure 10.18 illustrates the set of binary memory maps of Practical Example 10.1.

Free-Allocation Table. The free-allocation fields of the binary memory maps are summarized in the free-allocation table shown in Figure 10.19.

FIGURE 10.17 Practical Example 10.1, general memory map.

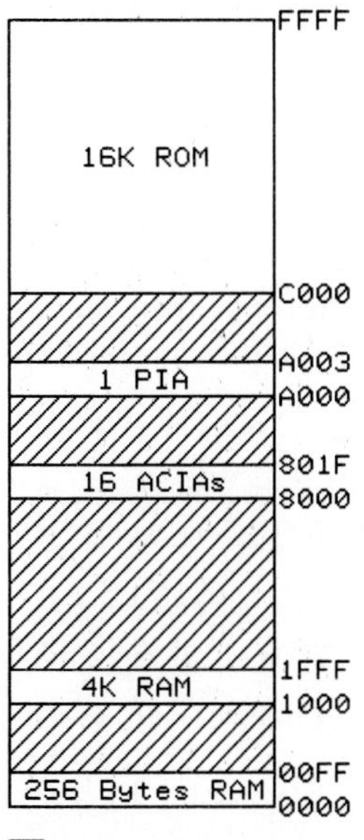

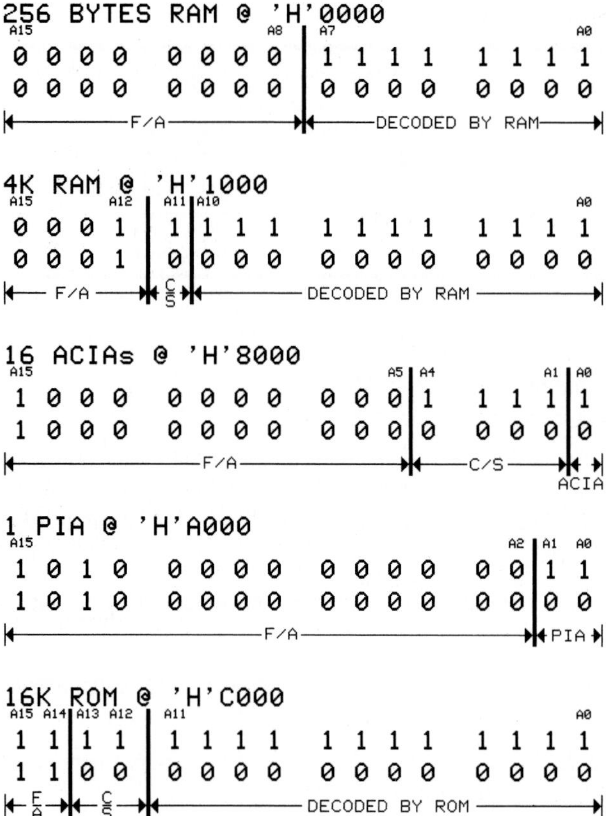

Figure 10.18 Practical Example 10.1, binary memory maps.

Chip-Select Tables. With the free-allocation logic now defined, we are ready to develop the chip-select tables for each block. Figure 10.20 illustrates the chip-select table for the ROM block.

This chip-select table is presented first, since it presents a new problem. This is the first case presented where there are not enough chip-select inputs to the chip to decode all of the address lines in the chip-select field and the free-allocation logic. Actually, this case

DEVICE	A15	A14	A13	A12	A11	A10	A9	A8	A7	A6	A5	A4	A3	A2	A1	A0	MINIMUM LOGIC
RAM0	(0)	0	0	(0)	0	0	0	0	0/1	0/1	0/1	0/1	0/1	0/1	0/1	0/1	$\overline{A15}$ $\overline{A12}$
RAM1	(0)	0	0	(1)	0/1	0/1	0/1	0/1	0/1	0/1	0/1	0/1	0/1	0/1	0/1	0/1	$\overline{A15}$ A12
ACIA	(1)	(0)	(0)	0	0	0	0	0	0	0	0	0/1	0/1	0/1	0/1	0/1	A15 $\overline{A14}$ $\overline{A13}$
PIA	1	(0)	(1)	0	0	0	0	0	0	0	0	0	0	0	0/1	0/1	$\overline{A14}$ A13
ROM	1	(1)	0/1	0/1	0/1	0/1	0/1	0/1	0/1	0/1	0/1	0/1	0/1	0/1	0/1	0/1	A14

FIGURE 10.19 Practical Example 10.1, free-allocation table.

FIGURE 10.20 Practical Example 10.1, ROM chip-select table.

ROM	C/S A13	C/S A12	OE_1	$\overline{OE_2}$
0	0	0	$\overline{A13}$	Q0
1	0	1	$\overline{A13}$	Q1
2	1	0	A13	Q0
3	1	1	A13	Q1

is more common than is implied here. The assignment of the chip-select inputs to the ROMs is rather straightforward:

1. Map address lines from the chip-select column to the chip-select inputs, leaving one chip-select input blank. In this example, A_{13} was mapped to OE_1.
2. Use a decoder, enabled by the free-allocation logic to drive the remaining chip-select inputs. The inputs to the decoder are the remaining address lines to be mapped, in this case the single address line, A_{12}. The outputs of the decoder will be used to drive the respective remaining chip-select input on each device.

Figure 10.21 illustrates a similar situation regarding decoding circuitry of the ACIA block. Note how the same method is used here. Address lines A_4 and A_3 are mapped over to CS_0 and CS_1, leaving CS_2 unassigned. Then, the remaining address lines A_2 and A_1 are decoded into the four individual permutations. Each decoder output is used to enable the respective device. Since the decoder is enabled with the free-allocation logic, the decoder enables the block of chips only when an address within the block is on the address bus.

ACIA	CHIP SELECT A4	A3	A2	A1	CS_0	$\overline{CS_1}$	$\overline{CS_2}$
0	0	0	0	0	$\overline{A4}$	$\overline{A3}$	Q0
1	0	0	0	1	$\overline{A4}$	$\overline{A3}$	Q1
2	0	0	1	0	$\overline{A4}$	$\overline{A3}$	Q2
3	0	0	1	1	$\overline{A4}$	$\overline{A3}$	Q3
4	0	1	0	0	$\overline{A4}$	A3	Q0
5	0	1	0	1	$\overline{A4}$	A3	Q1
6	0	1	1	0	$\overline{A4}$	A3	Q2
7	0	1	1	1	$\overline{A4}$	A3	Q3
8	1	0	0	0	A4	$\overline{A3}$	Q0
9	1	0	0	1	A4	$\overline{A3}$	Q1
10	1	0	1	0	A4	$\overline{A3}$	Q2
11	1	0	1	1	A4	$\overline{A3}$	Q3
12	1	1	0	0	A4	A3	Q0
13	1	1	0	1	A4	A3	Q1
14	1	1	1	0	A4	A3	Q2
15	1	1	1	1	A4	A3	Q3

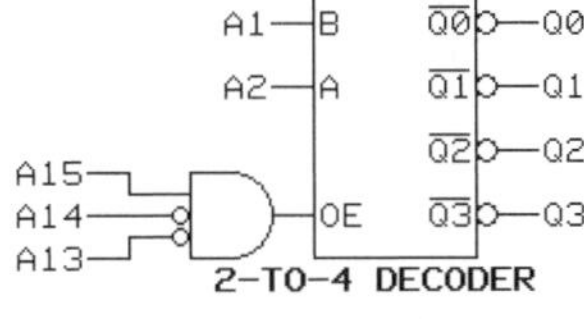

FIGURE 10.21 Practical Example 10.1, ACIA chip-select table.

256B RAM @'H'0000

RAM0	$\overline{CS_0}$	$\overline{CS_1}$
0	A15	A12

4K RAM @'H'1000

RAM1	C/S $\overline{A11}$	CS_0	CS_1	CS_2
0	0	$\overline{A11}$	A12	$\overline{A15}$
1	1	A11	A12	$\overline{A15}$

PIA @ A000

PIA	CS_0	$\overline{CS_1}$	$\overline{CS_2}$
0	A13	A14	0

FIGURE 10.22 Practical Example 10.1, RAM and PIA chip-select tables.

Figure 10.22 illustrates the remaining chip-select tables for Practical Example 10.1.

Schematic Diagram. With the chip-select tables in hand, we are now ready to draw a schematic diagram. Figure 10.23 illustrates that schematic diagram.

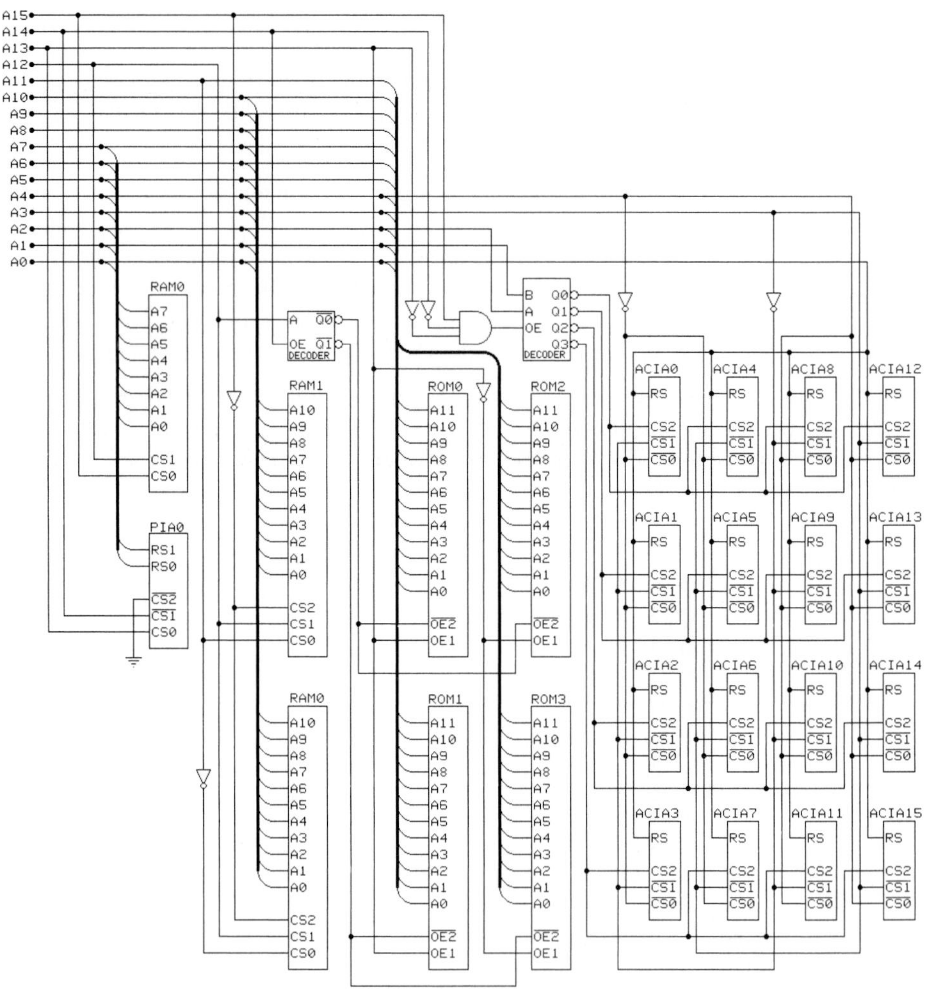

FIGURE 10.23 Practical Example 10.1, schematic diagram.

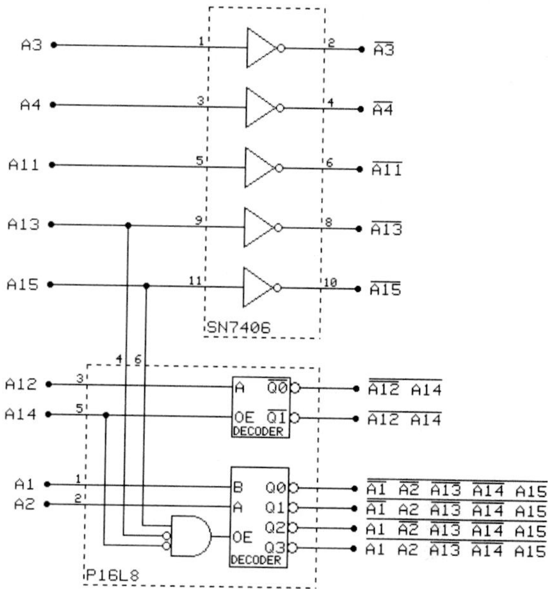

FIGURE 10.24 Practical Example 10.1, decoding circuitry.

CUPL Program. Casual observation of Figure 10.23 reveals that we need to implement two decoders with their functional inputs, and five inverters. Figure 10.24 is a summary of that decoding circuitry.

If this circuit were to be implemented in a single PAL chip, that chip would require 10 inputs and 11 outputs. The most complex PAL defined in this text has only 10 outputs, so more than one chip will have to be used. If an SN7406 inverter chip is used to implement the inverters in Figure 10.24, we can use a 16L8 PAL to implement the decoders and the three-input AND logic. If this is the case, we will be using only six inputs and six outputs of the PAL chip. Consider the following CUPL program solution. The documentation of the SN7406 hex inverter is provided for information only and is entered entirely as a set of comment statements.

```
/*************************************************************/
/* CUPL program to implement address bus decoding logic for */
/*                                                           */
/*   256B RAM   @ 'h'0000     RAM: 256x8, !CS0,!CS1          */
/*     4K RAM   @ 'h'1000     RAM:  2Kx8, OE1,!OE2           */
/*    16 ACIAs @ 'h'8000     ACIA:  2x8, CS0,!CS1,!CS2      */
/*      1 PIA   @ 'h'A000     PIA:   4x8, CS0,!CS1,!CS2      */
/*    16K ROM   @ 'h'C000     ROM:  2Kx8, CS0,CS1,CS2        */
/*************************************************************/
Name          prex-7-1;
Revision      01;
Date          04/07/96;
Designer      J. Carter;
Company       University of North Carolina, Charlotte;
Device P16L8;
```

```
/**********************************************************/
/*  ADDRESS BUS DECODE LOGIC                              */
/**********************************************************/
/* Inputs from the Address Bus */
Pin  1 =  A1;
Pin  2 =  A2;
Pin  3 = A12;
Pin  4 = A13;
Pin  5 = A14;
Pin  6 = A15;
/* Outputs to respective chip-select pins of memory & I/O   */
Pin 12 = !A12&A14;                    /* ROM0:OE2 ROM2:OE2 */
Pin 13 = A12&A14;                     /* ROM1:OE2 ROM3:OE2 */
Pin 14 = !A1&!A2&!A13&!A14&A15; /* CS2 of ACIA0,4,8,12     */
Pin 15 = !A1& A2&!A13&!A14&A15; /* CS2 of ACIA1,5,9,13     */
Pin 16 =  A1&!A2&!A13&!A14&A15; /* CS2 of ACIA2,6,10,14    */
Pin 17 =  A1& A2&!A13&!A14&A15; /* CS2 of ACIA3,7,11,15    */
/* The remaining decoding logic is implemented with an SN7406
   TTL inverter utilizing the following pin assignments:
   (Note that this entire program segment is a comment.)    */
/* Inputs from the Address Bus
Pin  1 =  A3;
Pin  3 =  A4;
Pin  5 = A11;
Pin  9 = A13;
Pin 11 = A15;                                               */
/* Output to respective chip-select pins of memory & I/O
Pin  2 =  !A3;          CS1 of ACIA 0,1,2,3,8,9,10,11
Pin  4 =  !A4;          CS0 of ACIA 0,1,2,3,4,5,6,7
Pin  6 = !A11;          CS0 of RAM0
Pin  8 = !A13;          CS0 of PIA0
Pin 10 = !A15;          CS2 of RAM0 and RAM1                */
```

Practical Exercise 10.1: Address Bus Decoding No. 1

Using the methodology described in this chapter, design the address bus decoding circuitry using the devices described in Table 10.3 for the following system configuration:

- 1024 bytes of high-speed (15 ns) static RAM at 'h'0000
- 2K of low-speed (60 ns) static RAM at memory address 'h'3000
- Four ACIAs at memory address 'h'5000
- Three PIAs at memory address 'h'6000
- 4K ROM at memory address 'h'7000

Include the general memory map, the set of binary memory maps, the free-allocation table, the chip-select tables, a detailed schematic diagram of the decoding circuitry (similar to Figure 10.24), and a CUPL program.

TABLE 10.3 Example devices.

Device	*Architecture*	*Chip-Select Labels*
2K low-speed RAM	2048 x 8 bits	CS_0, CS_1, CS_2
4K low-speed RAM	4096 x 8 bits	$!OE_1$
High-speed RAM	256 x 8 bits	$!CS_0$, $!CS_1$
4K ROM	4096 x 8 bits	OE_1, $!OE_2$
16K ROM	16384 x 8 bits	!OE
M6820 PIA (Parallel I/O)	4 x 8 bits	CS_0, $!CS_1$, $!CS_2$
M6850 ACIA (Serial I/O)	2 x 8 bits	CS_0, $!CS_1$, $!CS_2$
74LS165 Parallel-load eight-bit shift register	1 x 8 bits	!LD
74LS240 Octal buffer	1 x 8 bits	$!G_1$, $!G_2$

Practical Exercise 10.2: Address Bus Decoding No. 2

Using the methodology described in this chapter, design the address bus decoding circuitry using the devices described in Table 10.3 for the following system configuration:

- 8K of low-speed (60 ns) static RAM at memory address 'h'2000
- One ACIA at memory address 'h'1000
- Two PIAs at memory address 'h'1008
- Two buffers at memory address 'h'1010
- 16K ROM at memory address 'h'4000

Include the general memory map, the set of binary memory maps, the free-allocation table, the chip-select tables, a detailed schematic diagram of the decoding circuitry (similar to Figure 10.24), and a CUPL program.

Practical Exercise 10.3: Address Bus Decoding No. 3

Using the methodology described in this chapter, design the address bus decoding circuitry using the devices described in this chapter for the following system configuration:

- 1K of high-speed (15 ns) static RAM at memory address 'h'0400
- Two ACIAs at memory address 'h'0000
- Two PIAs at memory address 'h'0004
- Four buffers at memory address 'h'000C
- 2K ROM at memory address 'h'0800

Include the general memory map, the set of binary memory maps, the free-allocation table, the chip-select tables, a detailed schematic diagram of the decoding circuitry (similar to Figure 10.24), and a CUPL program.

Practical Exercise 10.4: Address Bus Decoding No. 4

Using the methodology described in this chapter, design the address bus decoding circuitry using the devices described in this chapter for the following system configuration:

- 768 bytes of high-speed (15 ns) static RAM at memory address 'h'0400
- Four ACIAs at memory address 'h'1000
- One PIA at memory address 'h'1010
- Three buffers at memory address 'h'1020
- 2K ROM at memory address 'h'2800

Include the general memory map, the set of binary memory maps, the free-allocation table, the chip-select tables, a detailed schematic diagram of the decoding circuitry (similar to Figure 10.24), and a CUPL program.

QUESTIONS AND PROBLEMS

Define each of the terms listed in 1–10.

1. Memory map
2. Memory-mapped I/O
3. Contiguous memory block
4. General memory map
5. Binary memory map
6. Chip-select field
7. Free-allocation field
8. Free-allocation table
9. Free-allocation logic
10. Chip-select table

11. What is the purpose of address bus decoding logic?
12. Is address bus decoding logic combinational, sequential, or a combination of both?
13. What types of logic devices may be used to implement address bus decoding logic?
14. What is the first step in designing address bus decoding logic?
15. List each of the design steps necessary to implement an address bus decoding scheme using programmable logic devices.
16. What is the purpose of the general memory map, and what information does it contain?
17. What is the purpose of a binary memory map, and what information does it contain?
18. How is a binary memory map constructed?
19. What is the purpose of the free-allocation table, and what information does it contain?
20. What is the relationship between a binary memory map and its relevant data on the free-allocation table?
21. How is the free-allocation logic function in a free-allocation table determined?
22. What is the purpose of free-allocation logic?
23. What is the purpose of the chip-select table, and what information does it contain?

24. What information from the free-allocation table must be included in the chip-select table?
25. How is the chip-select table constructed?
26. If the number of address lines in the chip-select field of a contiguous memory block is fewer than the number of chip-select inputs to a device to be allocated, what kind of address bus decoding circuitry will be needed?
27. If the number of address lines in the chip-select field of a contiguous memory block is *greater* than the number of chip-select inputs to a device to be allocated, what kind of address bus decoding circuitry will be needed?

To answer Problems 28–30, refer to Table 10.3, which describes the applicable devices.

28. How many 2K RAMs would be needed to implement a 16K x 8-bit contiguous memory block?
29. How many 16K ROMs would be needed to implement a 64K x 32-bit contiguous memory block?
30. How many 4K RAMs would be allocated at the same address if a 64-bit data word is used?

Refer to the following system configuration in answering Problems 31–38.

- 512 bytes of high-speed (15 ns) static RAM at memory address 'h'0000
- One ACIA at memory address 'h'0200
- One PIA at memory address 'h'0210
- Three 74LS240s at memory address 'h'0300
- 2K ROM at memory address 'h'0800

31. Draw the general memory map of the system.
32. How many address lines are required to implement the above system of devices? What is the total address range of this system?
33. Draw the binary memory maps of each contiguous block of this system.
34. How many high-speed RAMs are included in the system? How are they selected?
35. Draw the free-allocation table of this system.
36. Draw the chip-select tables of this system.
37. Draw a schematic diagram of the address decoding logic of this system.
38. Write a CUPL program that will implement the address bus decoding circuit using a programmable logic device.

Refer to the following system configuration in answering Problems 39–46.

- 16K of 4K RAMs at memory address 'h'0000
- Two ACIAs at memory address 'h'4000
- Two PIAs at memory address 'h'4010
- Four 74LS165s at memory address 'h'4020
- 32K of 16K ROMs at memory address 'h'8000

39. Draw the general memory map of the system.
40. How many address lines are required to implement the above system of devices? What is the total address range of this system?
41. Draw the binary memory maps of each contiguous block of this system.

42. How many RAMs are included in the system? How are they selected?
43. Draw the free-allocation table of this system.
44. Draw the chip-select tables of this system.
45. Draw a schematic diagram of the address decoding logic of this system.
46. Write a CUPL program that will implement the address bus decoding circuit using a programmable logic device.

APPENDIX A

Selected PLD Architectures

A.1 DEVICE LIST

The following is a partial list of some of the programmable logic devices supported by CUPL. Those marked with an asterisk (*) are documented in the following pages.

PAL

G16V8*	G16V8A	G16Z8	G18V10	G20RA10	G20V8*	G20V8A
G22V10*	P10H8*	P10L8*	P10P8	P10P8V	P12L10*	P12H6*
P12L6*	P12P10					
P12P6	P12P6V	P14H4*	P14L4*	P14L8*	P14P4	P14P4V
P14P8	P16C1*	P16H2*	P16H8	P16HD8	P16L2*	P16L6*
P16L8*	P16LD8	P16N8*	P16P4C	P16P6	P16P8	P16P8C
P16P8H	P16P8V	P16R4*				
P16R6*	P16R8*	P16RP4	P16RP6	P16RP6V	P16RP8	P16RP8V
P18L4*	P18P4	P20C1*	P20L10*	P20L2*	P20L8*	P20P2
P20P6	P20P8	P20R4*	P20R6*	P20R8*	P20RS10	P20RS4
P20RS8	P20S10	P22V10*				

FPLA/FPLS (Not in PAL Expert® library)

PLS105*					
PLS153	8 x 32 x 10	FPLA*	PLS167	14 x 48 x 6	FPLS*
PLS168	12 x 48 x 8	FPLS*	PLS405	16 x 64 x 8	FPLS*

PROM

RA10P4	1024 x 4	PROM*	RA10P8	1024 x 8	PROM
RA11P4	2048 x 4	PROM	RA11P8	2048 x 8	PROM
RA12P4	4096 x 4	PROM	RA12P8	4096 x 8	PROM
RA13P8	8192 x 8	PROM	RA5P8	32 x 8	PROM*
RA8P4	256 x 4	PROM*	RA8P8	256 x 8	PROM*
RA9P4	512 x 4	PROM*	RA9P8	512 x 8	PROM*

A.2 DEVICE LIBRARY DOCUMENTATION

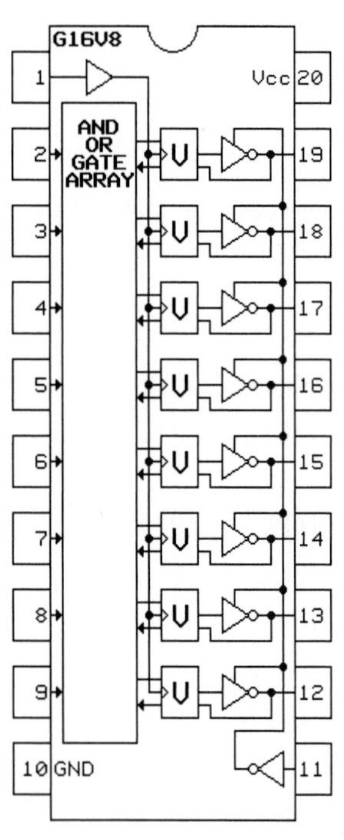

G16V8 Architecture (GAL)

Extensions: OE, D

AMD/MMI	PALCE16V8/4
AMD/MMI	PALCE16V8H/Q
LATTICE	GAL16V8
NATIONAL	GAL16V8
SGS-THOM.	GAL16V8
VLSI	VP16V8E

Input only: 2, 3, 4, 5, 6, 7, 8, 9
Input/Output: 12, 13, 14, 15, 16, 17, 18, 19
Gate array generates seven minterms per output.

This device emulates three different PAL architectures with their flexible output macro configuration. When this device mnemonic is used, the device parameters for the proper sub-mode are automatically selected according to the following:

A. Registered Mode. This mode is automatically chosen when the PLD source file has registered output. In the registered mode, specifying an output enable term for a registered output pin is not flagged as an error by the compiler or simulator. In this mode, the output enable control for registered pins is common to pin 11.

Input only: 2, 3, 4, 5, 6, 7, 8, 9
Input/Output: 12, 13, 14, 15, 16, 17, 18, 19

B. Complex Mode. This mode is automatically chosen when the PLD source file has an output enable term for a nonregistered pin and/or combinatorial feedback.

Input only: 1, 2, 3, 4, 5, 6, 7, 8, 9, 11
Output only: 12, 19
Input/Output: 13, 14, 15, 16, 17, 18

C. Simple Mode (Default). If none of the above are met, the device type defaults to the simple mode. In this mode, the input/output pins are configured as either input only or output only (that is, no feedback can occur).

Input only: 1, 2, 3, 4, 5, 6, 7, 8, 9, 11
Output only: 15, 16
Input/Output: 12, 13, 14, 17, 18, 19

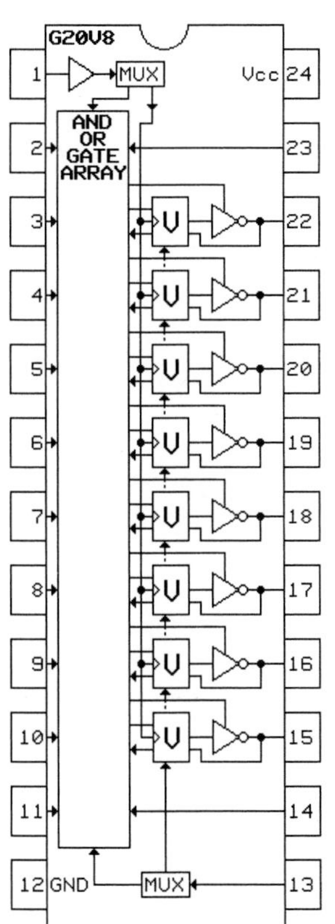

G20V8 Architecture (GAL)

Extensions: OE, D

AMD/MMI	PALCE20V8/4/PALCE20V8H/Q
LATTICE	GAL20V8/RAL20H2/RAL20H8
LATTICE	RAL14H8/RAL16H6/RAL18H4
NATIONAL	GAL20V8
NATIONAL	PAL14H8/20V8/PAL16H6/20V8
NATIONAL	PAL18H4/20V8/PAL20H2/20V8
NATIONAL	PAL20H8/20V8
SGS-THOM.	GAL20V8/RAL14H8/RAL16H6
SGS-THOM.	RAL20H2/RAL20H8/RAL18H4
VLSI	VP20V8E

Input only: 2, 3, 4, 5, 6, 7, 8, 9, 10, 11, 14, 23
Input/Output: 15, 16, 17, 18, 19, 20, 21, 22
Gate array generates seven minterms per output.

This device emulates three different PAL architectures with their flexible output macro configuration. When this device mnemonic is used, the device parameters for the proper sub-mode are automatically selected according to the following:

A. Registered Mode. This mode is automatically chosen when the PLD source file has registered output. In the registered mode, specifying an output enable term for a registered output pin is not flagged as an error by the compiler or simulator. In this mode, the output enable control for registered pins is common to pin 13.

Input only:	2, 3, 4, 5, 6, 7, 8, 9, 10, 11, 14, 23
Input/Output:	15, 16, 17, 18, 19, 20, 21, 22

B. Complex Mode. This mode is automatically chosen when the PLD source file has an output enable term for a nonregistered pin and/or combinatorial feedback.

Input only:	1, 2, 3, 4, 5, 6, 7, 8, 9, 10, 11, 13, 14, 23
Output only:	15, 22
Input/Output:	16, 17, 18, 19, 20, 21

C. Simple Mode (Default). If none of the above are met, the device type defaults to the simple mode. In this mode, the input/output pins are configured as either input only or output only (that is, no feedback can occur).

Input only:	1, 2, 3, 4, 5, 6, 7, 8, 9, 10, 11, 13, 14, 23
Output only:	18, 19
Input/Output:	15, 16, 17, 20, 21, 22

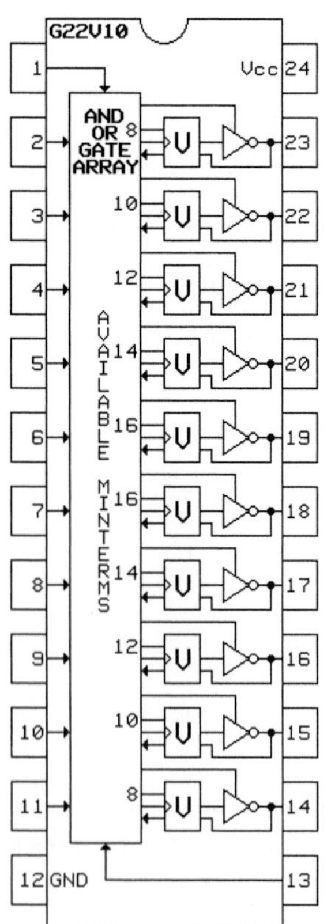

G22V10 Architecture (GAL)

Extensions: OE, D, AR, SP

LATTICE	GAL22V10(UES)
LATTICE	GAL22V10B
NATIONAL	GAL22V10

Input only:	1, 2, 3, 4, 5, 6, 7, 8, 9, 10, 11, 13
Input/Output:	14, 15, 16, 17, 18, 19, 20, 21, 22, 23

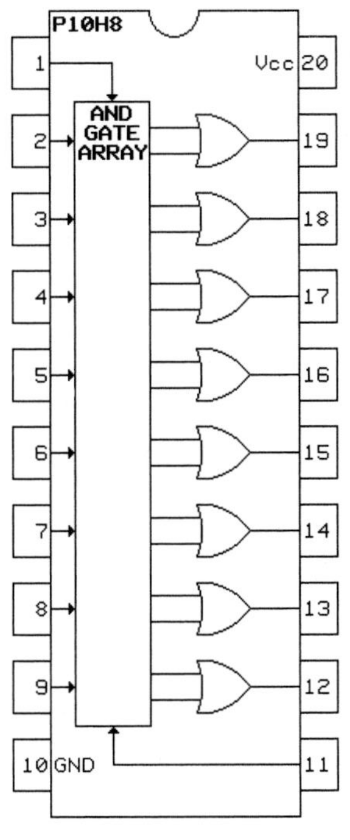

P10H8 Architecture (PAL)

AMD/MMI	PAL10H8/-2
LATTICE	RAL10H8
NATIONAL	PAL10H8/16V8, PAL10H8/A/A2
SGS-THOM.	RAL10H8
Input only:	1, 2, 3, 4, 5, 6, 7, 8, 9, 11
Output only:	12, 13, 14, 15, 16, 17, 18, 19

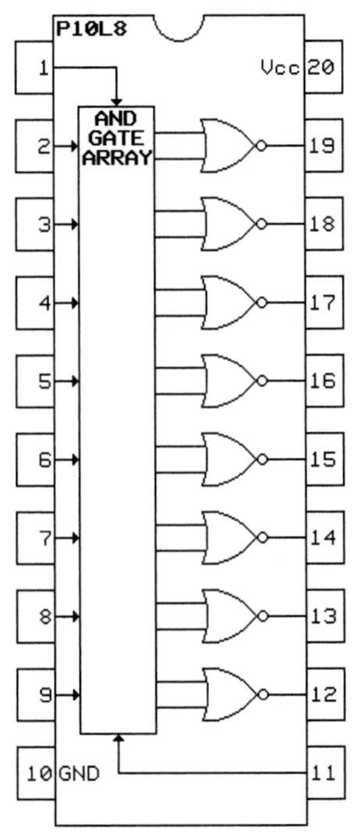

P10L8 Architecture (PAL)

AMD/MMI	PAL10L8/-2
LATTICE	RAL10L8
NATIONAL	PAL10L8/16V8, PAL10L8/A/A2
SGS-THOM.	RAL10L8
Input only:	1, 2, 3, 4, 5, 6, 7, 8, 9, 11
Output only:	12, 13, 14, 15, 16, 17, 18, 19

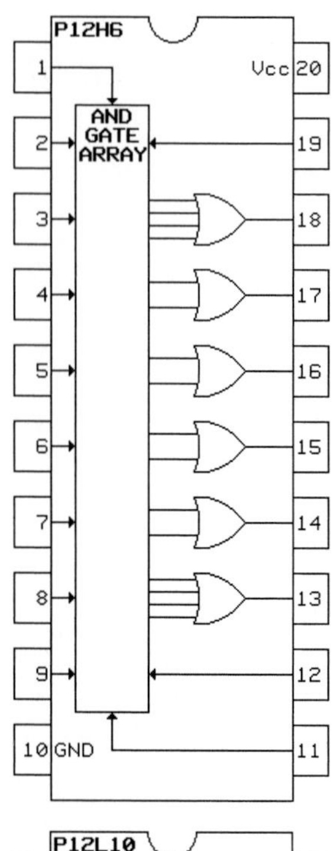

P12H6 Architecture (PAL)

AMD/MMI	PAL12H6/-2
LATTICE	RAL12H6
NATIONAL	PAL12H6/16V8
NATIONAL	PAL12H6/A/A2
SGS-THOM.	RAL12H6

Input only:	1, 2, 3, 4, 5, 6, 7, 8, 9, 11, 12, 19
Output only:	13, 14, 15, 16, 17, 18

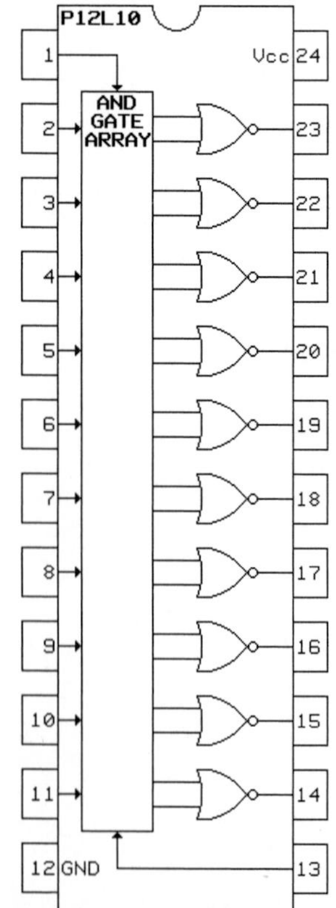

P12L10 Architecture (PAL)

AMD/MMI	PAL12L10
CYPRESS	PALC12L10
NATIONAL	PAL12L10

Input only:	1, 2, 3, 4, 5, 6, 7, 8, 9, 10, 11, 13
Output only:	14, 15, 16, 17, 18, 19, 20, 21, 22, 23

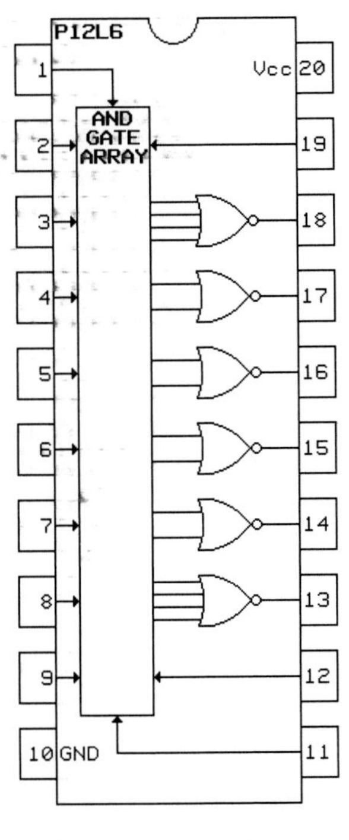

P12L6 Architecture (PAL)

AMD/MMI	PAL12L6/-2
LATTICE	RAL12L6
NATIONAL	PAL12L6/16V8
NATIONAL	PAL12L6/A/A2
SGS-THOM.	RAL12L6

Input only: 1, 2, 3, 4, 5, 6, 7, 8, 9, 11, 12, 19
Output only: 13, 14, 15, 16, 17, 18

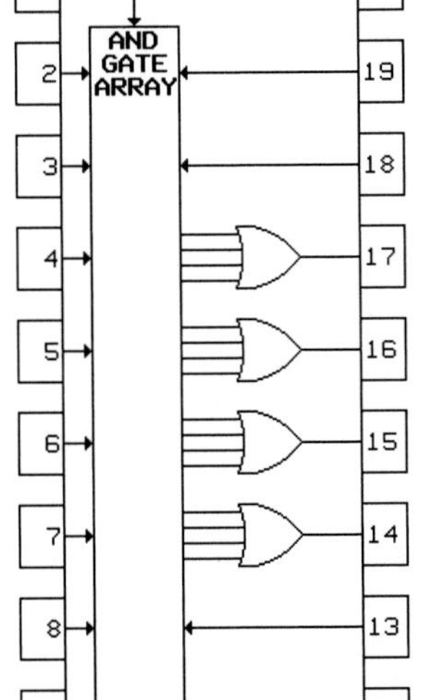

P14H4 Architecture (PAL)

AMD/MMI	PAL14H4/-2
LATTICE	RAL14H4
NATIONAL	PAL14H4/16V8
NATIONAL	PAL14H4/A/A2
SGS-THOM.	RAL14H4

Input only: 1, 2, 3, 4, 5, 6, 7, 8, 9, 11, 12, 13, 18, 19
Output only: 14, 15, 16, 17

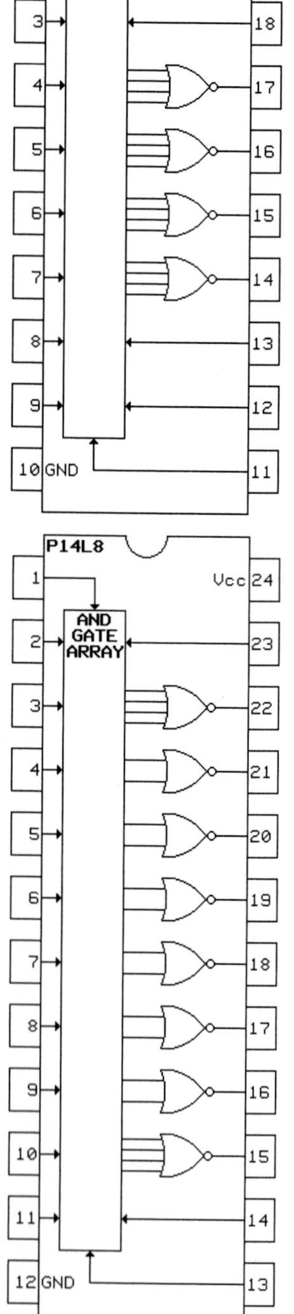

P14L4 Architecture (PAL)

AMD/MMI	PAL14L4/-2
LATTICE	RAL14L4
NATIONAL	PAL14L4/16V8
NATIONAL	PAL14L4/A/A2
SGS-THOM.	RAL14L4

Input only:	1, 2, 3, 4, 5, 6, 7, 8, 9, 11, 12, 13, 18, 19
Output only:	14, 15, 16, 17

P14L8 Architecture (PAL)

AMD/MMI	PAL14L8
CYPRESS	PALC14L8
LATTICE	RAL14L8
NATIONAL	PAL14L8
NATIONAL	PAL14L8/20V8
SGS-THOM.	RAL14L8

Input only:	1, 2, 3, 4, 5, 6, 7, 8, 9, 10, 11, 13, 14, 23
Output only:	15, 16, 17, 18, 19, 20, 21, 22

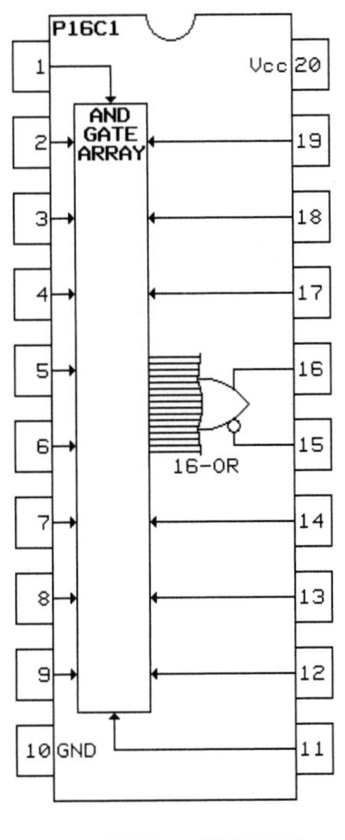

P16C1 Architecture (PAL)

AMD/MMI	PAL16C1/-2
LATTICE	RAL16C1
NATIONAL	PAL16C1/A/A2

Input only:	1, 2, 3, 4, 5, 6, 7, 8, 9, 11, 12, 13, 14, 17, 18, 19
Output only:	15, 16

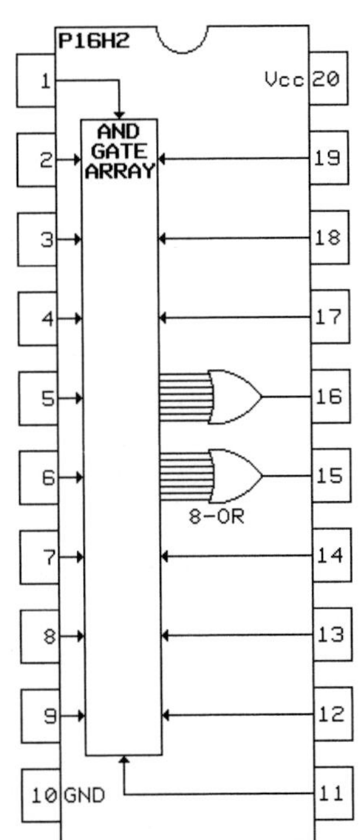

P16H2 Architecture (PAL)

AMD/MMI	PAL16H2/-2
LATTICE	RAL16H2
NATIONAL	PAL16H2/16V8, PAL16H2/A/A2
SGS-THOM.	RAL16H2

Input only:	1, 2, 3, 4, 5, 6, 7, 8, 9, 11, 12, 13, 14, 17, 18, 19
Output only:	15, 16

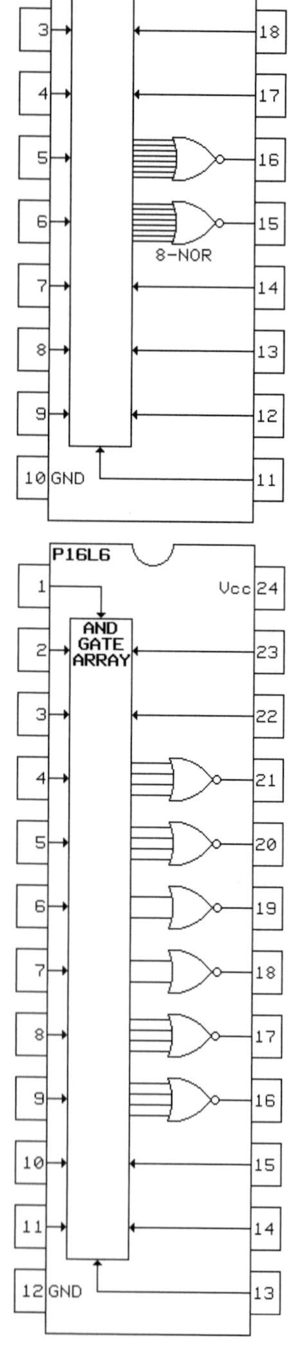

P16L2 Architecture (PAL)

AMD/MMI	PAL16L2/-2
LATTICE	RAL16L2
NATIONAL	PAL16L2/16V8
NATIONAL	PAL16L2/A/A2
SGS-THOM.	RAL16L2

Input only: 1, 2, 3, 4, 5, 6, 7, 8, 9, 11, 12, 13, 14, 17, 18, 19
Output only: 15, 16

P16L6 Architecture (PAL)

AMD/MMI	PAL16L6
CYPRESS	PALC16L6
LATTICE	RAL16L6
NATIONAL	PAL16L6
NATIONAL	PAL16L6/20V8
SGS-THOM.	RAL16L6

Input only: 1, 2, 3, 4, 5, 6, 7, 8, 9, 11, 12, 13, 14, 15, 22, 23
Output only: 16, 17, 18, 19, 20, 21

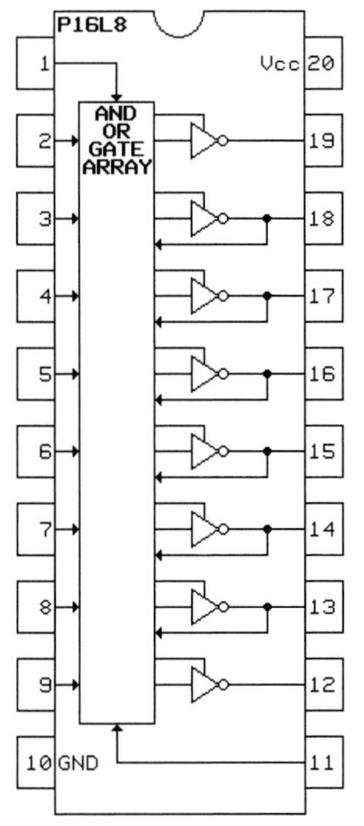

P16L8 Architecture (PAL)

Extensions: OE

AMD/MMI	AMPAL16L8/PAL16L8/A/A-2/A-4/5-7
AMD/MMI	PAL16L8B/B-2/B-4/BP/D/H-10/H15
AMD/MMI	PALC16L8Q,Z
CYPRESS	PAL16L8A/-2/PALC16L8
FAIRCHILD	F16L8
HARRIS	HPL16L8/HPL16LC8/HPL77209
LATTICE	RAL16L8
NATIONAL	PAL16L8/16V8/A/A2/B/B2/PAL16L8D/-7
PHILIPS	PLUS16L8
SAMSUNG	CPL16L8/L
SGS-THOM.	RAL16L8
SIGNETICS	PLHS16L8A/B/D/-7
SPRAGUE	SPL16LC8
TI	PAL16L8A/-2/-10/-12/15/2S/-7/-55

Gate array generates seven minterms per output.

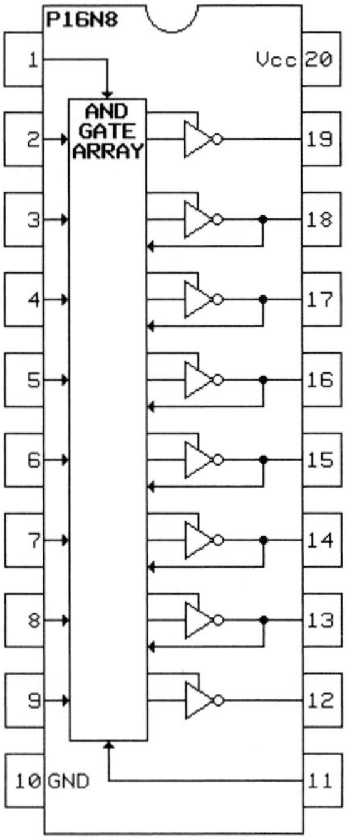

P16N8 Architecture (PAL)

Extensions: OE

PHILIPS	PHD16N8
SIGNETICS	PHD16N8
TI	TIBPAD16N8-7.5

Input only:	1, 2, 3, 4, 5, 6, 7, 8, 9, 11
Output only:	12, 19
Input/Output:	13, 14, 15, 16, 17, 18

One minterm available on each output.

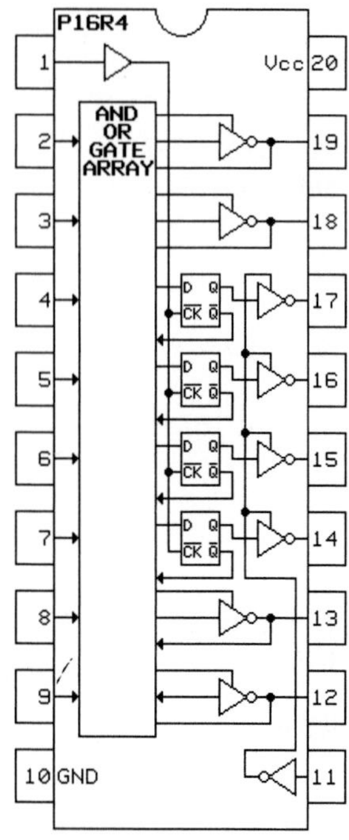

P16R4 Architecture (PAL)

Extensions: OE, D

AMD/MMI	AMPAL16R4
AMD/MMI	PAL16R4/A/A-2/A-4/-5/7/B/B-2/B-4/BP
AMD/MMI	PAL16R4D/H-10/H15/A/A-2/PALC16R4
AMD/MMI	PALC16R4Q/Z
FAIRCHILD	F16R4
HARRIS	HPL16R4
LATTICE	RAL16R4
NATIONAL	PAL16R4/16V8/A/A2/B/B2/D/-7
PHILIPS	PLUS16R4-7
SAMSUNG	CPL16R4/L
SGS-THOM.	RAL16R4
SIGNETICS	PLUS16R4D/-7
SPRAGUE	SPL16RC4
TI	PAL16R4A/-2/-10/-12/15/25/-7/-55

Gate array generates eight minterms per register, seven per logical output.

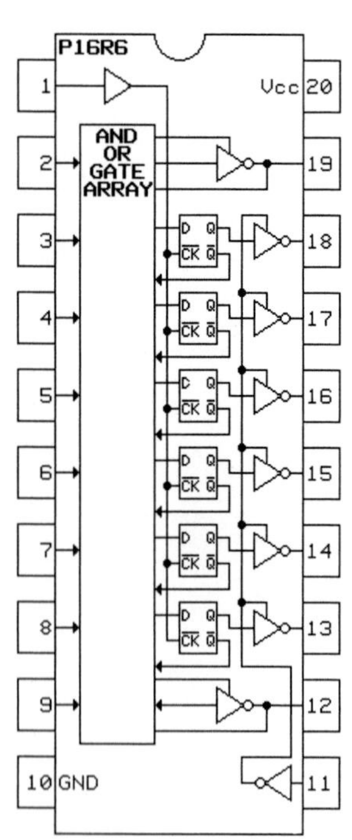

P16R6 Architecture (PAL)

Extensions: OE, D

AMD/MMI	AMPAL16R6
AMD/MMI	PAL16R6/A/A-2/A-4/-5/7/B/B-2/B-4/BP
AMD/MMI	PAL16R6D/H-10/H15/PALC16R6Q/Z
CYPRESS	PAL16R6A/A-2/PALC16R6
HARRIS	HPL16R6
LATTICE	RAL16R6
NATIONAL	PAL16R6/16V8/A/A2/B/B2/D/D/-7
SAMSUNG	CPL16R6/L
SGS-THOM.	RAL16R6
SIGNETICS/PHILIPS	PLUS16R6D/-7
SPRAGUE	SPL16RC6
TI	PAL16R6A/-2/-10/-12/15/25/-7
TI	TICPAL16R6-55

Gate array generates eight minterms per register, seven per logical output.

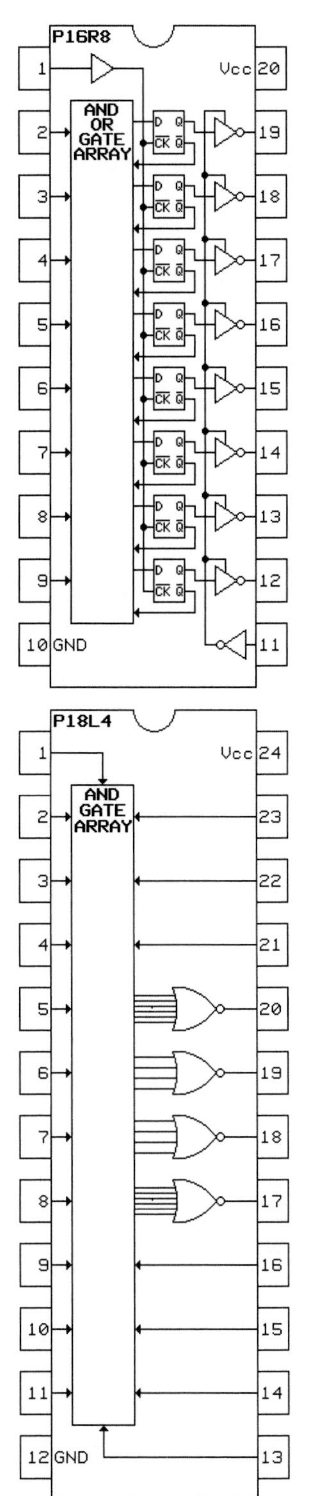

P16R8 Architecture (PAL)

Extensions: D

AMD/MMI	AMPAL16R8
AMD/MMI	PAL16R8/A/A-2/A-4/-5/7/B/B-2/B-4/BP
AMD/MMI	PAL16R8D/H1-/H15/PALC16R8Q/Z
CYPRESS	PAL16R8A/A-2/PALC16R8
FAIRCHILD	F16R8
HARRIS	HPL16R8
LATTICE	RAL16R8
NATIONAL	PAL16R8/16V8/A/A2/B/B2/D/-7
SAMSUNG	CPL16R8/L
SGS-THOM.	RAL16R8
SIGNETICS/PHILIPS	PLUS16R8D/-7
SPRAGUE	PL16RC8
TI	PAL16R8A/-2/-10/-12/15/25/-7
TI	TICPAL16R8-55

Gate array generates eight minterms per output.

P18L4 Architecture (PAL)

AMD/MMI	PAL18L4
CYPRESS	PALC18L4
LATTICE	RAL18L4
NATIONAL	PAL18L4
NATIONAL	PAL18L4/20V8
SGS-THOM.	RAL18L4

Input only: 1, 2, 3, 4, 5, 6, 7, 8, 9, 10, 11, 13, 14, 15, 16, 21, 22, 23
Output only: 17, 18, 19, 20

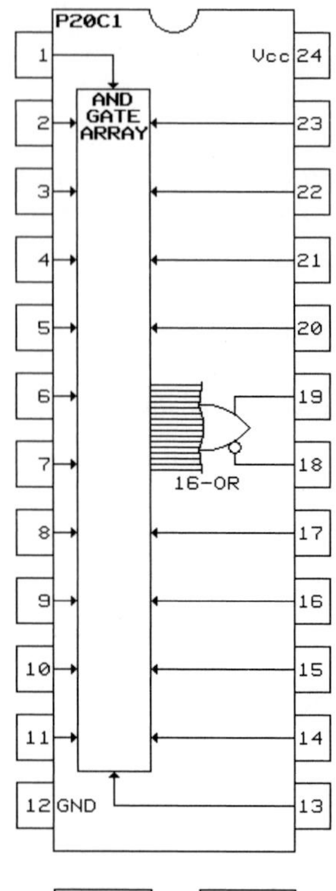

P20C1 Architecture (PAL)

AMD/MMI	PAL20C1
NATIONAL	PAL20C1

Input only:	1, 2, 3, 4, 5, 6, 7, 8, 9, 10, 11, 13, 14, 15, 16, 17, 20, 21, 22, 23
Output only:	18, 19

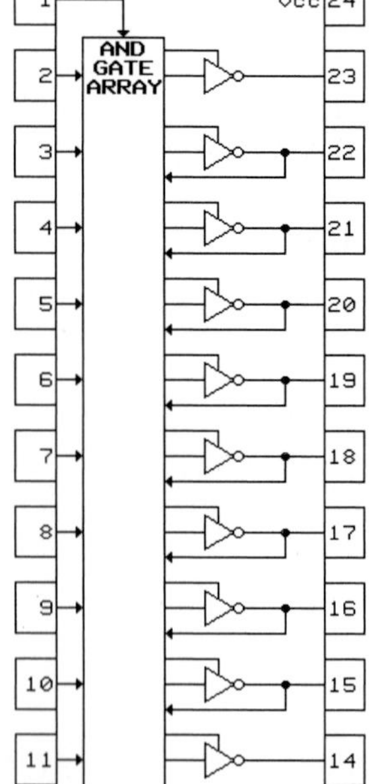

P20L10 Architecture (PAL)

Extensions: OE

AMD/MMI	AMPAL20L10
AMD/MMI	PAL20L10
AMD/MMI	PAL20L10A
CYPRESS	PALC20L10
NATIONAL	PAL20L10/A
SAMSUNG	CPL20L10
SAMSUNG	CPL20L10L
TI	TIBPAL20L10

Input only:	1, 2, 3, 4, 5, 6, 7, 8, 9, 10, 11, 13
Output only:	14, 23
Input/Output:	15, 16, 17, 18, 19, 20, 21, 22

Gate array generates seven minterms per output.

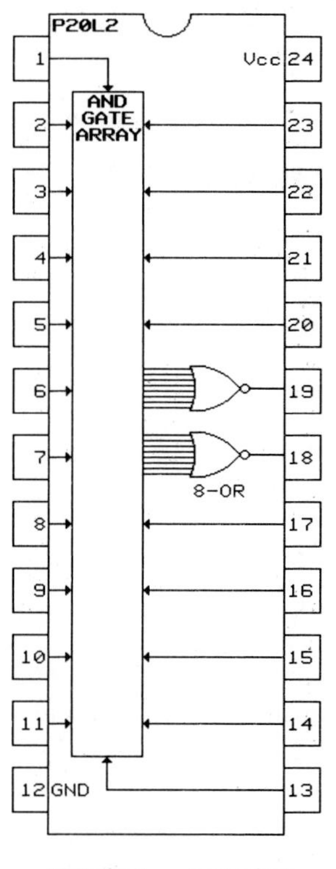

P20L2 Architecture (PAL)

AMD/MMI	PAL20L2
CYPRESS	PALC20L2
LATTICE	RAL20L2
NATIONAL	PAL20L2
NATIONAL	PAL20L2/20V8
SGS-THOM.	RAL20L2

Input only: 1, 2, 3, 4, 5, 6, 7, 8, 9, 10, 11, 13, 14, 15, 16, 17, 20, 21, 22, 23

Output only: 18, 19

P20L8 Architecture (PAL)

Extensions: OE

AMD/MMI	AMPAL20L8
AMD/MMI	PAL20L8/-5/7AMD/MMI/-10/A/A-2/B/B-2
AMD/MMI	PALC20L8
CYPRESS	PALC20L8
LATTICE	RAL20L8
NATIONAL	PAL20L8/20V8
NATIONAL	PAL20L8/A/B/D
SAMSUNG	CPL20L8/L
SGS-THOM.	RAL20L8
SIGNETICS	PLUS20L8D/-7
SPRAGUE	SPL20LC8
TI	PAL20L8A/ TIBPAL20L8-15/25

Input only: 1, 2, 3, 4, 5, 6, 7, 8, 9, 10, 11, 13, 14, 23,

Output only: 15, 22

Input/Output: 16, 17, 18, 19, 20, 21

Gate array generates seven minterms per output.

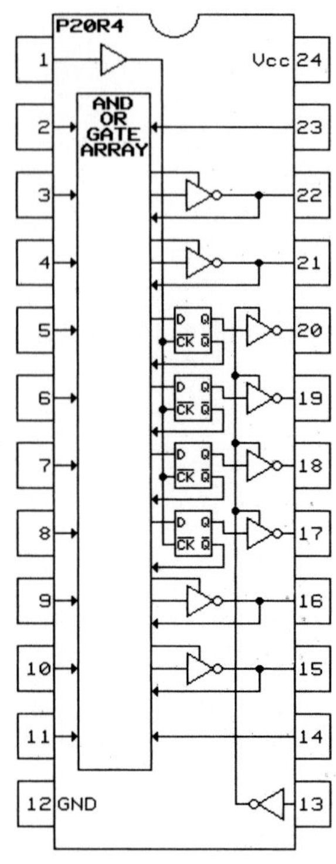

P20R4 Architecture (PAL)

Extensions: OE, D

AMD/MMI	AMPAL20R4
AMD/MMI	PAL20R4/-5/7/-10/A/A-2/B/B-2
AMD/MMI	PALC20R4Z
CYPRESS	PALC20R4
LATTICE	RAL20R4
NATIONAL	PAL20R4/20V8
NATIONAL	PAL20R4/A/B
NATIONAL	PAL20R4D
SAMSUNG	CPL20R4, CPL20R4L
SGS-THOM.	RAL20R4
SIGNETICS	PLUS20R4D/-7
TI	PAL20R4A/TIBPAL20R4-15/25

Input only: 2, 3, 4, 5, 6, 7, 8, 9, 10, 11, 14, 23
Output only: 17, 18, 19, 20
Input/Output: 15, 16, 21, 22
Gate array generates eight minterms per register, seven per logical output.

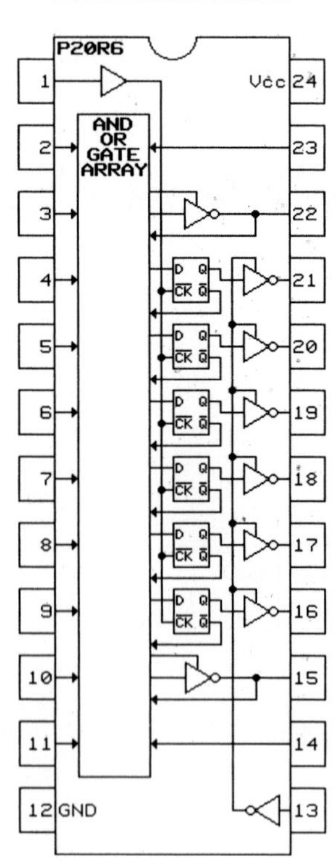

P20R6 Architecture (PAL)

Extensions: OE, D

AMD/MMI	AMPAL20R6
AMD/MMI	PAL20R6/-5/7/-10/A/A-2/B/B-2
AMD/MMI	PALC20R6Z
CYPRESS	PALC20R6
LATTICE	RAL20R6
NATIONAL	PAL20R6/20V8
NATIONAL	PAL20R6/A/B
NATIONAL	PAL20R6D
SAMSUNG	CPL20R6/L
SGS-THOM.	RAL20R6
SIGNETICS	PLUS20R6D/-7
TI	PAL20R6A
TI	TIBPAL20R6-15/25

Input only: 2, 3, 4, 5, 6, 7, 8, 9, 10, 11, 14, 23
Output only: 16, 17, 18, 19, 20, 21
Input/Output: 15, 22
Gate array generates eight minterms per register, seven per logical output.

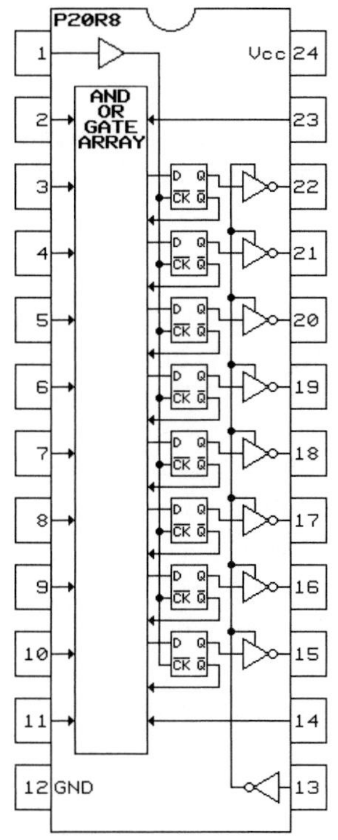

P20R8 Architecture (PAL)

Extensions: D

AMD/MMI	AMPAL20R8
AMD/MMI	PAL20R8/-5/7/-10/A/A-2/B/B-2
AMD/MMI	PALC20R8Z
CYPRESS	PALC20R8
LATTICE	RAL20R8
NATIONAL	PAL20R8/20V8
NATIONAL	PAL20R8/A/B
NATIONAL	PAL20R8D
SAMSUNG	CPL20R8
SAMSUNG	CPL20R8L
SGS-THOM.	RAL20R8
SIGNETICS	PLUS20R8D/-7
TI	PAL20R8A, TIBPAL20R8-15/25

Input only: 2, 3, 4, 5, 6, 7, 8, 9, 10, 11, 14, 23
Output only: 15, 16, 17, 18, 19, 20, 21, 22
Gate array generates eight minterms per expression.

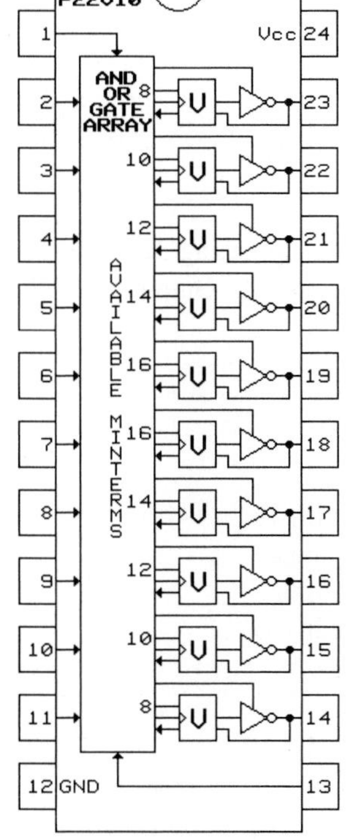

P22V10 Architecture (PAL)

Extensions: OE, D, AR, SP

AMD/MMI	AMPAL22V10, PALC22V10H/Q
AMD/MMI	PALCE22V10/4/H/Q/Z
ATMEL	AT22V10/L
CYPRESS	PAL22V10C, PALC22V10
GAZELLE	GA22V10
GOULD	PEEL22CV10
ICT	PEEL22CV10, A
SIGNETICS	PL22V10
TI	TIBPAL22V10/A, TICPAL22V10
TI	TICPAL22V10Z

Input only: 1, 2, 3, 4, 5, 6, 7, 8, 9, 10, 11, 13
Input/Output: 14, 15, 16, 17, 18, 19, 20, 21, 22, 23

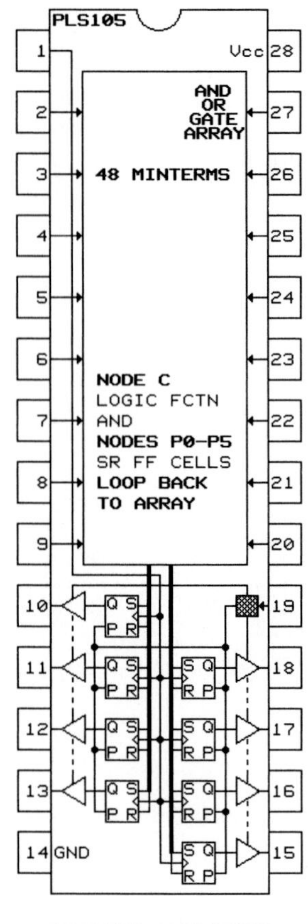

PLS105 Architecture (FPLA)

Extensions: S, R

All generated logic functions are capable of supporting 48 minterms with variables generated from all 16 inputs and seven internally generated nodes.

PHILIPS	PLS105/A
SIGNETICS	PLS105/A

Input only: 2, 3, 4, 5, 6, 7, 8, 9, 20, 21, 22, 23, 24, 25, 26, 27
Output only: 10, 11, 12, 13, 15, 16, 17, 18
Clock: Pin 1
Output Enable: Pin 19 (0 = OE, 1 = Preset of all flip-flops)

Internal nodes:	35	C	Combinational
	29–34	P0–P5	SR flip-flops

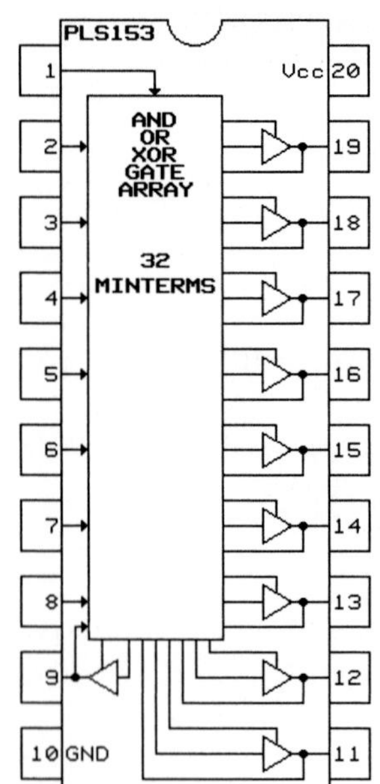

PLS153 Architecture (FPLS)

Extensions: OE

All generated logic functions are capable of supporting 32 minterms with variables generated from all inputs and outputs. Output assertion levels are programmable.

PHILIPS	PLS153/A
SIGNETICS	PLS153/A

Input only: 1, 2, 3, 4, 5, 6, 7, 8
Input/Output: 9, 11, 12, 13, 15, 16, 17, 18, 19

Internal nodes:	X0–X9	Output of AND logic
	S0–S9	Output of XOR logic
	D0–D9	OE of each output
	B0–B9	Outputs

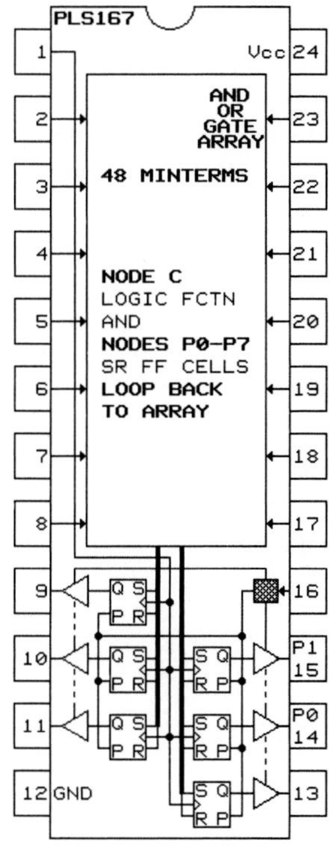

PLS167 Architecture (FPLS)

Extensions: S, R

All generated logic functions are capable of supporting 48 minterms with variables generated from all 14 inputs and nine internally generated nodes.

PHILIPS	PLS167/A
SIGNETICS	PLS167/A

Input only:	2, 3, 4, 5, 6, 7, 8, 17, 18, 19, 20, 21, 22, 23		
Output only:	9, 10, 11, 13		
Input/Output:	14, 15		
Clock: Pin 1			
Output enable: Pin 16 (0 = OE, 1 = Preset of all flip-flops)			
Internal nodes:	31	C	Combinational
	25–30	P2–P7	SR flip-flops

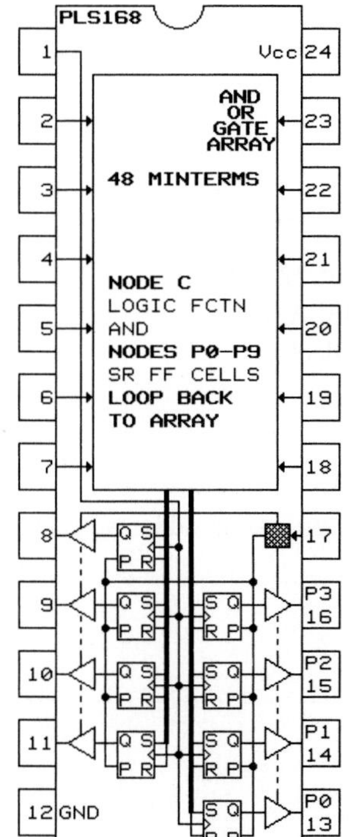

PLS168 Architecture (FPLS)

Extensions: S, R

All generated logic functions are capable of supporting 48 minterms with variables generated from all 12 inputs and 11 internally generated nodes.

PHILIPS	PLS168/A
SIGNETICS	PLS168/A

Input only:	2, 3, 4, 5, 6, 7, 18, 19, 20, 21, 22, 23		
Output only:	8, 9, 10, 11		
Input/Output:	13, 14, 15, 16		
Clock: Pin 1			
Output enable: Pin 16 (0 = OE, 1 = Preset of all flip-flops)			
Internal nodes:	31	C	Combinational
	25–30	P9–P4	SR flip-flops

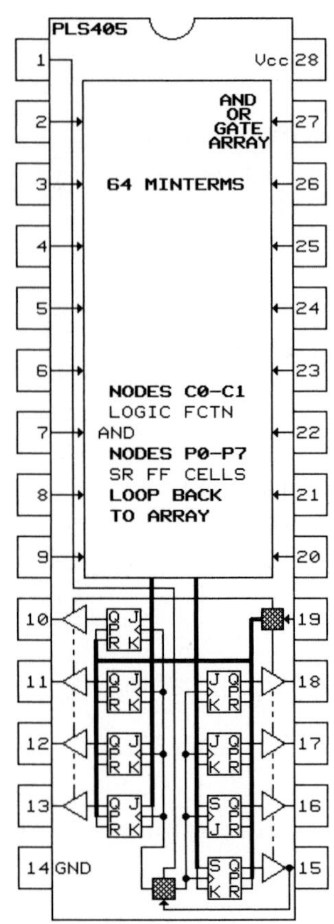

PLS405 Architecture (FPLA)

Extensions: S, R

All generated logic functions are capable of supporting 64 minterms with variables generated from all 16 inputs and 10 internally generated nodes.

PHILIPS	PLS405/A
SIGNETICS	PLS405/A

Input only: 2, 3, 4, 5, 6, 7, 8, 9, 20, 21, 22, 23, 24, 25, 26, 27
Output only: 10, 11, 12, 13, 15, 16, 17, 18
Clock: Pin 1
Output enable: Pin 19 (0 = OE, 1 = Preset of all flip-flops)
Internal nodes: 37–38 C0–C1 Combinational
29–36 P0–P7 SR flip-flops

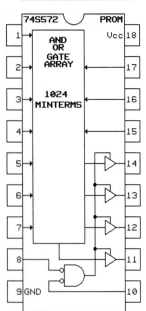

RA10P4 Architecture (PROM)

One-time field programmable 1024 x 4 architecture.

AMD/MMI	53/6352, 53/6353, AM27S32, AM27S33
HARRIS	HM7642, A, B, HM7643, A, B
NATIONAL	DM74S572, A (open collector)*
NATIONAL	DM74S573, A, B (tri-state)*
SIGNETICS	82S137/A/B
TI	TBP24S41, TBP24SA41

*Programmed by raising Vcc to +10.5 V, providing addressing and data information, one data bit at a time.

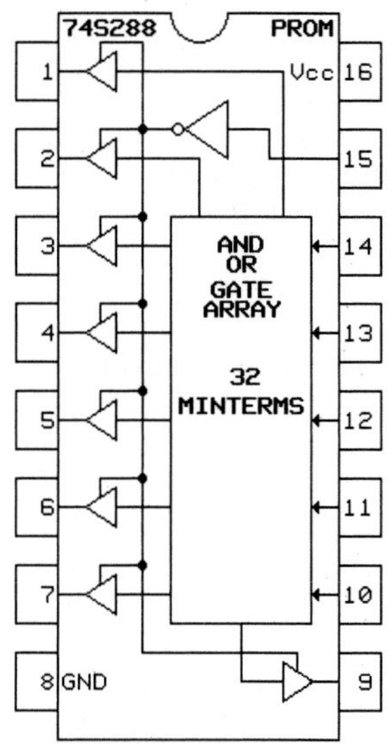

RA5P8 Architecture (PROM)

One-time field programmable 32 x 8 architecture.

AMD/MMI	53/6330, 53/6331, AM27S18, AM27S19
HARRIS	HM7602, HM7603
NATIONAL	DM74S188 (open collector)*
NATIONAL	DM74S288 (tri-state)*
SIGNETICS	82S123/A, 82S23/A
TI	TBP18S030, TBP18SA030

*Programmed by raising Vcc to +10.5 V, providing addressing and data information, one data bit at a time.

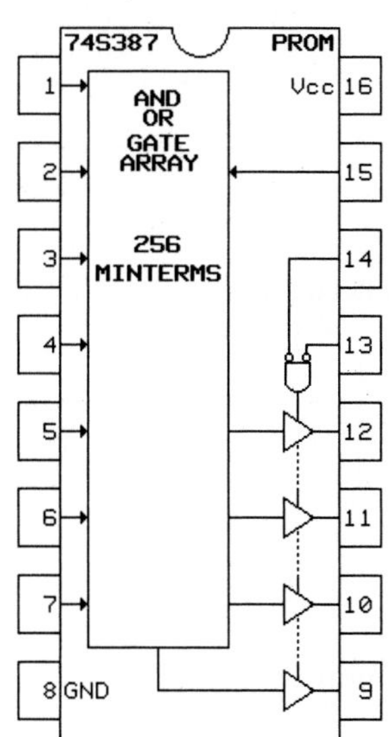

RA8P4 Architecture (PROM)

One-time field programmable 256 x 4 architecture.

AMD/MMI	53/6300, 53/6301, AM27S20, AM27S21
HARRIS	HM7610, A, B, HM7611, A, B
NATIONAL	DM74S287 (open collector)*
NATIONAL	DM74S387 (tri-state)*
SIGNETICS	82S126/A, 82S129/A
TI	TBP24S10, TBP24SA10

*Programmed by raising Vcc to +10.5 V, providing addressing and data information, one data bit at a time.

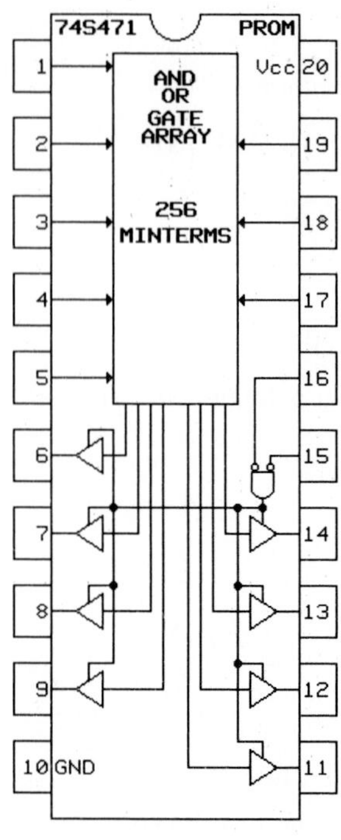

RA8P8 Architecture

One-time field programmable 256 x 8 architecture.

AMD/MMI	53/6308, 53/6309
NATIONAL	DM74LS471*
SIGNETICS	82S135
TI	TBP18S22, TBP18SA22

*Programmed by raising Vcc to +10.5 V, providing addressing and data information, one data bit at a time.

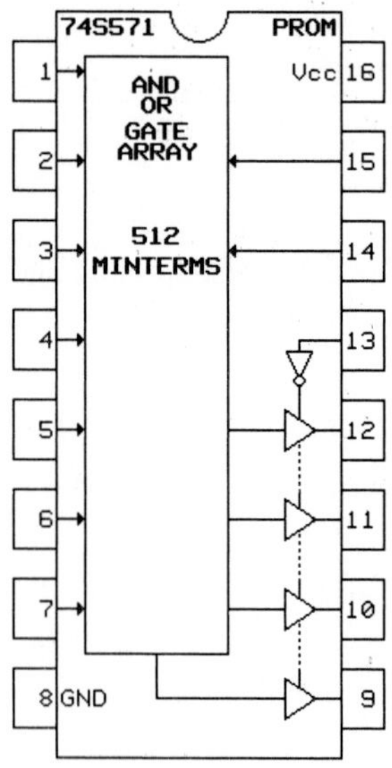

RA9P4 Architecture (PROM)

One-time field programmable 512 x 4 architecture.

AMD/MMI	53/6305, 53/6306
AMD/MMI	AM27S12, AM27S13
HARRIS	HM7620A, B, C
HARRIS	HM7621A, B, C
NATIONAL	DM74S570, A (o.c.)*
NATIONAL	DM74S571, A, B (t.s.)*
SIGNETICS	82S130/A, 82S131/A

*Programmed by raising Vcc to +10.5 V, providing addressing and data information, one data bit at a time.

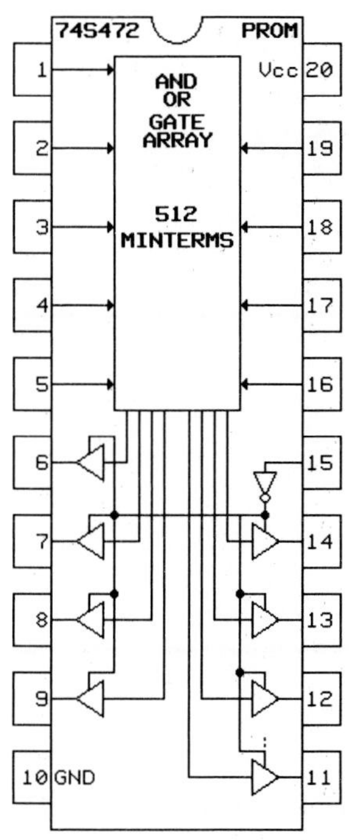

RA9P8 Architecture (PROM)

One-time field programmable

AMD/MMI	53/6348, 53/6349
AMD/MMI	AM27S28, AM27S29
HARRIS	HM7648, HM7649, HM7649A
NATIONAL	DM74S472, A, B*
NATIONAL	DM74S473, A*
SIGNETICS	82S147/A
TI	TBP28S42, TBP28SA42

*Programmed by raising Vcc to +10.5 V, providing addressing and data information, one data bit at a time.

APPENDIX B

CUPL PAL/PROM Device Library

B.1 PAL DEVICES

DEVICE CODE	DEVICE MNEMONIC	# OF PINS	# OF FUSES	# OF PTERMS
AMD/MMI				
AMPAL16H8/A	P16H8	20	2048	64
AMPAL16HD8/A	P16HD8	20	2048	64
AMPAL16L8A/B/L/AL/Q	P16L8	20	2048	64
AMPAL16LD8/A	P16LD8	20	2048	64
AMPAL16R4A/B/L/AL/Q	P16R4	20	2048	64
AMPAL16R6A/B/L/AL/Q	P16R6	20	2048	64
AMPAL16R8A/B/L/AL/Q	P16R8	20	2048	64
AMPAL20L10/B	P20L10	24	1600	40
AMPAL20L8A/B	P20L8	24	2560	64
AMPAL20R4A/B	P20R4	24	2560	64
AMPAL20R6A/B	P20R6	24	2560	64
AMPAL20R8A/B	P20R8	24	2560	64
AMPAL22V10/A	P22V10	24	5828	132
PAL10H8/-2	P10H8	20	320	16
PAL10H8/CE16V8	P10H8	20	320	16
PAL10L8/-2	P10L8	20	320	16
PAL10L8/CE16V8	P10L8	20	320	16
PAL10P8/CE16V8	P10P8	20	328	16
PAL12H6/-2	P12H6	20	384	16

Source: *Universal Compiler for Programmable Logic,* Logical Devices, Inc., Deerfield Beach, FL, 1991.

DEVICE CODE	DEVICE MNEMONIC	# OF PINS	# OF FUSES	# OF PTERMS
AMD/MMI (continued)				
PAL12H6/CE16V8	P12H6	20	384	16
PAL12L10	P12L10	24	480	20
PAL12L6/-2	P12L6	20	384	16
PAL12L6/CE16V8	P12L6	20	384	16
PAL12P6/CE16V8	P12P6	20	390	16
PAL14H4/-2	P14H4	20	448	16
PAL14H4/CE16V8	P14H4	20	448	16
PAL14H8/CE20V8	G20V8	24	560	20
PAL14L4/-2	P14L4	20	448	16
PAL14L4/CE16V8	P14L4	20	448	16
PAL14L8	P14L8	24	560	20
PAL14L8/CE20V8	P14L8	24	560	20
PAL14P4/CE16V8	P14P4	20	452	16
PAL14P8/CE20V8	P14P8	24	568	20
PAL16C1/-2	P16C1	20	512	16
PAL16H2/-2	P16H2	20	512	16
PAL16H2/CE16V8	P16H2	20	512	16
PAL16H6/CE20V8	G20V8	24	560	20
PAL16H8/CE16V8	P16H8	20	2048	64
PAL16L2/-2	P16L2	20	512	16
PAL16L2/CE16V8	P16L2	20	512	16
PAL16L6	P16L6	24	640	20
PAL16L6/CE20V8	P16L6	24	640	20
PAL16L8-5	P16L8	20	2048	64
PAL16L8-7	P16L8	20	2048	64
PAL16L8/A/A-2/A-4	P16L8	20	2048	64
PAL16L8/CE16V8	P16L8	20	2048	64
PAL16L8B	P16L8	20	2048	64
PAL16L8B-2/B-4	P16L8	20	2048	64
PAL16L8BP	P16L8	20	2048	64
PAL16L8D	P16L8	20	2048	64
PAL16L8H-10	P16L8	20	2048	64
PAL16L8H-15	P16L8	20	2048	64
PAL16P2/CE16V8	P16P2	20	514	16
PAL16P6/CE20V8	P16P6	24	646	20
PAL16P8/CE16V8	P16P8	20	2056	64
PAL16P8A	P16P8	20	2056	64
PAL16P8B	P16P8	20	2056	64
PAL16R4-5	P16R4	20	2048	64
PAL16R4-7	P16R4	20	2048	64
PAL16R4/A/A-2/A-4	P16R4	20	2048	64
PAL16R4/CE16V8	P16R4	20	2048	64

DEVICE CODE	DEVICE MNEMONIC	# OF PINS	# OF FUSES	# OF PTERMS
AMD/MMI (continued)				
PAL16R4B	P16R4	20	2048	64
PAL16R4B-2/B-4	P16R4	20	2048	64
PAL16R4BP	P16R4	20	2048	64
PAL16R4D	P16R4	20	2048	64
PAL16R4H-10	P16R4	20	2048	64
PAL16R4H-15	P16R4	20	2048	64
PAL16R6-5	P16R6	20	2048	64
PAL16R6-7	P16R6	20	2048	64
PAL16R6/A/A-2/A-4	P16R6	20	2048	64
PAL16R6/CE16V8	P16R6	20	2048	64
PAL16R6B	P16R6	20	2048	64
PAL16R6B-2/B-4	P16R6	20	2048	64
PAL16R6BP	P16R6	20	2048	64
PAL16R6D	P16R6	20	2048	64
PAL16R6H-10	P16R6	20	2048	64
PAL16R6H-15	P16R6	20	2048	64
PAL16R8-5	P16R8	20	2048	64
PAL16R8-7	P16R8	20	2048	64
PAL16R8/A/A-2/A-4	P16R8	20	2048	64
PAL16R8/CE16V8	P16R8	20	2048	64
PAL16R8B	P16R8	20	2048	64
PAL16R8B-2/B-4	P16R8	20	2048	64
PAL16R8BP	P16R8	20	2048	64
PAL16R8D	P16R8	20	2048	64
PAL16R8H-10	P16R8	20	2048	64
PAL16R8H-15	P16R8	20	2048	64
PAL16RP4/CE16V8	P16RP4	20	2056	64
PAL16RP4A	P16RP4	20	2056	64
PAL16RP6/CE16V8	P16RP6	20	2056	64
PAL16RP6A	P16RP6	20	2056	64
PAL16RP8/CE16V8	P16RP8	20	2056	64
PAL16RP8A	P16RP8	20	2056	64
PAL18H4/CE20V8	G20V8	24	560	20
PAL18L4	P18L4	24	720	20
PAL18L4/CE20V8	P18L4	24	720	20
PAL18P4/CE20V8	P18P4	24	724	20
PAL20C1	P20C1	24	640	16
PAL20H2/CE20V8	G20V8	24	560	20
PAL20H8/CE20V8	G20V8	24	560	20
PAL20L10	P20L10	24	1600	40
PAL20L10A	P20L10	24	1600	40
PAL20L2	P20L2	24	640	16

DEVICE CODE	DEVICE MNEMONIC	# OF PINS	# OF FUSES	# OF PTERMS
AMD/MMI (continued)				
PAL20L2/CE20V8	P20L2	24	640	16
PAL20L8	P20L8	24	2560	64
PAL20L8-10	P20L8	24	2560	64
PAL20L8-5	P20L8	24	2560	64
PAL20L8-7	P20L8	24	2560	64
PAL20L8/CE20V8	P20L8	24	2560	64
PAL20L8A/A-2	P20L8	24	2560	64
PAL20L8B	P20L8	24	2560	64
PAL20L8B-2	P20L8	24	2560	64
PAL20P2/CE20V8	P20P2	24	642	16
PAL20P8/CE20V8	P20P8	24	2568	64
PAL20R4	P20R4	24	2560	64
PAL20R4-10	P20R4	24	2560	64
PAL20R4-5	P20R4	24	2560	64
PAL20R4-7	P20R4	24	2560	64
PAL20R4/CE20V8	P20R4	24	2560	64
PAL20R4A/A-2	P20R4	24	2560	64
PAL20R4B	P20R4	24	2560	64
PAL20R4B-2	P20R4	24	2560	64
PAL20R6	P20R6	24	2560	64
PAL20R6-10	P20R6	24	2560	64
PAL20R6-5	P20R6	24	2560	64
PAL20R6-7	P20R6	24	2560	64
PAL20R6/CE20V8	P20R6	24	2560	64
PAL20R6A/A-2	P20R6	24	2560	64
PAL20R6B	P20R6	24	2560	64
PAL20R6B-2	P20R6	24	2560	64
PAL20R8	P20R8	24	2560	64
PAL20R8-10	P20R8	24	2560	64
PAL20R8-5	P20R8	24	2560	64
PAL20R8-7	P20R8	24	2560	64
PAL20R8/CE20V8	P20R8	24	2560	64
PAL20R8A/A-2	P20R8	24	2560	64
PAL20R8B	P20R8	24	2560	64
PAL20R8B-2	P20R8	24	2560	64
PAL20RS10	P20RS10	24	3338	80
PAL20RS4	P20RS4	24	3330	80
PAL20RS8	P20RS8	24	3338	80
PAL20S10	P20S10	24	3322	80
PAL22V10	P22V10	24	5828	132
PALC16L8Q	P16L8	20	2048	64
PALC16L8Z	P16L8	20	2048	64

DEVICE CODE	DEVICE MNEMONIC	# OF PINS	# OF FUSES	# OF PTERMS
AMD/MMI (continued)				
PALC16R4Q	P16R4	20	2048	64
PALC16R4Z	P16R4	20	2048	64
PALC16R6Q	P16R6	20	2048	64
PALC16R6Z	P16R6	20	2048	64
PALC16R8Q	P16R8	20	2048	64
PALC16R8Z	P16R8	20	2048	64
PALC20L8Z	P20L8	24	2560	64
PALC20R4Z	P20R4	24	2560	64
PALC20R6Z	P20R6	24	2560	64
PALC20R8Z	P20R8	24	2560	64
PALC22V10H	P22V10	24	5828	132
PALC22V10Q	P22V10	24	5828	132
PALCE16V8/4	G16V8	20	2194	64
PALCE16V8H/Q	G16V8	20	2194	64
PALCE20V8/4	G20V8	24	560	20
PALCE20V8H/Q	G20V8	24	560	20
PALCE22V10/4	P22V10	24	5828	132
PALCE22V10H/Q	P22V10	24	5828	132
PALCE22V10Z	P22V10	24	5828	132
ATMEL				
AT22V10/L	P22V10	24	5828	132
CYPRESS				
PAL16L8A	P16L8	20	2048	64
PAL16L8A-2	P16L8	20	2048	64
PAL16R4A	P16R4	20	2048	64
PAL16R4A-2	P16R4	20	2048	64
PAL16R6A	P16R6	20	2048	64
PAL16R6A-2	P16R6	20	2048	64
PAL16R8A	P16R8	20	2048	64
PAL16R8A-2	P16R8	20	2048	64
PAL22V10C	P22V10	24	5828	132
PALC12L10	P12L10	24	480	20
PALC14L8	P14L8	24	560	20
PALC16L6	P16L6	24	640	20
PALC16L8	P16L8	20	2048	64
PALC16R4	P16R4	20	2048	64
PALC16R6	P16R6	20	2048	64
PALC16R8	P16R8	20	2048	64
PALC18L4	P18L4	24	720	20
PALC20L10	P20L10	24	1600	40

DEVICE CODE	DEVICE MNEMONIC	# OF PINS	# OF FUSES	# OF PTERMS
CYPRESS (continued)				
PALC20L2	P20L2	24	640	16
PALC20L8	P20L8	24	2560	64
PALC20R4	P20R4	24	2560	64
PALC20R6	P20R6	24	2560	64
PALC20R8	P20R8	24	2560	64
PALC22V10	P22V10	24	5828	132
FAIRCHILD				
F16L8	P16L8	20	2048	64
F16P8	P16P8	20	2056	64
F16R4	P16R4	20	2048	64
F16R6	P16R6	20	2048	64
F16R8	P16R8	20	2048	64
F16RP4	P16RP4	20	2056	64
F16RP6	P16RP6	20	2056	64
F16RP8	P16RP8	20	2056	64
F20P8	P20P8	24	2568	64
GAZELLE				
GA22V10	P22V10	24	5828	132
GOULD				
PEEL22CV10	P22V10	24	5828	132
HARRIS				
HPL16H8	P16H8	20	2048	64
HPL16L8	P16L8	20	2048	64
HPL16LC8	P16L8	20	2048	64
HPL16P8	P16P8H	20	2056	64
HPL16R4	P16R4	20	2048	64
HPL16R6	P16R6	20	2048	64
HPL16R8	P16R8	20	2048	64
HPL16RC4	P16RP4	20	2056	64
HPL16RC6	P16RP6	20	2056	64
HPL16RC8	P16RP8	20	2056	64
HPL77209	P16L8	20	2048	64
HPL77216	P16P8H	20	2056	64
ICT				
PEEL22CV10	P22V10	24	5828	132
PEEL22CV10A	P22V10	24	5828	132

DEVICE CODE	DEVICE MNEMONIC	# OF PINS	# OF FUSES	# OF PTERMS
LATTICE				
GAL16V8	G16V8	20	2194	64
GAL16V8A	G16V8A	20	2194	64
GAL16V8B	G16V8A	20	2194	64
GAL20V8	G20V8	24	560	20
GAL20V8A	G20V8A	24	2706	64
GAL20V8B	G20V8A	24	2706	64
GAL22V10(UES)	G22V10	24	5892	132
GAL22V10B(UES)	G22V10	24	5892	132
RAL10H8	P10H8	20	320	16
RAL10L8	P10L8	20	320	16
RAL10P8	P10P8	20	328	16
RAL12H6	P12H6	20	384	16
RAL12L6	P12L6	20	384	16
RAL12P6	P12P6	20	390	16
RAL14H4	P14H4	20	448	16
RAL14H8	G20V8	24	560	20
RAL14L4	P14L4	20	448	16
RAL14L8	P14L8	24	560	20
RAL14P4	P14P4	20	452	16
RAL14P8	P14P8	24	568	20
RAL16C1	P16C1	20	512	16
RAL16H2	P16H2	20	512	16
RAL16H6	G20V8	24	560	20
RAL16H8	P16H8	20	2048	64
RAL16L2	P16L2	20	512	16
RAL16L6	P16L6	24	640	20
RAL16L8	P16L8	20	2048	64
RAL16P2	P16P2	20	514	16
RAL16P6	P16P6	24	646	20
RAL16P8	P16P8	20	2056	64
RAL16R4	P16R4	20	2048	64
RAL16R6	P16R6	20	2048	64
RAL16R8	P16R8	20	2048	64
RAL16RP4	P16RP4	20	2056	64
RAL16RP6	P16RP6	20	2056	64
RAL16RP8	P16RP8	20	2056	64
RAL18H4	G20V8	24	560	20
RAL18L4	P18L4	24	720	20
RAL18P4	P18P4	24	724	20
RAL20H2	G20V8	24	560	20
RAL20H8	G20V8	24	560	20
RAL20L2	P20L2	24	640	16

DEVICE CODE	DEVICE MNEMONIC	# OF PINS	# OF FUSES	# OF PTERMS
LATTICE (continued)				
RAL20L8	P20L8	24	2560	64
RAL20P2	P20P2	24	642	16
RAL20P8	P20P8	24	2568	64
RAL20R4	P20R4	24	2560	64
RAL20R6	P20R6	24	2560	64
RAL20R8	P20R8	24	2560	64
NATIONAL				
GAL16V8	G16V8	20	2194	64
GAL16V8-7	G16V8A	20	2194	64
GAL16VA/QS	G16V8A	20	2194	64
GAL20V8	G20V8	24	560	20
GAL20V8-7	G20V8A	24	2706	64
GAL20V8A/QS	G20V8A	24	2706	64
GAL22V10	G22V10	24	5892	132
PAL10H8/16V8	P10H8	20	320	16
PAL10H8/A/A2	P10H8	20	320	16
PAL10L8/16V8	P10L8	20	320	16
PAL10L8/A/A2	P10L8	20	320	16
PAL10P8/16V8	P10P8	20	328	16
PAL12H6/16V8	P12H6	20	384	16
PAL12H6/A/A2	P12H6	20	384	16
PAL12L10/A	P12L10	24	480	20
PAL12L6/16V8	P12L6	20	384	16
PAL12L6/A/A2	P12L6	20	384	16
PAL12P6/16V8	P12P6	20	390	16
PAL14H4/16V8	P14H4	20	448	16
PAL14H4/A/A2	P14H4	20	448	16
PAL14H8/20V8	G20V8	24	560	20
PAL14L4/16V8	P14L4	20	448	16
PAL14L4/A/A2	P14L4	20	448	16
PAL14L8/20V8	P14L8	24	560	20
PAL14L8/A	P14L8	24	560	20
PAL14P4/16V8	P14P4	20	452	16
PAL14P8/20V8	P14P8	24	568	20
PAL16C1/A/A2	P16C1	20	512	16
PAL16H2/16V8	P16H2	20	512	16
PAL16H2/A/A2	P16H2	20	512	16
PAL16H6/20V8	G20V8	24	560	20
PAL16H8/16V8	P16H8	20	2048	64
PAL16L2/16V8	P16L2	20	512	16
PAL16L2/A/A2	P16L2	20	512	16

DEVICE CODE	DEVICE MNEMONIC	# OF PINS	# OF FUSES	# OF PTERMS
NATIONAL (continued)				
PAL16L6/20V8	P16L6	24	640	20
PAL16L6/A	P16L6	24	640	20
PAL16L8/16V8	P16L8	20	2048	64
PAL16L8/A/A2/B/B2	P16L8	20	2048	64
PAL16L8D/-7	P16L8	20	2048	64
PAL16P2/16V8	P16P2	20	514	16
PAL16P6/20V8	P16P6	24	646	20
PAL16P8	P16P8	20	2056	64
PAL16P8/16V8	P16P8	20	2056	64
PAL16R4/16V8	P16R4	20	2048	64
PAL16R4/A/A2/B/B2	P16R4	20	2048	64
PAL16R4D/-7	P16R4	20	2048	64
PAL16R6/16V8	P16R6	20	2048	64
PAL16R6/A/A2/B/B2	P16R6	20	2048	64
PAL16R6D/-7	P16R6	20	2048	64
PAL16R8/16V8	P16R8	20	2048	64
PAL16R8/A/A2/B/B2	P16R8	20	2048	64
PAL16R8D/-7	P16R8	20	2048	64
PAL16RP4	P16RP4	20	2056	64
PAL16RP4/16V8	P16RP4	20	2056	64
PAL16RP6	P16RP6	20	2056	64
PAL16RP6/16V8	P16RP6	20	2056	64
PAL16RP8	P16RP8	20	2056	64
PAL16RP8/16V8	P16RP8	20	2056	64
PAL18H4/20V8	G20V8	24	560	20
PAL18L4/20V8	P18L4	24	720	20
PAL18L4/A	P18L4	24	720	20
PAL18P4/20V8	P18P4	24	724	20
PAL20C1/A	P20C1	24	640	16
PAL20H2/20V8	G20V8	24	560	20
PAL20H8/20V8	G20V8	24	560	20
PAL20L10/A	P20L10	24	1600	40
PAL20L2/20V8	P20L2	24	640	16
PAL20L2/A	P20L2	24	640	16
PAL20L8/20V8	P20L8	24	2560	64
PAL20L8/A/B	P20L8	24	2560	64
PAL20L8D	P20L8	24	2560	64
PAL20P2/20V8	P20P2	24	642	16
PAL20P8/20V8	P20P8	24	2568	64
PAL20R4/20V8	P20R4	24	2560	64
PAL20R4/A/B	P20R4	24	2560	64
PAL20R4D	P20R4	24	2560	64

DEVICE CODE	DEVICE MNEMONIC	# OF PINS	# OF FUSES	# OF PTERMS
NATIONAL (continued)				
PAL20R6/20V8	P20R6	24	2560	64
PAL20R6/A/B	P20R6	24	2560	64
PAL20R6D	P20R6	24	2560	64
PAL20R8/20V8	P20R8	24	2560	64
PAL20R8/A/B	P20R8	24	2560	64
PAL20R8D	P20R8	24	2560	64
RICOH				
EPL10P8A	P10P8V	20	664	32
EPL10P8B	P10P8V	20	664	32
EPL12P6A	P12P6V	20	786	32
EPL12P6B	P12P6V	20	786	32
EPL14P4A	P14P4V	20	908	32
EPL14P4B	P14P4V	20	908	32
EPL16P2A	P16P2V	20	1030	32
EPL16P2B	P16P2V	20	1030	32
EPL16P8B	P16P8V	20	2072	64
EPL16RP4B	P16RP4V	20	2072	64
EPL16RP6B	P16RP6V	20	2072	64
EPL16RP8B	P16RP8V	20	2072	64
SAMSUNG				
CPL16L8	P16L8	20	2048	64
CPL16L8L	P16L8	20	2048	64
CPL16R4	P16R4	20	2048	64
CPL16R4L	P16R4	20	2048	64
CPL16R6	P16R6	20	2048	64
CPL16R6L	P16R6	20	2048	64
CPL16R8	P16R8	20	2048	64
CPL16R8L	P16R8	20	2048	64
CPL20L10	P20L10	24	1600	40
CPL20L10L	P20L10	24	1600	40
CPL20L8	P20L8	24	2560	64
CPL20L8L	P20L8	24	2560	64
CPL20R4	P20R4	24	2560	64
CPL20R4L	P20R4	24	2560	64
CPL20R6	P20R6	24	2560	64
CPL20R6L	P20R6	24	2560	64
CPL20R8	P20R8	24	2560	64
CPL20R8L	P20R8	24	2560	64

DEVICE CODE	DEVICE MNEMONIC	# OF PINS	# OF FUSES	# OF PTERMS
SGS-THOMPSON				
GAL16V8	G16V8	20	2194	64
GAL16V8A	G16V8A	20	2194	64
GAL20V8	G20V8	24	560	20
GAL20V8A	G20V8A	24	2706	64
RAL10H8	P10H8	20	320	16
RAL10L8	P10L8	20	320	16
RAL10P8	P10P8	20	328	16
RAL12H6	P12H6	20	384	16
RAL12L6	P12L6	20	384	16
RAL12P6	P12P6	20	390	16
RAL14H4	P14H4	20	448	16
RAL14H8	G20V8	24	560	20
RAL14L4	P14L4	20	448	16
RAL14L8	P14L8	24	560	20
RAL14P4	P14P4	20	452	16
RAL14P8	P14P8	24	568	20
RAL16H2	P16H2	20	512	16
RAL16H6	G20V8	24	560	20
RAL16H8	P16H8	20	2048	64
RAL16L2	P16L2	20	512	16
RAL16L6	P16L6	24	640	20
RAL16L8	P16L8	20	2048	64
RAL16P2	P16P2	20	514	16
RAL16P6	P16P6	24	646	20
RAL16P8	P16P8	20	2056	64
RAL16R4	P16R4	20	2048	64
RAL16R6	P16R6	20	2048	64
RAL16R8	P16R8	20	2048	64
RAL16RP4	P16RP4	20	2056	64
RAL16RP6	P16RP6	20	2056	64
RAL16RP8	P16RP8	20	2056	64
RAL18H4	G20V8	24	560	20
RAL18L4	P18L4	24	720	20
RAL18P4	P18P4	24	724	20
RAL20H2	G20V8	24	560	20
RAL20H8	G20V8	24	560	20
RAL20L2	P20L2	24	640	16
RAL20L8	P20L8	24	2560	64
RAL20P2	P20P2	24	642	16
RAL20P8	P20P8	24	2568	64
RAL20R4	P20R4	24	2560	64
RAL20R6	P20R6	24	2560	64
RAL20R8	P20R8	24	2560	64

DEVICE CODE	DEVICE MNEMONIC	# OF PINS	# OF FUSES	# OF PTERMS
SIGNETICS				
PHD16N8	P16N8	20	512	16
PL22V10	P22V10	24	5828	132
PLHS16L8A	P16L8	20	2048	64
PLHS16L8B	P16L8	20	2048	64
PLUS16L8D/-7	P16L8	20	2048	64
PLUS16R4D/-7	P16R4	20	2048	64
PLUS16R6D/-7	P16R6	20	2048	64
PLUS16R8D/-7	P16R8	20	2048	64
PLUS20L8D/-7	P20L8	24	2560	64
PLUS20R4D/-7	P20R4	24	2560	64
PLUS20R6D/-7	P20R6	24	2560	64
PLUS20R8D/-7	P20R8	24	2560	64
SPRAGUE				
SPL16LC8	P16L8	20	2048	64
SPL16RC4	P16R4	20	2048	64
SPL16RC6	P16R6	20	2048	64
SPL16RC8	P16R8	20	2048	64
SPL20LC8	P20L8	24	2560	64
TEXAS INSTRUMENTS				
PAL16L8A/-2	P16L8	20	2048	64
PAL16R4A/-2	P16R4	20	2048	64
PAL16R6A/-2	P16R6	20	2048	64
PAL16R8A/-2	P16R8	20	2048	64
PAL20L8A	P20L8	24	2560	64
PAL20R4A	P20R4	24	2560	64
PAL20R6A	P20R6	24	2560	64
PAL20R8A	P20R8	24	2560	64
TIBPAD16N8-7.5	P16N8	20	512	16
TIBPAL16H8	P16H8	20	2048	64
TIBPAL16HD8	P16HD8	20	2048	64
TIBPAL16L8-10	P16L8	20	2048	64
TIBPAL16L8-12/15/25	P16L8	20	2048	64
TIBPAL16L8-5	P16L8	20	2048	64
TIBPAL16L8-7	P16L8	20	2048	64
TIBPAL16LD8	P16LD8	20	2048	64
TIBPAL16O2	P16L8	20	2048	64
TIBPAL16R4-10	P16R4	20	2048	64
TIBPAL16R4-12/15/25	P16R4	20	2048	64
TIBPAL16R4-5	P16R4	20	2048	64
TIBPAL16R4-7	P16R4	20	2048	64

DEVICE CODE	DEVICE MNEMONIC	# OF PINS	# OF FUSES	# OF PTERMS
TEXAS INSTRUMENTS (continued)				
TIBPAL16R6-10	P16R6	20	2048	64
TIBPAL16R6-12/15/25	P16R6	20	2048	64
TIBPAL16R6-5	P16R6	20	2048	64
TIBPAL16R6-7	P16R6	20	2048	64
TIBPAL16R8-10	P16R8	20	2048	64
TIBPAL16R8-12/15/25	P16R8	20	2048	64
TIBPAL16R8-5	P16R8	20	2048	64
TIBPAL16R8-7	P16R8	20	2048	64
TIBPAL20L10	P20L10	24	1600	40
TIBPAL20L8-15/25	P20L8	24	2560	64
TIBPAL20R4-15/25	P20R4	24	2560	64
TIBPAL20R6-15/25	P20R6	24	2560	64
TIBPAL20R8-15/25	P20R8	24	2560	64
TIBPAL22V10/A	P22V10	24	5828	132
TICPAL16L8-55	P16L8	20	2048	64
TICPAL16R4-55	P16R4	20	2048	64
TICPAL16R6-55	P16R6	20	2048	64
TICPAL16R8-55	P16R8	20	2048	64
TICPAL22V10	P22V10	24	5828	132
TICPAL22V10Z(T)	P22V10	24	5828	132
TICPAL22V10Z(ZP)	P22V10	24	5828	132
VLSI				
VP10P8	P10P8	20	328	16
VP12P6	P12P6	20	390	16
VP14P4	P14P4	20	452	16
VP16P2	P16P2	20	514	16
VP16P8	P16P8	20	2056	64
VP16RP4	P16RP4	20	2056	64
VP16RP6	P16RP6	20	2056	64
VP16RP8	P16RP8	20	2056	64
VP16V8E	G16V8	20	2194	64
VP20V8E	G20V8	24	560	20

B.2 PROM DEVICES

DEVICE CODE	DEVICE MNEMONIC	# OF PINS	ARRAY SIZE
AMD/MMI			
53/6300	RA8P4	16	256 x 4
53/6301	RA8P4	16	256 x 4

DEVICE CODE	DEVICE MNEMONIC	# OF PINS	ARRAY SIZE
AMD/MMI (continued)			
53/6305	RA9P4	16	512 x 4
53/6306	RA9P4	16	512 x 4
53/6308	RA8P8	20	256 x 8
53/6309	RA8P8	20	256 x 8
53/6330	RA5P8	16	32 x 8
53/6331	RA5P8	16	32 x 8
53/6348	RA9P8	20	512 x 8
53/6349	RA9P8	20	512 x 8
53/6352	RA10P4	18	1K x 4
53/6353	RA10P4	18	1K x 4
53/6380	RA10P8	24	1K x 8
53/6380JS	RA10P8	24	1K x 8
53/6380S	RA10P8	24	1K x 8
53/6381	RA10P8	24	1K x 8
53/6381JS	RA10P8	24	1K x 8
53/6381S	RA10P8	24	1K x 8
53/6388	RA11P4	18	2K x 4
53/6389	RA11P4	18	2K x 4
53/63S1641	RA12P4	20	4K x 4
53/63S1641A	RA12P4	20	4K x 4
53/63S1681	RA11P8	24	2K x 8
53/63S1681A	RA11P8	24	2K x 8
53/63S3281	RA12P8	24	4K x 8
53/63S3281A	RA12P8	24	4K x 8
AM27S12	RA9P4	16	512 x 4
AM27S13	RA9P4	16	512 x 4
AM27S18	RA5P8	16	32 x 8
AM27S180	RA10P8	24	1K x 8
AM27S181	RA10P8	24	1K x 8
AM27S184	RA11P4	18	2K x 4
AM27S185	RA11P4	18	2K x 4
AM27S19	RA5P8	16	32 x 8
AM27S190	RA11P8	24	2K x 8
AM27S191	RA11P8	24	2K x 8
AM27S20	RA8P4	16	256 x 4
AM27S21	RA8P4	16	256 x 4
AM27S28	RA9P8	20	512 x 8
AM27S29	RA9P8	20	512 x 8
AM27S32	RA10P4	18	1K x 4
AM27S33	RA10P4	18	1K x 4
AM27S37	RA10P8	24	1K x 8
AM27S40	RA12P4	20	4K x 4

DEVICE CODE	DEVICE MNEMONIC	# OF PINS	ARRAY SIZE
AMD/MMI (continued)			
AM27S41	RA12P4	20	4K x 4
AM27S43	RA12P8	24	4K x 8
AM27S49	RA13P8	24	8K x 8
HARRIS			
HM7602	RA5P8	16	32 x 8
HM7603	RA5P8	16	32 x 8
HM7610	RA8P4	16	256 x 4
HM7610A	RA8P4	16	256 x 4
HM7610B	RA8P4	16	256 x 4
HM7611	RA8P4	16	256 x 4
HM7611A	RA8P4	16	256 x 4
HM7611B	RA8P4	16	256 x 4
HM76160	RA11P8	24	2K x 8
HM76161	RA11P8	24	2K x 8
HM76161A	RA11P8	24	2K x 8
HM76161B	RA11P8	24	2K x 8
HM76164	RA12P4	20	4K x 4
HM76165	RA12P4	20	4K x 4
HM7620	RA9P4	16	512 x 4
HM7620A	RA9P4	16	512 x 4
HM7620B	RA9P4	16	512 x 4
HM7621	RA9P4	16	512 x 4
HM7621A	RA9P4	16	512 x 4
HM7621B	RA9P4	16	512 x 4
HM76321	RA12P8	24	4K x 8
HM7642	RA10P4	18	1K x 4
HM7642A	RA10P4	18	1K x 4
HM7642B	RA10P4	18	1K x 4
HM7643	RA10P4	18	1K x 4
HM7643A	RA10P4	18	1K x 4
HM7643B	RA10P4	18	1K x 4
HM7648	RA9P8	20	512 x 8
HM7649	RA9P8	20	512 x 8
HM7649A	RA9P8	20	512 x 8
HM76641	RA13P8	24	8K x 8
HM76641A	RA13P8	24	8K x 8
HM7680	RA10P8	24	1K x 8
HM7680P	RA10P8	24	1K x 8
HM7680R	RA10P8	24	1K x 8
HM7680RP	RA10P8	24	1K x 8
HM7681	RA10P8	24	1K x 8

DEVICE CODE	DEVICE MNEMONIC	# OF PINS	ARRAY SIZE
HARRIS (continued)			
HM7681A	RA10P8	24	1K x 8
HM7681P	RA10P8	24	1K x 8
HM7681R	RA10P8	24	1K x 8
HM7681RP	RA10P8	24	1K x 8
HM7684	RA11P4	18	2K x 4
HM7684P	RA11P4	18	2K x 4
HM7685	RA11P4	18	2K x 4
HM7685A	RA11P4	18	2K x 4
HM7685P	RA11P4	18	2K x 4
NATIONAL			
DM74LS471	RA8P8	20	256 x 8
DM74S188	RA5P8	16	32 x 8
DM74S287	RA8P4	16	256 x 4
DM74S288	RA5P8	16	32 x 8
DM74S387	RA8P4	16	256 x 4
DM74S472	RA9P8	20	512 x 8
DM74S472A	RA9P8	20	512 x 8
DM74S472B	RA9P8	20	512 x 8
DM74S473	RA9P8	20	512 x 8
DM74S473A	RA9P8	20	512 x 8
DM74S570	RA9P4	16	512 x 4
DM74S570A	RA9P4	16	512 x 4
DM74S571	RA9P4	16	512 x 4
DM74S571A	RA9P4	16	512 x 4
DM74S571B	RA9P4	16	512 x 4
DM74S572	RA10P4	18	1K x 4
DM74S572A	RA10P4	18	1K x 4
DM74S573	RA10P4	18	1K x 4
DM74S573A	RA10P4	18	1K x 4
DM74S573B	RA10P4	18	1K x 4
DM87S180	RA10P8	24	1K x 8
DM87S181	RA10P8	24	1K x 8
DM87S181A	RA10P8	24	1K x 8
DM87S184	RA11P4	18	2K x 4
DM87S185	RA11P4	18	2K x 4
DM87S185A	RA11P4	18	2K x 4
DM87S185B	RA11P4	18	2K x 4
DM87S190	RA11P8	24	2K x 8
DM87S190A	RA11P8	24	2K x 8
DM87S190B	RA11P8	24	2K x 8
DM87S191	RA11P8	24	2K x 8

DEVICE CODE	DEVICE MNEMONIC	# OF PINS	ARRAY SIZE
NATIONAL (continued)			
DM87S191A	RA11P8	24	2K x 8
DM87S191B	RA11P8	24	2K x 8
DM87S195	RA12P4	20	4K x 4
DM87S195A	RA12P4	20	4K x 4
DM87S195B	RA12P4	20	4K x 4
DM87S321	RA12P8	24	4K x 8
DM87S321A	RA12P8	24	4K x 8
SIGNETICS			
82S123/A	RA5P8	16	32 x 8
82S126/A	RA8P4	16	256 x 4
82S129/A	RA8P4	16	256 x 4
82S130/A	RA9P4	16	512 x 4
82S131/A	RA9P4	16	512 x 4
82S135	RA8P8	20	256 x 8
82S137/A/B	RA10P4	18	1K x 4
82S147/A	RA9P8	20	512 x 8
82S180	RA10P8	24	1K x 8
82S181/A/B	RA10P8	24	1K x 8
82S184	RA11P4	18	2K x 4
82S185/A/B	RA11P4	18	2K x 4
82S191/A/B/C	RA11P8	24	2K x 8
82S195	RA12P4	20	4K x 4
82S23/A	RA5P8	16	32 x 8
82S321	RA12P8	24	4K x 8
82S641	RA13P8	24	8K x 8
TEXAS INSTRUMENTS			
TBP18S030	RA5P8	16	32 x 8
TBP18S22	RA8P8	20	256 x 8
TBP18SA030	RA5P8	16	32 x 8
TBP18SA22	RA8P8	20	256 x 8
TBP24S10	RA8P4	16	256 x 4
TBP24S166	RA12P8	24	4K x 4
TBP24S41	RA10P4	18	1K x 4
TBP24S81	RA11P4	18	2K x 4
TBP24SA10	RA8P4	16	256 x 4
TBP24SA166	RA11P8	24	4K x 4
TBP24SA41	RA10P4	18	1K x 4
TBP24SA81	RA11P4	18	2K x 4
TBP28S166	RA11P8	24	2K x 8
TBP28S42	RA9P8	20	512 x 8

DEVICE CODE	DEVICE MNEMONIC	# OF PINS	ARRAY SIZE
TEXAS INSTRUMENTS (continued)			
TBP28S86A	RA10P8	24	1K x 8
TBP28SA166	RA11P8	24	2K x 8
TBP28SA42	RA9P8	20	512 x 8
TBP28SA86A	RA10P8	24	1K x 8

INDEX